The Illustrated Encyclopedia of Pterosaurs

The Illustrated Encyclopedia of Pterosaurs

DR PETER WELLNHOFER

a Salamander book

Published by Salamander Books Limited
LONDON

A Salamander Book

Published by Salamander Books Ltd,
129/137 York Way,
London N7 9LG,
United Kingdom

ISBN 0 86101 566 5

Distributed in the UK by
Hodder & Stoughton Services,
P.O. Box 6,
Mill Road,
Dunton Green,
Sevenoaks,
Kent TN3 2XX

All correspondence concerning the content of this volume should be addressed to Salamander Books Ltd.

Credits

Editor:
Philip de Ste. Croix

Designers:
Roger Hyde, Jill Coote, John Heritage

Colour pterosaur artwork:
John Sibbick © Salamander Books Ltd

Skeletal artwork and black and white pterosaur restorations:
© Peter Wellnhofer

Diagrams:
© Peter Wellnhofer and Geoff Denney © Salamander Books Ltd

Text translation from German:
Michael Robinson

Index:
Stuart Craik

Filmset:
SX Composing Ltd, England

Colour and monochrome reproductions:
Scantrans PTE Ltd, Singapore and Contemporary Litho Ltd, England

Printed in Hong Kong

The Author

Born in 1936 in Munich, Bavaria, Dr Peter Wellnhofer is Hauptkonservator at the Bavarian State Collection of Palaeontology and Historical Geology, an institution that houses some of the finest pterosaur fossil remains in the world. He has worked in this field of research for more than 20 years and is recognised as a leading authority worldwide. Apart from his numerous scientific publications, he also prepared the volume on pterosaurs for the *Encyclopedia of Palaeoherpetology*, and is the author of an illustrated booklet on pterosaurs and early birds from Solnhofen, *Solnhofer Plattenkalk: Urvögel und Flugsaurier*.

The Artist

John Sibbick undertook his art training in the early 1970s at Guildford Art School, Surrey, where he studied graphic design and, latterly, illustration. He subsequently spent four years in various art studios in London 'learning the ropes' before going fully freelance. He has always been interested in prehistoric animals, and has worked on a number of books concerned with dinosaurs and their contemporaries including the much acclaimed companion volume to this, *The Illustrated Encyclopedia of Dinosaurs*. Apart from his publishing commissions, he also works regularly for galleries and museums, and has recently been involved in producing artwork for use in television documentary programmes about prehistoric life in the world before man.

Acknowledgements

Having devoted a quarter of a century of my scientific career to the study of those extinct flying saurians called pterosaurs, I began to consider writing a book about these fascinating creatures from a past so long ago. I felt the need for such a documentation in order to summarize our present knowledge, destined not only for specialists in the field but also for an increasingly interested public audience. Since the popular *Dragons of the Air* published by Harry Govier Seeley in 1901, no such book had been written, despite the growing number of similar books on dinosaurs. With regard to the many fossil pterosaur finds and to the intensive research on pterosaurs, a tremendous amount of new data have accumulated since the turn of the century. So, this book seemed to be long overdue. However, it was not until Dr Colin Pennycuick, Professor of Biology at the University of Miami, suggested decisively that I should write a book on pterosaurs, that I decided to pursue the idea. For this, and his continuous interest in this project I want to thank him especially.

During the preparation of the book I enjoyed in many ways the support of colleagues all over the world. For supplying photographs, information, advice and for allowing access to museum specimens in their care I am deeply obliged to Natalia Bakhurina, Moscow, Christopher Bennett, Lawrence, Kansas, Dr José Bonaparte, Buenos Aires, Martyn Cowley, Simi Valley, California, Prof Dr Dong Zhiming, Beijing, China, Dr Burkard Engesser, Basel, Switzerland, Karl A. Frickhinger, Gräfelfing, Germany, Rolf. B. Hauff, Holzmaden, Germany, Prof Dr Dietrich Herm, Munich, Germany, Alexander W. A. Kellner, Rio de Janeiro, Brazil, Dr Heinz A. Kollman, Vienna, Austria, Dr Theo Kress, Solnhofen, Germany, Prof Dr Wann Langston, Jr, Austin, Texas, Dr John G. Maisey, New York, Dr Helmut Mayr, Munich, Germany, Dr Angela Milner, London, Dr Ralph Molnar, Fortitude Valley, Queensland, Australia, Dany Oppliger, Basel, Switzerland, Prof Dr John H. Ostrom, New Haven, Connecticut, Dr Kevin Padian, Berkeley, California, Prof Dr Giovanni Pinna, Milan, Italy, Dr Georg Plodowski, Frankfort, Germany, Matt B. Smith, Bozeman, Montana, David Unwin, Reading, England, Dr Günter Viohl, Eichstätt, Germany, Prof Dr Frank Westphal, Tübingen, Germany, and Dr Rupert Wild, Stuttgart, Germany.

Without the assistance of the technical staff palaeontological research would hardly be possible. This is the place to thank the various people in my institution who have contributed to this book by their skilled work in preparing, casting and photographing fossil specimens. My thanks for their help over many years go to Leonhard Bimmer, Franz Höck, Renate Liebreich and Ernst Schmieja, all in Munich.

I would also like to express my gratitude to the editor of this book, Philip de Ste. Croix, for the smooth co-operation over the last three years, for his numerous suggestions in order to improve the text and the illustrations, and for taking my proposals seriously into account. My gratitude is extended to the artist, John Sibbick, who, in his outstanding colour restorations, managed to bring the various pterosaurs to life again. They are among the best life restorations of pterosaurs I know.

Finally, I want to thank my wife Ingrid for her patience over the last years and for her understanding that her husband had to spend most of his spare time preparing drawings, photographs and text for this book.

Dr Peter Wellnhofer

CONTENTS

Left: Pterosaurs went extinct about 65 million years ago; it goes without saying that no human being has ever seen a live pterosaur. We must, therefore, rely upon scientific study of their fossil remains to form an impression of this fascinating group of animals, and to allow us to make informed deductions about their likely appearance and life styles. Despite the relative scarcity of fossil evidence, palaeontologists have been able to build up quite a detailed picture of a number of these 'flying dragons' of primeval times. Some people have even been stimulated to make three-dimensional flying reconstructions of pterosaurs, such as this model of *Quetzalcoatlus* which is about to undergo a flight test.

Earth keeps a firm hold on its creatures. From earliest times, all life has been subject to the inescapable force of gravity. Myths of the ancient world like the legend of Daedalus and Icarus[1], the early aircraft experiments of the tailor of Ulm[2] or Otto Lilienthal[3] show us how man has always dreamed of conquering these forces, of releasing himself from the pull of the Earth with mechanical aids, and flying.

In just the same way biological evolution did not come to a stop when confronted with air as a possible habitat: it conquered it in an astonishing number of ways. Two thirds of all species living today are capable of flight. The largest group are of course insects, by far the largest class within the animal kingdom, with more than three quarters of a million living species. The earliest representatives of flying insects in the history of the Earth are known as fossils in Carboniferous strata about three hundred million years old.

Apart from insects, only vertebrates, in the shape of birds, bats and pterosaurs, have succeeded in penetrating the air as an 'ecological niche' by achieving active flight. All three groups solved the problem of overcoming gravity independently. Of these three groups it was the pterosaurs who were the first vertebrates to do this successfully, and that was over 200 million years ago, in the Triassic period, at the beginning of the Mesozoic Era.

Flying makes extreme demands on the animal body, on skeleton and structure and on physiology. A flying animal is of course always heavier than air, but it must be as light as possible. On the other hand its increased energy requirement demands an appropriate respiratory and circulatory system and a high metabolic rate. Complicated steering manoeuvres in the air are only possible with suitably developed nerve and brain structures. An ability to fly actively does however offer enormous advantages: increased mobility in the search for food and in avoiding enemies, flexibility when living conditions are unfavourable, extension of territory over a considerably extended habitat.

Despite differences in detail, a generally uniform body plan is to be found in all vertebrates capable of active flight. Obviously the only possible way in which tetrapods, or four-footed vertebrates, could fly was by their forelimbs changing into wings. So their flying organs were not an absolutely free gift, they came at the price of the loss of the forelegs.

It is tempting to interpret early adaptations of reptiles to the medium of air as a first, careful move towards acquiring the ability to fly, as in *Daedalosaurus*[4] in the Permian, or *Kuehneosaurus*[5] and *Icarosaurus*[6] in the Triassic. These reptiles, the size of lizards, had flaps of skin between their elongated ribs, which allowed them more or less passive downward gliding flight over a limited distance. A surviving example of this principle of locomotion is the genus *Draco*[7], the so-called dragon of South-East Asia.

In 1971 the Moscow zoologist A.G.Sharov discovered a hitherto unknown reptile in deposits from the late Triassic in Kirghizia (USSR), that he called *Podopteryx* (leg-flier), and classified in the reptile order of pseudo-

Above: Hours of painstaking research are needed before a convincing portrait of the pterosaur as it appeared in life can be attempted. Here the author is seen reconstructing the head of *Pterodaustro*, the extraordinary 'flamingo pterosaur' from Argentina.

1 According to Greek legend Daedalus was an Athenian architect and sculptor. He built an artificial labyrinth for the Minotaur for King Minos of Crete. In order to escape from captivity on Crete he made wings for himself and his son Icarus, using birds' feathers held together with wax. With their help they rose into the air and flew off in the direction of home. Despite his father's warning that he should not fly too high, proud young Icarus soared up so far that the heat of the sun melted the wax in his wings, and he plummetted into the sea. His unhappy father found him dead on the shore of a nearby island, on which he buried him, and called the island Icaria. It is one of the Dodecanese Islands in the Aegean.

2 In 1811 A. Berblinger, a tailor, tried to fly over the Danube at Ulm in a glider he had built himself. The attempt failed.

3 Otto Lilienthal (1848-1896), an engineer, studied birds' flight and constructed gliders with cambered wings, in which he conducted flying experiments over distances of up to 1,000ft (300m). In August 1896 he crashed in a monoplane glider, and was killed. The Wright brothers later took up this pioneer of flight's experiments in America.

4 Carroll, R. L., 1978. *A gliding reptile from the Upper Permian of Madagascar.* Palaeontographica Africana, 21: 143-159.

5 Robinson, P.L., 1962. *Gliding lizards from the Upper Keuper of Great Britain.* Proceedings of the Geological Society, London, 1601: 137-146.

6 Colbert, E.H., 1966. *A gliding reptile from the Triassic of New Jersey.* American Museum Novitates, 2246: 1-23.

7 Flying dragons of the genus *Draco* can spread their rib wings and glide for a long distance from tree to tree. To increase the gliding area the tail is also somewhat broader at the base, and there are flaps of skin at the side of the chin. In the case of *Draco* gliding flights of up to 200ft (60m) have been observed.

Left: Big Bend National Park in West Texas. The park was so-named because the Rio Grande river makes a great U-turn in this part of Texas, its waters cleaving through deep-cut canyon walls for over 100 miles (160km). The slopes in the distance in this picture are late Cretaceous flood plain sediments which have been exposed by the action of erosion. These strata have yielded vertebrate fossils, including dinosaurs and the giant pterosaurs of the latest Cretaceous. It was here that the fossil remains of the largest known pterosaur, *Quetzalcoatlus*, were first discovered in 1971 by Douglas Lawson. The wingspan of this largest flying creature of all time was about 39ft (12m). A half size model of it is seen on the opposite page, about to undertake a flight test in Death Valley.

suchids.[8] Later this genus was renamed *Sharovipteryx* in his honour. The fossil had survived in such good condition that skin imprints could still be made out after 240 million years. This led Sharov to the conclusion that between the extremely elongated hind legs and the long tail, and also on the sides of the body, a flight membrane could be spread that must have given the animal good possibilities of gliding flight.

But all these experiments did not allow active, that is sustained, flight in the open air with complete control over steering and direction. This decisive evolutionary step was not achieved until the appearance of pterosaurs, who practised it successfully throughout the entire Mesozoic, thus for over 150 million years.

Without fossil finds and without palaeontology, the study of life in the Earth's past, we would have no knowledge of this large, interesting, indeed fascinating group of animals which became extinct, along with other saurians, at the end of the Cretaceous, 65 million years ago. This also means that no human being has ever seen a live pterosaur, of which the largest had a wing-span of about 39 feet (12m). We can only form an impression of them by scientific analysis of their remains: fossilized skeletal bones, imprints of their skin and possibly tracks that they left behind on soft ground, and by making comparisons with other flying vertebrates that we can study more closely, namely birds and bats.

In doing this it quickly becomes clear that pterosaurs were different in many respects, and in some ways unique. Phylogenetically they are not closely related to birds and bats, which of course are mammals, and pterosaurs also occupy a special position vis-à-vis reptiles as we know them today, like tortoises, snakes, lizards, crocodiles and the tuatara. If we include the geology of deposits in which pterosaurs have been found and their geographical distribution in our considerations, we can gradually build up a picture of this 'dragon' of primeval times, fit it together like a mosaic, and after millions of years bring these creatures back to life in our mind's eye.

If this book succeeds in doing this for the reader then it has fulfilled its purpose. At the same time we should always be aware that fossil finds – often chance finds – only represent a small section of the actual prehistoric living world, and that palaeontologists cannot give definite answers to many questions. And so in the last resort pterosaurs will always have their riddles and retain some secrets. But that does not make them any the less fascinating; indeed it leaves room for the imagination to play its part.

8 Sharov, A.G., 1971. *New flying reptiles from the Mesozoic deposits of Kazakhstan and Kirghizia.* Trudy Akademia Nauk SSSR, Paleontological Institute, 130: 104-113 (in Russian).

Above: This photograph shows a group of models of *Rhamphorhynchus*, a long-tailed pterosaur that populated the coasts and islands of the Solnhofen lagoon in that part of the world that is now Bavaria in the Late Jurassic about 150 million years ago. These models are on display in Palaeontological Museum of the Bayerische Staatssammlung für Paläontologie und historische Geologie in Munich, a collection that houses many of the fossils from the Solnhofen limestone. Pterosaurs are so well documented in Solnhofen that it must be considered the most important site in which they have been found in the world. The unique quality of the fossilization has led to the preservation of complete skeletons and even sometimes the documentation of soft parts like flight membrane prints.

INTRODUCTION TO PTEROSAURS

In the early years of this century, four English explorers, led by the famous London zoologist Professor George Edward Challenger, discovered living primeval creatures in a faraway corner of South America, creatures which had hitherto been believed to have become extinct at least from the Mesozoic. On an inaccessible plateau, completely cut off from the outside world, Professor Challenger, his colleague Professor Summerlee, traveller and sportsman Lord John Roxton and Ed Malone, reporter on the *Daily Gazette*, also met living pterosaurs from the Jurassic. Malone's reports provide a clear record of the adventures and dangers faced by the members of the expedition.

Malone describes their first encounter with a pterosaur like this: '. . . we saw, at the distance of a mile or so, something which appeared to be a huge grey bird flap slowly up from the ground and skim smoothly off, flying very low and straight, until it was lost among the tree-ferns.' On the same evening they were to make even closer acquaintance with a pterosaur: this is the only thing it could possibly have been. The men were sitting around the camp fire roasting an aguti, a small pig-like animal, that Lord Roxton had killed for their supper. But let Ed Malone tell us himself: 'The night was moonless, but there were some stars, and one could see for a little distance across the plain. Well, suddenly out of the darkness, out of the night, there swooped something with a swish like an aeroplane. The whole group of us were covered for an instant by a canopy of leathery wings, and I had a momentary vision of a long, snake-like neck, a fierce, red greedy eye, and a great snapping beak, filled, to my amazement, with little gleaming teeth. The next instant it was gone – and so was our dinner. A huge black shadow, twenty feet across, skimmed up into the air; for an instant the monster's wings blotted out the stars, and then it vanished over the brow of the cliffs above us.'

Finally the expedition succeeded in capturing a living creature and bringing it back to London in a big crate. Professor Challenger reported on his discovery of living pterosaurs at a meeting of the Royal Zoological Society. The meeting was not prepared to believe him; the professor gave a sign and a large, rectangular crate was brought on to the platform. He pushed back the lid, and a moment later an in-

Above: 'The next instant it was gone -- and so was our dinner.' Zdeněk Burian's illustration of a scene in Conan Doyle's *The Lost World* when Professor Challenger's party of explorers meet a living pterosaur on a mountain plateau in South America.

describably loathsome creature appeared with a scratching, rattling sound, unfolded two leathery wings to a span of ten feet and rose into the air to circle slowly under the ceiling of Queen's Hall, in which the meeting was taking place. Everyone was staring at this display as if transfixed, when suddenly Professor Challenger yelled: 'The window! For Heaven's sake close the window!' But it was too late. The next second the creature reached the opening, forced its body through it and disappeared.

The last record of the London *pterodactyl*, as it was later called, is found in the log of the *Friesland*, a postal steamer of the Holland-America Line. This contains an entry saying that next morning at 9 o'clock, ten miles off Start Point, a strange creature, 'something between a flying goat and a monstrous bat', flew past at enormous speed in a south-westerly direction. Its homing instinct had certainly set it on the right course. It probably never reached its goal, so this last living pterosaur perished somewhere out in the wastes of the Atlantic.

Below: The last glimpse of the London pterodactyl as observed by the crew of the Dutch-American steamer, S.S. *Friesland*. This reconstruction of a scene from *The Lost World* is again by Zdeněk Burian.

The reader will certainly have noticed that this story is far too good to be true. Its author, Sir Arthur Conan Doyle, creator of the legendary Sherlock Holmes, published it in 1912 under the title *The Lost World*.[9] It is remarkable in two respects.

Firstly, Conan Doyle describes the appearance, way of life and behaviour of animals that disappeared from the face of the earth 65 million years ago. Thus no human being can ever have seen them alive; all we have are fossils, remains of the creatures that have been turned to stone. Conan Doyle does not describe these strange primeval creatures as a scholar, but as an imaginative writer, but thoroughly convincingly and realistically.

Secondly, Conan Doyle has his prehistoric pterosaurs survive on a high plateau somewhere in South America, more precisely in the Amazon region, thus in Brazil. It is quite amazing that in 1970, that is to say 60 years after Conan Doyle wrote the story, fossilized pterosaur bones actually were found on an extensive high plateau in north-eastern Brazil, the Araripe plateau. The Brazilian geologist Llellewyn Price described the find and called the species *Araripesaurus castilhoi*.[10] Pterosaur fossils from the Araripe plateau have continued to appear, representing several different genera, some with wing spans of up to 20ft (6m). Some of them actually must have looked as Conan Doyle described them: '. . . with a long, snake-like neck, a great snapping beak, filled with little, gleaming teeth'. The Araripe plateau in north-eastern Brazil became one of the most important sources of Cretaceous pterosaurs, documenting the highly diverse development of this group of animals about 100 million years ago.

In fact the first discoveries of pterosaur fossils were made as early as the 18th century, at the time when people started to collect natural objects and display them in natural history collections. Thus the first pterosaur find, a small fossil on a slab of Solnhofen lithographic limestone from Eichstätt in Bavaria, appeared in the natural history collection of the Palatine prince, Karl Theodor, in Mannheim. The supervisor of this collection, Cosimo Alessandro Collini, probably noticed that this was an unusual fossil skeleton, but was undecided in his interpretation of what sort of a creature it was. It was not until 1801 that the great Parisian anatomist Georges Cuvier recognized that it was a reptile that could fly. It was neither a bat nor a bird, as others thought, but a hitherto entirely unknown type of creature that he called *Pterodactylus* (=flight finger), because the wing membrane was stretched over a single, albeit very long, digit.

Genuine scientific research on pterosaurs began with Cuvier. The small *Pterodactylus* from Eichstätt was followed by many other finds, not only in Solnhofen lithographic limestone, which also provided the famous primeval bird *Archaeopteryx*, but from many other

9 Conan Doyle, A., 1912. *The Lost World*. A new edition of this fascinating book was published in paperback by Puffin Books and Penguin Books in 1981.

10 Price, L.I., 1971. *A Presença de Pterosauria no Cretáceo Inferior da Chapada do Araripe, Brasil.* Anais Academia Brasiliana de Ciencias, 43 (suppl.): 352-461.

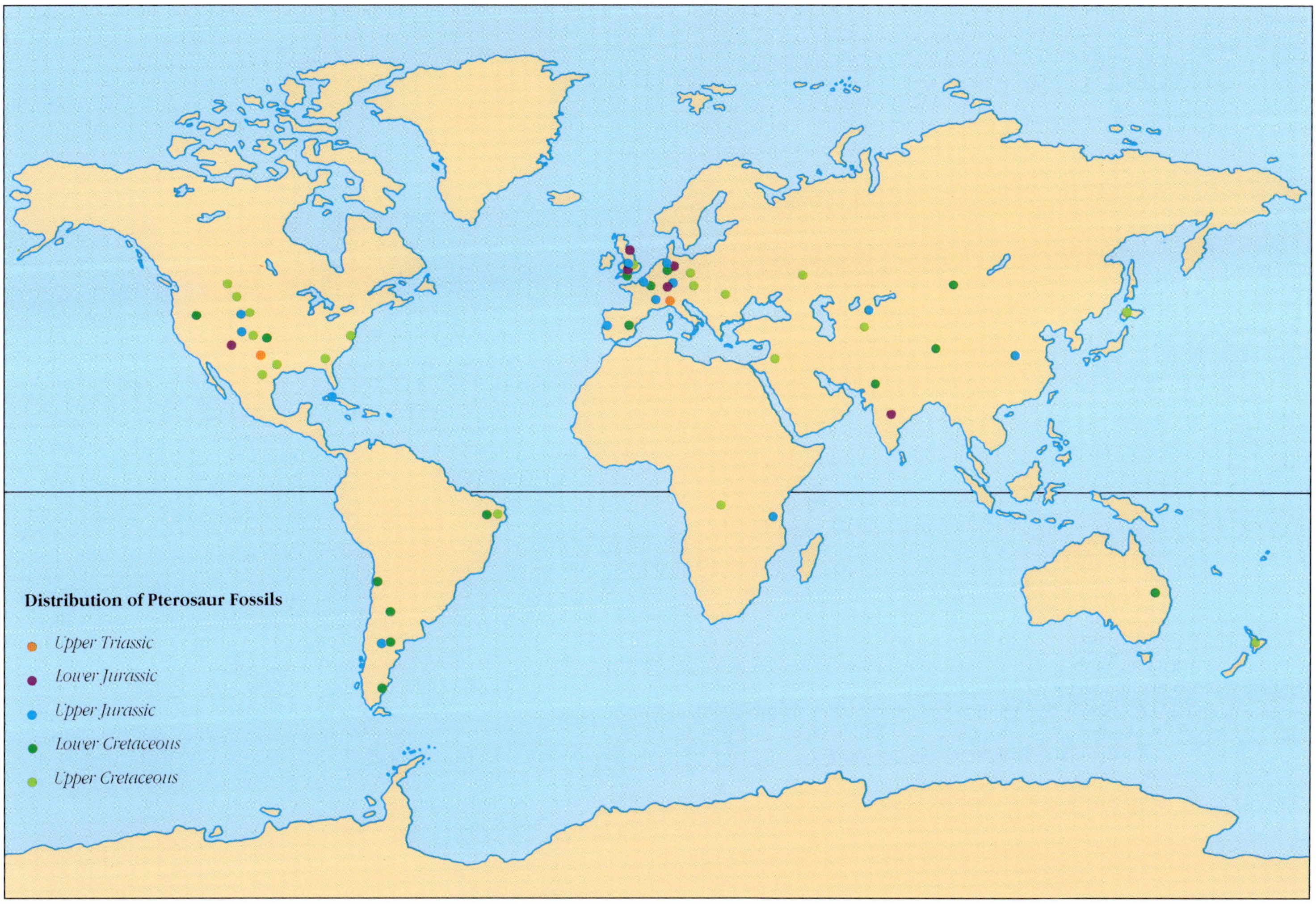

Above: The distribution of fossil finds indicates that pterosaurs once lived on all continents except Antarctica. Only sedimentary rocks of the Mesozoic era are likely to yield pterosaur fossils. This fact is reflected by the distributional pattern shown here, which also reflects the accessibility of the available fossil sites around the globe.

Below: Fossil wing bones of a pterosaur discovered in a split nodule from the Lower Cretaceous Santana Formation of the Araripe Plateau in north-eastern Brazil. This specimen, named as *Araripesaurus castilhoi*, was the first scientific record of pterosaurs from this area of South America, where, prophetically, Conan Doyle set *The Lost World*.

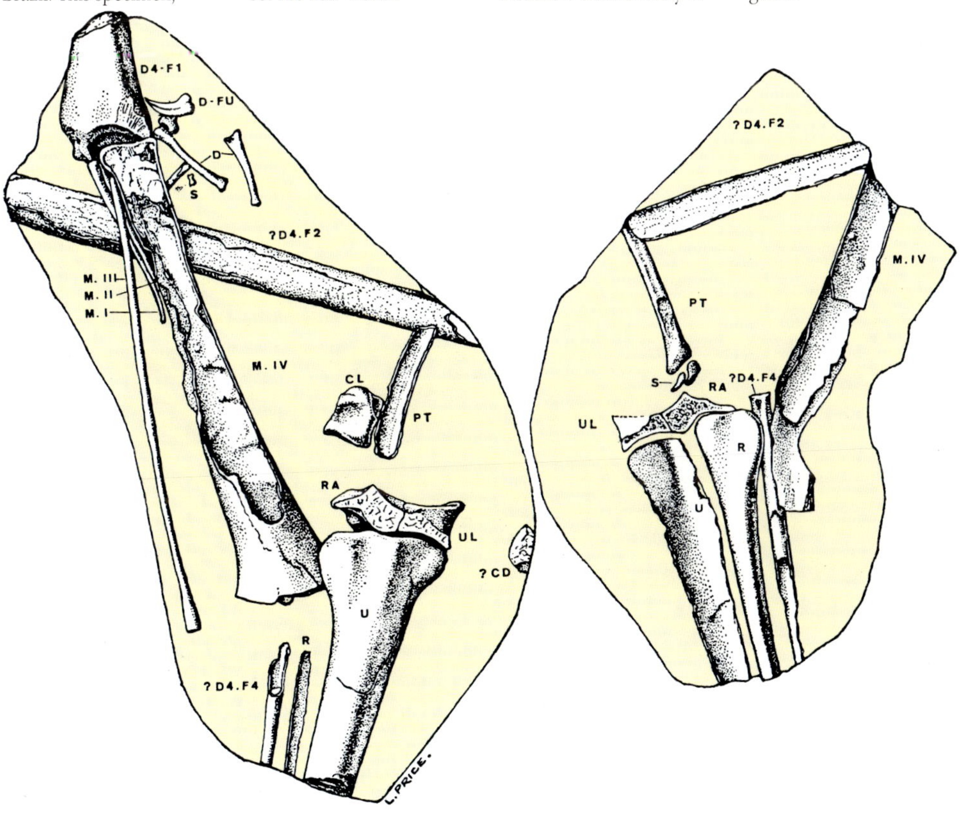

geological formations as well, from the late Triassic (220 million years ago) to the late Cretaceous (65 million years ago).

They appeared in an extraordinary variety of forms, ultimately distributed world-wide. Pterosaurs lived on all continents with the exception of the Antarctic, and included genera from the size of a sparrow up to the largest flying creatures of all time, with a wing span of 39ft (12m). Despite the wide range of different forms, pterosaurs maintained their special characteristics to the end: a unique wing structure and extreme lightness of skeleton.

Scientific research has not been at a standstill since Cuvier. Generations of palaeontologists and zoologists since that time have tried to find out what sort of creatures pterosaurs were, what they looked like, where and how they lived, how they moved, how they flew, fed and reproduced themselves, what their flight membranes and body covering were like, whether they were warm-blooded or not, how they evolved and developed phylogenetically, what role they played in the prehistoric animal world and finally why they became extinct 65 million years ago, without successors, leaving birds to take over domination of the air.

Scientific research on these and many other questions is in full swing. It will probably never reach an ultimate conclusion, as new fossil finds are always being made, and also because human curiosity, the driving force in any field of science, will never be extinguished. Thus attempts have recently been made to reconstruct the aerodynamics of the pterosaur wing using mathematical and physical methods, and to test the results by building flying models.

Today pterosaurs can be considered as one of the most thoroughly investigated groups of fossil reptiles. Interesting answers have been provided to many of the questions. Others could only be answered by making assumptions, as documentation available to palaeontologists in fossil form is by its very nature incomplete. Unfortunately we are not in the position of Conan Doyle's Professor Challenger, and able to observe living pterosaurs. However, if we allow our imaginations to be inspired and stimulated by Conan Doyle, this and our scientifically based knowledge unite to convince us of their fascination.

Fossilization

It is the palaeontologist's task to draw conclusions from fossils about life in prehistoric times. The word fossil comes from the Latin verb *fodere*, to dig, and originally meant anything that had been dug up, or discovered in the ground. Today, by fossils we understand remains of organisms, but also traces of creatures' activities. As a rule fossils are found in older deposits from the Earth's past, usually in sedimentary rocks, and consist of the hard parts of the creature that survived decay. A close connection with geology is thus of fundamental importance to palaeontology. Fossils are not only biological, but also historical documents of the development of life on Earth. Thus as far as zoology and botany are concerned, fields of research concerned with the modern world of plants and animals, this represents an enormous broadening of our knowledge of life forms, as plants and animals of earlier ages, now long since extinct, can be included. Today palaeontology has an overview of the development of organized life on the higher level over a period of almost 600 million years.

But along with this broader knowledge of the formal variety of life in general there is also

Left: This *Pterodactylus* from the Solnhofen lithographic limestone of Bavaria has a broken wing finger (see also the close-up photograph of the fracture below). This accident probably happened while the creature was alive, and as a result it would not have been able to spread its wing membrane which relied upon the extension of the wing finger for its proper functioning. With one wing hanging limp, flight would have been impossible and the search for food severely impaired. It is quite likely that this individual starved to death, and its corpse was then washed into the lagoon where the process of fossilization began. The drawing below shows in detail the arm and hand bones, and the broken finger, which is circled in the photograph above.

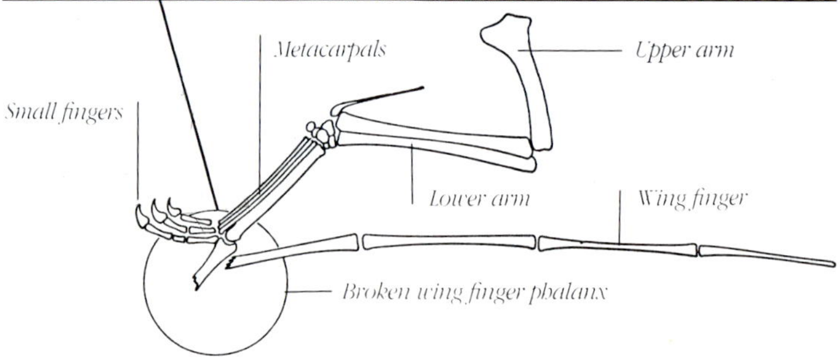

Fossilization (above and below)
This sequence of drawings shows the first of the stages by which a dead pterosaur may have been turned into a fossil. In the first drawing the carcass is seen floating in the sea. After some time it sinks to the sea floor where it is embedded in soft mud. At this stage the soft parts, muscles, skin and connective tissue decay completely, and water currents may disturb and eventually dislocate parts of the skeleton.

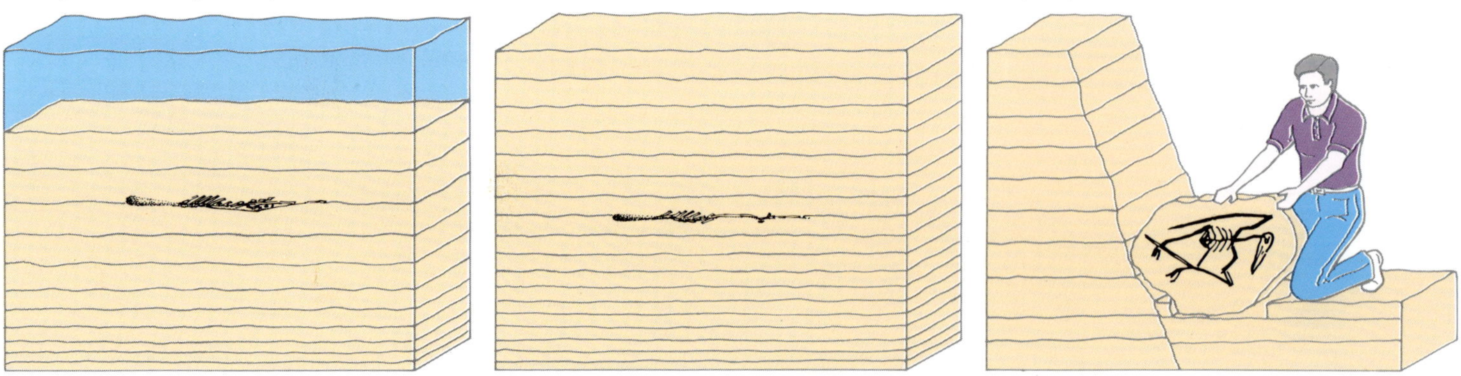

Steady sedimentation will enclose and preserve the bones, and, as the overburden increases, also compress the skeletal remains. In the course of compaction, dehydration, and mineralisation the sediment is transformed into hard rock, for example flaggy limestone, and the pterosaur bones will be petrified, i.e. fossilized (middle drawing above). Finally, if the quarryman is lucky enough, he will find the fossil pterosaur on a slab of limestone.

an associated loss of information. A palaeontologist cannot experiment like a zoologist. Direct research into physiological processes is not available to him. Only in rare cases can he make statements about soft parts, body covering and colouring. This is a result of the fossilization process itself.

After death a living organism usually breaks down into inorganic substances, and no trace of it remains. Thus fossilization represents an exception to the general rule, and several factors have to be present simultaneously to make the process possible. This is particularly true of pterosaurs, which have very lightly constructed and fragile skeletons. One of the most important prerequisites was that the pterosaur corpses had to be quickly covered up and protected from exposure to the air, so that natural processes of decomposition were checked. As pterosaurs were land creatures, they are likely to have died on land in most cases, and so only had a chance of becoming fossils if their bodies were washed into a river or lake, or if they fell into a sand bank or sank into mud.

By far the most frequent pterosaur finds are made in marine strata, that is to say in rocks which originated in prehistoric marine deposits. Most of the known species of pterosaur probably lived near the coast and fed on fish or other aquatic organisms. The cause of their death was probably natural only in the rarest cases, in other words they did not often die of old age. Pterosaurs found as fossils could have died of disease, parasites, poisoning or wounds. Many were probably themselves prey to larger predatory saurians, and were eaten. It is quite conceivable that flight membrane injuries left them unable to fly, and were thus life threatening. Pterosaur fossils have been found in Solnhofen limestone in Bavaria with the first wing phalanx broken. That can only have happened while the creature was alive. These pterosaurs had thus, for whatever reason, had an accident or been attacked, and broken the flight digit that spread the flight membrane. This meant that one wing hung limp, and the creatures could no longer fly in search of food. They had to starve.

In many cases we can assume that pterosaurs were forced into water in a tropical storm and drowned, or were carried from the mainland by rivers into the sea near the coast. In both cases the corpse sank to the sea bed after drifting for a longer or shorter period, then was finally covered in mud and subsequent sedimentation of increasingly massive strata. In the course of this, soft parts, muscles, skin and connective tissue decomposed.

Water and oxygen cause decay. If there is no oxygen, putrefaction sets in. In rare cases soft parts also survived as fossils: their decomposition was delayed, meaning that they could be infiltrated by mineral substances during the fossilization process. This preserves a detailed image of parts of the body not usually capable of fossilization. In the case of some pterosaurs we are therefore familiar with imprints of flight membranes, of webs between the long toes, of a throat pouch, of respiratory tubes and of the outline of the body.

Evidence of a hair-like body covering caused a sensation. Sometimes it is in the form of an imprint, or hair has even been preserved itself in some cases.[11] In general, fossil pterosaur skeletons have only come down to us incomplete or as fragments. The smaller a pterosaur was, the greater its chance of being completely

11 Sharov, A.G., op. cit. (8).

Below: The skeleton of this juvenile *Pterodactylus antiquus* from Solnhofen is completely preserved. Its undisturbed position shows that the carcass was quickly embedded in the sediment together with its soft parts. However, wing membranes, muscles, skin and tendons have decayed later, leaving no trace.

Above: This *Rhamphorhynchus* from Solnhofen still shows the impressions of wing membranes preserved in much detail, as well as the imprint of a terminal vane on the tail. The fossil documentation of soft parts like these requires very favourable conditions, and a fine grained sediment.

fossilized. If the size of the body was greater than that of a pigeon or a seagull, we generally only find incomplete skeletal remains or individual parts of the skeleton, like isolated skulls, wing bones, leg bones, pelvises or shoulder blades. The reason for this is that pterosaur corpses, rather like the corpses of birds, drift around in the water for a long time.[12] Their bodies were light, their bones hollow, and partially filled with air. The heavy parts, above all the head, wings and hind legs of the dead pterosaur hung down and were the first items to be detached from the body after decomposition of the soft parts. They sank to the sea bed and were embedded individually. The other parts of the body followed, until the whole skeleton was spread over a large area of the sea floor.

In the case of complete skeletons one frequently observes that the pterosaur's neck is bent back, and is often drawn backwards so strongly that the skull is above the pelvis. This position also occurs in modern bird corpses, especially when they have dried out on land to become a 'mummy'. This is caused by contraction of the muscles of the neck, which draws the neck backwards. This might mean that fossilized pterosaur skeletons with heads bent back belonged to creatures that died on land and dried out there, thus becoming mummified. They must then have been washed in this condition by a river into the sea or a lake.

However, this bending back of the neck could also be explained by the fact that a decrease in tension of the neck muscles after death caused elastic ligaments between the neural spines of the cervical vertebrae to draw the neck back automatically. One also observes

12 Schäfer, W., 1962. *Aktuo-Paläontologie nach Studien in der Nordsee*. Frankfurt am Main. Schäfer investigated post-mortem history of seabirds in detail. His findings can also be applied to pterosaurs.

Left: This *Rhamphorhynchus longicaudus* presumably lost its head when its carcass was floating in the waters of the Solnhofen lagoon. If a dead pterosaur drifted around on the surface of the water for a long period of time, it must have been quite usual for parts of the carcass to disintegrate and eventually get lost. The skull would normally become detached first, and then the wings and legs. In this specimen, only the skull is missing; the rest of the skeleton has remained intact.

Below and below right: This skeleton of *Pterodactylus elegans* from the Solnhofen limestone of Bavaria is seen with its head and neck strongly bent backward. Such a posture is commonly observed in fossil pterosaurs, and is caused by the pull of ligaments between the neck vertebrae after the tension of the musculature has ceased to operate after the animal has died. This peculiar bending of the neck can also be seen in fossil and extant birds, especially when the carcass has been mummified on dry land. Such an example is shown in the drawing of the Common Tern (p. 13).

this in many birds who sleep with their heads turned back into their feathers. They do not need any muscular activity to do this, it is enough to relax the neck muscles so that the interspinal ligaments can exert a backward pull.

If the skeletal remains finally sank to the bottom of a lake or sea, they could possibly still have been disarticulated by currents and gradually covered and completely embedded by continuing sedimentation. The skeleton was crushed by pressure from the layers of sediment, which at first contained a lot of water, then gradually became more and more compacted and heavy. In the course of further compacting petrifaction, or mineralization, gradually occurred, and this also affected the bones of the skeleton.

The bones were impregnated with circulating mineral solutions. This explains why fossil bones and teeth generally consist of inorganic, mineral substances rather than organic ones, and thus are 'petrified'. In this way the internal structure of the bones is often preserved, often down to very fine details.

The kind of mineralization that occurs is dependent on the surrounding milieu. Limestone is formed from limy marine ooze, of the kind that occurs in the Upper Jurassic lagoon of Solnhofen. Fossil bones then consist of calcite or calcium carbonate ($CaCo_3$). Sandy deposits turn into sandstone; the bones are then usually silicified, that means they consist of very hard quartz-like silicic acid (SiO_2). Clayey ooze becomes shale or slate. Here too bones can occur in calcified form.

A special form of fossil preservation is that in calcareous nodules or concretions. Famous examples are fish fossils from the Santana formation in Brazil.[13] Concretions are concentrated accumulations of materials that were originally evenly distributed. They can form as a result of the presence of decomposing organic substances. Pterosaur bones are preserved three-dimensionally in the concretions of the Santana formation; these fossils are not compressed and fragmented, as is usually the case. Often complete wings, vertebral columns and whole skulls are found in a concretion, and in rare cases remains of wing membranes as well. The concretion surrounds the fossils and protects them at the same time. Thus extremely thin-walled and hollow pterosaur bones are protected from the distorting pressure of the rocks above them. However, a fossil enclosed in a solid limestone concretion is extremely difficult to prepare and requires a great deal of patience, skill and experience from the preparator concerned.

Fossilization is only possible in sedimentary rock. In magmatic rock, formed from red-hot molten magma in the depths of the earth's crust or from volcanic lava, no fossils can survive. Thus when looking for fossils we can happily ignore rocks like granite or basalt. We restrict ourselves to strata formed at some point in the history of the Earth as sedimentary rocks, whether on land or at sea. In the case of pterosaurs these were mainly deposits formed in lakes and seas.

Also with respect to pterosaurs we must always be aware that only a small percentage of all organisms has survived in fossil form, probably well below 1 per cent.[14] Thus a large proportion of individual finds of a species or genus shows that on many occasions we are dealing with random finds and owe our knowledge of them merely to collector's luck. Primarily there is little chance of fossilization in areas affected by erosion, in uplands in other words, as most remains are carried away and destroyed.

Thus in the case of pterosaurs we have to reckon with enormous gaps in the fossil record, especially for forms that inhabited higher inland areas. A group of animals that lived on Earth for as long as birds have today, in other words at least 150 million years, must have occupied as many habitats as modern birds. We know that pterosaurs lived all over the world. It can hardly be assumed that they lived exclusively near to coasts and shores, where their fossils have been found in sedimentary lowland areas. Additionally one must take into consideration that only very small sections of the sedimentary rocks in which pterosaur fossils potentially could occur are on the surface of the Earth today. To a large extent they were eroded in the course of the history of the Earth or are concealed under more recent deposits.

However, although we only know a small proportion of the pterosaurs that once lived on Earth, it is astonishing how much information we nevertheless have about them. Indeed, they

Above: By contrast with the previous specimens, this *Rhamphorhynchus* skeleton has been disarticulated, and some elements may have already been lost when it sank to the sea bed. There, water currents scattered the bones before they were embedded.

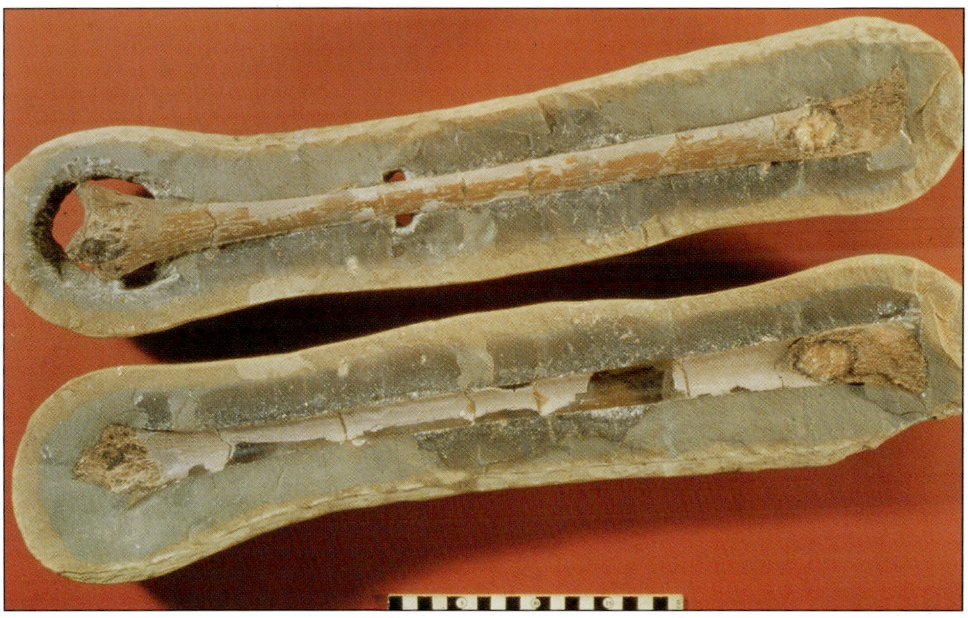

Below: The famous fossil concretions of the Santana formation in Brazil contain mostly fishes, but rarely also skeletal remains of pterosaurs. Here the first wing phalanx of a large pterosaur, *Araripesaurus dehmi*, is preserved three-dimensionally.

13 Martill, D.M., 1988. *Preservation of Fish in the Cretaceous Santana Formation of Brazil.* Palaeontology, 31 (1): 1-18.

14 The number of plant and animal species to have lived since the Cambrian, i.e. in the last 590 million years, is estimated at about one thousand million. Some even suggest 1·6 thousand million fossil species. Of these about 130,000 are described. An estimated 4·5 million plant and animal species live on the Earth today. (E. Kuhn-Schnyder, 1977. *Die Geschichte des Lebens auf der Erde.* Mitteilungen der Naturforschenden Gesellschaft des Kantons Solothurn, 27.)

Left: Since Roman times the Solnhofen quarries in Bavaria have been worked by hand to excavate the bright yellow limestone for tiles, steps, and latterly printing stones. Careful removal of the slabs often reveals fossils here.

Above: Bones preserved in calcareous nodules at the Santana formation on the Araripe plateau are not compressed and so may be preserved three-dimensionally. Careful preparation has revealed these pterosaur wing bones.

can be said to be one of the best-researched and best-known fossil reptiles of all.

Excavation and Preparation

It is not really possible to look specifically for pterosaur fossils. They are far too rare for that. Even in the well-known sites where finds have been made, like Solnhofen in Bavaria, on the Araripe Plateau in Brazil or in the Kansas Chalk, pterosaurs are always rare and valued finds. The relatively high number of specimens from these sites is due to intensive quarrying and tireless collecting over a long period.

Pterosaur bones are extremely light in structure. The long bones are hollow like those of birds, and have extremely thin walls. The vertebrae of larger species also have cavities. The skull is often only a structure of bands surrounding large openings and apertures. This means that skeletal bones of a pterosaur were often very fragile, and were damaged or rubbed away by coarser sediments like sand and gravel. And indeed we only find pterosaurs in fine-grained sediments like limestone, shale and marl, or in calcareous concretions, but very rarely in coarse sandstones. Here the delicate, fragile bones scarcely had a chance of fossilizing.

If the remains of a skeleton are discovered on a stratum of rock, the first thing to be attempted is carefully to loosen the rock containing the fossil from the stratum in which it is set without breaking or losing bones. Also it is essential that care is taken that all the parts of the fossil are completely excavated. It has frequently occurred that hours have to be spent looking for a missing piece of stone that was mistakenly thrown away, but still contained parts of the fossil. Usually fossil remains are enclosed in rock and thus protected. Remains of skeletons can then only be recognized by irregular protrusions on the surface of the rock or from cross-sections of bones on the edges of the rock.

If the fossilized bones are brittle and crumbly, then they have to be fixed and hardened on site. Shellac or thin artificial resin lacquers are used for hardening them. To guarantee safe transport to the laboratory, protective plaster bandages may be used or a plaster jacket, and possibly polyurethane foam as well, may have to be wrapped around the fossil remains. It is advisable to sketch or photograph the position of the skeletal remains beforehand. It is often sensible to take a compass bearing in order to determine the distribution of the fossils. This is important if the position of the fossils was affected by water currents: their position can be used to work out the direction of the current and thus the likely position of further fossil remains.

After discovering and excavating fossils, the next task is to preserve the remains, i.e. remove them from the rock, which is usually hard, by suitable methods, and uncover them. This is done both with the traditional tools of hammers and chisels of various sizes and degrees of fineness, and nowadays often with the aid of

Left: Quarry 'Mina Pedra Branca' at the Araripe plateau in Brazil. The quarry is operated for the gypsum at the bottom of the wall; above it is a section of the early Cretaceous Santana formation.

electrically driven vibrating needles, with an air abrasive or in suitable cases with various acids (acetic acid or formic acid), which eat away the rock.[15] This work requires expert knowledge, a craftsman's art and above all patience and instinctive skill. This preparation is the work of a team of experts without whose technical assistance palaeontology as a science would not be possible.

X-ray technology has often proved a valuable aid for reliable preparation work.[16] An X-ray of a suitable fossil will show up all the details of a skeleton, even if it is still enclosed in rock itself. The preparator can use X-ray photography as an aid to uncovering the bones of a pterosaur skeleton, which are often very small and delicate.

Thus the palaeontologist's working methods are often based on the morphology, shape and form of the hard parts of a skeleton that may be preserved in fossil form. Fortunately under particularly favourable conditions the structure of soft parts has also survived, usually as an imprint of the body skin or the flight membrane. With the aid of ultra-violet light, organic fossil remains can be made visible against the inorganic surface of the rock. Organic substances fluoresce in yellow-green light, so that (for example) soft parts, horn beaks or horn claws, become visible. Even the shape of the brain is known in the case of some pterosaurs, for example if fine ooze was forced into the empty brain case of the skull and the ooze petrified along with the skull. A natural endocast produced in this way is a faithful copy of the brain, as it is possible to work on the assumption that the pterosaur brain filled the entire brain cavity, as is the case with birds.

Above: The weak outlines of bone in this thin slab of shale indicate a pterosaur skeleton within. The X-ray photograph (below) reveals the complete skeleton of *Dorygnathus*, and guides the preparator in his efforts to uncover the bones.

Above right: Sometimes the brain cavity of fossil vertebrates was filled with sediment resulting in an endocast of the brain case which duplicates the size and shape of the brain itself. This is *Parapsicephalus*, a Liassic pterosaur.

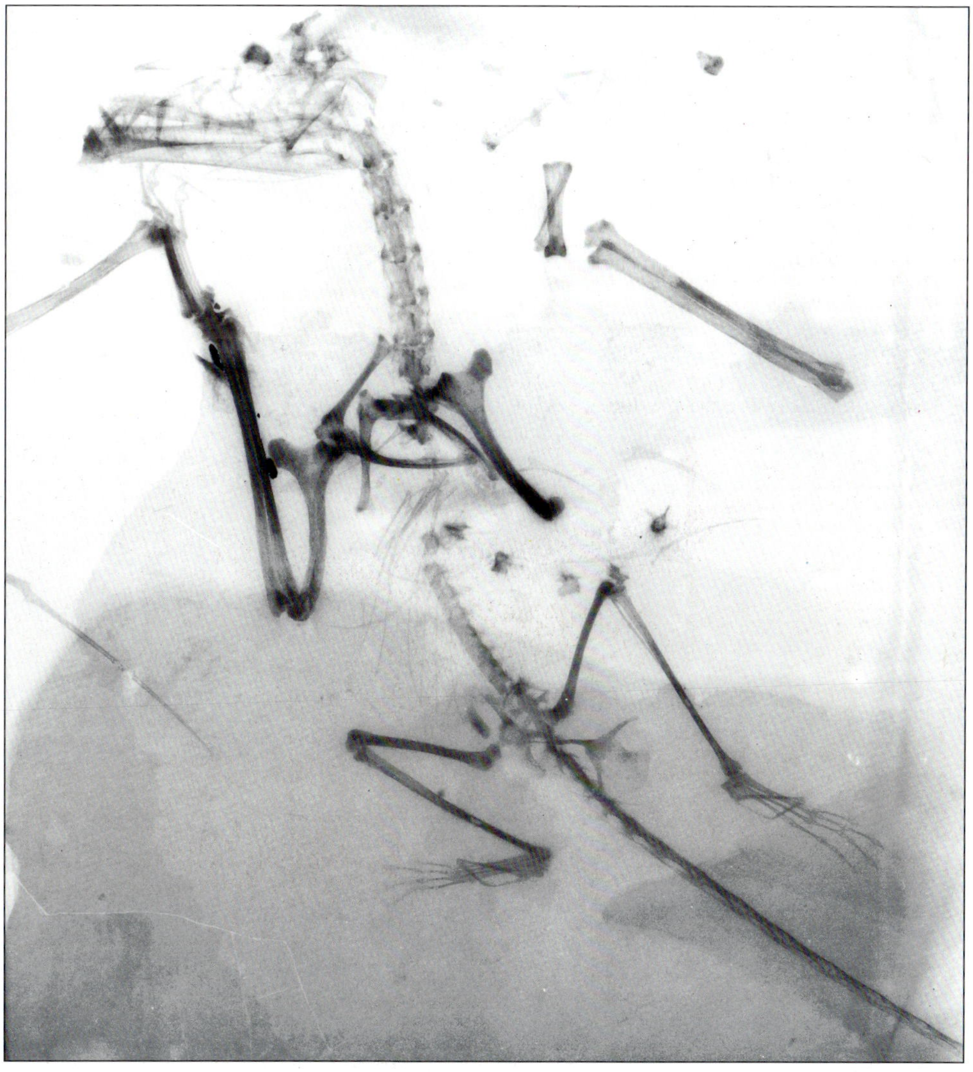

Dating the Fossils

When we work on pterosaur fossils, we know that we have before us the fossilized remains of living animals that really did exist in the mists of time. They were creatures of flesh and blood, subject to the same biological life principles as we are. Palaeontology as the study of the living world of the geological past is, however, not just a biological, but also an historical discipline. It does not only investigate animal and plant organisms in previous ages, but also follows the changes they underwent in the course of time, that is to say their development or evolution. This is only possible if fossil finds can be dated.

Fossils are generally found in geological bodies, usually in sedimentary deposits. Thus palaeontology's connection with geology is just as important as its connection with biology.

15 Acid preparation is particularly successful if the matrix surrounding the fossil is limestone, which dissolves in acid (usually acetic or formic acid). The most suitable fossils are those which are silicified in limestone, as silicon dioxide (SiO_2) is not attacked by the preparation acids. This procedure was used very successfully for the preparation of fish and pterosaurs from Solnhofen limestone and the limestone concretions of the Brazilian Santana formation. Usually 10 per cent acetic acid is used (CH_3COOH), but 8-10 per cent formic acid (CHCOOH) is better. The concretion is left in the acid for 6-30 hours, and is then treated with water and neutralized with soda lye. The bones thus released from the rock are dried, then coated with diluted polystyrol adhesive or paraffin. They are then put back in the acid. This process is repeated for as long as is necessary to free the remains of the skeleton from the rock adhering to them. This can take several weeks.

16 Fossils were X-rayed soon after the discovery of X-rays by Wilhelm C. Röntgen in 1896. The best results are achieved with thin fossil slabs in which there is a high contrasting density between the fossil and the surrounding rock, as for example in the case of pyritic fossils in slate slabs. Pioneers in the field of X-ray investigation of fossils were the Bonn physicist Walter M. Lehmann and the Erlangen physical chemist and palaeontologist Wilhelm Stürmer.

On the question of the age of fossils therefore, geological methods have to be applied. This produces the concept of relative geological time, which works with *relative* determination of the age of rocks and the fossils they contain. Beyond this geology has, by various physical methods, also established *absolute* dating by year numbers, refining and extending it down to the present day.

Relative Dating

Long before it was possible to calculate the absolute age of rocks using radioactive isotopes, a system of relative dating was used. This is based on the so-called *stratigraphic principle*, first established by the Danish doctor Nicolaus Steno in 1669. This simply means that when sedimetary rocks are formed the upper strata are bound to be more recent than the ones underneath them, as they were deposited at a later date. In an undisturbed sequence of rocks, known to geologists as a *profile*, there is thus an age sequence from bottom to top, from the older rocks to the more recent ones.

This essentially obvious regularity became extraordinarily important for palaeontology and historical geology. At a very early stage, particularly in England, France and Germany, scholars began to classify rocks according to their relative position and age, to divide them into geological epochs and formations and to give them names. Geological periods like *Cambrian* or *Devonian* are named after landscapes in Wales and South-West England, the *Permian* after a Russian province, and the *Jurassic* after the Jura range in Switzerland. *Ordovician* and *Silurian* are derived from the names of Celtic tribes which once lived in Wales. *Carboniferous* denotes the presence of coal, and Carboniferous deposits throughout the world contain a great deal of coal. The *Cretaceous* is associated with chalk via the Latin. This leaves the *Triassic*, the trinity, because of its three subdivisions, *Bundsandstein*, *Muschelkalk* and *Keuper*, and *Tertiary*[17] and *Quaternary*, the third and fourth systems. At first only four *systems* were identified.

Later the six older periods, Cambrian, Ordovician, Silurian, Devonian, Carboniferous and Permian were grouped together and classified as the Palaeozoic ('ancient life') era, the middle three, Triassic, Jurassic and Cretaceous were termed the Mesozoic ('middle life') era and the two latest, the Tertiary and Quaternary, became the Cenozoic ('recent life') era. The period before the Palaeozoic is called the Pre-Cambrian.

This geological history without dates was principally worked out from fossils, from the point around 1800 when the English engineer William Smith[18] observed in the course of canal-building in the Midlands that fossils of animals and plants do not occur randomly in rock, but that each stratum contained characteristic fossils distinct from those above and below. Strata that occur at various, often widely separated points, must be of the same age if they contain the same fossils. Later, when observations were refined, it was recognized that many species of fossilized flora or fauna remain the same over several strata, but that others only occur at a particular level. This led to the conclusion that some species must have had longer and others shorter existences. It was the short-lived species that were of greatest significance for the establishment of relative geological age. If at the same time they were widely distributed, they could be used as time markers and for dating and correlating strata that were a considerable distance apart. These are then known as *index fossils*.

Now that it was possible to use the terms *older* and *younger* in geological terms, it was noticed that only simpler, more 'primitive' organisms occurred in the older strata, and more complicated, more 'progressive' organisms in the younger strata, but never the other way round. This can only lead to the conclusion that a development, an *evolution* of life took place in the course of geological history, which in the last resort led from unicellular organisms to man. Thus the relative age of species preserved as fossils in rock can be determined from their evolutionary level. This *biostratigraphy* is very refined today. It uses various plant and animal groups as index fossils. The smallest unit of time in this system is the *biozone*, which is determined by the lifespan of an index fossil species. Its beginning is marked by the first and its end by the last appearance of the index species.

Unfortunately it is not possible to make such refined zone classifications for pterosaurs. Finds are far too rare for this, and they obviously do not develop as quickly as the ammonites, for example. These marine, squid-like molluscs with spiral shells are outstanding index fossils, especially in the Jurassic and Cretaceous. But if we apply a somewhat coarser time-grid, it is perfectly possible to make state-

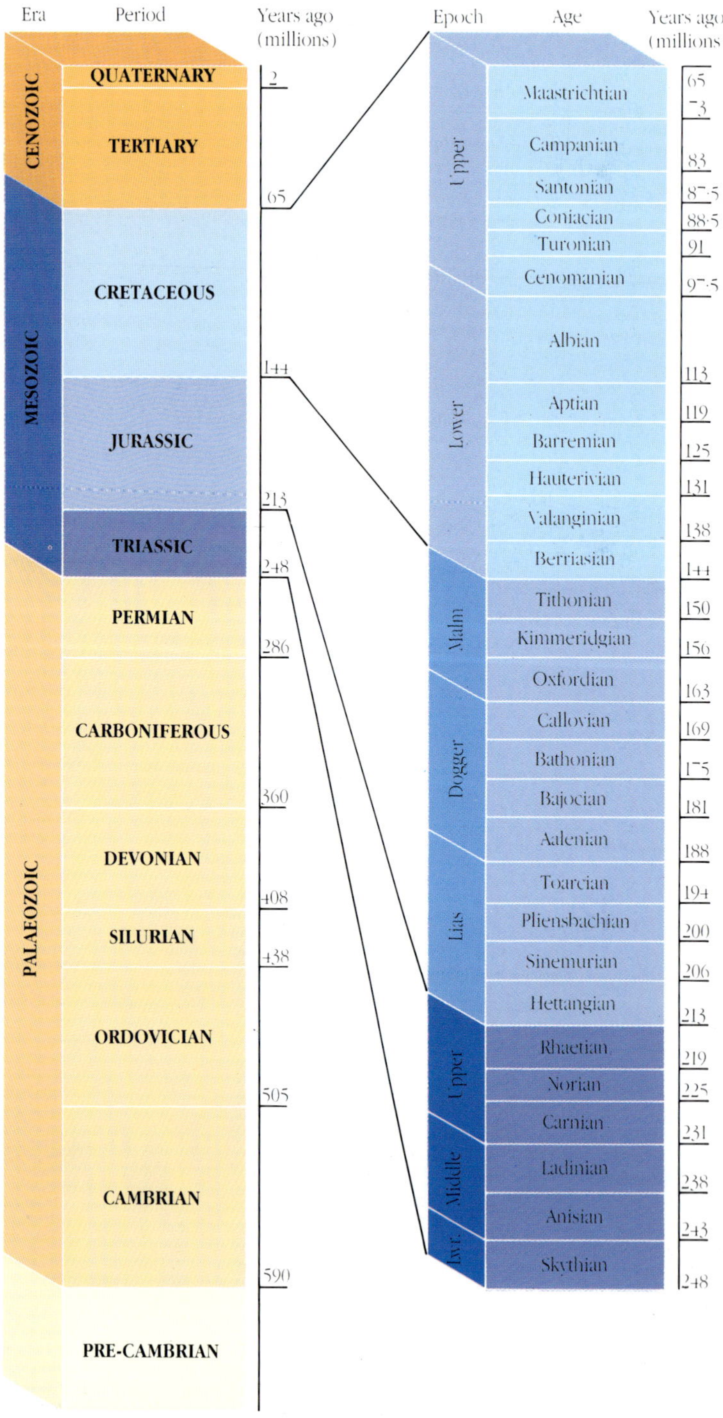

The Mesozoic Era (left)
This diagram shows how geologists and palaeontologists classify rocks according to their relative position and age, and so can construct a stratigraphic sequence which is basically a step-by-step history of the Earth's formation. The major divisions are called eras, and shown here are the Palaeozoic era (590-248 million years ago), the Mesozoic era (248-65 million years ago) and the Cenozoic era (65 million years ago to the present day). The time before the Palaeozoic is called the Pre-Cambrian. These eras are then divided into smaller units called periods, as can be seen from the main diagram (far left). In this book we are concerned with the Mesozoic, which was the age of the pterosaurs. This era has been pulled out in detail (near left) to enable the reader to see the further sub-divisions which scientists recognise during this span of time. The three Mesozoic periods – Triassic, Jurassic and Cretaceous – are split into epochs, and each epoch in turn encompasses a number of ages which, refined with absolute dates, form a relatively exact calendar of Mesozoic life history.

Relative Dating (right)
This geological profile shows different sedimentary rocks deposited at different times. The lower the position of the rock in the profile, the earlier it was deposited and hence the older are the fossils it contains.

17 The name Tertiary is derived from Montes tertiarii, a concept introduced by the Italian geologist Giovanni Arduino in 1759 for deposits at the foothills of the Alps in northern Italy. At that time the older strata were called Montes primarii (pre-Cambrian and Palaeozoic) and Montes secundarii (Mesozoic). The name Quaternary (the fourth age) was introduced by the French geologist Jules Desnoyers in 1829 to cover the post-Tertiary period.

18 William Smith (1769-1839), English geologist and engineer, established stratigraphy, the science of the age sequence of rock strata, in 1816-1819. *Strata identified by organized fossils.* 4 parts.

ments about age, as indeed it is for other fossil reptiles as well. We know for example that short-tailed pterosaurs, the Pterodactyloidea, first appear in the Upper Jurassic, but not in any older strata. On the other hand long-tailed pterosaurs (Rhamphorhynchoidea) only occur until the Upper Jurassic, and not at all in the Cretaceous. Thus if remains of long-tailed pterosaurs are found, then the Cretaceous can confidently be excluded from our dating estimates. Equally a short-tailed pterosaur can never occur in Triassic deposits, but is typical of the Upper Jurassic or Cretaceous. In most cases more precise relative dating is possible by identifying accompanying fauna in the same deposit and from the geological situation.

Absolute Dating

Measured by the standards of human age, the age of the Earth, today thought to be about five thousand million years, is not conceivable to us. And yet the enormously long periods in which the Earth and life upon it have developed are the daily bread of palaeontologists and geologists. These periods are part of the reality of the world, just as the gigantic distances to the stars and galaxies in space, which astronomers no longer measure by earthly standards, in miles, but only in light-years. A light-year is the distance travelled by light at a speed of 186,000 miles per second (300,000 km/sec) in a year. That is over 6·2 million million miles (10 million million km).

As the notion of a light-year contains a time component, it is possible at least in our minds to make a connection with the history of the Earth. We receive light from distant stars that was emitted millions of years ago in some cases. We do not even know whether these light sources far out in space still exist today or have already been extinguished. On the other hand for example a hypothetical observer in a group of galaxies in the Coma Berenices constellation between Virgo and the Great Bear looking at the Earth today through an enormously powerful telescope, could see living pterosaurs and dinosaurs. The light from the Earth would have taken from the Cretaceous until today to arrive.

Above: Fossil preparation is a delicate art. Here the preparator uses a hammer and fine chisel to chip away the rock containing the skull of a large pterosaur from the Santana formation of Brazil. For fine details a binocular microscope is also a necessary tool.

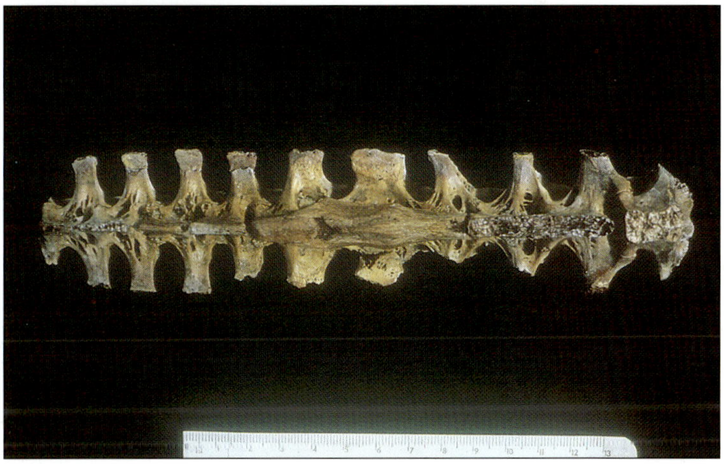

Right: If the chemical conditions of the rock and fossil are right, acid preparation can be applied. These vertebrae were treated with acetic acid.

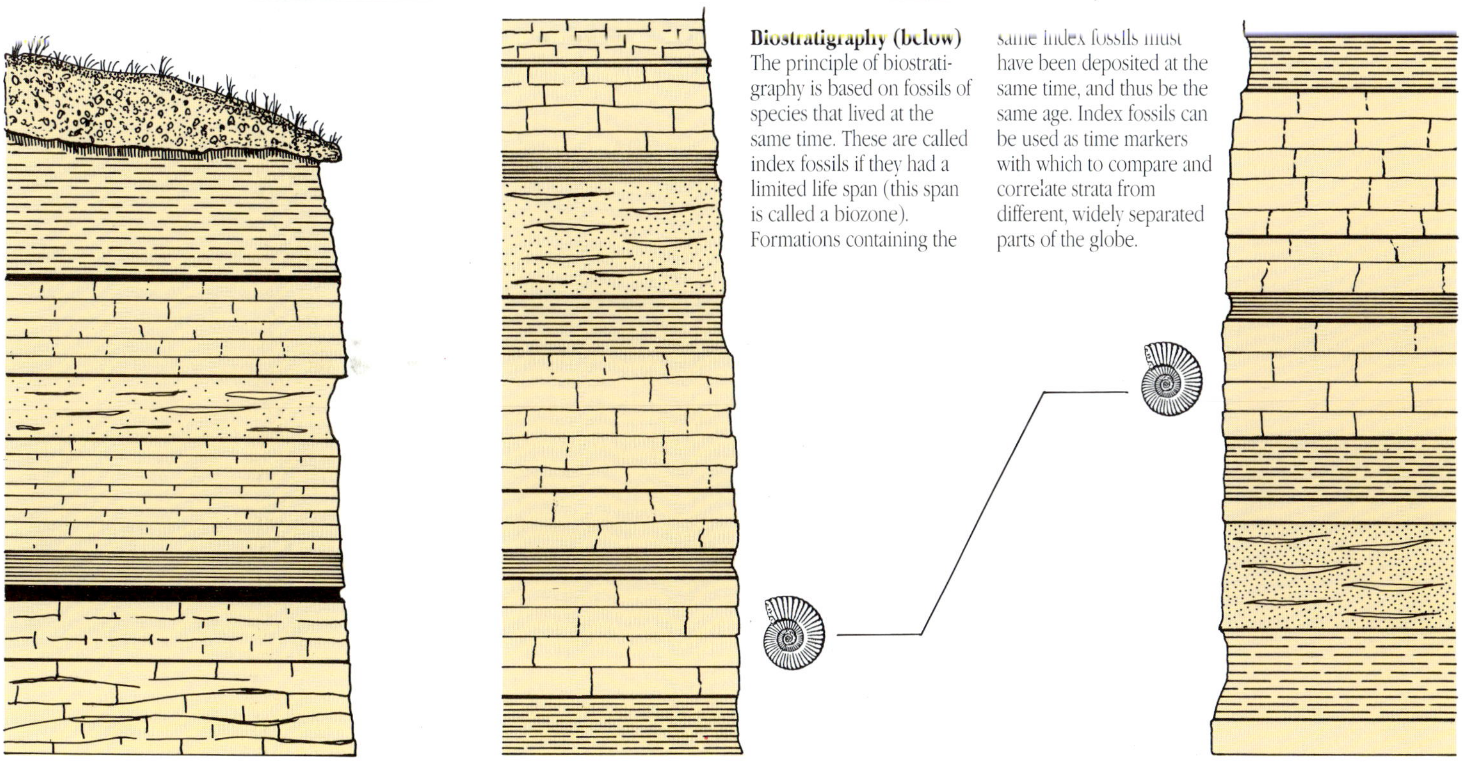

Biostratigraphy (below) The principle of biostratigraphy is based on fossils of species that lived at the same time. These are called index fossils if they had a limited life span (this span is called a biozone). Formations containing the same index fossils must have been deposited at the same time, and thus be the same age. Index fossils can be used as time markers with which to compare and correlate strata from different, widely separated parts of the globe.

How do geologists arrive at such high geological ages? In the last resort rocks and the fossils occurring in them just do not reveal their absolute age.

In 1650, Archbishop James Ussher of Armagh worked out from biblical evidence that the Creation took place on Sunday 23 October in the year 4004 BC, at nine o'clock in the morning. The calculation made in 1862 by William Thomson (the eminent British physicist who later became Lord Kelvin), working on the time it would take the Earth to cool from its red-hot molten condition, is founded on a physical basis. He arrived at an age for the Earth of between 20 and 400 million years.

But it was not until the early twentieth century that physicists Ernest Rutherford in England and Bertram B. Boltwood in America discovered that it was possible to use radioactive elements as a means of measuring time and determining the age of rocks. Many rocks contain small quantities of radioactive minerals. The principle is based on the fact that in the course of time radioactive elements decay to form stable end products and in a way that is completely constant and independent of any outside influence. The measurement of the rate of decay is the so-called half-life, which means the time it takes a radioactive isotope to go through half the decay and transformation process. From the measured relationship between original substance, the parent isotope, and end product, the daughter isotope, it is possible to calculate the time at which this atomic clock started to tick. This reveals the date of the rock's origin, and thus its age.

The isotopes uranium 238 and 235, thorium 232, rubidium 87 and potassium 40 are used as suitable radioactive elements today. Uranium 238 for example decays through various stages to form gaseous helium and the stable end product lead 206. So if rock containing uranium is examined, its age can be calculated from the ratio of remaining uranium to the newly formed lead. In this case the half-life is 4,510 million years. This means that after 4,510 million years half the uranium 238 has decayed to form lead 206.

For a dating up to a maximum of 50,000 years the radio-carbon method is used, working with carbon C14, with a half-life of only 5,570 years. The C14 method is based on the fact that in the structure of living organisms a type of carbon that has become radioactive in the upper layers of the atmosphere, the carbon isotope C14, occurs alongside normal carbon. On the death of the organism, i.e. the plant or animal, the radioactive carbon incorporated into bone or woody tissue begins to decay, as no new C14 is taken up from this point. The date at which decay began, and thus the age of the specimen, can be calculated if the very weak radio-active emission is measured. Because of its limited range the C14 method is of particular use for human prehistory.

Another method of numerical dating is dendrochronology, which has a span of a few thousand years, and works by counting annual rings of old trees and fossilized timber. Varve chronology uses the seasonal rhythm of meltwater deposits in areas in which glaciers and inland ice melt. The so-called varves, alternate light and dark layers of lake clay showing fine annual banding, were counted, and by combining and overlapping profiles from North Germany to North Sweden and America, a continuous sequence can be established for the past 10,000 years, thus dating the post ice age.

Evolution of Life in Time

Absolute dating methods at our disposal today, which are continually being refined, have shown that rocks in which we find the first traces of the first, simple life are about 3,500 million years old, that the first multicellular organisms appeared about 2,000 million years ago, that genuine plants existed a thousand million years ago, and 590 million years ago, in other words at the beginning of the Palaeozoic Era, in the Cambrian, it is possible to observe a rapid spread of a great variety of organisms. These were exclusively marine life forms. Thus life came from the sea.

The first vertebrates were jawless, fish-like sea creatures which appear for the first time in 450 million year old strata of the Ordovician in the fossil record. Evidence of the first fish is not found until deposits which are 420 million years old. Other important evolutionary stages during the rise of the vertebrates were the first appearance of land-based tetrapods, of the first amphibians 370 million years ago, in the Devonian, and of the first reptiles over 300 million years ago, in the Carboniferous.

So reptiles (from the Latin *repere*, to crawl) flourished for the first time at the beginning of the Mesozoic, in the Triassic. Their success was

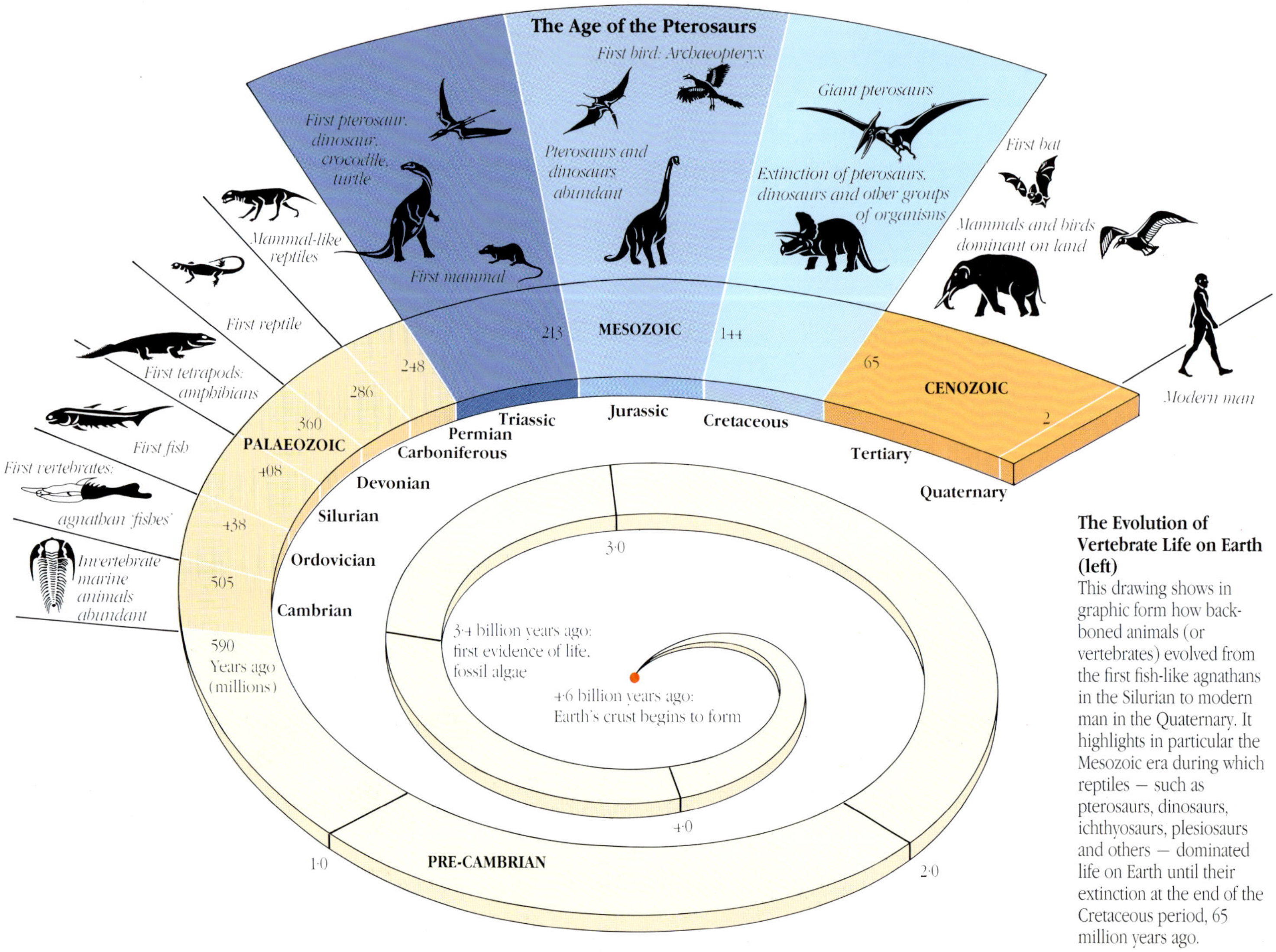

The Evolution of Vertebrate Life on Earth (left)
This drawing shows in graphic form how backboned animals (or vertebrates) evolved from the first fish-like agnathans in the Silurian to modern man in the Quaternary. It highlights in particular the Mesozoic era during which reptiles – such as pterosaurs, dinosaurs, ichthyosaurs, plesiosaurs and others – dominated life on Earth until their extinction at the end of the Cretaceous period, 65 million years ago.

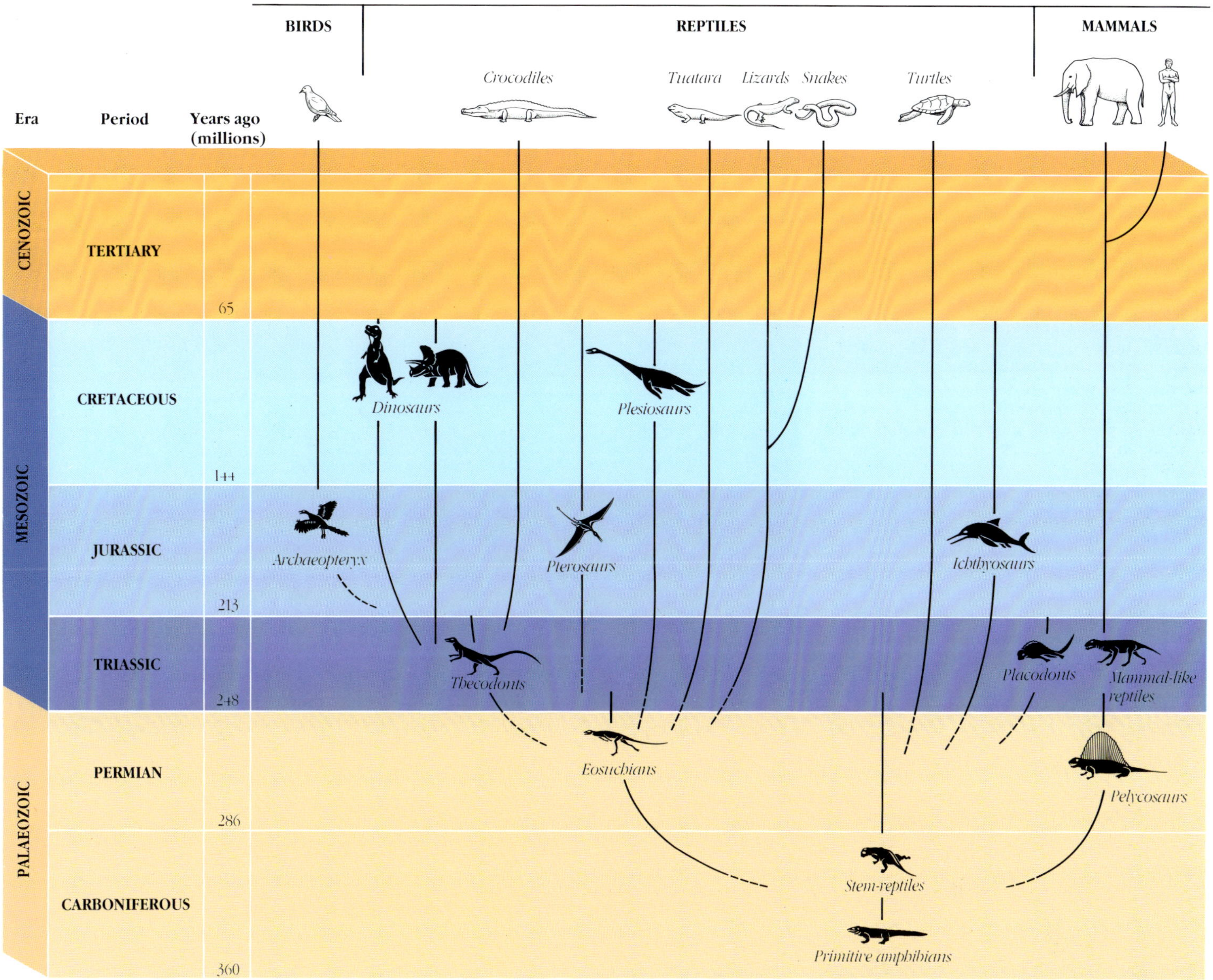

The Age of Reptiles (above)
The Mesozoic era was the age of the reptiles in general. Of many different groups only few escaped the great extinction at the end of the Cretaceous, 65 million years ago, which accounted for dinosaurs, pterosaurs, plesiosaurs and ichthyosaurs. Those that have survived until today are: crocodiles, the tuatara (*Sphenodon*), lizards, snakes, and turtles. Birds as well as mammals are considered to be descendants of different reptilian ancestors.

based on the fact that they were finally able to emancipate themselves from the water: reptiles lay their eggs on dry land. The eggs are large, with big yolks, and protected by a hard shell. Thus, unlike amphibians, they do not have to go through a larval stage in water. Their skin was protected against drying out. Thus they could make use of dry land as a new and extensive habitat.

Evolution grabbed this opportunity with great suddenness, as even as early as the Triassic all orders of reptiles, including pterosaurs, plesiosaurs, dinosaurs, placodonts and others were established. Some of these saurians returned to life in the water, with special adaptations, like for example the ichthyosaurs and plesiosaurs.

There is evidence of the earliest pterosaurs in the late Triassic. Their fossil remains were discovered in rocks in the Alpine foothills near Bergamo in northern Italy. At the same time it was proved that the pterosaurs were the *first* vertebrates to adapt to a life of active flight. They dominated the air without competition until the Late Jurassic, for almost 70 million years. Their first competitors were not to appear until 150 million years ago, with the first birds, represented by *Archaeopteryx* in the lithographic limestone of Solnhofen.

Throughout the Mesozoic, i.e. during the Triassic, Jurassic and Cretaceous, reptiles were absolute lords of the earth. For this reason the Mesozoic is also known as the Age of the Reptiles. In absolute figures their dominance lasted for over 180 million years. Certainly most of them became extinct either before or at the end of the Cretaceous, among them the pterosaurs. Latterly these reached enormous dimensions, with wing spans of 39ft (12m), making them the largest flying creatures of all times.

Pterosaurs first flourished in the Jurassic. The earliest pterosaur skeleton finds in Lower Jurassic strata on the Dorset coast of southern England are about 200 million years old. Finds at Holzmaden in Württemberg are somewhat younger, about 190 million years, but indicate a greater range. In the Upper Jurassic, about 150 million years ago, we see not only considerable evolution of a great variety of pterosaurs, but also world-wide distribution. The most important finds were made in the region of Solnhofen and Eichstätt in Bavaria. But pterosaur fossils have been found in Upper Jurassic strata in France, England, Portugal, Russia, East Africa, Cuba, the United States and China as well.

After long-tailed pterosaurs, the Rhamphorhynchoidea, became extinct in the late Jurassic, short-tailed pterosaurs, the Pterodactyloidea, developed considerably in the Cretaceous. The earliest Pterodactyloidea appeared in the Upper Jurassic, but they evolved to the largest extent in the early Cretaceous, about 135 million years ago. The strangest pterosaurs, like *Pterodaustro* for example, a creature with filter dentition from the Cretaceous in Argentina, lived in this period.

Other pterosaurs, some of them enormous, lived later, in the Upper Cretaceous in China, Australia and Brazil. The pterosaur fossils described in the last century from the Greensand in Cambridge in England are somewhat more recent. Giant pterosaurs of the genus *Pteranodon*, found by O.C. Marsh's fossil hunters in the Kansas Chalk of the United States, can be dated to about 80 million years ago.

Geologically the most recent and also the largest known pterosaur, the famous *Quetzalcoatlus*, lived in the region which is now West Texas only towards the end of the Cretaceous, about 65 million years ago.

At the turn of the Cretaceous to the Tertiary, 65 million years ago, the reptiles' age of splendour was over. Of the 17 orders of reptiles only four survived, crocodiles, snakes and lizards, tortoises and the tuatara. Subsequently mammals and birds evolved in their place. During the Mesozoic they had eked out an undistinguished existence, and are scantily represented in the fossil record. At the beginning of the Cenozoic, in the Tertiary, mammals also produced an actively flying representative, the bat. Here the oldest fossil find is *Icaronycteris*, from 50 million year old lake deposits in Wyoming, USA. The last scion of this evolution of the vertebrates, only roughly sketched here, is man, the genus *Homo*. He appeared late, that means a mere two million years ago, and modern man only at the beginning of the last ice age, about 70,000 years ago.

We now want to present the enormous timespans of the geological past and the evolution of life on Earth in a form which it is easier for our minds to grasp. To do this we want to represent the total duration of the history of the Earth from the moment in which it came into being until the present day as a single 24 hour day. Thus the beginning of the day is the formation of the Earth 4,600 million years ago: 00.00 hours. The present moment is reached at the last stroke of midnight: 24.00 hours. Thus an hour represents 191·7 million years, a minute 3,194,000 years and a second still lasts for 53,240 years.

The most significant dates in the history of the Earth and the evolution of vertebrates can be read from this clock like a timetable of life:

00.00 h
Formation of the Earth
03.07 h
Formation of the Earth's crust
06.15 h
First signs of life
13.34 h
First multicellular organisms
20.55 h
Beginning of the Cambrian
21.39 h
First vertebrates
22.04 h
First land-based tetrapods (amphibians)
22.26 h
First saurians (reptiles)
22.51 h
First pterosaurs
23.13 h
First birds (*Archaeopteryx*), long-tailed pterosaurs become extinct
23.35 h
Short-tailed pterosaurs flourish
23.39 h
Pterosaurs, dinosaurs and others become extinct
37 seconds before 24.00 h
Appearance of early man (*Homo*)
2 seconds before 24.00 h
Appearance of modern man (*Homo sapiens*)

Dragons of the Myths

If we pursue the history of the investigation of pterosaurs, the flying saurians of prehistoric times, there is a natural link in our minds with the myths and legends of dragons. These were usually also seen as winged lizards or snakes.

For 16th and 17th century scholars dragons were still a reality. For example, in the Schlangenbuch (Snake Book) by the famous Swiss naturalist and town doctor Conrad Gessner, dating from 1589, there is a chapter called 'Von den Tracken', in which he describes and illustrates various dragons. He also describes a battle between a Swiss called Winkelried and a dragon which took place near the Swiss village of Wyler.[19] The scholarly Jesuit father Athanasius Kircher provided a picture of this fight in his great work on natural history *Mundus Subterraneus* (The world below the Earth) in 1678. According to this the dragon had a long neck and tail, four legs, and wings. The Viennese palaeontologist Othenio Abel suggested in this context that the drawing could have been based on fossil reptile finds, possibly long-necked plesiosaurs from the Jurassic strata of Württemberg.

At a time when even naturalists believed in fabulous creatures and monsters, the discovery of fossil bones and remains of skeletons in caves must have reinforced ideas of dragons. Old names like dragon's cave, dragon's rock or dragon's stone still occur on modern maps.

Another picture of a dragon by Athanasius Kircher has come down to us. It was said to have been killed by a knight of St. John on the island of Rhodes. This picture is also said to have been the inspiration for Schiller's ballad 'The Fight with the Dragon'.

19 According to an ancient legend, in the early days of Swiss settlement a cruel dragon lived above the village of Wyler, and drove out men and cattle. A countryman called Winkelried, who had been banished for murder, offered to kill the dragon, in exchange for which he would be allowed to return. After he had defeated the dragon he held the bloody sword high in the air and the dragon's blood dripped on to his body, causing his death. Bölsche, W., 1929, *Drachen. Sage und Naturwissenschaft*, Stuttgart (Kosmos).

Below: The Swiss countryman Winkelried killing a dragon that had menaced the village of Wyler. This illustration appeared in *Mundus Subterraneus* by the Jesuit scholar Athanasius Kircher in 1678. It was conceivably based upon fossil finds.

Left: Every year in August a traditional festival called the '*Drachenstich*' (the slaying of the dragon) takes place in the small town of Furth im Wald in eastern Bavaria, Germany. A huge fire-breathing, winged dragon, which is actually mounted on wheels and powered by an engine, is ritually 'killed' by a prince on horseback during the performance. Such celebrations recall Christian allegories such as that of St George and the Dragon which express the triumph of good over evil. St Michael, St Margaret, St Sylvester and St Martha are all depicted as slaying dragons.

Right: This 18th century sculpture depicts a dragon as part of the fountain in front of the pilgrimage church of Loreto in Italy. Again it reminds us that in the Christian tradition the dragon has always figured as the symbol of the devil.

Even in the early seventeenth century such legends persisted, like that of the flying dragon on Mount Pilatus near Lucerne in Switzerland. It is reported that the creature flew out of a cave on Mount Pilatus in 1619, and flapped across the valley with slowly beating wings. Athanasius Kircher also created an imaginative picture of this, and it was still being printed on the map of Switzerland in Mattaeus Seutter's Atlas of the World in 1730.

There is still a ceremony of 'dragon-slaying' in a festival in the small town of Furth im Wald in eastern Bavaria. This involves a large, fire-spewing winged dragon being killed by a knight. When one remembers that even in our century the teeth of fossilized mammals were sold as 'dragons' teeth' by Chinese chemists, this gives a sense of how deeply notions of dragons were rooted in the peoples of Eastern Asia.

In China dragons were symbols of the might of the Emperor and of happiness for thousands of years, whilst in the culture of the West the dragon was usually a symbol of evil and the devil. Pictures in which the Archangel Michael, St. George or Christ himself are portrayed as dragon-killers are ancient in origin and very widespread. Fights with dragons are a recurrent myth in various forms amongst many peoples. They always deal with a god fighting with a dragon, for example Indra, Apollo, Hercules, Jason, Thor or Siegfried. The well-known Greek legend of Cadmus tells how he killed a dragon, from whose teeth the dragon's seed, armed men, sprang up.

Ancient notions of dragons suggest they had a snake-like body, two legs and bat's wings. A second pair of legs was not added until the 16th century.[20] These different types of dragons are impressively presented in Conrad Gessner's 1589 *Book of Snakes*. There are similar pictures in Sebastianus Munsterus' *Cosmographia Universa*, printed in Basel in 1544. They may have been the model for the creator of the Lindwurm Monument in Klagenfurt in Austria. In fact this sculpture was based on a fossil find from the region, the skull of an ice-age woolly rhinoceros. Clearly this was thought to be the skull of a prehistoric dragon.[21]

Below: The Lindwurm Monument in Klagenfurt dates back to the late 16th century. The head is based on the skull of an ice-age woolly rhinoceros which had been found in 1335 near the town, and taken to be the skull of a dragon that had lived nearby.

However, none of these concepts of dragons has anything to do with pterosaur fossil finds. Even when they really were based on fossilized remains, in many cases those of ice-age cave bears, the wings are always represented as bird- or bat-like. Actual pterosaur fossils could hardly have given rise to dragon legends, as the significance of such remains was disputed as late as the early nineteenth century, when scientific investigation had already begun.

20 Smith, G.E., 1919. *The Evolution of the Dragon*. London.

21 Abel, O., 1939. *Vorzeitliche Tierreste im Deutschen Mythus, Brauchtum und Volksglauben*. Jena.

Round about 1757 Karl Theodor, the Elector Palatine, discovered a number of assorted shells in a cupboard in his palace in Mannheim. He was so fascinated by their colours and shapes that he decided to start his own collection of natural objects. In the 18th century it was quite the fashion for princes and noblemen to collect minerals, zoological and botanical treasures, fossils, ethnological items and above all natural rarities and curiosities. So Karl Theodor established a *Naturalienkabinett* (cabinet of natural objects), an act which was in keeping with his desire to instruct; in other words it was a service to the public. Today we would call such a collection a natural history museum, although at first it occupied only two rooms in the east wing of the Mannheim palace, later extended to four galleries containing mineralogical and palaeontological collections, and also specimens of extant vertebrates, birds, insects and rare plants.

In 1764 the Elector handed over direction of the development of the *Naturalienkabinett* to historian and naturalist Cosimo Alessandro Collini (1727-1806).[1] He was to be the first man to examine a pterosaur fossil and write a scientific treatise on it; this appeared in 1784 in the Acta Academiae Theodoro-Palatinae in Mannheim.[2] The object of this description was an exceptionally attractive fossil found in the limestone quarries at Eichstätt in Bavaria. We know that Elector Karl Theodor received fossils for his natural history collection from Graf Friedrich Ferdinand of Pappenheim, a small town not far from the famous quarries of Solnhofen and Eichstätt. The fossil must have arrived in Mannheim between 1767 and 1784, as Collini did not list it in his 1767 catalogue of the collection.

Like many 18th century naturalists, Collini no longer thought that fossils were mineral formations, freaks of nature or even objects left behind after the Great Flood. He realized they were remains of living things from earlier epochs and compared them with modern plants and animals. However, faced with the fossilized vertebrate of Eichstätt he got into considerable difficulty: there was nothing in the animal kingdom comparable with the object in front of him, a remarkable skeleton with a long, toothed snout, claws on hands and feet and long, thin bones for the forelimbs.

1 Cosimo Alessandro Collini was born on 14 October 1727 in Florence, first studied law in Pisa, then went to Switzerland in 1749 and Prussia in 1750, to Frederick the Great's Berlin, intending to study Pure Science and undertake historical research. He met a number of scholars here, and in 1752 became secretary to the French writer, philosopher and historian Voltaire, who was living in Berlin at the time, and considered to be the embodiment of the Age of Enlightenment. In 1756 Collini left Voltaire and took a post as tutor in Strasbourg. During this period he became interested in natural science, especially zoology and anatomy, with which he concerned himself intensively. On Voltaire's recommendation he became an official at the Mannheim court of the Elector Palatine, Karl Theodor, in 1760, and was subsequently appointed supervisor of the natural history collection. At the same time he became a member of the Mannheim Academy of Science and later published several essays on animal fossils that had found their way into the Mannheim natural history collection.

2 Collini, C.A., 1784. *Sur quelques Zoolithes du Cabinet d'Histoire naturelle de S.A.S.E. Palatine et de Bavière, à Mannheim*. Acta Academiae Theodoro-Palatinae Mannhein, 5, pars physica: 58-103.

Left: Karl Theodor (1733-1799), Elector of the Palatinate and Bavaria, who founded the Mannheim Natural History Collection in the middle of the 18th century. The first known pterosaur fossil specimen was housed in this *Naturalienkabinett.* This engraving was made in Paris in 1768.

Despite this Collini provided an astonishingly precise reproduction of the fossil, a copper engraving by Verhelst, which has proved an important basis for later research. Strangely enough he thought it possible that the right hand and right foot were not part of the creature, as they were not positioned at the point where they actually belonged. But he rightly recognized that the creature's arms could be folded, and expressed a supposition that a membrane could have been attached to them.

In an attempt to categorize this fossilized vertebrate, he established that it was impossible to place it in any known group of animals. He expressly excluded it from the class of birds, a view later put forward by Professor Friedrich Blumenbach[3] of Göttingen in particular. As far as the long folding arms, tail, hind legs and feet were concerned he recognized a similarity with bats, but rejected this interpretation because of the long, toothed beak. Finally he thought the most reasonable supposition was that the animal was a sea creature, but without being able to interpret the long legs with any clarity.

Collini concluded his treatise with a remarkable observation for its period, which modern palaeoecologists would fully endorse. He writes: 'In order to understand the function of all parts of its body, one must also know its habitat, its diet, its enemies, the other circumstances of its natural behaviour and the way in which the species survived and reproduced.'

3 Johann Friedrich Blumenbach (1752-1840) was Professor of Medicine in Göttingen and in his time a prominent exponent of comparative anatomy and anthropology. Thus he was also very interested in fossil forms, and described and named the ice-age woolly rhinoceros and the mammoth. His principal work is the *Handbuch der Naturgeschichte*, which appeared in many editions.

Above: Cosimo Alessandro Collini (1727-1806), the first keeper of the Mannheim Natural History Collection established by Karl Theodor. In 1784 Collini published a description of the first known pterosaur, without, however, coming to a definite conclusion about its true identity as a flying reptile.

Georges Cuvier

Seventeen years passed before attention again turned to the still nameless Eichstätt fossil. The Parisian anatomist Georges Cuvier (1769-1832),[4] realized that this creature must have been a reptile.[5] He recognized the long bones on the forelimbs as highly elongated phalanges of one digit of the hand, on which a flight membrane must have been mounted. Naturally this membrane had not survived as a fossil. Thus the creature was capable of flight, a flying saurian comparable to *Draco* (dragon), the modern flying lizard. Cuvier based his classification of the creature within the zoological system solely on Collini's description and illustration of 1784, he had never seen the fossil itself. About 50 years later pterosaur fossils were found in the Jurassic Solnhofen limestone with imprints of the flight membrane surviving on the surface of the rock. This completely confirmed Cuvier's assumption. The creature itself remained nameless, until in a later, detailed treatise written in 1809 Cuvier placed it in a reptilian genus of its own and gave it the name *'Ptero-dactyle'*, meaning roughly 'flight finger'.[6]

4 Georges Cuvier (1769-1832) was born on 23 August 1769 as Georg Küfer in Mömpelgard, then part of Württemberg, now Montbéliard in the Département of Doubs, France, in the same year as Napoleon, whom he later served, and Alexander von Humboldt, the eminent naturalist, whose close friend he was to become. He established himself as the foremost comparative anatomist of his day, and was honoured with many administrative and political appointments, notably: in 1796 member of the *Institut National*, 1808 University council member, 1811 Chevalier of the Légion d'Honneur, 1814 made a councillor of state by Napoleon and chancellor of the University, in 1818 he refused an offer of the interior ministry and became a member of the Académie Française. He was made a baron by the king in 1819 and finally a Peer of France in 1832. He died unexpectedly of cholera in May of the same year. Cuvier used his influence to promote science in schools. He also made the Musée d'Histoire Naturelle and Jardin des Plantes in Paris the leading scientific research institution of its day.

5 Cuvier, G., 1801. *Reptile volant. Extrait d'un ouvrage etc.*, an 9:6; Paris.

6 Cuvier, G., 1809. *Mémoire sur le squelette fossile d'un Reptil volant des environs d'Aichstedt, que quelques naturalistes ont pris pour un oiseau, et donc nous formons un genre de Sauriens, sous le nom de Ptero-Dactyle.* Annales du Musée d'Histoire Naturelle Paris, 13:424,

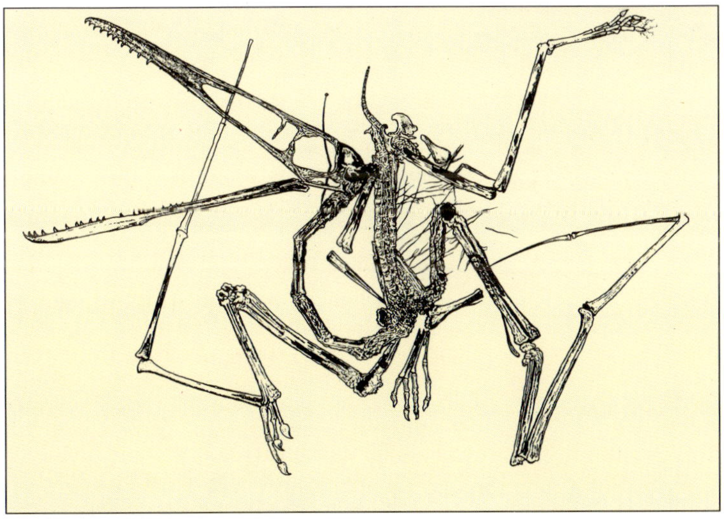

Above and left: The actual specimen of *Pterodactylus antiquus* from Eichstätt which Collini studied is seen above, while to the left is reproduced the original copper plate engraving by Verhelst which appeared in Collini's 1784 treatise entitled *Sur un animal fossile d'un genre particulier.*

Right: Georges Cuvier (1769-1832) is regarded as the father of vertebrate palaeontology. Cuvier recognised the reptilian nature of the Eichstätt fossil in his study of 1801.

Cuvier's particular academic strength lay in the field of comparative anatomy. He is considered the father of scientific vertebrate palaeontology. He came from modest circumstances. His father commanded the guard at the castle of Mömpelgard. Cuvier's interest in science was established at an early stage. In 1784, the same year that Collini described his mysterious Eichstätt fossil, the first pterosaur, Cuvier went as a fourteen-year-old student to the Hohe Karlsschule in Stuttgart to study economics. Even at this time he took every opportunity to collect plants and insects, and to immerse himself in the scientific works of the French zoologist, Leclerc de Buffon, and the Swedish botanist Carl Linnaeus.

After his period at the Karlsschule, Cuvier took a job as a residential tutor in Normandy. This enabled him to spend the dangerous years of the French Revolution by the sea, where he again took up the study of plants, insects and sea creatures and found fossils, which stimulated him to compare them with living specimens. He also maintained a lively correspondence with distinguished naturalists in Paris, who inevitably recognized the talents of the young scientist and offered him a professorship at the *École centrale* of the Panthéon. Thus in 1795 he went to Paris, where the political situation was slowly calming down, and embarked upon his remarkable career.

His first great publication on comparative anatomy appeared as early as 1800. His 'correlation principle' states that the shape of missing parts of a creature can be deduced from the shape of each isolated piece. As fossil skeletons are often fragmentary and incomplete, Cuvier's principle of correlation was of extraordinary significance for vertebrate palaeontology in general. Cuvier's precise knowledge of skeletal structure in living reptiles also meant that he was able to classify the Eichstätt vertebrate skeleton as a reptile.

He dealt with this interesting fossil in detail in his treatise of 1809, which appeared in the annals of the Natural History Museum in Paris. In the meantime other interpretations had appeared, such as that of Friedrich Blumenbach, who argued that it was a waterfowl. In this examination Cuvier again had to rely on Collini's first description of 1784, as it was thought that the original specimen was missing. Cuvier was in fact aware that Karl Theodor's natural history collection had been transferred to Munich, but it seemed that no-one could find it there. We now know that the specimen really was in Munich, but the Munich anatomist Samuel Thomas von Soemmerring (1755-1830) was working on the fossil at the time and preparing a scientific treatise of his own about it.

But how did the exhibit get to Munich? For political reasons, closely connected with the line of succession of the house of Wittelsbach. In 1777 the old Bavarian Wittelsbach line died out with the Bavarian Elector Max III Joseph. The succession passed to his cousins in the Palatinate, to Karl Theodor von Pfalz-Neuburg-Sulzbach, who moved his residence from Mannheim to Munich in the following year. It was not until 1802, after his death and under the regency of his successor, Elector Max IV Joseph, that the Mannheim natural history collection was brought to Munich and with it the rare fossil of the Eichstätt pterosaur, which was still unique. It was handed over to the natural history collection of the Bavarian Academy of Science, into the care of Samuel Thomas von Soemmerring.[7]

Soemmerring

Soemmerring was one of the most versatile scholars of his time. He was an anatomist and surgeon, but also undertook palaeontological research, and concerned himself with astronomy and technical problems. Thus he was the inventor of an electro-chemical telegraph that is still kept and shown in the Deutsches Museum in Munich. In Munich Soemmerring became a member of the Bavarian Academy of Science, where he was made director of the natural history collection, which had considerably increased in size only a few years earlier through the acquisition of Karl Theodor's Mannheim collection. The collection included among its many zoological specimens, plants, minerals, rocks and fossils the fossil described by Collini and recognized by Cuvier as a flying reptile, the first *Pterodactylus*.

This fossil must have immediately roused Soemmerring's interest and stimulated him to examine this controversial vertebrate skeleton himself and produce an interpretation of his own. He presented the results of his studies on 27 December 1810 in the form of a lecture to the mathematics and physics class at the Academy of Science. At that time the Academy was housed in the former Jesuit monastery at St Michael's in Neuhauserstrasse in Munich, where the Bavarian state scientific collections, the successors of the Academy collection, were housed until 1944.

The lengthy title of Soemmerring's lecture was: 'About an *Ornithocephalus* or the unknown creature of the prehistoric world, whose fossil skeleton was described by C. Collini in the fifth volume of the Actorum Academiae Theodoro-Palatinae, along with a pictorial representation in actual size, and which skeleton is at present to be found in the Royal Academy of Science in Munich.' In his lecture, later printed in the Memoirs of the Academy in 1812,[8] Soemmerring addressed Collini's views critically. He did not see Cuvier's work of 1809 until 1811, and he considered his interpretation of the Eichstätt fossil in an appendix.

Left: Samuel Thomas von Soemmerring (1755-1830) restudied the Eichstätt pterodactyl, but concluded that it was a mammal which he named *Ornithocephalus*.

Above: The Academy of Sciences in Munich as it appeared in 1826, shortly after Soemmerring had become keeper of its Natural History Collections.

His analysis comes to the conclusion that the creature must be a mammal, a hitherto unknown kind of bat. This was a view similar to that held by Professor Johann Hermann of Strasbourg, who had pointed out the fossil in the Mannheim collection to Cuvier, and had also suggested that he thought it was a mammal, to be classified between bats and birds.

Soemmerring too was convinced that the creature was a transitional form between the mammal and the bird class. He even speaks of a 'graduated sequence' of animals between flying mammals and actual birds. It seems that Soemmerring was an early adherent of the ideas of Jean Baptiste de Lamarck, whose theory of evolution assumed a gradual sequence of living things, becoming increasingly complex. Thus Lamarck's view was in complete contrast with that of Cuvier, who worked on the basis that there were times when the entire living world was extinguished by catastrophe, and replaced with a completely new one.

Soemmerring called the Eichstätt fossil *Ornithocephalus antiquus*, which means 'old bird-head'. Although, unlike Cuvier, he had the advantage of being able to study the original fossil, he made serious mistakes, especially in his interpretation of the forelimbs. For example, he did not recognize the bones of the upper arm for what they were, but thought they were breast bones. This meant that he identified the actual lower arm bones as those of the upper arm, and the long bones of the middle hand as those of the lower arm. But he correctly established that the hand had four digits and that the highly elongated fourth digit with its four long phalanges must have been a 'mast' to stretch a flight membrane. He thought the creature was a bat-like mammal, closest to the modern Indian flying fox. Independently of Cuvier he also thought it probable that it fed on insects which it caught in flight. Soemmerring wrote of the habitat and environment of *Ornithocephalus* 'that at that time the present Danube region must have had a hot, southern Indian climate.'

In 1817 Soemmerring also described the second pterosaur fossil[9] from Jurassic limestone slabs, also from the Eichstätt region. He gave it the name *Ornithocephalus brevirostris* (=short-snouted bird-head), as it was a species with a short snout. He thought that it was even more bat-like than the first and compared it with the parti-coloured bat *Vespertilio murinus*. He felt his earlier view was reinforced by

---

7 Samuel Thomas von Soemmerring (1755-1830) was born on 28 January 1755 in Mainz. He lectured in anatomy and surgery at the Collegium Carolinum in Kassel, from 1784 he was Professor of Anatomy and Physiology at the university in his home town of Mainz. From 1795-1805 he worked as a doctor in Frankfurt and was a founder member of the Senckenbergische Naturforschende Gesellschaft. In 1805 he was appointed Professor of Anatomy and Surgery at the University of Munich and was personal physician to Maximilian Joseph, the first King of Bavaria. In 1819 he settled in Frankfurt again, and died there in 1830.

8 Soemmerring, S.T.v., 1812. *Über einen Ornithocephalus*. Denkschriften der Akademie der Wissenschaften München, math.-phys. Classe, 3: 89-158.

9 Soemmerring, S.T.v., 1817. *Über einer Ornithocephalus brevirostris der Vorwelt*. Denkschriften der Akademie der Wissenschaften München, math.-phys. Classe, 6: 89-104.

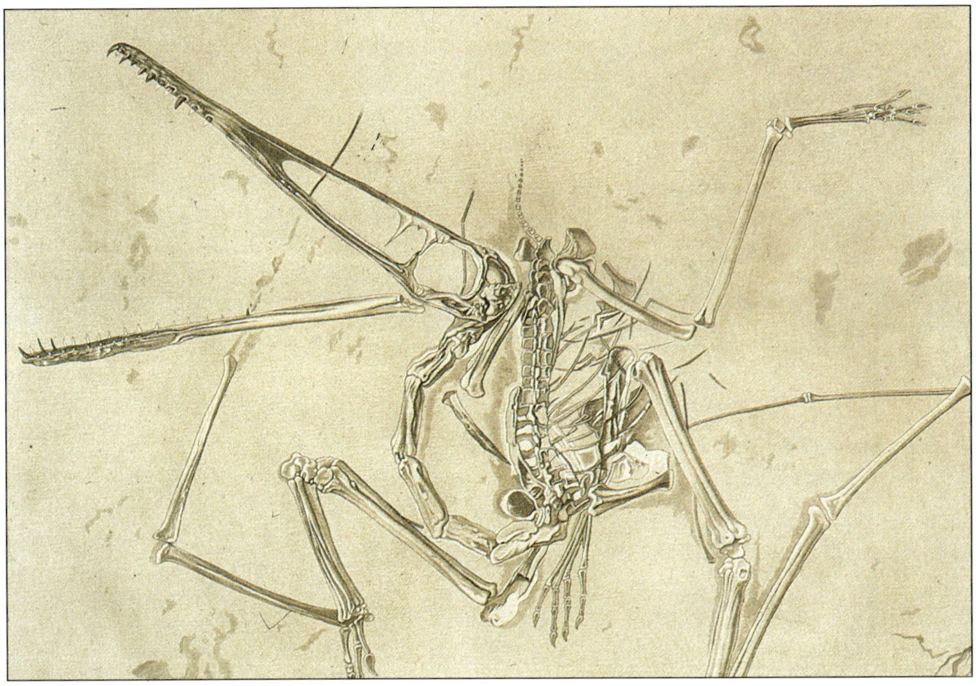

Above: In 1817 Soemmerring described a second pterodactyl from near Eichstätt. Because of the short beak he named it *Ornithocephalus brevirostris*. This specimen, long thought to be lost, was rediscovered in a school collection in 1969.

Above: The Eichstätt pterodactyl as illustrated in Soemmerring's treatise of 1812. He considered the skeleton to be that of a bat-like mammal.

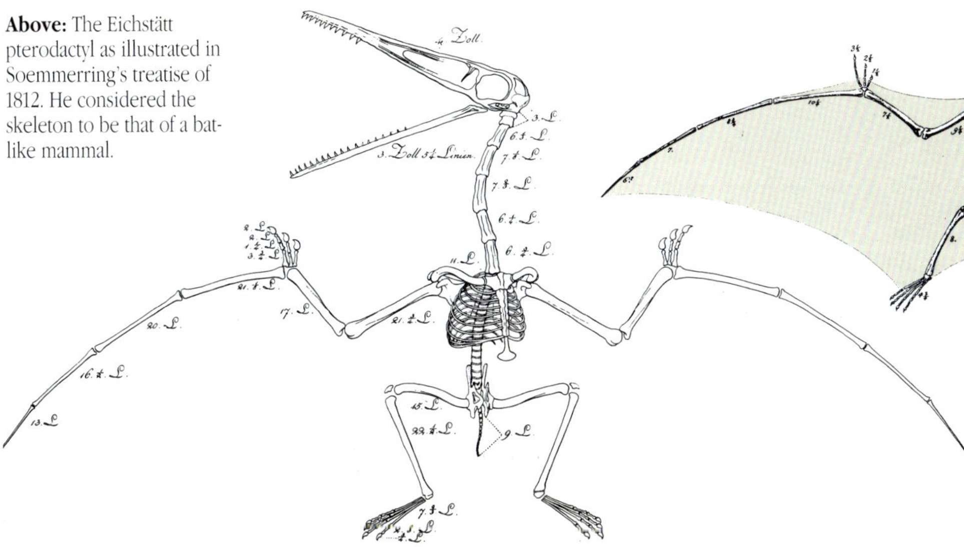

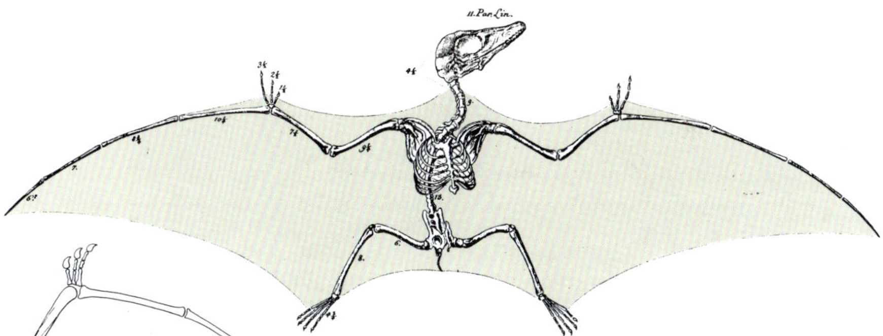

Left: The first skeletal restoration of a pterosaur as produced by Soemmerring in his 1812 treatise. His interpretation of the forelimb bones, however, was seriously in error.

Above: A skeletal restoration of *'Ornithocephalus' (=Pterodactylus) brevirostris* by Soemmerring, 1817, drawn on the basis of the second known pterodactyl from the Jurassic limestone of Bavaria. By marking the outlines of the wing membranes, Soemmerring gave it an even more bat-like appearance.

this and classified both forms as a new genus of bat, supposed to be similar to the flying fox.

Soemmerring's authority was apparently so great that of the German naturalists only Lorenz Oken (1779-1851) sided with Cuvier.[10] In 1819, in the magazine Isis, edited by Oken, Cuvier had called the Eichstätt pterosaur first known as *Ptero-dactyle* by its correct name *Pterodactylus* and given it the specific name *longirostris* (=long-snouted). According to international rules for the naming of zoological genera and species priority is given to

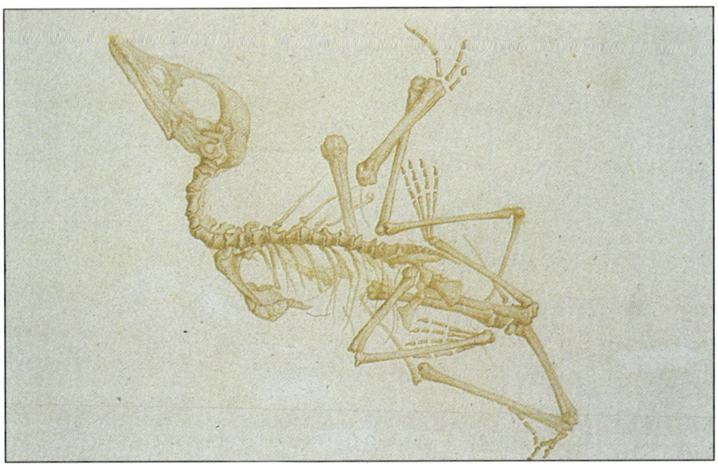

Left: Soemmerring's drawing of the *'Ornithocephalus' brevirostris* fossil pictured above. This appeared in his treatise of 1817. Today, this specimen is recognised as a juvenile individual of a larger species of the genus *Pterodactylus*. It is a baby pterosaur with a wing span of only 10in (25cm).

Cuvier's *Pterodactylus*, but not to his specific name *longirostris*. Here the name *antiquus*, given by Soemmerring seven years earlier, is valid. Thus the first two pterosaur species should correctly be called in scientific terms *Pterodactylus antiquus* and *Pterodactylus brevirostris*. Later it turned out, however, that the species identified as *brevirostris* was simply a young animal of the species *antiquus*, and thus had to carry the same specific name as the adult individual, namely *antiquus*.

In the meantime, Cuvier had started to bring out his major work *Recherches sur les Ossemens fossiles* (Examinations of fossil bones) in 1812, which was to be a standard work on vertebrate palaeontology for a long time. He deals with *Pterodactylus* and its position in the zoological system in great detail in this work. With the aid of a cast and some detailed drawings that he had received from colleagues travelling through Munich, he once again took it upon himself to speak out against Soemmerring's bat

10 Lorenz Oken (1779-1851) was Professor at the University of Jena, but had seen the original *Pterodactylus* fossil in Munich himself. He wrote in 1819 in the magazine Isis, of which he was the editor: 'Consequently each individual piece of bone of this creature says loudly and clearly that it is an amphibian, and indeed of the order of lizards.' A year earlier he had taken Cuvier's side in the quarrel between Cuvier and Soemmerring, with the words: 'Truly, we never thought that we would find ourselves defending Cuvier! On other occasions we are always fleet to deny him something. But in this case! No! That is too bad!' The nature of these differences of opinion with Cuvier is clear if one compares Oken's nature philosophy with Cuvier's doctrine of catastrophe and his views on the constancy of species.

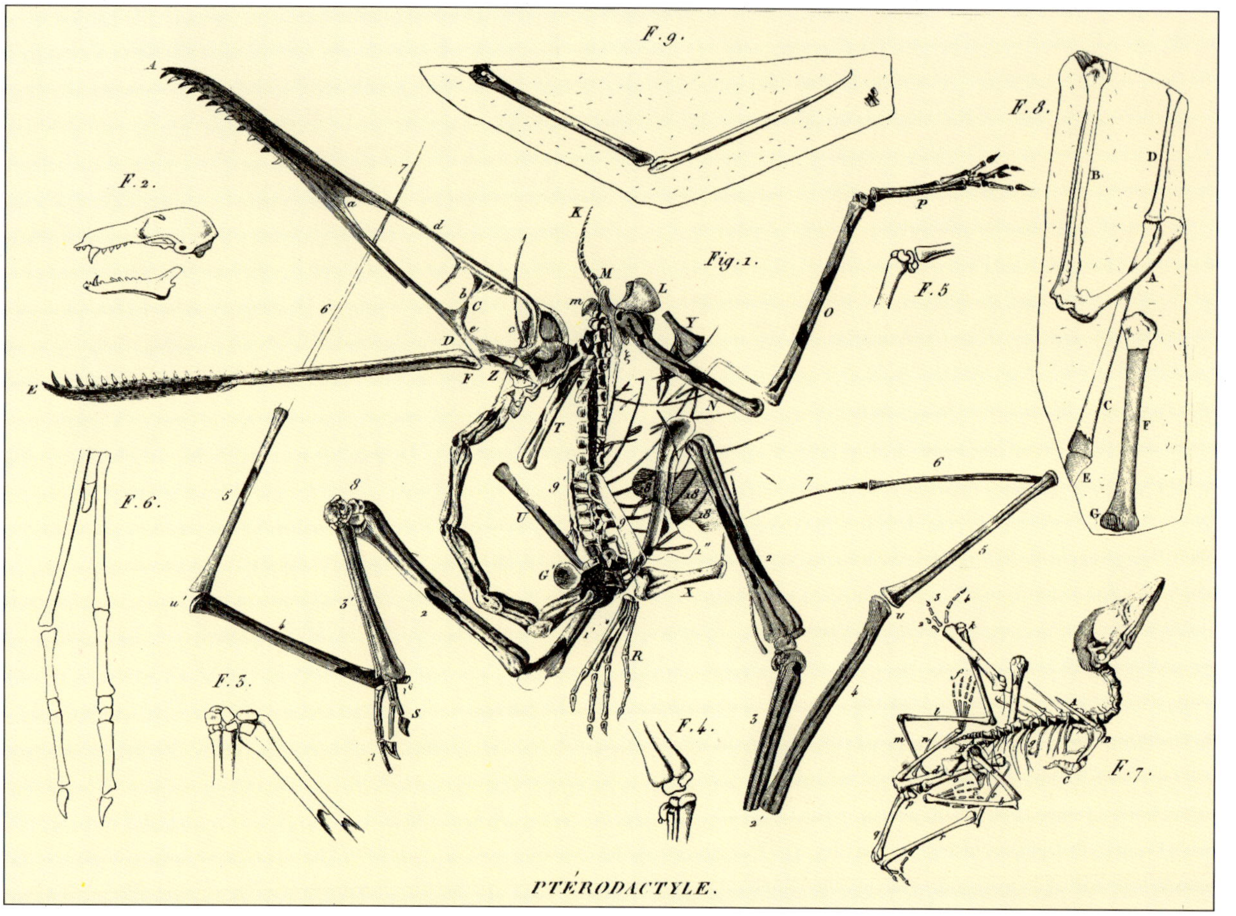

Left: A plate from Cuvier's fundamental work *Recherches sur les Ossemens fossiles* showing the Eichstätt pterodactyl, *Pterodactylus antiquus*, in the centre, with several drawings of osteological details (F.3-6), *Pterodactylus brevirostris* (F.7), and two more fragmentary pterosaur fossils (F.8 and 9). Cuvier also included a drawing of the skull of a bat, *Pteropus minimus* (F.2), taken by Soemmerring to be very similar to the skull of *Pterodactylus*. Cuvier, however, could demonstrate its great difference.

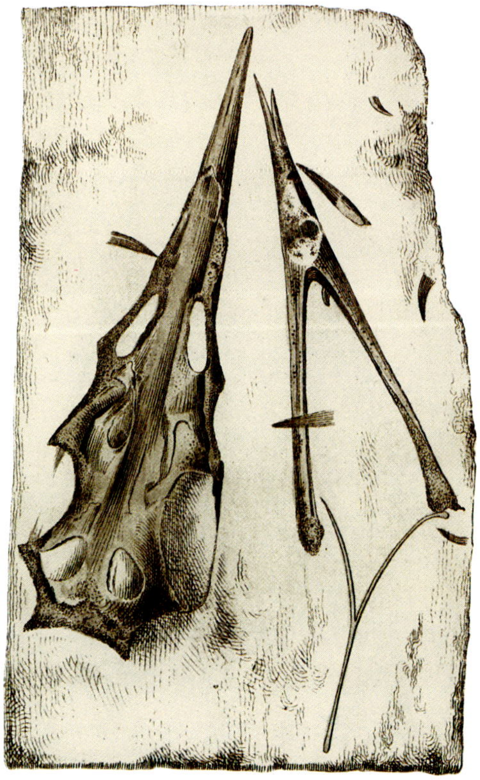

Below: Soemmerring thought that this fossil skull from the Solnhofen limestone belonged to a sea bird. In 1831 Professor August Goldfuss of Bonn University recognised it as belonging to a pterosaur. He named it *Pterodactylus muensteri*. Later it was realised that in fact it represented a completely new type of pterosaur, the rhamphorhynchids.

interpretation. In particular he pointed out the great differences between *Pterodactylus* and bats, in which it is possible to make a distinction in the dentition between incisors, molars and canines. But in *Pterodactylus*, as in reptiles, the teeth are simple, pointed and uniform. He also stressed other reptilian characteristics of the skeleton, especially in the vertebrae, ribs and pelvis. The skeleton of the foot in particular was completely different from that of birds and bats, and much more like that of a lizard. The short tail, which Soemmerring considered batlike, did not prove anything, as some reptiles, tortoises for example, also had short tails. Of course neither Cuvier nor Soemmerring could have known at the time that long-tailed pterosaurs would later be discovered; this would have emphasized lizard-like characteristics all the more.

Because of *Pterodactylus*' long hindlimbs, Cuvier assumed that the creature could only stand upright on its hind legs, and that the forelimbs could be folded back like birds' wings, and had not been an aid to standing. Finally Cuvier summed up his comparative anatomical investigation of *Pterodactylus* as follows: these creatures were reptiles that flew using a membrane supported by a single digit of the four-fingered hand and which could hang from tree branches with their three other, smaller digits. They could perhaps also crawl on the ground, but could only stand upright on their hindlimbs. They had large heads with jaws full of small, pointed teeth that were only suitable for catching insects and other small creatures. Cuvier thought these flying reptiles were the most extraordinary of living things, and that if one had met them alive they would seem the strangest creatures in nature. Scientific proof, using methods of comparative anatomy and Cuvier's authority, supported this assessment of *Pterodactylus* as a flying reptile, a pterosaur, and it is still recognized today.

Other Finds – Other Ideas

Soemmerring clung to his bat hypothesis to the end. In 1820 he was still describing some hollow bones which without doubt came from a Solnhofen pterosaur as the fossil remains of a genus of large bat.[11] In the same year the Munich zoologist Johann Baptist von Spix (1781-1826) published a short treatise[12] on two phalanges from a pterosaur, also from Solnhofen, which he said were those of the southern Asian red-necked fruit bat. Incidentally in the same year, 1820, Spix had returned from a journey of exploration in Brazil. He had undertaken this with the botanist Carl Friedrich Philipp von Martius (1794-1868) and had returned with rich zoological and botanical collections. The two Munich naturalists had found large quantities of fish fossils in the north-eastern province of Ceará; in the 1970s pterosaur fossil remains were discovered here for the first time.

In 1825 the great Bayreuth fossil collector and palaeontologist Georg Graf zu Münster (1776-1844) asked Soemmerring's opinion about a presumed *Pterodactylus* head from Solnhofen that had come into his possession. In November of the same year Soemmerring replied: 'The Ornitholith, which is as rare as it is valuable, seems to belong to a particular genus of waterfowl which could have been similar to *Larus tridactylus* (a seagull) and *Colymbus* (a diver).' From that time this skull was counted as the first remains of a bird from the Jurassic, long before the discovery of *Archaeopteryx*, the so-called *Urvogel*, or primordial bird, in 1861. Later Graf Münster sent a cast of this fossil slab to Georg August Goldfuss (1782-1848), Professor of Natural History at the University of Bonn. In his *Beiträge zur Kenntnis verschiedener Reptilien der Vorwelt* (Contributions to the knowledge of various pre-

11 Soemmerring, S.T.v., 1829. *Über die fossilen Reste einer großen Fledermausgattung, welche sich zu Karlsruhe in der Großherzoglichen Sammlung befinden*. Denkschriften der Akademie der Wissenschaften München, math.-phys. Classe, 6: 105-112.

12 Spix, J.B.v., 1820. *Über ein neues, vermutlich dem Pteropus Vampirus Linn. zugehöriges Petrefikat aus dem Solenhofer Kalkbruch in Bayern*. Denkschriften der Akademie der Wissenschaften München, math.-phys. Classe, 6: 59-68.

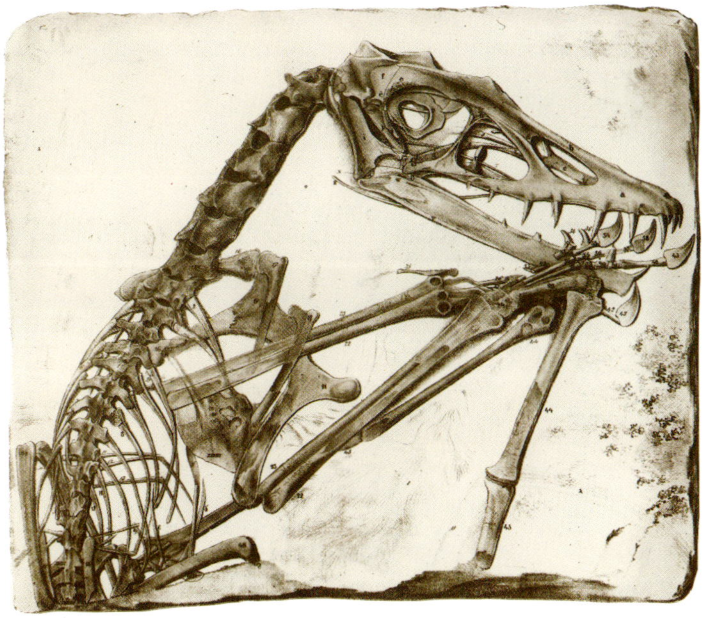

Left: The first more complete skeleton of a long-tailed pterosaur was also described by Goldfuss in 1831. The specimen from the Solnhofen limestone was first named *'Pterodactylus' crassirostris*, because its long tail was not yet discovered. Today its correct scientific name is *Scaphognathus crassirostris*. To give an idea of scale, its skull is only 4·57in (11·6cm) long.

Bottom of page: A rather strange life restoration of pterodactyls was presented by Edward Newman in 1843 He thought that pterosaurs were warm-blooded and covered by a coat of fur, and so concluded that they could not have been reptiles. Instead, he restored them as flying marsupials with rather elegant little ears.

the unbridled imagination of a Chinese artist and less like a representation of a product of nature that really existed.' Goldfuss also believed that he had seen signs of hair on his specimen of *Pterodactylus crassirostris*. This was later seen to have been a mistake. Certain proof of a covering of hair for pterosaurs was successfully established, but not until 1927, by the Munich palaeontologist Ferdinand Broili.

The Munich zoologist Johann Georg Wagler had a completely different view of the position of *Pterodactylus* in the animal kingdom. In his *Natural System of Amphibians*[14] he combined it with other fossil groups like ichthyosaurs and plesiosaurs in a class of vertebrates of its own that he called Gryphi, placing them between birds and mammals. For this reason Wagler called *Pterodactylus* an arm griffin. He thought it had been hairless and that its feet were fin-shaped and covered with a kind of sheath of strong skin, but that some claws protruded beyond their outer edge 'to hold the female during the act of copulation'. Wagler had a strange idea of the way of life of these creatures. He thought they were aquatic, and used their wings as paddles in the water.

The English zoologist Edward Newman imagined pterodactyls quite differently in 1843.[15] He thought they were furry flying creatures with pouches, marsupials in other

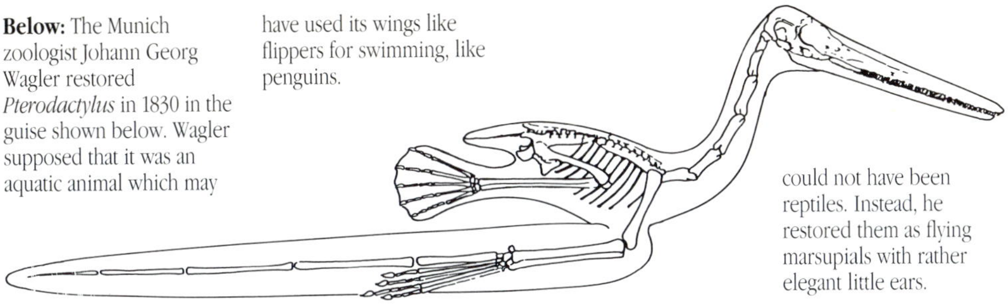

Below: The Munich zoologist Johann Georg Wagler restored *Pterodactylus* in 1830 in the guise shown below. Wagler supposed that it was an aquatic animal which may have used its wings like flippers for swimming, like penguins.

14 Wagler, J.G., 1830. *Natürliches System der Amphibien*. München, Stuttgart, Tübingen.

15 Newman, E., 1843. *Note on the Pterodactyle Tribe considered as Marsupial Bats*. Zoologist, 1: 129-131.

historic reptiles) of 1831 Goldfuss suggested that this could be a pterosaur skull.[13] In honour of Graf Münster he immediately named it *Ornithocephalus (Pterodactylus) muensteri*.

Certainly Soemmerring's mistake was understandable. In comparison with the *Pterodactylus* species *antiquus* and *brevirostris* the skull in the Münster collection was very different. Goldfuss on the other hand could compare it with the *Pterodactylus crassirostris* he had just acquired from Solnhofen, later given the generic name *Scaphognathus* (=tub jaw), whose skull was much more like Graf Münster's fossil. Münster wrote later, after further preparation and freeing of the skull bones from the rock: 'the toothless, spiniform point of the beak justifies the assumption of a subgenus of the genus *Pterodactylus* or perhaps, on closer acquaintance with the remaining bones, the formation of a new genus of these most wondrous creatures of the prehistoric world.' It was later established that this was the first long-tailed pterosaur find (=Rhamphorhynchoidea). All previous finds had been short-tailed pterosaurs (=Pterodactyloidea).

Goldfuss expressed views about the way of life of pterosaurs that seem very modern for 1831. He also thought they were flying reptiles which 'only used their claws to cling on to cliffs, crevices or trees, if they were available, and to climb up steep walls. Like bats they could fly and probably hovered above the surface of the water to catch insects and probably aquatic creatures as well.' And yet the Bonn professor could not escape the fascination exerted by these flying reptiles, and wrote at the end of his treatise: 'The image of this creature always seems more like a painting created by

13 Goldfuss, A., 1831. *Beiträge zur Kenntnis verschiedener Reptilien der Vorwelt*. Nova Acta Academiae Leopoldinae, 15: 61-128.

words. For this reason they have pretty little ears in his reconstruction drawing. The assumption that they must have been hairy also arose from the theory that these active flying animals were warm-blooded, and thus needed a covering of hair for reasons of heat conservation, like bats and other mammals.

Until 1827 all known pterosaur fossils came from the lithographic limestone in the quarries of Solnhofen and Eichstätt in Bavaria. Then finds were announced at other sites. The Proceedings of the Geological Society in London contained a description of a new species of *Pterodactylus* from early Jurassic strata from Lyme Regis, on the coast of Dorset.[16] These skeletal remains from the Blue Lias were dealt with by William Buckland (1784-1856), Reader in Geology at the University of Oxford. He was one of the 'fathers of British geology', and had described the first dinosaur, *Megalosaurus*, in England. Buckland's field was not just the geology of various parts of England and fossil reptiles in southern England, but also the structure of the Alps and fossil mammals of the Ice Age. He called the new species of pterosaur *Pterodactylus macronyx*, after the great claws on its digits. It later became clear that the genus was new as well. Richard Owen, later to be director of the Natural History Museum in London, gave it the name *Dimorphodon* (=two-form tooth) because of the double shape of its teeth. It was a flying creature with a wing span of 4·6ft (1·4m).

Buckland had acquired the fossil from Mary Anning (1799-1847), one of the first professional fossil collectors. She spent her whole life in the little town of Lyme Regis, on the Dorset coast of southern England. It is said that she was struck by lightning as a baby, and that her nurse was killed in the accident. She was a bright child, never married, and earned a living by collecting and selling fossils. The tongue-twister 'she sells sea shells sitting on the sea shore' is said to have been inspired by her. As an eleven-year-old girl she found the first articulated skeleton of a Jurassic ichthyosaur in the fossil-rich cliffs near her home town, and later the first complete plesiosaur skeleton. In December 1828 she discovered the first pterosaur in England, which she sent to Professor Buckland in Oxford. Mary Anning's customers included the King of Saxony, as well as many of the great palaeontologists and geologists of her time. The king bought fossils from her for his natural history collection in Dresden.

To be strictly accurate, the first English pterosaur remains had already been described by Gideon Mantell (1790-1852), but he thought they were the bones of a bird.[17] They were very fragmented, hollow bones originating from

Above: Dean William Buckland (1784-1856). Originally a scholar at Corpus Christi College, Oxford, he was appointed a Reader in Geology in 1818. In 1829 he described the first English pterosaur, *'Pterodactylus' (=Dimorphodon) macronyx.*

Above: Mary Anning (1799-1847) was one of the first professional fossil collectors. She made a living out of collecting and selling fossils from the Dorset coast near her home town of Lyme Regis. It was she who sold the specimen of *Dimorphodon* pictured left to William Buckland in 1828.

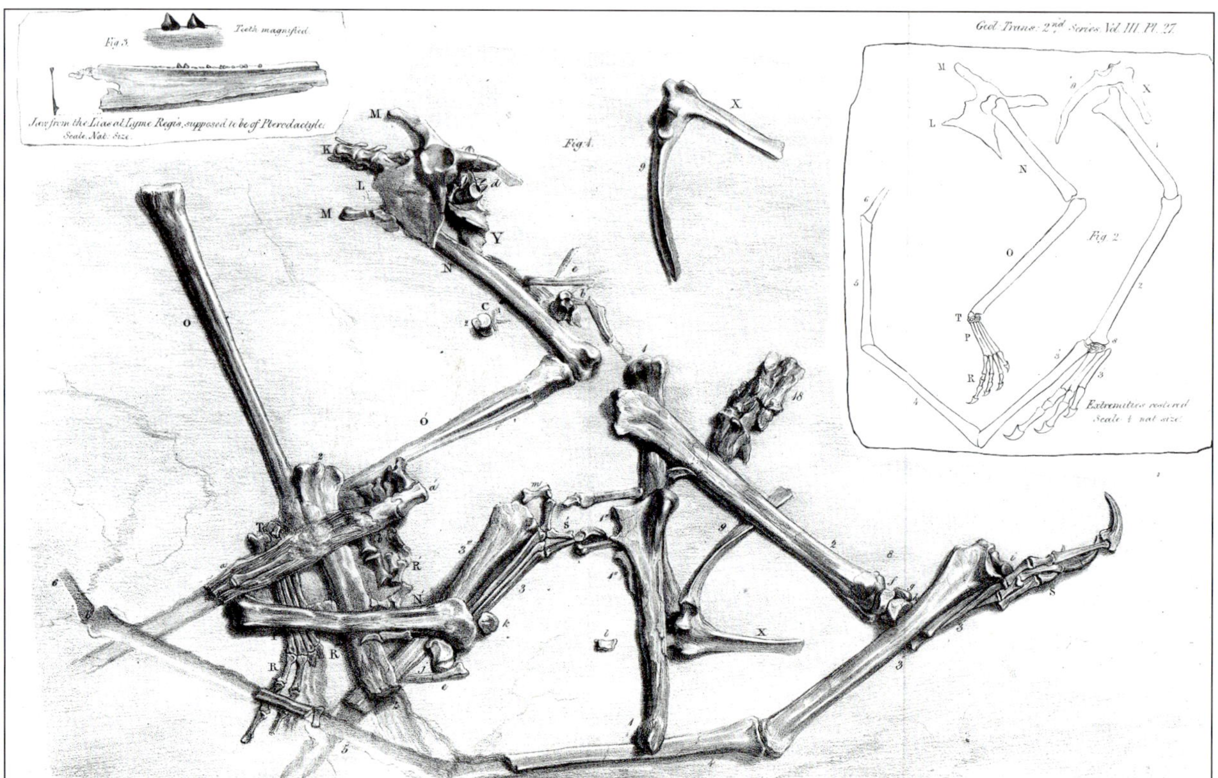

Left: The skeletal remains of the first Jurassic pterosaur in England were discovered by Mary Anning in the cliffs of the 'Blue Lias' near Lyme Regis. The skeleton was somewhat disarticulated and the skull was missing. The elements preserved were merely the incomplete wing bones and hind legs including the basin and shoulder girdle. This is the first illustration of *Dimorphodon macronyx* by Buckland in the 1835 Transactions of the Geological Society London.

16 Buckland, W., 1829. Proceedings of the Geological Society London, 1: 127.

17 Mantell, G.A., 1827. *Illustrations of the Geology of Sussex*. London.

Above: Gideon Mantell (1790-1852), a country doctor from Lewes in Sussex, actually described the first pterosaur remains found in England in 1827, but he mistakenly identified the hollow bones as those of a bird. He had collected them from the Wealden clay of Tilgate Forest in Sussex. Mantell is more famous for his discovery of the first remains of the dinosaur *Iguanodon* — some teeth which had originally been picked up by his wife, Mary Ann, near quarries in the Cuckfield area of Tilgate Forest.

Right: A plate from Mantell's *Illustrations of the Geology of Sussex*, published in 1827. Mantell described these remains as 'bones of birds'; only later were they recognised as pterosaur bones.

Below: Carl Theodori (1788-1857), founder and keeper of the *Petrefaktensammlung* (fossil collection) in the castle of Banz on the river Main in Upper Franconia. In 1830 he announced the discovery of the first pterosaur bones from the Lias in Germany.

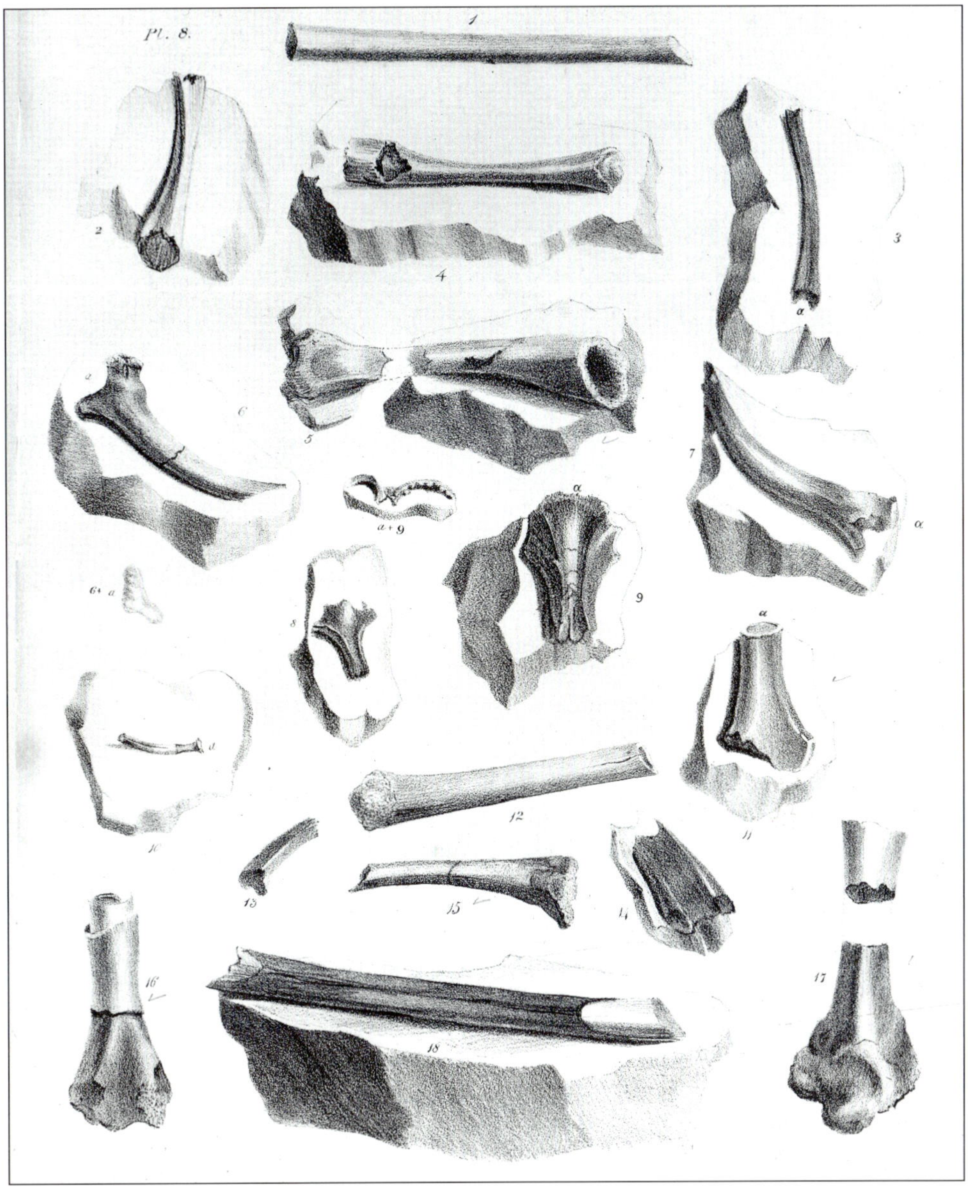

Lower Cretaceous strata in Tilgate Forest in Sussex. Mantell was a country doctor in Lewes, a small town in Sussex. He was a great lover and collector of fossils from the area around his home and wrote books and treatises on fossils and the geology of Sussex. In 1822 his wife Mary Ann had found the first remains of dinosaurs on English soil. They were only single teeth, which Mantell described as *Iguanodon* (=iguana tooth) because of their shape. Most of the more complete *Iguanodon* skeletons later discovered came from the same strata in Tilgate Forest as the pterosaur bones.

Pterosaur fossils from the Lias had also been found in Germany at the same time as the English finds. In 1830 Carl Theodori (1788-1857) reported the discovery of isolated bones from pterosaur skeletons around the monastery of Banz near Staffelstein on the upper Main in Franconia. Theodori, born in Landshut in Bavaria, was private secretary and chancellery councillor of Herzog Wilhelm in Bavaria, who in 1814 had acquired the former Benedictine monastery of Banz high above the Main, north of Bamberg, after it had been dissolved in the secularization of 1803. Herzog Wilhelm wanted to use the Brothers Dientzenhofer's magnificent monastery buildings with their splendid church, a superb example of Main-Franconian baroque, as a summer residence for himself and his family. For this reason Theodori spent the summer months here each year and began

Left: The baroque cloister of Banz in the Main valley, formerly a Benedictine monastery, became the castle of Wilhelm, Duke of Bavaria, in 1814. His secretary, Carl Theodori, collected fossils from the neighbourhood, and assembled a celebrated collection in the castle. This collection is still housed in Banz. It was refurbished in 1988 and exhibits many interesting specimens including a giant ichthyosaur skull, 6·6ft (2m) in length.

Below: A plate from Carl Theodori's memoir of 1852 illustrating the Liassic pterosaur bones from Banz, comprising jaw bones, vertebrae, ribs, upper arm, and shoulder girdle of a long-tailed pterosaur. Theodori named this as *'Pterodactylus' banthensis*. This species, the name is derived from Banz, was later assigned to a separate genus *Dorygnathus* (=spear jaw).

a systematic geological investigation of the environs of Banz, especially the Lias and the Dogger, the Lower Jurassic rocks that formed the basis of the Banz hills. With the former Benedictine father and priest Augustin Geyer he assembled a famous collection of fossils from the immediate neighbourhood, and this was displayed in the monastery, now palace, of Banz as the 'Petrefaktensammlung'.[18]

Theodori was in active contact with the leading naturalists of his time. The Banz collection was visited by Buckland, Owen, Goldfuss and Leopold von Buch, the preeminent German geologist at the time. The famous Frankfurt vertebrate palaeontologist Hermann von Meyer also visited the Banz collection in summer 1830 and later wrote: '. . . I really succeeded in finding remains of *Pterodactylus macronyx*, which had been discovered in England shortly before, for the Banz region.' Theodori himself had described the Banz pterosaurs as *Pterodactylus banthensis*,[19] thus distinguishing them from the English finds in Lyme Regis. Later they were even separated as a different genus called *Dorygnathus* (=spear jaw).

Fossil pterosaur remains were also found later in Liassic strata in Württemberg. The first find was reported by Albert Oppel (1831-1865), later to be Professor of Palaeontology in Munich. It was a lower jaw recovered from the

18 The collection is still there today. It was completely refurbished in 1988 and reopened in new rooms. The finest object in the little museum was and is a huge, 6·6ft (2m) long fossilized skull of an ichthyosaur, discovered in 1842 during building work in nearby Unnersdorf and then described by Theodori as *Ichthyosaurus trigonodon*. Of course Theodori suffered from the seclusion and isolation of Banz, especially from the lack of a library, essential for his scientific work.

19 Theodori, C., 1830. *Knochen von Pterodactylus aus der Liasformation von Banz*. Frorieps Notizen für Natur und Heilkunde, 632: 101.

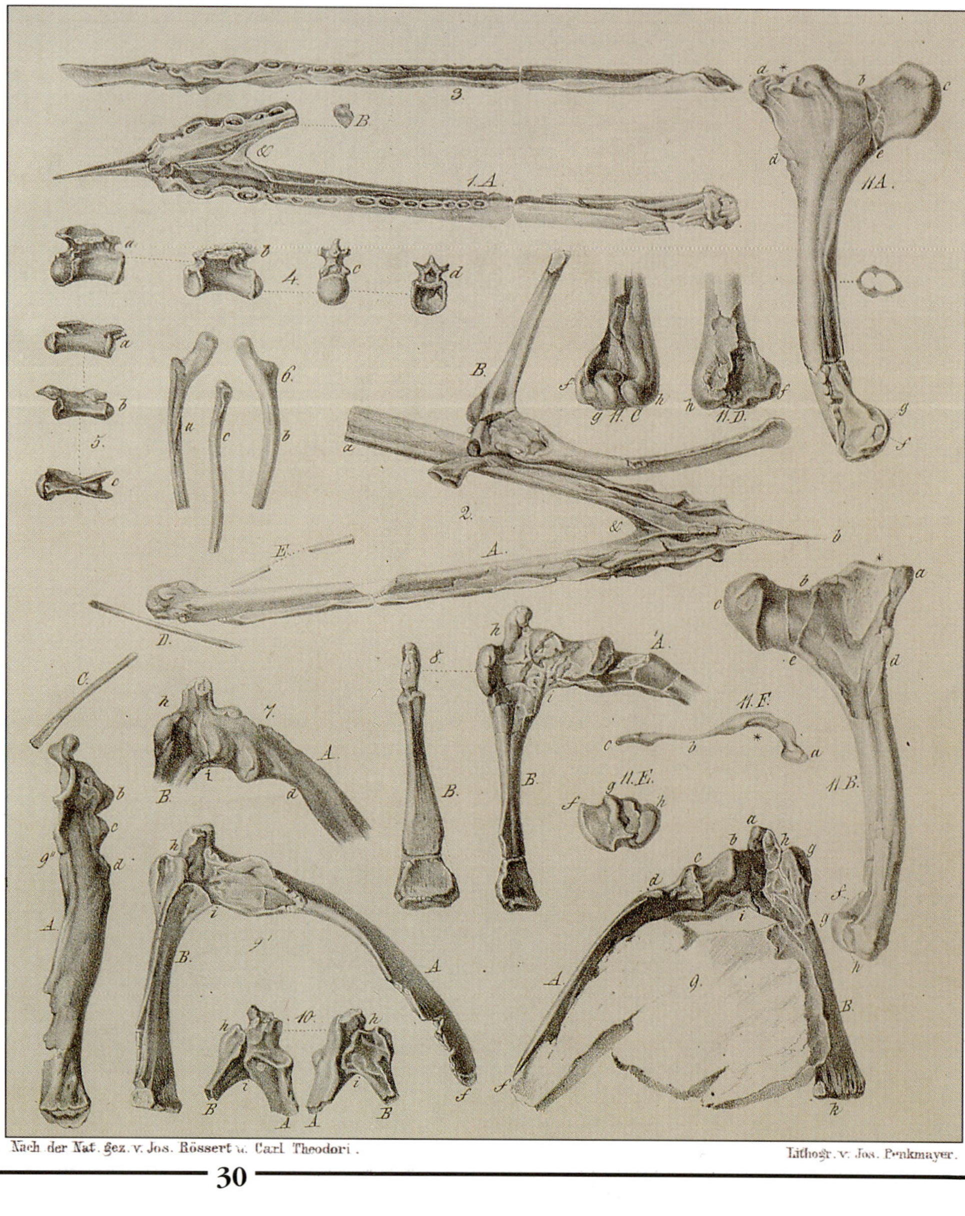

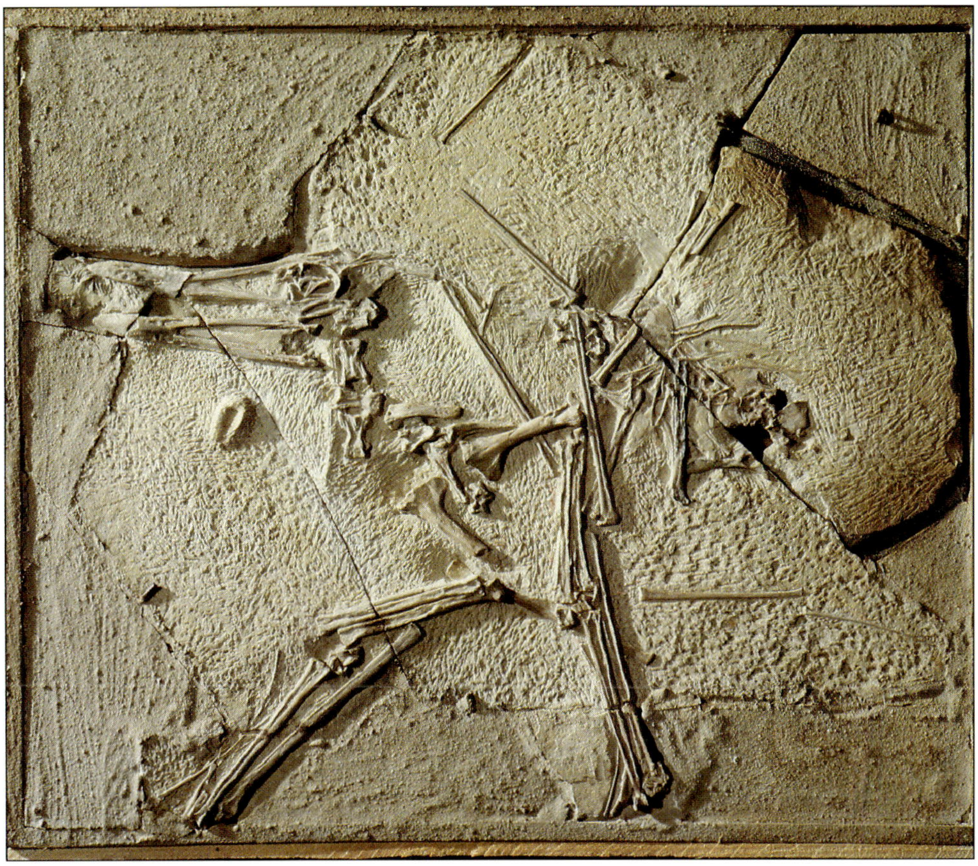

Lias at Wittberg near Metzingen in 1856, and he classified it as a *Pterodactylus* of the Banz species, *Pterodactylus banthenis*.[20] Two years later the Tübingen professor Friedrich August Quenstedt (1809-1889), a trail-blazing geological explorer, particularly of the Jurassic formation in Württemberg, reported on a *Pterodactylus liasicus* from the same site.[21] There were a few wing bones, scattered and embedded. Quenstedt thought the English, Franconian and Swabian pterosaur finds were of different ages, but saw them as examples of evolution of the species in geological time sequence. This shows that he too was a supporter of the ideas of the French zoologist Jean Baptist de Lamarck, as he writes: 'Then proof will also emerge that there were no crises in creation, but generally speaking only gradual development.' Here he was clearly taking up a position opposed to the catastrophe approach supported by Georges Cuvier. In fact it later emerged that the English Lyme Regis Lias pterosaurs are older than the Swabian and Franconian finds. But the latter two are the same age, dating from the late Liassic.

It was also Quenstedt who made known the first Swabian pterosaur find, in a letter to the Heidelberg professor Heinrich Georg Bronn, editor of the 'Neues Jahrbuch für Mineralogie' of 29 July 1854. 'At last I can tell you something about the first *Pterodactylus württembergicus* from the limestone slabs of our White Jurassic.' The fossil came from the limestone quarries of Nusplingen in the Schwäbische Alb, where the rocks are similar in age and origin to the Solnhofen limestone slabs. A year later Quenstedt described this pterosaur skeleton as *Pterodactylus suevicus*, from an animal with a wing span of about 4·3ft (1·3m).[22]

Above: The first pterosaur from Württemberg was announced by August Quenstedt as *Pterodactylus suevicus* in 1845. The skeleton was discovered in a limestone quarry near Nusplingen. Today this species is assigned to *Gallodactylus*, a genus first described from late Jurassic limestone from France.

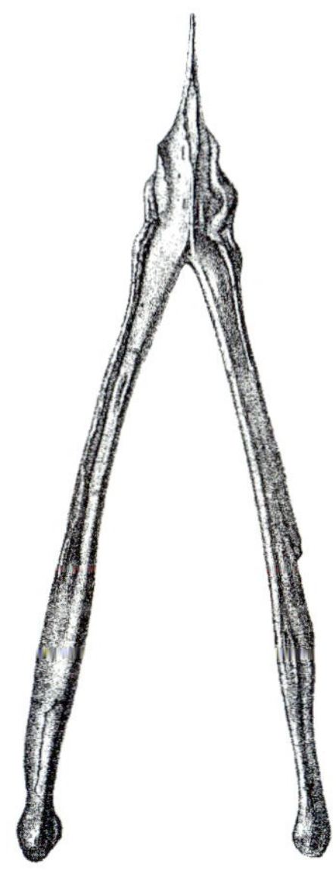

Right: This mandible was collected by Albert Oppel in 1856 near Bad Boll, not far from Holzmaden. The illustration is from von Meyer's monograph of 1859-60 in which this specimen is assigned to *Dorygnathus banthensis*.

Later more skeletal remains were discovered at Nusplingen quarry, but the principal area for finds in Württemberg was the region of Holzmaden, Ohmden, Bad Boll and Zell near Kirchheim/Teck on the northern edge of the Schwäbische Alb. Black bituminous shale from the Upper Liassic had been quarried there for years. It was also called *Posidonienschiefer* after the *Posidonia* shells that frequently occurred there.

Here again it was Albert Oppel who was the first to report a pterosaur find at Bad Boll, a lower jaw,[23] which he sent to the Frankfurt palaeontologist Hermann von Meyer for investigation. More complete pterosaur skeletons from Holzmaden were first described from 1895 by the Stuttgart professor Felix Plieninger. He was able to present evidence of two different genera, *Dorygnathus* and *Campylognathus* (=crooked jaw),[24] both long-tailed varieties. It was also Plieninger who introduced the division of the pterosaur order into two sub-orders, Rhamphorhynchoidea (for long-tailed forms) and Pterodactyloidea (for short-tailed forms).

Solnhofen and Eichstätt in Bavaria and Holzmaden in Württemberg have remained the most important sites for pterosaur finds until today. The Solnhofen limestone slabs continued to provide new species. Thus in 1839 Graf Münster reported a '*Pterodactylus* from the Solnhofen limestone quarries", which differed from other species in respect of its long, thin tail. For this reason he called it *Pterodactylus longicaudus*, long-tailed *Pterodactylus*. It was the first indication of a second type of pterosaur, which could be distinguished from short-tailed forms, with a few very short caudal vertebrae, by its long tail with a large number of vertebrae.[25]

A New Genus

In 1846 Hermann von Meyer[26] had the opportunity of examining another long-tailed pterosaur from Solnhofen. Because of the 'beak-like projection on the snout' he gave this new genus the name *Rhamphorhynchus* (=beak-snout). Hermann von Meyer (1801-1869), the actual founder of vertebrate palaeontology in Germany, also gave the first summary of pterosaurs of the Upper Jurassic, in a magnificent folio volume which he published in 1859-60 under the title *Die Reptilien aus dem Lithographischen Schiefer des Jura in Deutschland und Frankreich*. At the time lithographic limestone meant limestone slabs of the Solnhofen limestone type, which were used in lithography, a stone-printing process invented by Alois Senefelder in 1796, because they were particularly fine-grained and dense.

There were also fossiliferous late Jurassic limestone slabs with fossil fish, small lizards, crocodiles and turtles in south-western France, near the little village of Cerin in the Département of Ain. However, only two bones suggested the existence of pterosaurs, an upper arm and a shin bone from a *Pterodactylus* with a wing span of about 6·5ft (2m), named *Pterodactylus cerinensis* by von Meyer.

20 Oppel, A., 1856. *Die Juraformation*. Jahreshefte des Vereins für Vaterländische Naturkunde in Württemberg, 12.

21 Quenstedt, F.A., 1858. *Über Pterodactylus liasicus*. Jahreshefte des Vereins für Vaterländische Naturkunde in Württemberg, 14: 299-336.

22 Quenstedt, F.A., 1855. *Über Pterodactylus suevicus im lithographischen Schiefer Württembergs*. Tübingen.

23 Oppel, A., 1858. *Die geognostische Verbreitung der Pterodactylen*. Jahreshefte des Vereins für Vaterländische Naturkunde in Württemberg, 14: 55.

24 Plieninger, F., 1895. *Campylognathus Zitteli. Ein neuer Flugsaurier aus dem oberen Lias Schwabens*. Palaeontographica, 41: 193-222.

25 Münster, G. zu, 1839. *Über einige neue Versteinerungen in den lithographischen Schiefern von Baiern*. Neues Jahrbuch für Mineralogie etc.: 676-682.

26 Hermann von Meyer (1801-1869) was born in Frankfurt on 3 September 1801. He was an extremely versatile researcher and the outstanding vertebrate palaeontologist of 19th century Germany. He was 'Bundescassier' to the Deutscher Bundestag in Frankfurt. From 1851-1852 he was director of the museum of the Senckenbergische Naturforschende Gesellschaft in Frankfurt. He described pterosaurs from the Upper Jurassic from Solnhofen in Bavaria, Nusplingen in Württemberg and Cerin in France in a magnificently presented monograph: *Die Reptilien aus dem Lithographischen Schiefer des Jura in Deutschland und Frankreich*, 1859-1860.

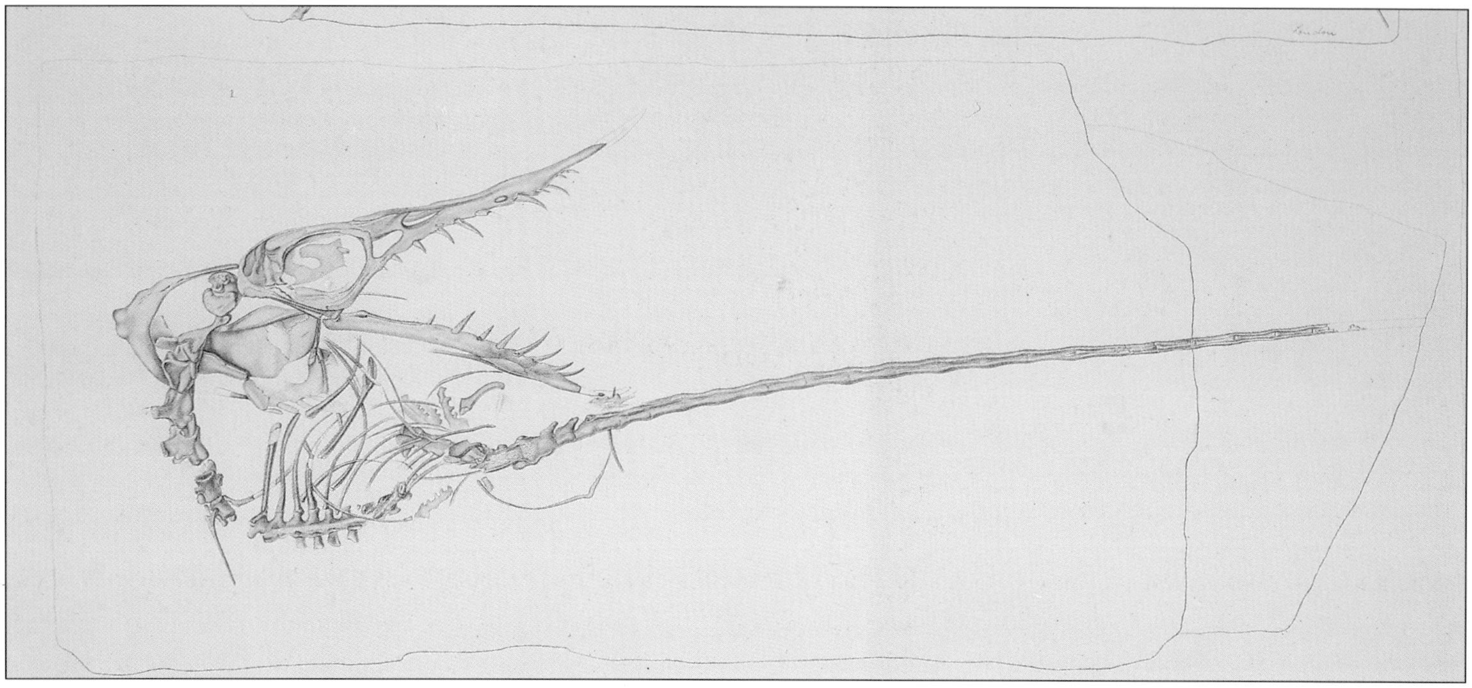

In Hermann von Meyer's great monograph there is also a description of skeletal remains found in Jurassic limestone slabs from Riedenburg in the Altmühl valley, about 25 miles (40km) east of Eichstätt. Because of its different, strong feet this species was given the name *Pterodactylus crassipes*, the thick footed. The fossil was later acquired by the Teyler Museum in Haarlem, Holland. In 1970 it was examined there by John H. Ostrom, Professor of Geology at Yale University, New Haven, USA. Professor Ostrom was more than a little surprised to discover that the sparse fragments of the hindlimbs and the hand did not come from a *Pterodactylus*, indeed they were not even part of a pterosaur, but came from a bird, the famous primordial bird *Archaeopteryx*. At that time there were only three specimens of this rare fossil, deposited in London, Berlin and Maxberg near Solnhofen.

John Ostrom discovered indistinct impressions on the rock surface of the limestone slab which had already confused Hermann von Meyer. Meyer wrote: 'I have never, on any other *Pterodactylus*, seen phenomena that one is inclined to ascribe to the folds of the flight membrane as clearly as on this specimen. But even here I cannot think them to be what I would dearly like to call them, because, apart from the fact that the flight digit is nowhere near them, they are not sharp and definite enough for folds of skin, particularly in a compressed condition. Therefore the reason for this phenomenon must be unevenness of the bed and movement of the water, which have caused this formation.'

Under oblique light these weak impressions on the surface of the stone showed up quite clearly for John Ostrom as imprints of feathers. Comparison of the skeletal remains with other specimens of prehistoric birds removed any remaining doubts. The *Pterodactylus crassipes* in the Teyler Museum in Haarlem was not a pterosaur, but an *Archaeopteryx*. Thus this primordial bird, the fourth specimen, had been woken like Sleeping Beauty after 110 years.[27]

For many years the curator of the Palaeontological Museum in Munich, Johann Andreas

27 Ostrom, J.H., 1970. *Archaeopteryx: Notice of a 'New' Specimen*. Science, 170: 537-538.

Above: This illustration of a *Rhamphorhynchus gemmingi* from Solnhofen appears in Hermann von Meyer's magnificent monograph *Die Reptilien aus dem Lithographischen Schiefer des Jura in Deutschland und Frankreich*. It shows the well preserved skull and typical long tail very nicely. The wings were obviously lost before the skeleton was embedded in the limy mud. Today this specimen is housed in the Teyler Museum, Haarlem.

Below: This is a plate from E.T. Newton's description of *'Scaphognathus' purdoni*, a fossil skull of a pterosaur discovered by the Rev D. W. Purdon near Whitby in Yorkshire. It shows the skull from both sides (Figs. 2 and 3) and from behind (Fig. 5). The preserved brain cast is seen from the left (Fig. 6) and from above and behind (Fig. 7). For comparison a skull of the Tuatara (*Sphenodon*) (Figs. 10 and 11) and a back view of the skull of a fowl (Fig. 12) are also included.

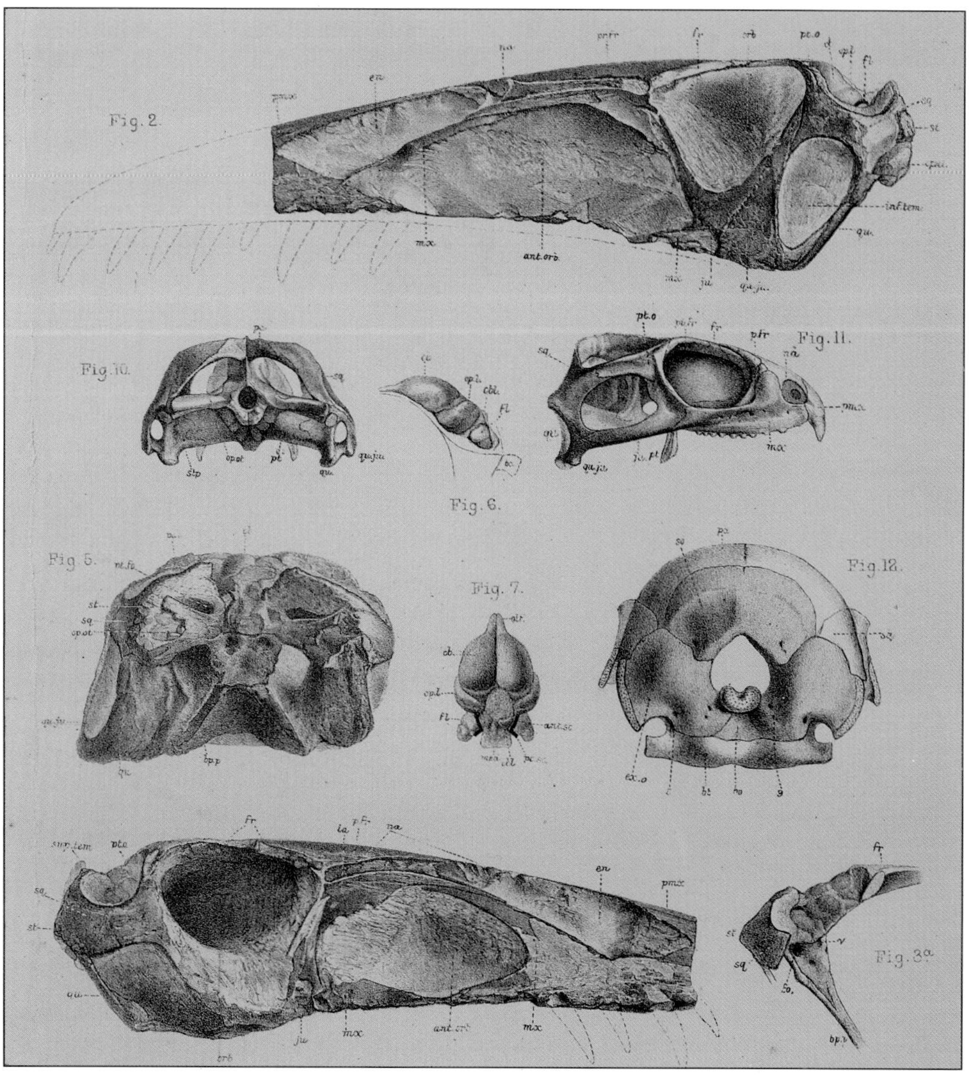

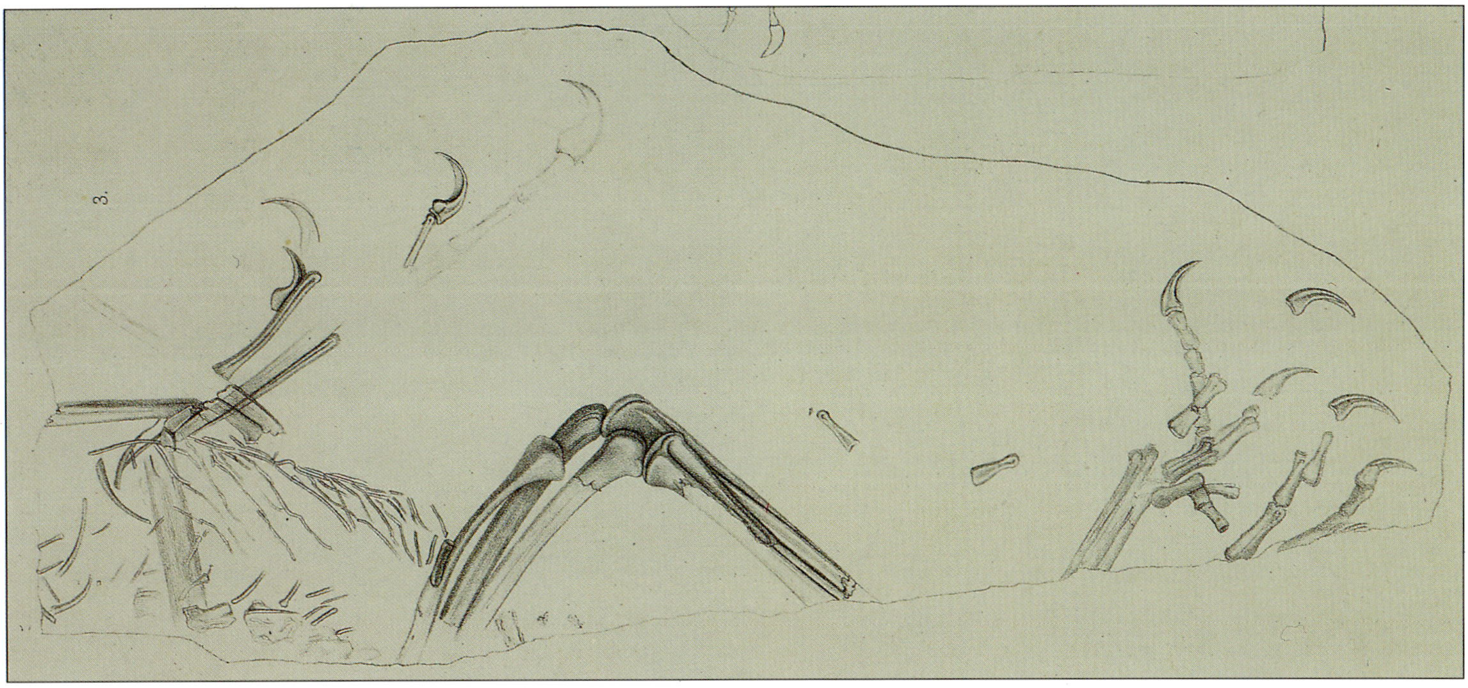

Wagner (1797-1861), devoted himself to investigating Solnhofen pterosaurs.[28] He described a number of new species. Many of these specimens came to light as a result of the purchase of Dr Carl Haeberlein's collection. He was a country doctor from Pappenheim, and the same Carl Haeberlein who in 1862 sold a large collection of Solnhofen fossils, including the first *Archaeopteryx* skeleton, to the British Museum (Natural History) in London. Interestingly Wagner saw pterosaurs as a transitional phase between reptiles and birds, although he was a convinced supporter of the Biblical creation story and violently rejected Darwin's ideas even in the last year of his life, 1861.

Remains of pterosaurs from strata of the Middle and Upper Jurassic had also come to light in England, for example from the so-called Stonesfield Slate in Oxfordshire, called *Pterodactylus bucklandi* by Hermann von Meyer in 1832, and in 1874 from the Kimmeridge Clay on the coast of Dorset in southern England. In the 1880s the Reverend D.W. Purdon found a fossil skull in the Liassic rocks of Whitby in Yorkshire, which he sent to the Geological Survey in London for investigation. There it was described by E.T. Newton in 1888, who named the species *Scaphognathus purdoni*, in honour of its finder. The genus *Scaphognathus* had been established by Andreas Wagner in 1861.

The fossil discovered by the Reverend Purdon was an unusually well-preserved pterosaur skull, in which the petrified contents of the brain capsule could be seen through an opening in the top of the skull. This kind of preservation is called an endocast, and it occurs when ooze is pressed into the hollow parts of the skull after the soft parts have decayed. This becomes firm, along with the surrounding sediment and also turns into stone. After removing the bones of the skull, Newton had a faithful cast of the brain that had controlled the pterosaur's life functions 180 million years ago. It showed remarkable similarity to the brain of a bird and was clearly different from the brain of a reptile.

In 1844, after Gideon Mantell himself recognized that the bones he had found in Tilgate Forest, thought to be bird bones, were in fact pterosaur bones, skeletal remains of pterosaurs were found at other Cretaceous sites in England, for example in 1851 in the Cambridge Greensand, which provided numerous specimens in subsequent years, although they were always isolated bones. Two researchers in particular devoted themselves to scientific investigation of these finds: Sir Richard Owen (1804-1892), director of the British Museum (Natural

Above: Also figured in his great monograph of 1859-60, Hermann von Meyer assigned these remains from the Solnhofen limestone to a new species of pterodactyl, as *Pterodactylus crassipes*. It was not until 1970 that it was recognised as another specimen of the famous primordial bird *Archaeopteryx* by John H. Ostrom of Yale University. The fossil is also housed in the Teyler Museum, Haarlem, Netherlands.

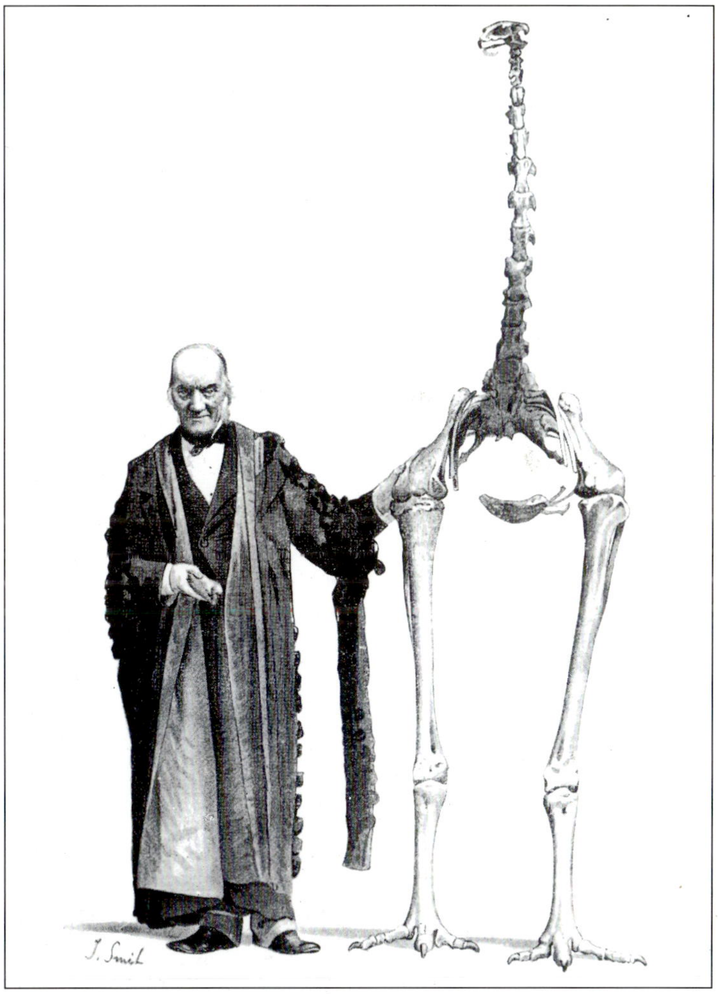

Right: Sir Richard Owen (1804-1892) with a mounted skeleton of the fossil giant bird *Dinornis*. In his time Owen was a leading comparative anatomist, and eventually became the first superintendent of the British Museum (Natural History) in London. In several monographs Owen described the English pterosaurs, which he regarded as cold-blooded reptiles. His view conflicted with that of Harry Govier Seeley, who insisted that pterosaurs must have been warm-blooded, and a violent controversy between the two scientists ensued.

28 Johann Andreas Wagner (1797-1861) was born in 1797 in Nuremberg. He was the first lecturer in zoology at the University of Erlangen, then went to Munich in 1836, became curator of the state palaeontological collection and was appointed to the first chair of palaeontology in Germany. He published numerous works, including material on fish and reptiles from Solnhofen, Tertiary primates from Greece and ice-age mammals from caves in Franconia. He was a profound believer in the Biblical creation story and was moved to oppose Charles Darwin's 'adventurous' ideas in the last years of his life. Thus for him the primeval bird *Archaeopteryx*, found in 1861, was not a bird, but a reptile, which he called *Griphosaurus* (=puzzle lizard), despite its feathers.

History) in London, and Harry Govier Seeley (1839-1909), Professor of Geology at King's College London. In 1901 Seeley published the first complete and popular book on pterosaurs, *Dragons of the Air*.

Seeley published a whole series of essays on pterosaurs (his Saurornia), which he later called Ornithosauria, following the Italian ornithologist Prince Charles Bonaparte. (But the name *Pterosauria* has priority, as it was suggested by the Darmstadt palaeontologist Johann Jakob Kaup as early as 1834). Seeley was concerned in particular with lavish finds of pterosaur bones in the Cretaceous Greensand of Cambridge. He gives an idea of the sites at which the bones were found in 1901 (p.34), when he writes: 'To give some idea of their abundance, it may be stated that they were mostly gathered during two or three years, as a matter of business, by an intelligent foreman of washers of the nodules of phosphate of lime, which, in commerce, are named coprolites. He soon learned to distinguish Pterodactyle bones from other fossils by their texture, and learned the anatomical names of bones from specimens in the University Museum. This workman, Mr Pond, brought together not only the best of the remains at Cambridge, but most of those at the museums at York and in London, and the thousands of less perfect specimens in public and private collections.'

In 1864, by analogy with birds, he insisted that pterosaurs, which he called Saurornia, must have been warm-blooded, which in his view justified separation from the reptile class.[29] Seeley felt that pterosaurs (he now used the name Ornithosauria) could not have been genuine reptiles. He wrote in *Dragons of the Air*. '. . . Ornithosaurs may now with more accuracy be described as dinosaurian. The dinosaurs, like the pterodactyles, must be re-

29 Seeley, H.G., 1864. *On the pterodactyle as evidence of a new subclass of Vertebrata (Saurornia)*. Reports of the British Association of Scientists, 34th meeting: p.69.
Seeley, H.G., 1870. *The Ornithosauria. An elementary study of the bones of Pterodactyles*. Cambridge University Press.
Seeley, H.G., 1901. *Dragons of the Air*. New York. Reprinted 1967 by Dover Paperback. This was the first popular book devoted exclusively to pterosaurs as a group. In it Seeley summarized knowledge about pterosaurs as it stood at the time.

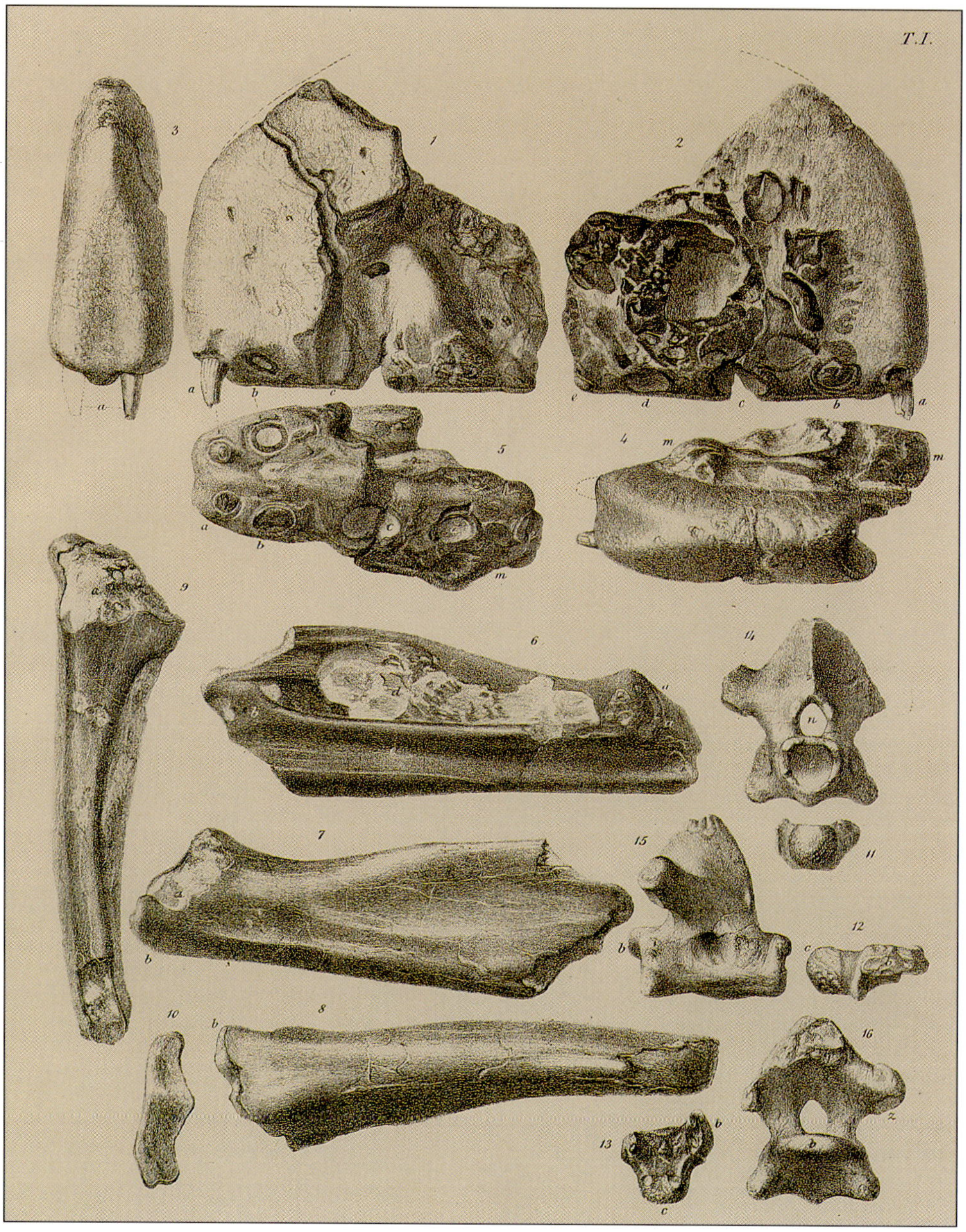

Above: A plate from one of Owen's monographs on the *Fossil Reptilia* depicting jaw fragments and vertebrae of *Criorhynchus* from the Cambridge Greensand.

Below: A life restoration of *Dimorphodon macronyx* from the Lias of Lyme Regis by H.G. Seeley, as seen in his popular book *Dragons of the Air* (1901).

Right: The Yale expedition of 1870 with O.C. Marsh standing in the centre. It was this party that discovered the first American pterosaur remains in Kansas. Because the West was still Indian territory, standard equipment included not only geological hammer, but also rifle and bowie-knife.

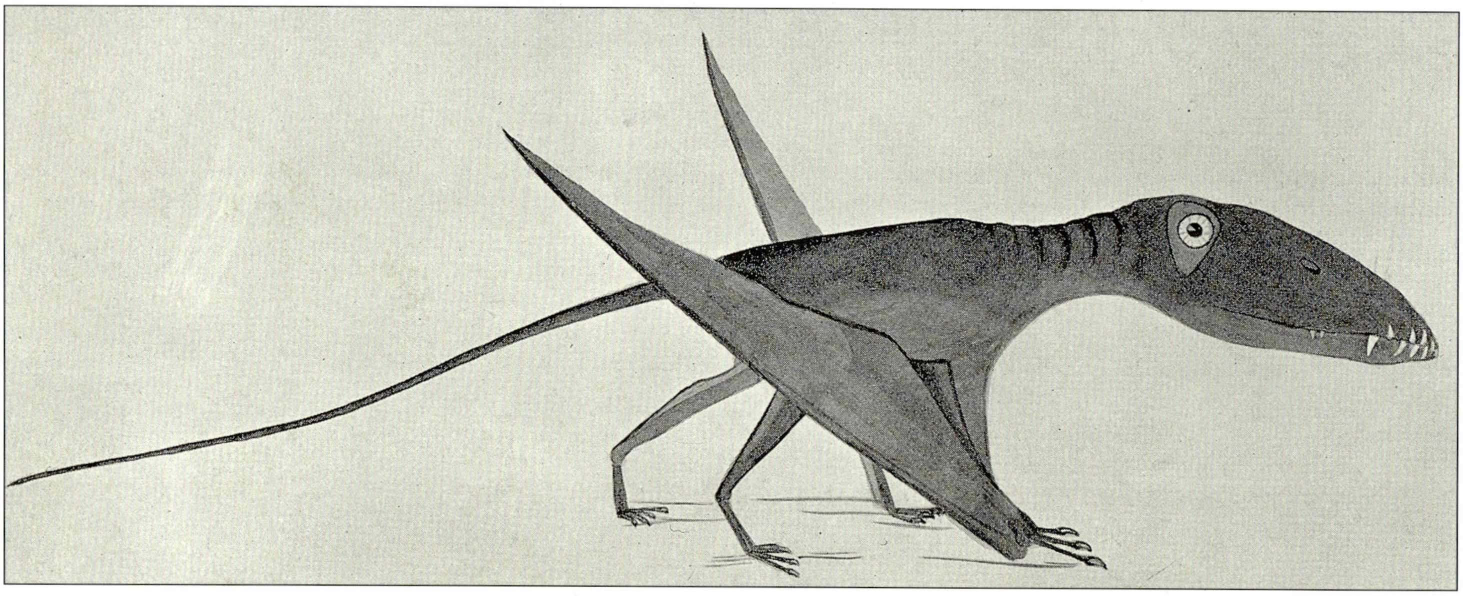

garded as intermediate in some respects between reptiles and birds.' Seeley then suggested a position of their own for pterosaurs, warm-blooded in his opinion, in the system of vertebrates. They were to be classified parallel with birds between reptiles and mammals.

Richard Owen's opinions were quite different from Seeley's.[30] He believed for example that the size of pterosaurs was limited from the outset. He thought they were cold-blooded and thus physiologically inferior to warm-blooded flying creatures like birds and bats. No larger pterosaurs were in fact known until about 1840. But in 1845 the secretary of the London Palaeontographical Society, John Scott Bowerbank, described a new species of pterosaur from chalk strata in Kent, whose wing span must have been at least eight to nine feet (2·5m). And somewhat later he discovered fossil bones of an even larger species, which must have had a span of almost 16·5ft (5m) with wings extended. For this reason Bowerbank called the species *Pterodactylus giganteus*.[31] This did not mean that Owen changed his views about the physiology of pterosaurs; no, he now declared, as a convinced supporter of Biblical teaching on creation, that God's power as a creator had been greater in the Earth's past, and thus went beyond calculations based upon contemporary nature.

Above: Edward Drinker Cope (1840-1897) was one of the pioneers of North American palaeontology, especially in exploring the Western territories for fossil vertebrates. His rivalry with Marsh is legendary.

A violent controversy developed between Owen and Seeley about the warm-bloodedness and physiology of pterosaurs. In 1870 Owen had received new skeletal remains of the Liassic pterosaur *Dimorphodon* from Lyme Regis in Dorset and compared its osteological characteristics with modern crocodiles and lizards in particular.[32] For Owen, as an anti-evolutionist, pterosaurs must have been reptilian and cold-blooded; they could not be warm-blooded and equally ranked with birds. The decisive fact was not that pterosaurs were actively flying animals, but that there was no evidence that they had an insulating skin covering to conserve warmth, in other words no feathers or hairs. In Seeley's eyes Owen was going against physiological principles by separating metabolic level and an ability to fly actively. Owen's reaction to much later evidence that pterosaurs had hair would have been interesting.

Immediately after the appearance of Owen's monograph Seeley published a critical statement.[33] For him pterosaurs were not reptiles, but closely related to birds. Like these they had a high metabolic level and similar physiology. This was also the case for lungs, heart and brain. He supported this on the basis of his excellent knowledge of the osteology of English Cretaceous pterosaurs. But essentially the quarrel between Seeley and Owen was not about the interpretation of skeletal structures, but a debate on the validity of species transformations and evolutionary progress.

American Discoveries

In the 1870s much bigger pterosaurs were discovered, the largest of all flying creatures. These were found in America. Investigation of these giant pterosaurs is closely associated with Edward Drinker Cope (1840-1897) and Othniel Charles Marsh (1831-1899), both pioneers of geological and palaeontological exploration of the North American West, but also bitter rivals. Their principal service was the discovery and exploitation of famous dinosaur remains. Cope was a member of the Academy of Science in Philadelphia, Marsh was a professor at Yale College in New Haven.

In summer 1870 O.C. Marsh mounted an expedition to the Rocky Mountains to look for fossils. One of his companions was Colonel William F. Cody, better known as Buffalo Bill. In November, on the way back, Marsh's group began to explore the Cretaceous strata of this region, the Niobrara Chalk of western Kansas,

30 It was Richard Owen (1804-1892) who introduced the name *Dinosauria* in 1841, although only three specimens of these giant reptiles were known from English Jurassic and Cretaceous strata. After the appearance of Darwin's book *The Origin of Species* in 1859 he became a leading anti-Darwinist, opposing Darwin's theory, which stated that animal and plant species on Earth evolved one from another by a gradual process of transformation, and thus were the result of evolution controlled by natural selection. Only the best-adjusted organisms survive in the 'struggle for life'. Owen described the first skeleton find of the primeval bird *Archaeopteryx*, bought by the British Museum from the Pappenheim doctor Carl Haeberlein in 1862, as a primitive, but genuine bird. Unlike the Darwinists he did not see it as a 'missing link' in evolution from reptile to bird.

31 Bowerbank, J.S., 1846. *On a new species of Pterodactyle of the upper Chalk*. Quarterly Journal of the Geological Society London, 1(5):7.

32 Owen, R., 1870. *Monograph of the fossil Reptilia of the Liassic Formations. III*. Monographs of the Palaeontographical Society London, pp.41-81.

33 Seeley, H.G., 1870. *Remarks on Professor Owen's Monograph on Dimorphodon*. Annals and Magazine of Natural History, (4), 6: 129.

from their camp on the Smoky Hill River. Kansas was in fact E.D. Cope's territory, and he had recently received fossils of mosasaurs and plesiosaurs from the same strata. Cope, in fact, had never visited the area himself. He had collectors, his 'bone hunters', who sent material to Philadelphia for him. By contrast, Marsh went to explore the West himself. At that time this was quite a dangerous adventure, as the Indians could attack at any moment.

After reaching the Smoky Hill River, Marsh's group immediately found three new species of mosasaur, great marine lizards, including a skeleton almost 33ft (10m) long. Some problematical bones were also eventually excavated. They were long, hollow and thin-walled, reminscent of bird bones. But their joints were the same shape as those of the English pterodactyls, although these were much smaller. The fragment of a metacarpal bone was 6·7in (17cm) long, from which Marsh calculated a wing span for this truly gigantic flying creature of 20ft (6·1m).[34]

In the following summer (1871) Marsh's hopes of more complete pterosaur finds on the Smoky Hill River were fulfilled. He found the other half of the metacarpal bone he had come across the previous year, and this gave an overall length of 16in (40cm). The first phalanx of the flight digit was 16·5in (42cm) long. These new finds confirmed Marsh's first calculations of overall size. Somewhat higher in the Chalk profile he discovered the wing bones of an even bigger creature, whose wing span he estimated at about 21·6ft (6·6m). He named the two forms in 1872 as two new species of *Pterodactylus*, *Pterodactylus ingens* (the huge) and *Pterodactylus occidentalis* (the western).[35]

It was a triumph for Marsh to have discovered and named the first pterosaurs on American soil, also the largest so far known. His rival Cope had not been inactive in the meantime. Immediately after Marsh's departure at the end of the 1871 season, Cope travelled to Kansas and found the same sites on the Smoky Hill River. He started to dig feverishly to establish his own collection of Cretaceous reptiles, so that he could publish before Marsh if possible. On 1 March 1872 he presented a paper to the American Philosophical Society, describing his Kansas pterosaurs.[36] But Marsh's mouthpiece, the American Journal of Science, which he edited, had published the description of *Pterodactylus occidentalis* and *Pterodactylus ingens* five days earlier, and thus secured priority for Marsh.

The Kansas pterosaurs were not only gigantic, but surprisingly their jaws were toothless, in complete contrast with pterosaurs found in the slightly older English Cretaceous deposits. But above all they had a long, thin bone crest at the rear of the head, which must have made them look very peculiar. In 1876 Marsh called the American form *Pteranodon* (=toothless flyer), thus separating them from the European genera.[37] He also distinguished a second genus

Above: Othniel Charles Marsh (1831-1899) discovered the first pterosaur bones in the New World in the autumn of 1870. These were fragments of a wing bone in the Niobrara Chalk of Kansas.

from the Kansas Chalk, the so-called Niobrara Formation, which was smaller and had no bone crest on the skull. He called it *Nyctosaurus* (=naked reptile). In subsequent years the Kansas sites were so productive that George F. Eaton, curator of Osteology and Vertebrate Palaeontology at the Peabody Museum at Yale College in New Haven, listed a collection of pterosaur bones from a total of 465 individual specimens.[38]

One of the most active researchers into North American Cretaceous pterosaurs was Samuel Wendell Williston (1852-1918). In his younger years he had been one of the most successful of O.C. Marsh's 'bone hunters' in the famous Western dinosaur sites. In 1882 he became geological assistant at Yale College, in 1887 Professor of Anatomy at Yale, in 1890 Professor of Geology at the University of Kansas and in 1902 Professor of Palaeontology at the University of Chicago. It was in his Kansas period that Williston was most concerned with pterosaur finds from the Niobrara Chalk. He wrote a total of nine essays on the 'Kansas Pterodactyls'[39], about the genera *Pteranodon* and *Nyctosaurus*, which had earlier been named by Marsh. Williston provided highly detailed observations about the skeleton of these pterosaurs, on the basis of considerably more complete skeletal material. He also believed that the genus *Ornithostoma*, described in England by Seeley, was identical with the American *Pteranodon*, a question that has still not been conclusively settled. In 1902 he reported on a remarkable specimen of *Nyctosaurus*, found shortly before in the Niobrara Chalk.[40] It was so complete that many new anatomical characteristics were observed, like for example the peculiar joints of the hind legs, from which he concluded that the creature must have walked most awkwardly on the ground. From this specimen, now in the Field Museum in Chicago, Williston provided an outstanding skeletal reconstruction of *Nyctosaurus* with a wing span of 7·9ft (2·4m); this is still reproduced in some modern textbooks.

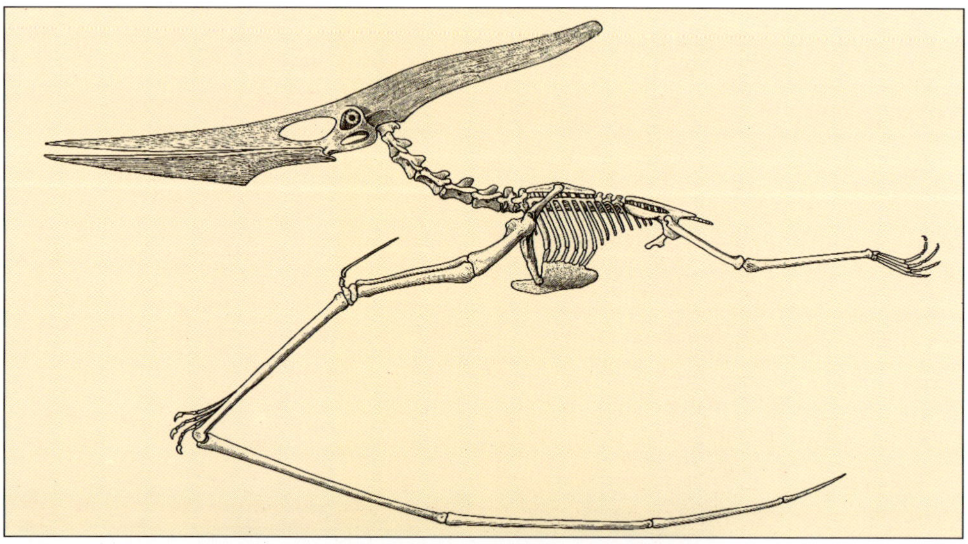

Above: George F. Eaton published a monograph on the *Osteology of Pteranodon* in 1910, on the basis of Marsh's finds. For the first time he reconstructed the skeleton of this giant pterosaur.

Left: The first evidence of pterosaurs in America was this metacarpal bone, the fragments of which were discovered by O.C. Marsh.

34 Marsh, O.C., 1871. *Note on a new and gigantic species of Pterodactyle*. American Journal of Science, 1: 472.

35 Marsh, O.C., 1872. *Discovery of additional remains of Pterosauria etc.* American Journal of Science, 3: 241.

36 Cope, E.D., 1872. *Two new Ornithosaurians from Kansas*. Proceedings of the American Philosophical Society, 12: 420-422.

37 Marsh, O.C., 1876. *Principal characters of American Pterodactyles*. American Journal of Science, 12: 479.

38 Eaton, C.F., 1910. *Osteology of Pteranodon*. Memoires of the Connecticut Academy of Arts and Sciences, 2: 1-38.

39 Williston, S.W., 1892-1893. *Kansas Pterodactyls*. Kansas University Quart., I, II.

40 Williston, S.W., 1902. *On the skeleton of Nyctodactylus, with restoration*. American Journal of Anatomy, 1: 297-305.

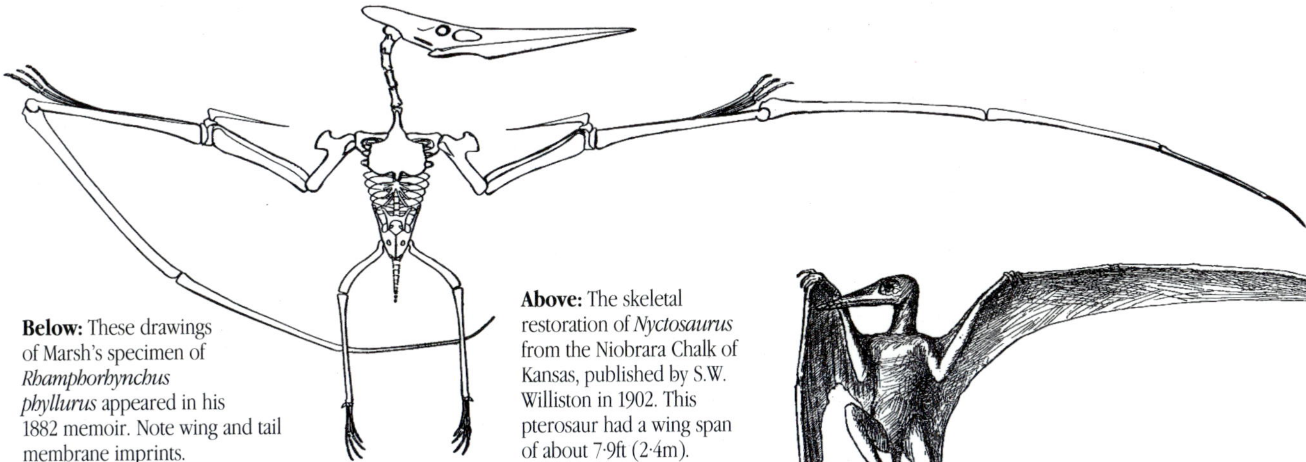

Below: These drawings of Marsh's specimen of *Rhamphorhynchus phyllurus* appeared in his 1882 memoir. Note wing and tail membrane imprints.

Above: The skeletal restoration of *Nyctosaurus* from the Niobrara Chalk of Kansas, published by S.W. Williston in 1902. This pterosaur had a wing span of about 7·9ft (2·4m).

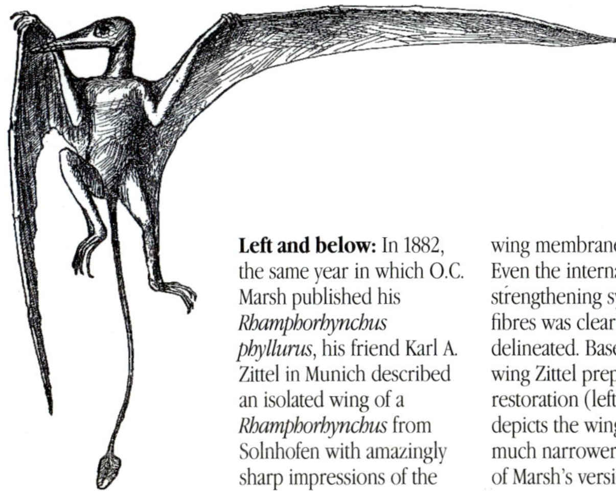

Left and below: In 1882, the same year in which O.C. Marsh published his *Rhamphorhynchus phyllurus*, his friend Karl A. Zittel in Munich described an isolated wing of a *Rhamphorhynchus* from Solnhofen with amazingly sharp impressions of the wing membrane preserved. Even the internal strengthening system of fibres was clearly delineated. Based on this wing Zittel prepared the life restoration (left) which depicts the wings as being much narrower than those of Marsh's version.

Above: Karl Alfred von Zittel (1839-1904), Professor of Geology and Palaeontology in Munich, was a leading 19th century palaeontologist. Under his direction, the Bavarian State Collection became one of the most important collections in Europe.

In 1873 O.C. Marsh heard of the discovery of a Solnhofen pterosaur with wing membrane imprints. Through a friend he snapped up this sensational find from under the noses of German museum curators, for a thousand dollars. It was a magnificently preserved *Rhamphorhynchus*, to which Marsh gave the name *phyllurus* (=leaf-tail) in his description published in 1882.[41] *Rhamphorhynchus*' flight membrane was unknown until this time, but in this specimen it was imprinted on the surface of the rock with complete clarity. Surprisingly Marsh also found a skin imprint at the end of the long tail. This showed a rhomboid, terminal tail vane, which he oriented vertically, and interpreted as a steering rudder.

In his reconstruction drawing of the creature in life he also assumed the existence of another flight membrane between the tail and the hind legs, looking like the uropatagium of a bat. Apparently Marsh was influenced by Owen to make this restoration, who had shown a similar flight membrane for his *Dimorphodon*. However, there is no evidence in the fossil material itself for a bat-like 'uropatagium' of this kind in either *Rhamphorhynchus* or *Dimorphodon*.

By chance Karl Alfred Zittel[42] also described a *Rhamphorhynchus* wing membrane from the State Collection in Munich in the same year, 1882. Only an isolated wing had been preserved, but there was a very detailed imprint of the wing membrane, not only in outline, but with the internal strengthening system of stiff fibres, which were embedded very densely in the membrane. Zittel concluded from this find that *Rhamphorhynchus*' wings must have been very narrow, not broad and including the hind legs, as is the case with bats.

A particular controversy developed around the question of how the terminal tail vane was oriented in long-tailed pterosaurs, vertically or horizontally? Was it to be interpreted as intended to control height or sideways movement? Marsh oriented it, as we have seen, vertically, as he had noticed a slight asymmetry in the outline. Later for aerodynamic reasons it was seen as a height control and oriented horizontally. But several indications suggest that Marsh was right after all.

In Germany the so-called lithographic limestone from the many quarries in the area of Solnhofen, Eichstätt and even further east as far as Kelheim at the confluence of the Altmühl and Danube rivers continued to provide many discoveries which led to new perceptions about pterosaurs of the Jurassic. Because of the extremely fine grain of the sediment these strata preserve soft parts that are usually unable to survive very well. Thus in 1908 Karl Wanderer observed imprints above and below the skull of a *Rhamphorhynchus* from the Dresden Museum that suggested a soft crest on the head and a throat pouch.[43] Also he was the first to find numerous dots like needle pricks on the flight membrane prints, which were later (1927) established as being caused by hair folli-

41 Marsh, O.C., 1882. *The wings of Pterodactyles.* American Journal of Science, 3, no. 16: 223.

42 Zittel, K.A., 1882. *Über Flugsaurier aus dem lithographischen Schiefer Bayerns.* Palaeontographica, 29: 47-80.

43 Wanderer, K., *Rhamphorhynchus Gemmingi H.v. Mayer. Ein Exemplar mit teilweise erhaltener Flughaut aus dem Kgl. Mineralog.-Geol Museum zu Dresden.* Palaeontographica, 55: 195-216.

cles by the Munich palaeontologist Ferdinand Broili.[44] This meant that pterosaurs were probably warm-blooded, on the basis of direct evidence from the fossil, confirming the view of Seeley and others. In the same publication Broili also reported on the observation of webs between the toes of a Solnhofen *Rhamphorhynchus*. This confirmed another of Seeley's assumptions, that pterosaurs could swim.

It was also in 1927 that the important palaeoneurologist Tilly Edinger published her findings on the brain structure of pterosaurs.[45] Brain endocasts had been known since 1888, when Newton wrote about the skull of *Scaphognathus purdoni* from the Whitby Lias. But some Solnhofen pterodactyls and rhamphorhynchids showed evidence of brains in the form of endocasts. Tilly Edinger's research showed that, even in the Upper Jurassic, pterosaurs had developed brains that were more like those of birds than the brain of their contemporary, the 'primordial bird' *Archaeopteryx*. Thus the pterosaur brain was by no means reptile-like and small, as in modern crocodiles and lizards, but closer to that of a bird in shape and size. This was an important prerequisite for flight control and steering. As for Seeley and many palaeontologists after him, this was both a sign of a high metabolic rate and of warm-bloodedness, and thus a reason for not classifying pterosaurs as reptiles, but for placing them in a separate class of vertebrates.

If pterosaurs were still alive today and had not become extinct 65 million years ago, we would certainly never group these hairy, warm-blooded and possibly intelligent creatures with crocodiles, lizards or tortoises, but give them a place of their own in the zoological system of vertebrates. But this problem will be discussed in the following chapter.

Above: Tilly Edinger (1897-1967) was the founder of modern palaeoneurology, a science for the investigation of brain structures in fossil animals. She studied in detail the structure of pterosaur brains.

Above: D.M.S. Watson (1886-1973) was one of the first palaeontologists to regard pterosaurs as highly efficient mechanisms. He studied their flight behaviour from an aerodynamic point of view.

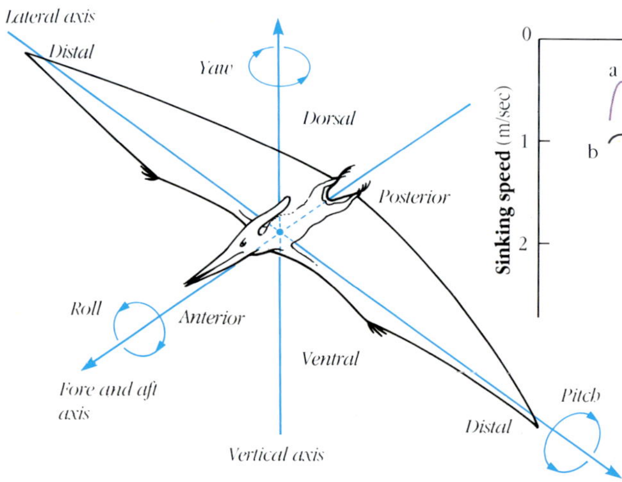

Above: This diagram (based on Bramwell and Whitfield's monograph on the *Biomechanics of* Pteranodon) shows the three axes of a flying object as applied to *Pteranodon*. The flying animal is regarded like an aircraft requiring the necessary control of motion about the roll, pitch and yaw axes.

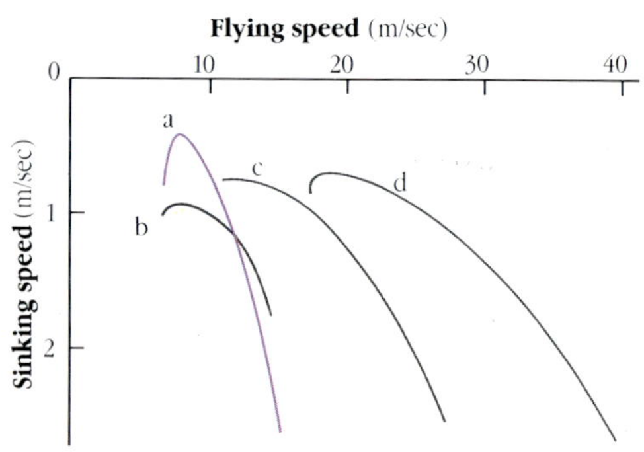

Above: These polar curves plot flying speed against sinking speed for *Pteranodon* (a), a falcon (b), an albatross (c) and a glider (d). It can be seen that the best flying speed for *Pteranodon* was about 26ft/sec (8m/sec) when the sinking speed was at its minimum.

Researching Pterosaur Flight

Stimulated by the beginnings of air travel with motor-driven aircraft in the early years of the century, interest was aroused in the flight characteristics of pterosaurs from the technical and aerodynamic point of view. Typically one of the first publications on this fascinating subject did not appear in a zoological or palaeontological magazine, but in the official organ of the Aeronautical Society of Great Britain, the Aeronautical Journal of October 1914. The essay by E.H. Hankin and D.M.S. Watson[46] was called 'On the Flight of *Pterodactyls*' and dealt with the flight of *Pteranodon*, the giant pterosaur from the Kansas Chalk. But they also used English skeletal material, with which they could test the movements of the individual wing joints. The pterosaur bones from the Cambridge Greensand were particularly suitable for this, as they were completely uncrushed. This was how they summed up their judgement of *Pteranodon*: 'Other flying animals can walk, run or swim, besides fly. But in the case of the higher pterodactyls their structure is such that it is difficult to understand how they can have had any other means of progression than flying. With a body little larger than that of a cat they had a span of wing asserted in some cases to have reached 21 feet or more.' And they added: 'The weakness of the flapping muscles makes it highly probable that their habitual mode of flight was by soaring, rather than by flapping.'

Subsequently *Pteranondon* in particular was the object of biomechanical and aerodynamic analyses, most recently by Cherrie D. Bramwell and George R. Whitfield[47] of the University of Reading in 1974. They calculated that Pteranodon, with a wing span of about 23ft (7m) and weighing 36lb (16·6kg) would have had a maximum flight speed of about 26ft/sec (8m/sec), and reached a lowest flight speed of 22ft/sec (6·7m/sec). This extremely slow flight also made it possible to land on the ground without danger. Bramwell and Whitfield were also of the opinion that an increase of average wind speed when the climate deteriorated at the end of the Cretaceous might have been one of the factors that brought about the extinction of the pterosaurs.

Finally scholars were not content with calculating the flight of pterosaurs theoretically, but even simulated it with flying models. The first attempt was by the German zoologist and behavioural researcher Erich von Holst.[48] In 1957 he built a flapping flight model of the Jurassic pterosaur *Rhamphorhynchus* with a wing span of 4ft (1·2m). He wanted to find out whether the terminal, rhomboid tail vane on the long tail of this pterosaur was originally oriented vertically or horizontally. The model, made of Japanese paper and balsa wood and driven by a rubber band motor, did in fact fly by flapping its own wings, but only with a horizontal tail vane functioning as a height control.

44 Broili, F., 1927. *Ein Rhamphorhynchus mit Spuren von Haarbedeckung*. Sitzungsberichte der Bayerischen Akademie der Wissenschaften München: pp.49-67.

45 Edinger, T., 1927. *Das Gehirn der Pterosaurier*. Zeitschrift für Anatomie und Entwicklungsgeschichte, 82 (1/3): 105-112.

46 Hankin, E.H. and Watson, D.M.S., 1914. *On the Flight of pterodactyls*. Aeronautical Journal, 18: 324-335.

47 Bramwell, C.D. and Whitfield, G.R., 1974. *Biomechanics of Pteranodon*. Philosophical Transactions of the Royal Society of London, B. Biol. Sci., 267: 503-592.

48 Holst, E.v., 1957. *Der Saurierflug*. Paläontologische Zeitschrift, 33: 15-22.

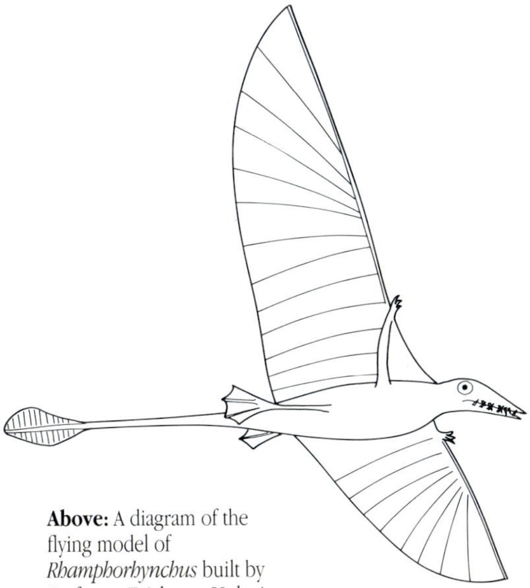

Above: A diagram of the flying model of *Rhamphorhynchus* built by Professor Erich von Holst in 1957.

Above: Stephen Winkworth launches his radio-controlled gliding model of *Pteranodon* into the air in 1985. The model flew successfully above the cliffs of the Dorset coast in southern England.

Below: The half-size model of *Quetzalcoatlus* built by Dr Paul MacCready and his team at AeroVironment Inc. in Monrovia, California. This model actually flapped its 18ft (5·5m) span wings during flight.

In 1985 an Englishman called Stephen Winkworth built a gliding model of a *Pteranodon* with built-in radio control.[49] And finally the American aeronautical engineer Paul MacCready[50] built a flapping flight model of the giant Texan pterosaur *Quetzalcoatlus*, half actual size, but still with a wing span of 18ft (5·5m). After successful test flights in Death Valley in California the model *Quetzalcoatlus* crashed in May 1986 in front of a large number of spectators in Washington.

In the last few decades it has become increasingly clear that the pterosaurs of the Mesozoic were an extraordinarily varied and successful group of animals. They have been found on all continents with the exception of the Antarctic. In the last few years further finds have been made in Europe and North America, and skeletal remains have also been found in South America, Asia, Australia and New Zealand. The first Chinese pterosaurs were particularly significant. They were found in the Lower Cretaceous of the province of Sinkiang, and described by C.C. Young in 1964.[51] There was also the strange pterosaur from the Lower Cretaceous in Argentina with filter dentition made up of hundreds of long, thin, bristle-shaped 'teeth',[52] and last but not least the excellently preserved pterosaur bones in the limestone concretions of the Santana Formation on the Araripe Plateau in Brazil, which has developed into one of the world's most important pterosaur localities.[53]

Among the most important discoveries of recent times are two superlatives, the largest pterosaur and the oldest pterosaur. In 1975 Douglas Lawson, then a student at the University of Texas in Austin, while doing field work for his master's thesis, found the skeletal remains of the largest pterosaur so far known, which he called *Quetzalcoatlus northropi*, (after the Mexican god Quetzalcoatl).[54] The fossils came from deposits from the Upper Cretaceous in the Big Bend National Park in West Texas. A reconstruction of the skeleton resulted in a wing span of 36 to 39·4ft (11 to 12m). What would Richard Owen have said about that?

Quetzalcoatlus is not just the largest pterosaur so far found, but also the geologically most recent pterosaur known to us. The oldest was discovered in 1973 in limestone of the Upper Triassic in the Bergamo Alpine foothills in Italy.[55] The Italian geologist Rocco Zambelli called it *Eudimorphodon* (=true two-form tooth), following the previous oldest, which Owen had christened *Dimorphodon* (=two-form tooth).

All these finds were documents of evolution and mosaic stones in the overall picture that we have gradually been able to form of this fascinating animal group called pterosaurs. The future will certainly bring further discoveries, and present us with a number of unexpected and surprising forms. Such is the nature of scientific inquiry, and we shall then, as always, have enthusiastic collectors and researching scientists to thank for this.

49 Winkworth, S., 1985. *Pteranodon flies again*. New Scientist: 32-33.
50 MacCready, P., 1985. *The Great Pterodactyl Project*. Engineering and Science, 49 (2): 18-24.
51 Young, C.C., 1964. *On a new pterosaurian from Sinkiang, China*. Vertebrata Palasiatica, 8: 221-256. Description of *Dsungaripterus* from the Lower Cretaceous in Sinkiang.
52 Bonaparte, J.F., 1970. *Pterodaustro guinazui gen. et sp. nov.. Pterosaurio de la formación Lagarcito, provincia de San Luis, Argentina*. Acta Geologica Lilloana, 10 (10): 207-226.
Sanchez, T.M., 1973, *Redescription del Craneo y Mandibulas de Pterodaustro guinazui Bonaparte (Pterodactyloidea, Pterodaustridae)*. Ameghiniana, 10 (4): 313-325.
53 Price, L.I., 1971. *Presenca de Pterosauria no Cretáceo Inferior da Chapada do Araripe, Brasil*. Annals of the Brazilian Academy of Sciences, 43 (suppl.): 451-461.
54 Lawson, D.A., 1975. *Pterosaur from the Latest Cretaceous of West Texas. Discovery of the Largest Flying Creature*. Science, 187: 947-948.
55 Zambelli, R., 1973. *Eudimorphodon ranzii gen. nov., sp. nov., uno pterosauro triassico*. Rendiconti Istituto Lombard. Scienze, B, 107: 27-32.
Wild, R., 1978. *Die Flugsaurier (Reptilia, Pterosauria) aus der Oberen Trias von Cene bei Bergamo, Italien*. Bolletino Società Paleontologica Italiana, 17 (2): 176-257.

WHAT ARE PTEROSAURS?

Ever since 1801, when Georges Cuvier defined the Eichstätt *Pterodactylus* as a flying reptile, pterosaurs have been thought of as reptiles. The modern class Reptilia includes tortoises and turtles, the tuatara, which is now only found in New Zealand, lizards and snakes (Squamata) and crocodiles. These are the only surviving four of about 17 orders of reptiles that lived on earth in the Mesozoic Era, the age of the reptiles. They included creatures as diverse as dinosaurs, ichthyosaurs, placodonts, mammal-like reptiles, the pterosaurs themselves, and several others. For this reason it is difficult to give a definitive description of the class as a whole.

Reptiles form a common group with birds and mammals, the Amniota: in their embryonic stage they are protected by a fluid-filled sac, the amnion, not present in amphibians. Because they produce eggs with large yolks and hard shells, reptiles are able to live exclusively on dry land, and their young emerge fully developed from eggs laid there. Unlike amphibians such as frogs and newts, they do not need to spend the early stages of their life as larvae with gills. The hard shell prevents the amniotic egg from drying out. This 'invention' about 300 million years ago was the essential step needed to enable the reptiles successfully to conquer dry land, and to evolve in the astonishing way they did. Their descendants, birds and mammals, and by extension we human beings as well, owe their existence to this crucial step.

We could make it easy for ourselves and define reptiles as Amniota that are neither birds nor mammals. Modern reptiles are easily distinguishable from both amphibians and mammals. They are cold-blooded, so their body temperature rises or falls with the ambient air temperature. Their skin is horny, usually scaly. Their hearts are incompletely chambered, and so oxygen-rich and oxygen-poor blood is not as strictly separated in their circulatory system as it is in warm-blooded creatures such as birds and mammals. Like birds, reptiles lay eggs with hard shells and large yolks. Amphibians' eggs are soft, with small yolks, and have to be spawned in water, while embryonic development of mammals takes place in the mother's body. Young mammals are born alive and fed with milk from special glands in the mother.

We shall quickly run into difficulties if we try to apply the above list of modern reptile characteristics to pterosaurs. Even though the only evidence that is usually available to us is their fossilized skeletons, we can still draw indirect conclusions from these about pterosaur physiology, and thus establish the regular and recurring features of their lives. The fact that pterosaurs were flying creatures implies a high energy requirement, comparable to the needs of other flying vertebrates like birds and bats, with a significantly higher metabolic rate than reptiles. This can only be 'financed' by warm-bloodedness. In order to maintain a high and constant body temperature birds have a body covering of feathers, and mammals hair or fur. This is intended to prevent loss of body heat, so primarily has an insulating function. Protection is a secondary function, and in the case of birds feathers are also an aid to flight. Pterosaurs had a body covering of hair, as fossil finds have shown, and so they must have been warm-blooded.

This characteristic alone makes pterosaurs different from modern, cold-blooded reptiles like crocodiles or lizards. It is also possible that they had hearts with four chambers, which prevented a mixture of oxygen-rich and oxygen-poor blood circulating in the body. Their young may possibly also have been born alive, like the ichthyosaurs, and fed for as long as they remained in the nest. In this respect too pterosaurs would be different from typical modern reptiles.

We cannot really avoid the question: 'Are pterosaurs reptiles at all?' Formerly many scholars would have answered no. They wanted to establish a class of vertebrates especially for pterosaurs, placed between reptiles and birds, and between reptiles and mammals, or even

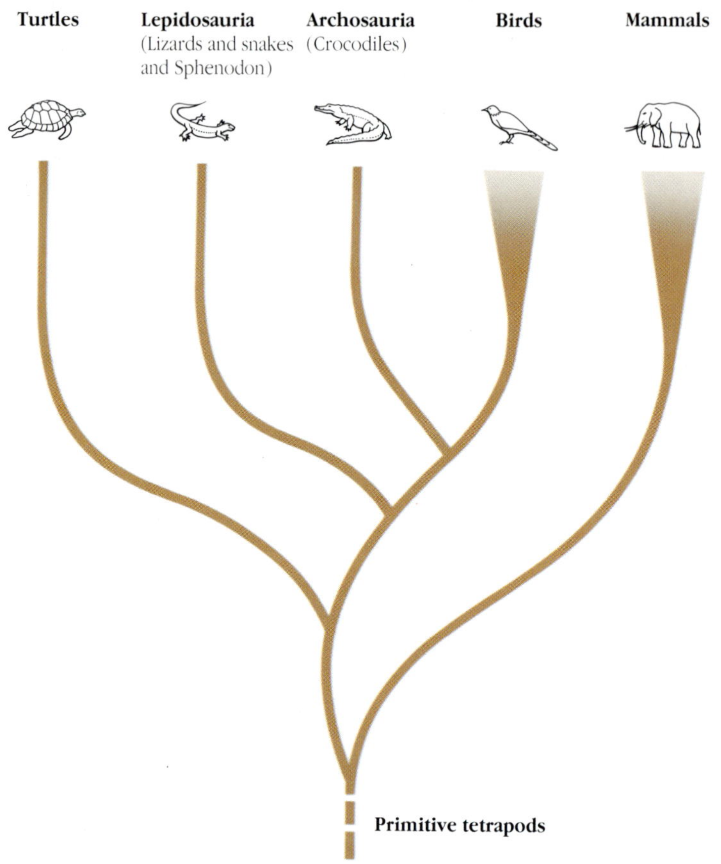

The Relationships of Amniotes (left)
Amniotes are vertebrates that develop within an amnion, a fluid-filled sac that protects the embryo before birth. The diagram takes the form of a family tree that shows in a simplified way the phylogenetic relationships of amniotes that evolved from a single stock of primitive tetrapods. About 300 million years ago, during the Upper Carboniferous, three major lineages can be distinguished: one that gave rise to mammals, a second to turtles, and a third to other reptilian groups and to birds. The class Reptilia includes all the lineages in solid colour. Reptiles can be defined as amniotes that lack the specialized characters of birds and mammals. Within modern fauna, crocodiles share a more recent common ancestry with birds than they do with lizards and snakes.

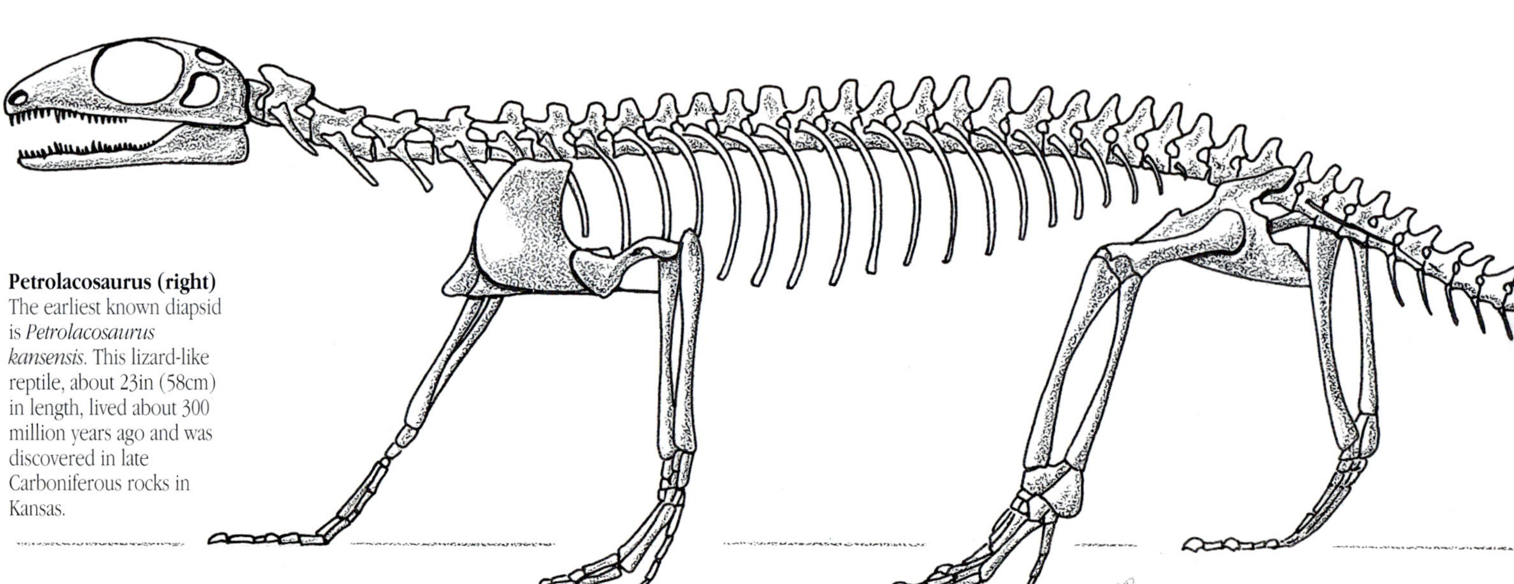

Petrolacosaurus (right)
The earliest known diapsid is *Petrolacosaurus kansensis*. This lizard-like reptile, about 23in (58cm) in length, lived about 300 million years ago and was discovered in late Carboniferous rocks in Kansas.

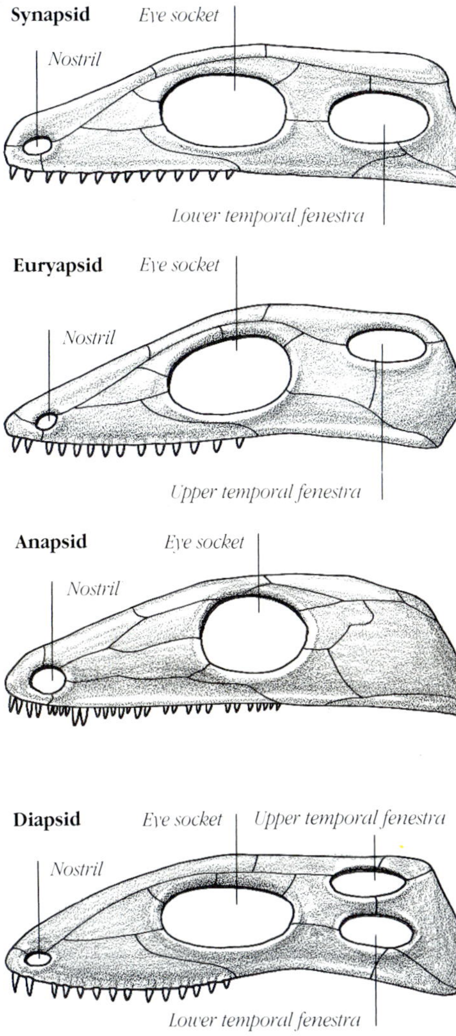

Reptile Classification
The subclasses of reptiles are recognized by the pattern of skull openings behind the eye socket, called temporal fenestrae. The synapsids have a single temporal fenestra low down on the side of the skull. The euryapsids also have one temporal fenestra but it is higher in position. Anapsids have no temporal opening at all, while diapsids have two, an upper and a lower one. During the Mesozoic diapsids were particularly abundant; they included the dinosaurs and pterosaurs in their number.

between birds and mammals. However, the problem can also be solved by defining reptiles in a different way, within a much broader framework. Such a definition could also accommodate warm-blooded creatures like pterosaurs, and certain dinosaurs like the coelurosaurs, also thought to have been warm-blooded. Despite this the Reptilia, if we include the fossil groups, are a motley crew, probably without a common, natural phylogeny, thus not monophyletic.

The absence or presence of certain openings in the skull, like the temporal openings behind the eye socket, is considered a significant pointer when reconstructing the phylogeny and classification of reptiles. A distinction is made here between **anapsids** without a temporal opening (tortoises for example), **synapsids** with only one, lower temporal opening (the mammal-like reptiles), **euryapsids** with only one, upper temporal opening (ichthyosaurs, for example) and finally **diapsids** with two temporal openings, an upper and a lower. According to this scheme pterosaurs are diapsids, as they have *two* temporal openings on each side of the skull behind the eye sockets.

The earliest diapsid reptile known is from the Upper Carboniferous, or Pennsylvanian system, about 260 million years ago. It was a lizard-sized creature called *Petrolacosaurus*.[1] There are two groups of diapsid reptiles, lepidosaurs and archosaurs. The lepidosaurs include lizards, snakes and tuataras, and the extinct rhynchosaurs and eosuchians. Crocodiles and alligators are the only remaining archosaurs. In the geological past they also included dinosaurs and a large and extremely diverse group of reptiles, the thecodonts.

Archosaurs and pterosaurs have *one* particular characteristic in common. They have another aperture in the skull wall between the eye and nose sockets, a so-called preorbital opening. Thus pterosaurs can be classified as archosaurs. Their closest living relatives are crocodiles and alligators. Their origin and family relationships will be discussed in more detail below.

Origin and Relationships

We have already established that from an osteological point of view pterosaurs can be classified as diapsids, and then as archosaurs because of their characteristic skull openings. According to the classic system established by the American vertebrate palaeontologist Alfred Sherwood Romer,[2] archosaurs include dinosaurs, crocodiles, pterosaurs and thecodonts. As the oldest archosaurs are the Triassic thecodonts, the origin of pterosaurs has been thought to lie with these, usually terrestrial, reptiles.

The Tübingen reptile specialist Friedrich Freiherr von Huene investigated the skeletal remains of a small reptile from Lossiemouth in Scotland[3] in 1914. The fossils came from the Upper Triassic Elgin sandstone and had already been named *Scleromochlus taylori* by Arthur Smith Woodward of the British Museum in London in 1907.[4] Although the bones had survived only as prints in the coarse sandstone, by no means complete and very fragmented, Huene risked a reconstruction of the entire skeleton of this tiny reptile which was only about 9in (23cm) long. He classified it as a pseudosuchid, a sub-order of the thecodonts, and because of its long hind legs suggested that *Scleromochlus* was a creature that lived in trees and could jump from branch to branch or tree to tree. Huene even thought it possible that *Scleromochlus* was a gliding creature, with folds of skin on its forelimbs that could be used as wings. He therefore assumed that it represented a stage of development that must be postulated as a direct precursor of pterosaurs: a climbing, jumping creature living in trees, which gradually developed a flight membrane like that of the parachute animals.[5] Essentially Huene was suggesting an arboreal origin for pterosaurs, in other words they were descended from creatures that lived in trees, in which an active ability to fly had developed from a passive parachute and gliding phase.

The view that pterosaurs were descended from Triassic thecodonts survived for a long time, and has appeared in textbooks. In *Scleromochlus* there is a great disproportion between the fore- and hindlimbs: the hindlimbs being considerably longer than the forelimbs. In pterosaurs exactly the reverse is the case. Their forelimbs, the wings, are considerably longer and more powerful than the hindlimbs.

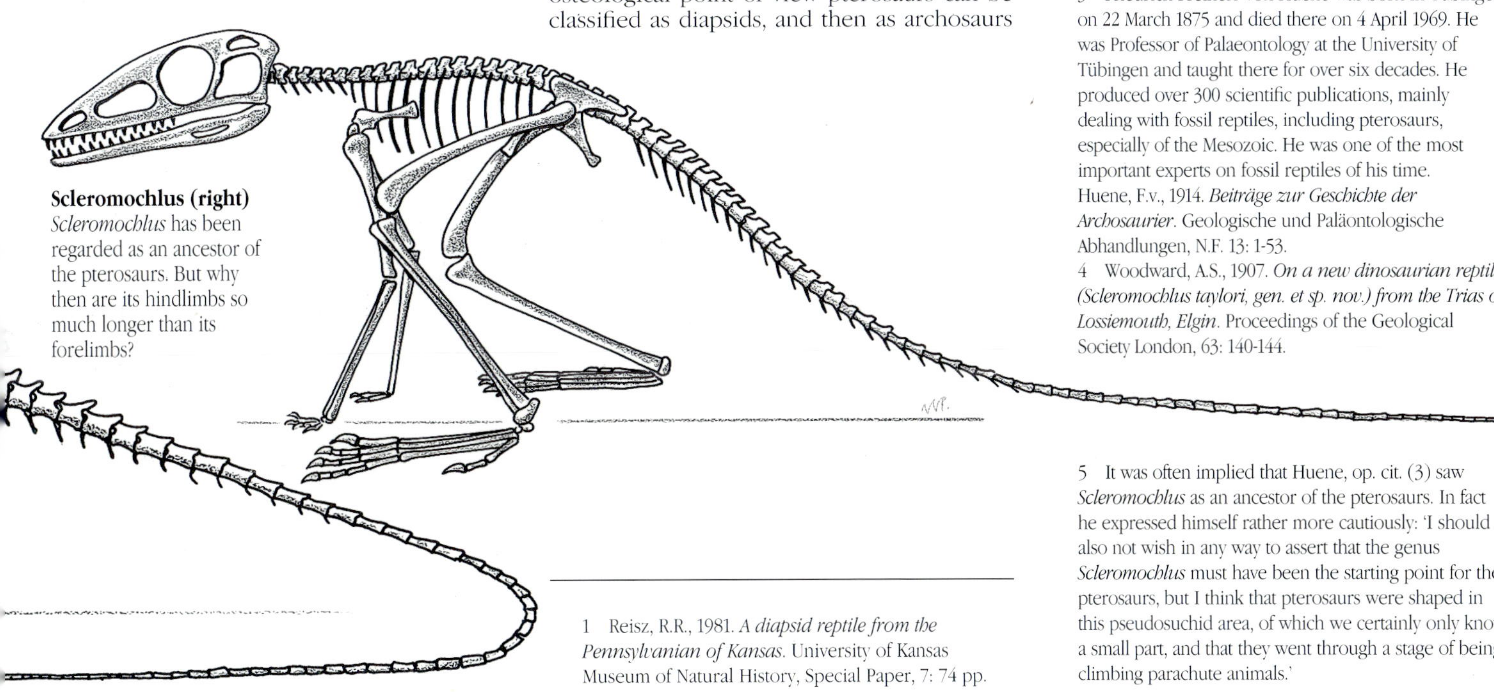

Scleromochlus (right)
Scleromochlus has been regarded as an ancestor of the pterosaurs. But why then are its hindlimbs so much longer than its forelimbs?

1 Reisz, R.R., 1981. *A diapsid reptile from the Pennsylvanian of Kansas*. University of Kansas Museum of Natural History, Special Paper, 7: 74 pp.

2 Romer, A.S., 1956. *Osteology of the Reptiles*. Chicago.

3 Friedrich Freiherr von Huene was born in Tübingen on 22 March 1875 and died there on 4 April 1969. He was Professor of Palaeontology at the University of Tübingen and taught there for over six decades. He produced over 300 scientific publications, mainly dealing with fossil reptiles, including pterosaurs, especially of the Mesozoic. He was one of the most important experts on fossil reptiles of his time. Huene, F.v., 1914. *Beiträge zur Geschichte der Archosaurier*. Geologische und Paläontologische Abhandlungen, N.F. 13: 1-53.

4 Woodward, A.S., 1907. *On a new dinosaurian reptile (Scleromochlus taylori, gen. et sp. nov.) from the Trias of Lossiemouth, Elgin*. Proceedings of the Geological Society London, 63: 140-144.

5 It was often implied that Huene, op. cit. (3) saw *Scleromochlus* as an ancestor of the pterosaurs. In fact he expressed himself rather more cautiously: 'I should also not wish in any way to assert that the genus *Scleromochlus* must have been the starting point for the pterosaurs, but I think that pterosaurs were shaped in this pseudosuchid area, of which we certainly only know a small part, and that they went through a stage of being climbing parachute animals.'

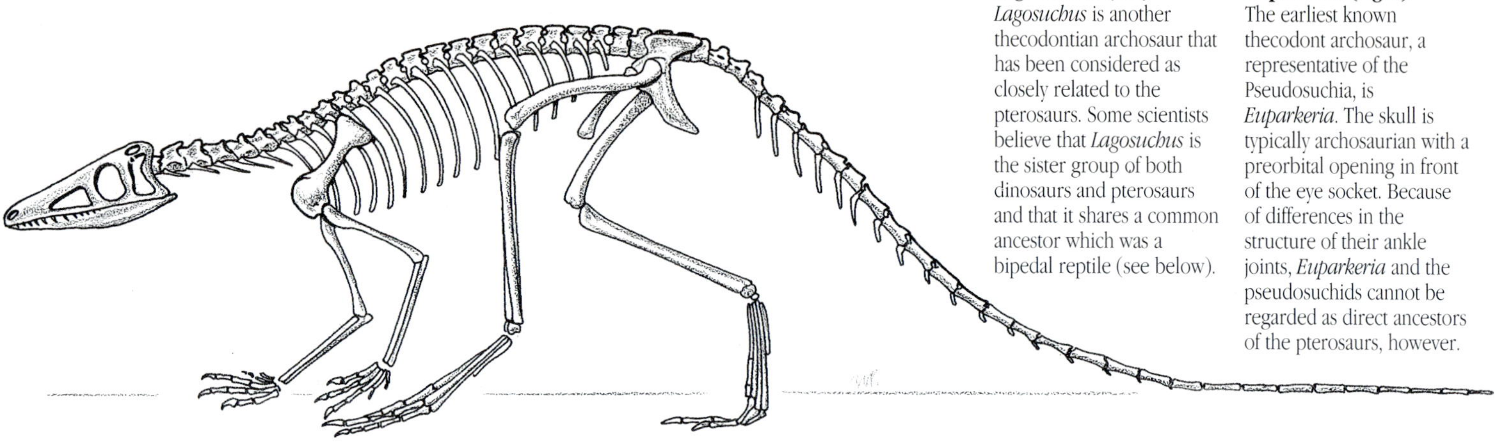

Lagosuchus (left)
Lagosuchus is another thecodontian archosaur that has been considered as closely related to the pterosaurs. Some scientists believe that *Lagosuchus* is the sister group of both dinosaurs and pterosaurs and that it shares a common ancestor which was a bipedal reptile (see below).

Euparkeria (right)
The earliest known thecodont archosaur, a representative of the Pseudosuchia, is *Euparkeria*. The skull is typically archosaurian with a preorbital opening in front of the eye socket. Because of differences in the structure of their ankle joints, *Euparkeria* and the pseudosuchids cannot be regarded as direct ancestors of the pterosaurs, however.

Direct predecessors of the pterosaur must therefore have shown a tendency in this direction. No thecodont fossil has limb proportions distorted in this way, and thecodonts also differ from pterosaurs in other skeletal characteristics. It is thus not possible to consider any particular thecodont as a pterosaur ancestor.

However, *one* fossil has been found that was considered to relate closely to the pterosaurs. It is a reptile about 12in (30·5cm) long with extremely long, slender limbs. It was found in the Chañares formation (Middle Triassic) in Argentina. This small thecodont, *Lagosuchus*, like *Scleromochlus* also has hindlimbs which are much longer than its forelimbs.[6] But it seems to have close affinities with dinosaurs, and to represent an intermediate stage between them and the original thecodonts. *Lagosuchus* has also been even more closely related to pterosaurs.[7] Kevin Padian of the University of Berkeley, California, assumes that *Lagosuchus* is related both to dinosaurs and pterosaurs. He believes that the common ancestor of

6 Bonaparte, J.F., 1975. *Nuevos materiales de Lagosuchus talampayensis Romer (Thecodontia-Pseudosuchia) y su significado en el origen de los Saurischia*. Acta Geologica Lilloana, 13, 1: 5-90; Tucuman.

7 Padian, K., 1984. *The Origin of Pterosaurs*. Third Symposium on Mesozoic Terrestrial Ecosystems (Reif, W.-E. and Westphal, F., eds.), pp.163-168; Tübingen, Attempto Verlag.

Family Relationships
This cladogram shows the hypothetical phylogenetic relationships between the pterosaurs, dinosaurs, *Lagosuchus*, and ornithosuchids as suggested by Kevin Padian of Berkeley University. In this concept dinosaurs and pterosaurs are regarded as sister-groups sharing a common ancestor with *Lagosuchus* and the ornithosuchids.

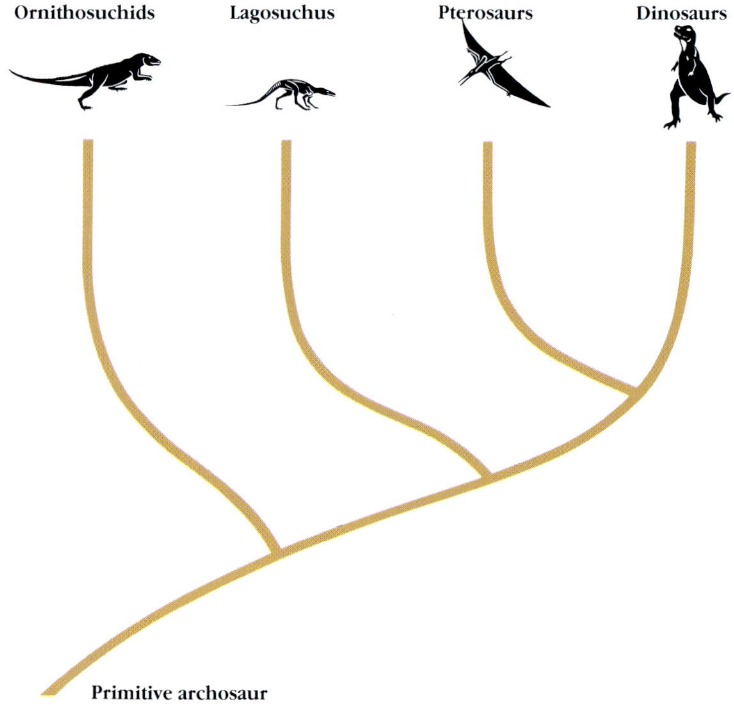

Below: This life restoration of *Lagosuchus talampayensis* suggests that this small archosaur might have been able to run bipedally. The extreme lengthening of the hind legs exhibits a tendency of limb proportion which is in marked contrast to the pterosaurs in which the forelimbs elongated and turned into wings.

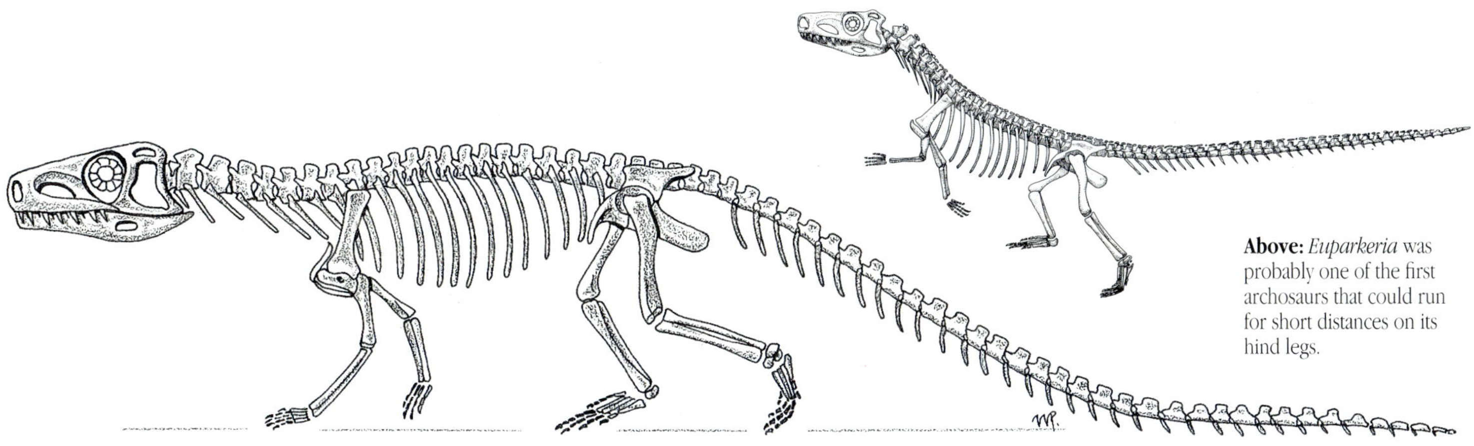

Above: *Euparkeria* was probably one of the first archosaurs that could run for short distances on its hind legs.

pterosaurs and dinosaurs was a biped with fully erect stance and gait. Expressed phylogenetically, *Lagosuchus* would be the sister group of both pterosaurs and dinosaurs, which would also themselves be sister groups. As an alternative to this scheme, pterosaurs could be seen as a sister-group of *Lagosuchus* and the dinosaurs, which would themselves be sister groups. In saying this, Padian presupposes that pterosaurs were bipeds and like birds could move along the ground on two legs. Consequently pterosaur flight would not have developed from the trees down, but from the ground up, a view which is disputed. According to this phylogenetic scheme *Lagosuchus*, dinosaurs, pterosaurs and the thecodont *Ornithosuchus* would have a common ancestor. The ornithosuchids were large, long reptiles (up to 13ft, 4m) of the Upper Triassic, partially capable of biped movement.[8]

It is interesting to note that as early as 1900 the German zoologist M. Fürbringer,[9] and H. G. Seeley in his *Dragons of the Air* of 1901, both put forward the view that pterosaurs could be traced back to dinosaur ancestors. But here too the fossil record presents no candidate as ancestor for the pterosaurs.

The earliest pterosaurs in the fossil record date from the Upper Triassic. In order to find out anything about their ancestors we have to look at appropriate fossil finds from the Middle and Lower Triassic, perhaps even the Permian. Of the thecodonts, only the second group, the pseudosuchians, are candidates here. Their earliest representative is the genus *Euparkeria*, a small reptile from the Uppermost Lower Triassic in South Africa, a creature afforded a central place in the evolution of archosaurs. *Euparkeria* is one of the first land animals which might have been able to walk on two legs, although it has a specialized ankle joint of crurotarsal or crocodiloid shape. In this type of joint the hinge articulation between leg and foot has a kink to allow a sideways twisting movement of the foot. Crocodiles also have an ankle of this kind, while dinosaurs, birds and pterosaurs have an ankle joint in which the joint articulation runs in a straight line, transversely between the tarsal bones. This is called a mesotarsal ankle joint. It therefore seems questionable that pterosaurs should originate from Triassic pseudosuchians. Certainly the pseudosuchid *Scleromochlus* seems to have a mesotarsal ankle joint[10] and in *Lagosuchus* the ankle has been interpreted as an intermediate form between pseudosuchid and dinosaur.[11]

Precise analysis of skeletal characteristics of the Triassic pterosaurs *Eudimorphodon* and *Peteinosaurus* from the Italian Alps led Rupert Wild of the Naturkundemuseum in Stuttgart to the conclusion that pterosaurs must be descended not from thecodonts, but, like the rest of the archosaurs, directly from eosuchians of the Permian or the Lowest Triassic.[12] The Triassic pterosaurs are already completely developed flying saurians with all the special adaptations to active flight and typical combinations of distinguishing features in their anatomy that

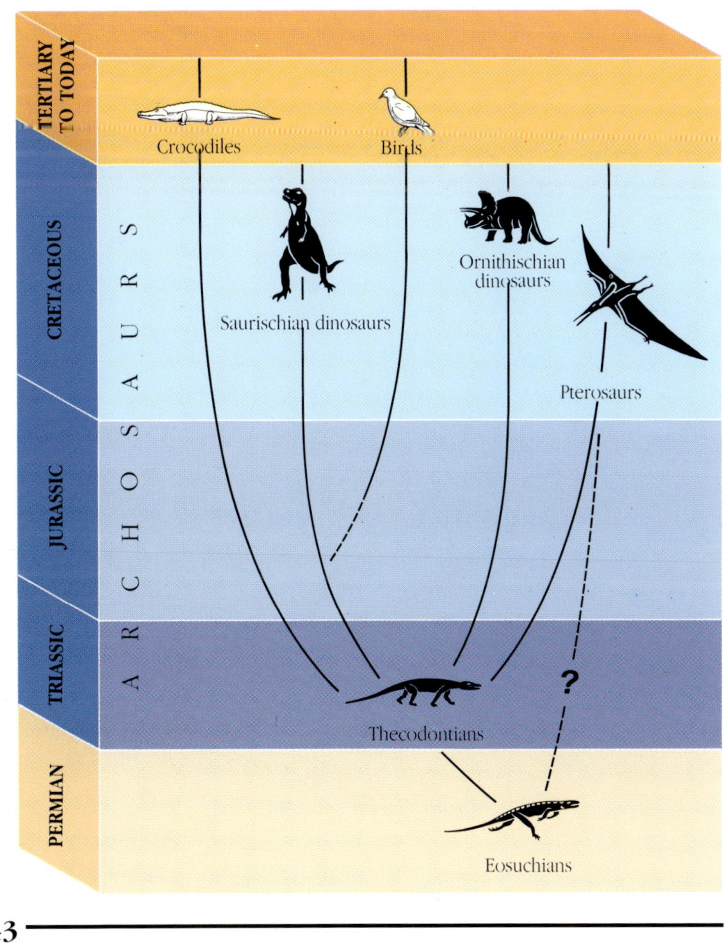

Probable Lines of Descent (right)
As a result of his studies on the oldest known pterosaurs – the Triassic forms *Eudimorphodon* and *Peteinosaurus* from the Italian Alps – Rupert Wild of the Stuttgart Natural History Museum concluded that the pterosaurs cannot be direct descendants of the Thecodontia, the ancestral stock of dinosaurs and crocodiles, but that more likely both have a common ancestor in the Eosuchia, a diapsid stem-group of small Permian reptiles which gave rise to the lepidosaurs as well as to the archosaurs as shown in this diagram. Although fossil evidence of intermediate forms is not known, we may suppose that the origin of the pterosaurs, and the beginnings of their evolution from land-bound reptiles to active fliers, lies in early Triassic or even late Permian times, perhaps as much as some 250 million years ago.

8 A similar view is put forward by Gauthier and Padian (1985). They relate *Euparkeria*, Ornithosuchidae, *Lagosuchus* and pterosaurs in this order more closely to dinosaurs (including birds) and use the taxon Ornithosuchia for this combination.

Gauthier (1986) argues similarly by including dinosaurs, pterosaurs, *Lagosuchus* and the Triassic Herrerasauridae in a new taxon Ornithodira, and also categorizes these, together with Ornithosuchidae, as Ornithosuchia ('birds and archosaurs closer to birds than to crocodiles'). Ornithosuchia and Pseudosuchia ('crocodiles and archosaurs closer to crocodiles than to birds') together make up the Archosauria, according to Gauthier (1986) 'all the descendants of the most recent common ancestor of crocodiles and birds.'

Gauthier, J. and Padian, K., 1985. *Phylogenetic, Functional and Aerodynamic Analyses of the Origin of Birds and their Flight*. The Beginnings of Birds, Proceedings of the International Archaeopteryx Conference Eichstätt 1984: 185-197.

Gauthier, J., 1986. *Saurischian Monophyly and the Origin of Birds*. The Origin of Birds and the Evolution of Flight (K. Padian, ed.). Memoires of the California Academy of Science, 8: 1-55.

9 Fürbringer, M., 1900. *Zur vergleichenden Anatomie des Brust-Schulterapparates und der Schultermuskeln. 5. Teil. Reptilien*. Jenaische Zeitschrift für Naturwissenschaften, 34, N.F. 27: 215-718; Jena.

10 Padian, K., op. cit. (7), p.165.

11 Bonaparte. J.F., op. cit. (6).

12 Wild, R., 1983. *Uber den Ursprung der Flugsaurier*. Weltenburger Akademie, Erwin Rutte-Festschrift, pp. 231-238; Kelheim/Weltenburg.
Wild, R., 1984. *Flugsaurier aus der Obertrias von Italien*. Naturwissenschaften, 71: 1-11; Springer-Verlag.

Learning To Fly

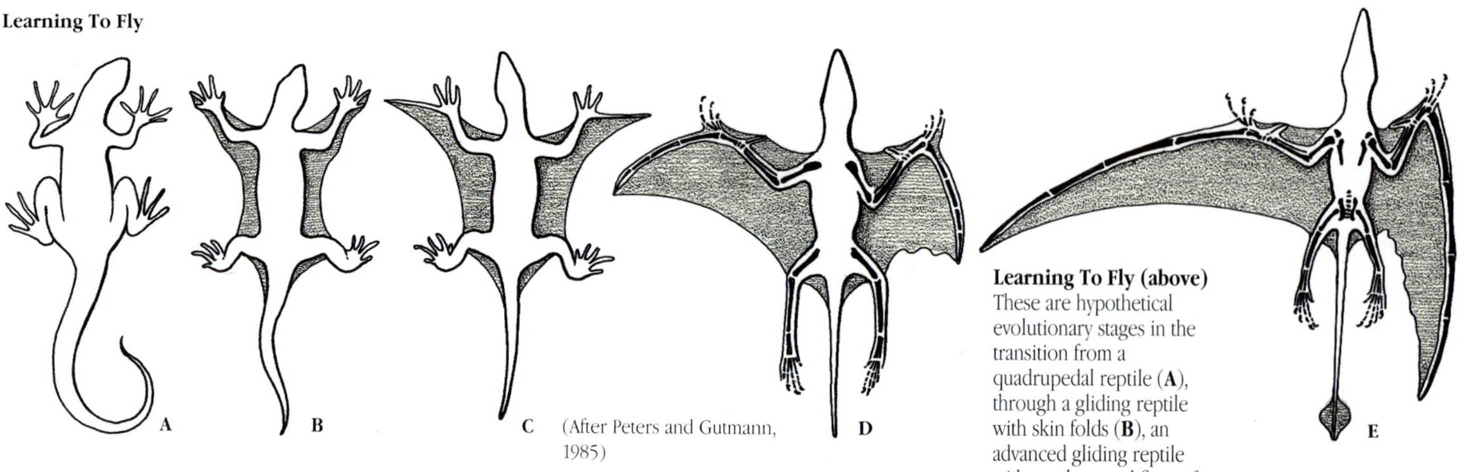

(After Peters and Gutmann, 1985)

Learning To Fly (above) These are hypothetical evolutionary stages in the transition from a quadrupedal reptile (**A**), through a gliding reptile with skin folds (**B**), an advanced gliding reptile with an elongated finger for steering (**C**), a protopterosaur with an enlarged wing finger supporting a wing membrane for flapping (**D**), to a developed pterosaur with greatly elongated wing finger supporting a long wing membrane and a terminal tail membrane for steering and counterbalance (**E**).

we find in all pterosaurs. Thus the earliest forms in the fossil record are not transitional or intermediate links.

Additionally, these oldest known pterosaurs themselves represent various lines of development. They can be classified as three different families of Ramphorhynchoidea, the long-tailed pterosaurs, with at least one of them already seeming very specialized. But that means that their origin is to be found much further back in the geological past than had hitherto been assumed. The beginning of their development from small, land-bound reptiles via a presumed arboreal gliding stage to large-scale active flapping flight must have taken place in the Permian, about 250 million years ago, and so we should direct our attention there. The eosuchians present themselves as a potential ancestor group, small diapsid reptiles now generally held to come from the same ancestral stock as all later archosaurs and lepidosaurs.

The general structure of the eosuchians can be seen from *Heleosaurus*, a reptile about 20in (51cm) long from the Upper Permian in South Africa. It still had vestigial skin armour, but already showed a tendency to walk on two legs. Certain osteological features of the skull of *Eudimorphodon*, the best-known of the Triassic pterosaurs, were also found by Rupert Wild in eosuchians, but not in thecodonts. It is not known whether *Heleosaurus* had a preorbital opening in the skull, a feature that distinguishes archosaurs from other diapsids.

Other reptile groups are also descended from the eosuchians, the prolacertilians, for example. *Tanystropheus* is also one of these, the giraffe-necked saurian which in its juvenile stage had teeth with many cusps, very similar to the teeth of the Triassic pterosaur *Eudimorphodon*. Overall Triassic pterosaurs and eosuchians+prolacertilians have more features in common than pterosaurs and thecodonts. This means that the origin of pterosaurs probably lay with Permian eosuchians or certain transitional forms between eosuchians and prolacertilians, but not with thecodonts. Pterosaurs and thecodonts are branches of the diapsids which developed independently of one another from Upper Permian or Lower Triassic eosuchians.

For this reason Michael Benton of the University of Bristol has declared that pterosaurs were not archosaurs.[13] He classifies them with archosaurs and a few other groups of different early diapsids as archosauromorphs. But if pterosaurs are no longer to be considered archosaurs, then the question arises of the meaning and evaluation of the preorbital opening in the skull, because in Robert Carroll's view 'the (most primitive) archosaurs can be distinguished from other early diapsids only by a single character, the *antorbital fenestra*.'[14] Modern crocodiles, although they are archosaurs, in contrast to Jurassic and Cretaceous crocodiles have *no* preorbital fenestra, but it is assumed that some eosuchians still had an opening of this kind. This is obviously a matter of definition, and it can only be satisfactorily resolved as a result of more complete fossil finds in the future.

Pterosaurs can be regarded as an independent branch of the diapsids and assigned, following Kuhn-Schnyder and Rieber,[15] to a subclass of their own, Pterosauromorpha. In order to explain their phylogenetic relationship with the rest of the archosaurs we still need more thorough investigation and more detailed analysis of characteristics.

According to Rupert Wild a small, tree-climbing reptile can be reconstructed as a hypothetical pre-pterosaur or propterosaur, in which a tendency to enlargement of the body surface proceeded from a lateral band of skin. This skin extended from the rear of the forelimbs and a fourth digit of the hand that grew longer and longer with evolution, to the flanks of the body and to the upper legs. At the same time the fifth digit of the hand regressed, while the first three digits retained their normal length and had sharp, hook-shaped claws for climbing.

The formation of a flight membrane was of selective advantage for these climbing creatures, as it checked free fall and made a safe landing on the ground possible. At the same time airspace was increasingly used, both to escape from enemies and for locomotion and enlargement of the radius of action, with the associated exploitation of new sources of food over a large area. The flight membrane area gradually grew larger, while the fourth digit lengthened to twenty times the length of

13 Benton, M.J., 1985. *Classification and phylogeny of the diapsid reptiles*. Zoological Journal of the Linnean Society, 84: 97-194. Benton (Nature 296: 306-307, 1982) first reintroduced the sub-class Diapsida (Osborn 1903), but classified all diapsid reptiles with the exception of *Petrolacosaurus* as Neodiapsida. They were regarded as sister groups. According to this scheme the Neodiapsida include the Archosauromorpha (=Pterosauria, Rhynchosauria, Prolacertiformes and Archosauria) and the Lepidosauromorpha (=Younginiformes and Lepidosauria). In a note to his 1985 essay dated July 1985 Benton announced, however: 'I now accept that the pterosaurs are archosaurs, and a close sistergroup of the Dinosauria, as argued by Padian (1983, 1984) and Gauthier (thesis). M.J. Benton'.

14 Carroll, R., 1987. *Vertebrate Paleontology and Evolution*. pp.698. W.H. Freeman and Co., New York.

15 Kuhn-Schnyder, E. and Rieber, H., 1984. *Paläozoologie*. pp.390. Thieme Verlag, Stuttgart, New York.

Below: *Heleosaurus scholtzi* reveals the general body form of an eosuchian reptile. It was a lizard-like animal about 20in (51cm) long. Here, too, a tendency to bipedal locomotion is discernible.

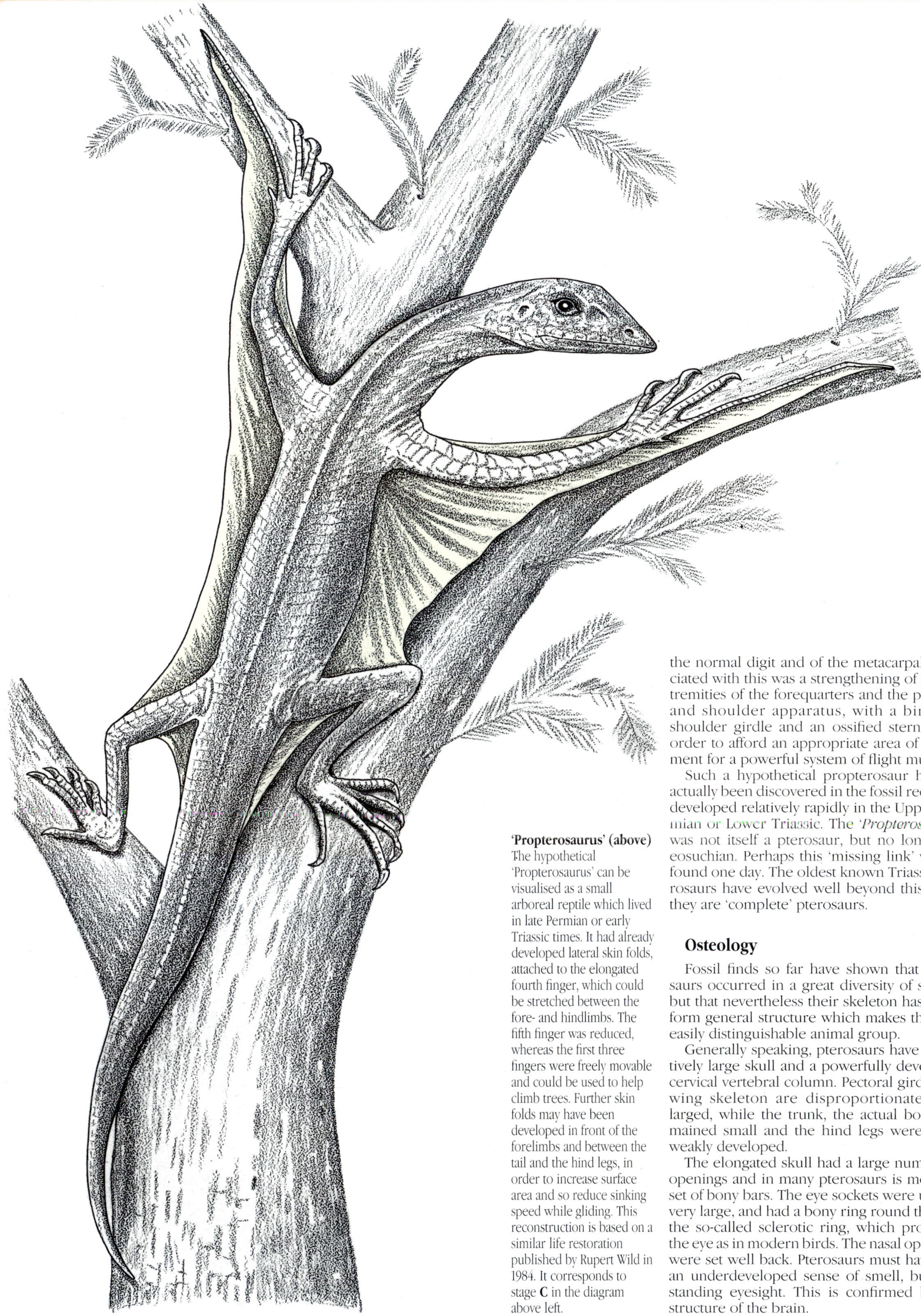

'Propterosaurus' (above)
The hypothetical 'Propterosaurus' can be visualised as a small arboreal reptile which lived in late Permian or early Triassic times. It had already developed lateral skin folds, attached to the elongated fourth finger, which could be stretched between the fore- and hindlimbs. The fifth finger was reduced, whereas the first three fingers were freely movable and could be used to help climb trees. Further skin folds may have been developed in front of the forelimbs and between the tail and the hind legs, in order to increase surface area and so reduce sinking speed while gliding. This reconstruction is based on a similar life restoration published by Rupert Wild in 1984. It corresponds to stage **C** in the diagram above left.

the normal digit and of the metacarpal. Associated with this was a strengthening of the extremities of the forequarters and the pectoral and shoulder apparatus, with a bird-like shoulder girdle and an ossified sternum, in order to afford an appropriate area of attachment for a powerful system of flight muscles.

Such a hypothetical propterosaur has not actually been discovered in the fossil record. It developed relatively rapidly in the Upper Permian or Lower Triassic. The '*Propterosaurus*' was not itself a pterosaur, but no longer an eosuchian. Perhaps this 'missing link' will be found one day. The oldest known Triassic pterosaurs have evolved well beyond this stage: they are 'complete' pterosaurs.

Osteology

Fossil finds so far have shown that pterosaurs occurred in a great diversity of shapes, but that nevertheless their skeleton has a uniform general structure which makes them an easily distinguishable animal group.

Generally speaking, pterosaurs have a relatively large skull and a powerfully developed cervical vertebral column. Pectoral girdle and wing skeleton are disproportionately enlarged, while the trunk, the actual body, remained small and the hind legs were more weakly developed.

The elongated skull had a large number of openings and in many pterosaurs is merely a set of bony bars. The eye sockets were usually very large, and had a bony ring round the eye, the so-called sclerotic ring, which protected the eye as in modern birds. The nasal openings were set well back. Pterosaurs must have had an underdeveloped sense of smell, but outstanding eyesight. This is confirmed by the structure of the brain.

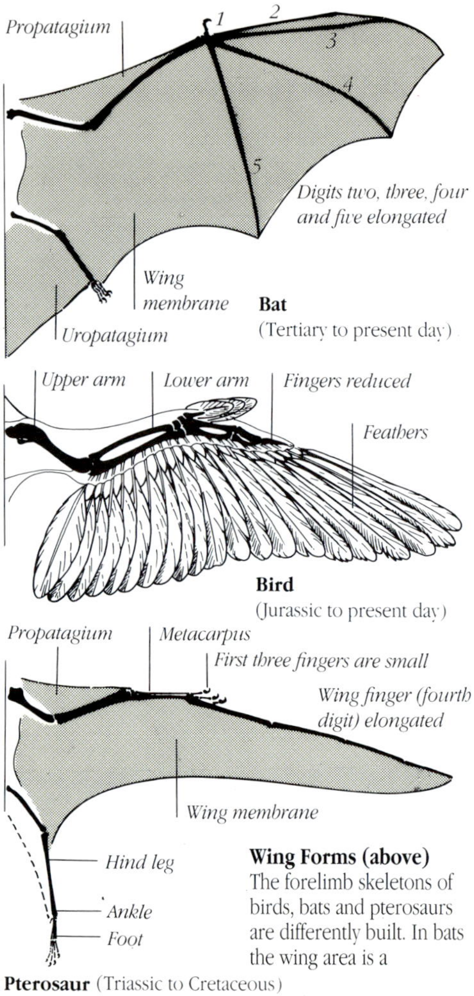

Wing Forms (above)
The forelimb skeletons of birds, bats and pterosaurs are differently built. In bats the wing area is a membrane between the four fingers and the hind leg. In birds the hand area is greatly reduced and fused. The wing area is formed by feathers. In pterosaurs a wing membrane was stretched between the elongated fourth finger and the body and upper leg.

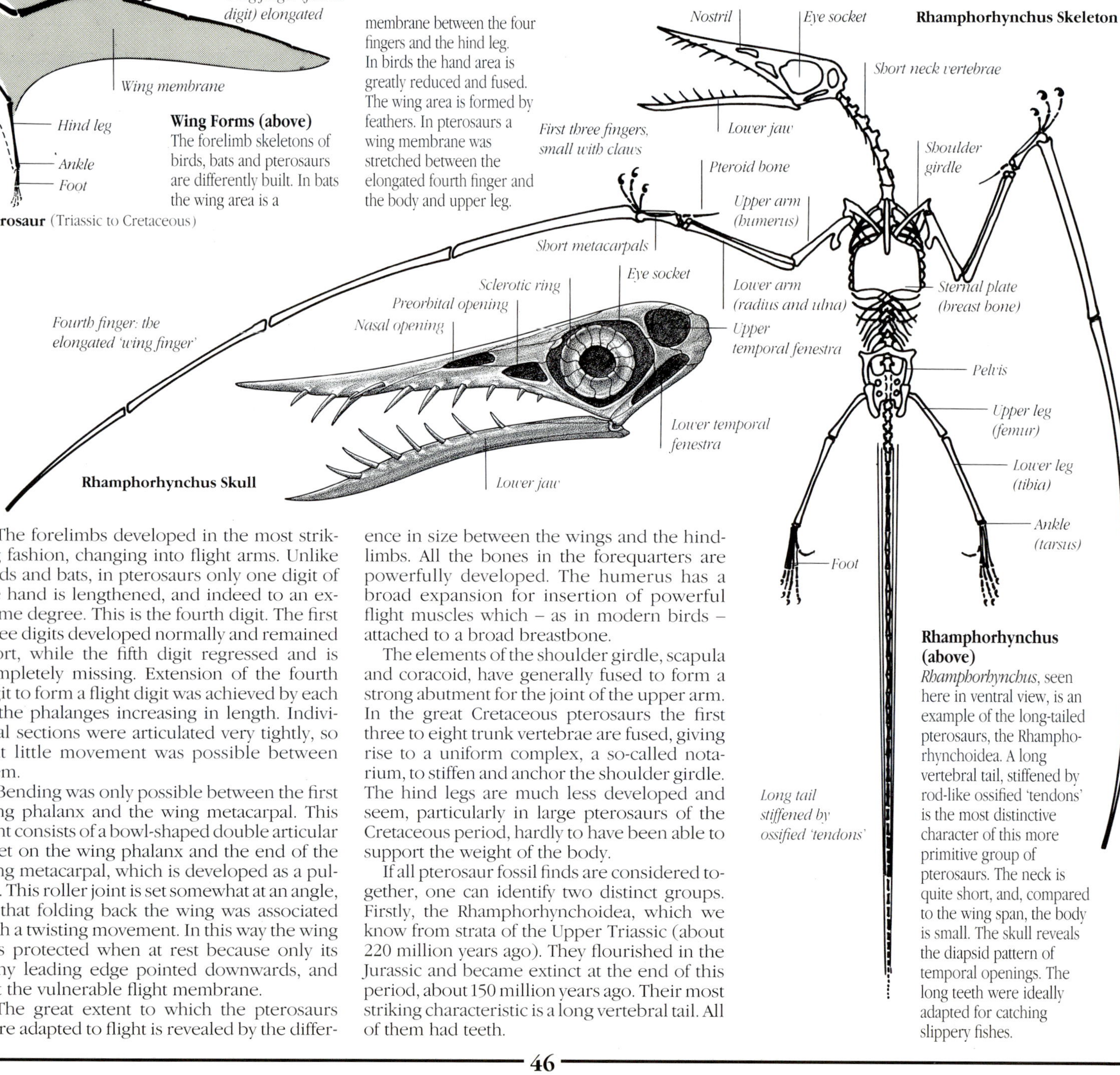

Rhamphorhynchus Skull

The forelimbs developed in the most striking fashion, changing into flight arms. Unlike birds and bats, in pterosaurs only one digit of the hand is lengthened, and indeed to an extreme degree. This is the fourth digit. The first three digits developed normally and remained short, while the fifth digit regressed and is completely missing. Extension of the fourth digit to form a flight digit was achieved by each of the phalanges increasing in length. Individual sections were articulated very tightly, so that little movement was possible between them.

Bending was only possible between the first wing phalanx and the wing metacarpal. This joint consists of a bowl-shaped double articular facet on the wing phalanx and the end of the wing metacarpal, which is developed as a pulley. This roller joint is set somewhat at an angle, so that folding back the wing was associated with a twisting movement. In this way the wing was protected when at rest because only its bony leading edge pointed downwards, and not the vulnerable flight membrane.

The great extent to which the pterosaurs were adapted to flight is revealed by the difference in size between the wings and the hindlimbs. All the bones in the forequarters are powerfully developed. The humerus has a broad expansion for insertion of powerful flight muscles which – as in modern birds – attached to a broad breastbone.

The elements of the shoulder girdle, scapula and coracoid, have generally fused to form a strong abutment for the joint of the upper arm. In the great Cretaceous pterosaurs the first three to eight trunk vertebrae are fused, giving rise to a uniform complex, a so-called notarium, to stiffen and anchor the shoulder girdle. The hind legs are much less developed and seem, particularly in large pterosaurs of the Cretaceous period, hardly to have been able to support the weight of the body.

If all pterosaur fossil finds are considered together, one can identify two distinct groups. Firstly, the Rhamphorhynchoidea, which we know from strata of the Upper Triassic (about 220 million years ago). They flourished in the Jurassic and became extinct at the end of this period, about 150 million years ago. Their most striking characteristic is a long vertebral tail. All of them had teeth.

Rhamphorhynchus (above)
Rhamphorhynchus, seen here in ventral view, is an example of the long-tailed pterosaurs, the Rhamphorhynchoidea. A long vertebral tail, stiffened by rod-like ossified 'tendons' is the most distinctive character of this more primitive group of pterosaurs. The neck is quite short, and, compared to the wing span, the body is small. The skull reveals the diapsid pattern of temporal openings. The long teeth were ideally adapted for catching slippery fishes.

Left: For preparation and study of smaller pterosaur specimens a binocular microscope has to be used. Fine details, such as teeth, skull sutures, or small bones are sometimes smaller than one millimetre in length, as in this *Rhamphorhynchus* skeleton exposed on a slab of Solnhofen limestone.

Pterodactylus (right) *Pterodactylus*, seen in ventral view, is an example of the short-tailed pterosaurs, the Pterodactyloidea. Here the tail is reduced, neck and skull are elongated, and the elements of the wing skeleton have different proportions. In the Upper Cretaceous this advanced group of pterosaurs evolved into toothless and giant forms with wing spans of up to 39ft (12m).

Pterodactylus Skeleton

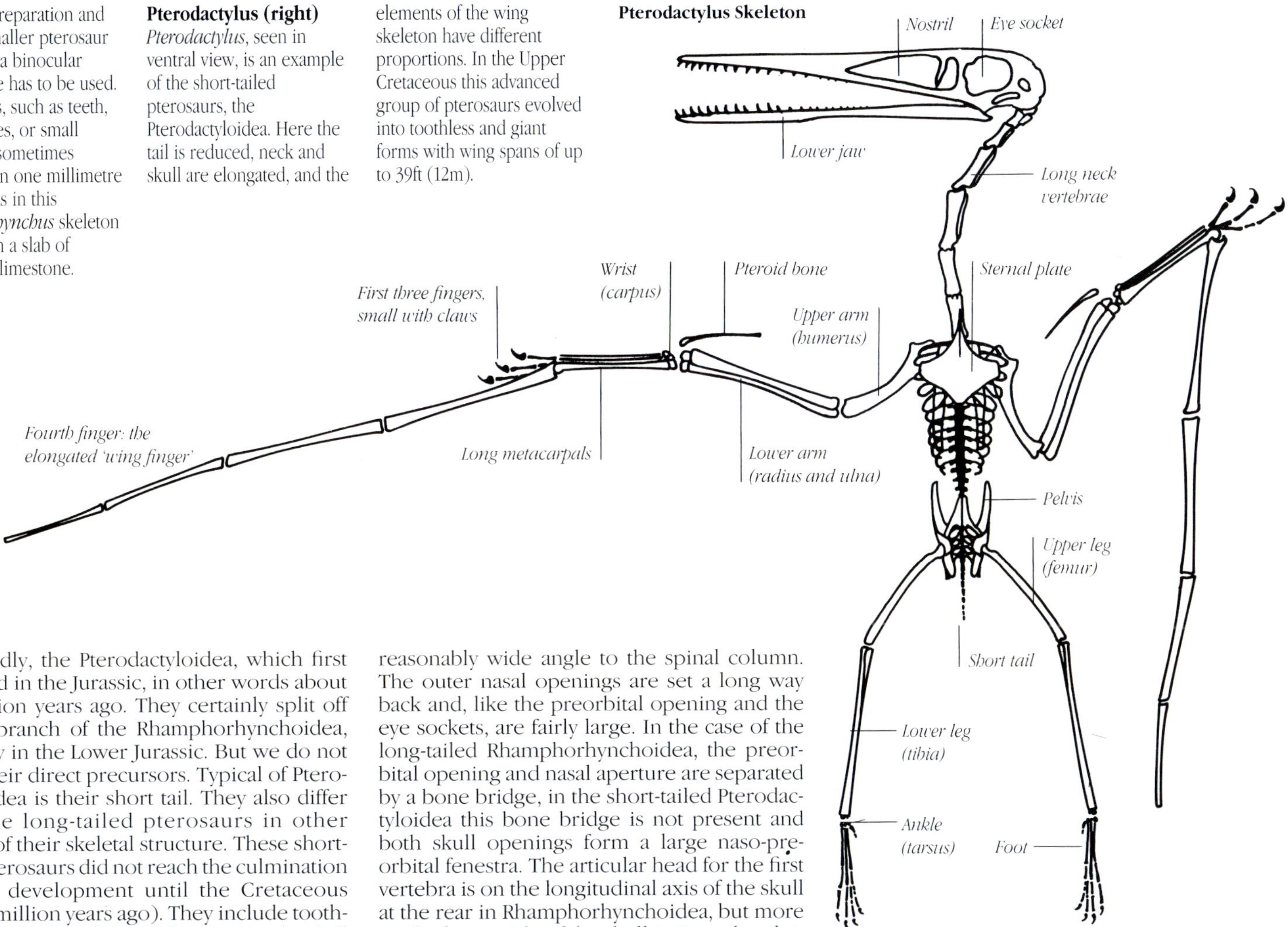

Secondly, the Pterodactyloidea, which first appeared in the Jurassic, in other words about 150 million years ago. They certainly split off from a branch of the Rhamphorhynchoidea, probably in the Lower Jurassic. But we do not know their direct precursors. Typical of Pterodactyloidea is their short tail. They also differ from the long-tailed pterosaurs in other aspects of their skeletal structure. These short-tailed pterosaurs did not reach the culmination of their development until the Cretaceous (144-65 million years ago). They include toothless forms and the largest flying animals of all times, with wing spans of about 40ft (12·2m). They became extinct at the end of the Cretaceous, and were the last of the pterosaurs.

In order to appreciate better the anatomy of the pterosaur, we shall now consider in turn the basic components of its skeleton, beginning with the head.

The Skull

As a rule the pterosaur skull is relatively large, low and very elongated, and set at a reasonably wide angle to the spinal column. The outer nasal openings are set a long way back and, like the preorbital opening and the eye sockets, are fairly large. In the case of the long-tailed Rhamphorhynchoidea, the preorbital opening and nasal aperture are separated by a bone bridge, in the short-tailed Pterodactyloidea this bone bridge is not present and both skull openings form a large naso-preorbital fenestra. The articular head for the first vertebra is on the longitudinal axis of the skull at the rear in Rhamphorhynchoidea, but more on the lower side of the skull in Pterodactyloidea. This meant that Rhamphorhynchoidea carried their heads in a rather extended position whereas in the latter the head was carried at a distinct angle to the neck.[16]

The teeth are set in the jaw in the same manner as thecodonts, that is to say they have individual sockets. The teeth may extend around the full length of the jaw, or only occupy part of it. In Rhamphorhynchoidea the teeth are either inclined forwards, or upright. The front teeth are usually longer and more curved, compressed at the sides, smooth and pointed. They become smaller towards the back of the jaw. In *Rhamphorhynchus* the long front teeth, which point forward and outwards, fit alternately into one another. Together with the toothless ends of the jaw, which were equipped with pointed horn beaks, they made an outstanding

16 Wellnhofer, P., 1978. *Pterosauria* in *Encyclopedia of Paleoherpetology*, (ed. P. Wellnhofer), part 19: 82 pp. Gustav Fischer Verlag, Stuttgart, New York.

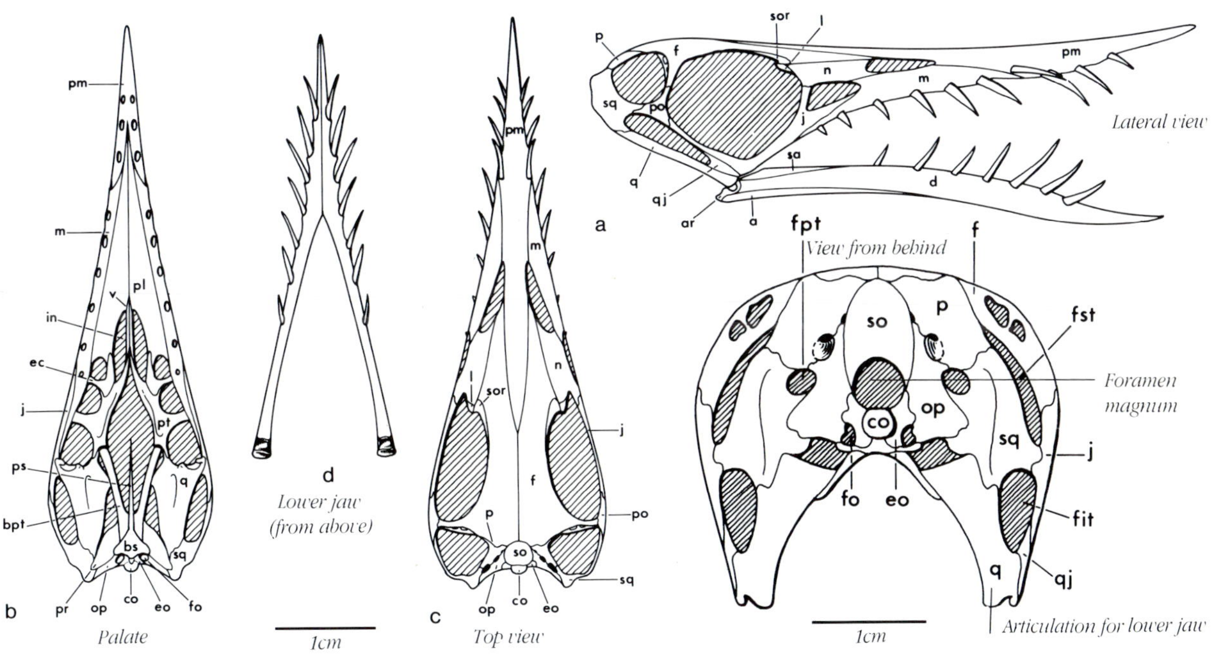

The Skull of Rhamphorhynchus (left) The drawings reveal the cranial osteology of *Rhamphorhynchus* as presented in a scientific monograph. The abbreviations refer to the scientific names of various elements of the skull. Reconstructions like this are the result of painstaking studies of many fossil specimens. The skull is shown in lateral, upper and palatal views, and from the back (near left), while the lower jaw is seen in lateral and upper view. The occipital condyle (co) forms the ball and socket articulation with the first neck vertebra. The opening above it (the foramen magnum) leads from the brain to the spinal cord.

Above: The picture of *Pterodactylus kochi* shows the position of the skull in relation to the neck that is typical of short-tailed pterodactyloids. The point of articulation (occipital condyle) is located underneath the braincase; in rhamphorhynchoids it is in a more posterior position.

prehensile organ for catching fish.[17] In some forms the toothless end of the lower jaw is laterally compressed and juts out beyond the upper jaw. Probably these pterosaurs used it to plough through the surface of the water when catching fish, like the black skimmer.[18]

Eudimorphodon is the only Triassic pterosaur with both robust front teeth and a large number of small, tightly-packed teeth with three and five cusps, in the upper and the lower jaw. In *Peteinosaurus* and *Dimorphodon* a large number of small teeth are set closely behind a few large front teeth.

Most Pterodactyloidea have complete sets of teeth, like *Pterodactylus* and *Ornithocheirus*. By contrast with Rhamphorhynchoidea they all have smaller, more even teeth. Reduction of dentition occurs in two ways: either only the front part of the jaw has teeth, as for example in *Gallodactylus*, or the points of the jaw become toothless, as in *Germanodactylus* and *Dsungaripterus*. Some Cretaceous pterosaurs are completely toothless. They probably had horn beaks with sharp edges, like birds.

Among Pterodactyloidea the genera *Ctenochasma*, *Gnathosaurus* and *Pterodaustro* had teeth which were very long and set close together, and used to filter planktonic food out of the water.

One often finds the small bone plates which formed the sclerotic ring preserved fossilized in the eye socket. This ring of bone supports the cornea from the inside in modern birds. It must have had the same function in pterosaurs, as they had large eyes. The pterosaurs' sclerotic ring consists of 12 to 20 thin, overlapping bone plates according to genus.

Variously shaped crests on the skull are particular features of many Jurassic and Cretaceous pterosaurs. Some pterosaurs have a long, low, median crest in the middle of the skull, like *Germanodactylus*, *Gnathosaurus*, *Ctenochasma* and *Dsungaripterus*, for example. Others have shorter or longer crests starting at the rear of the skull, the so-called supraoccipital crest, like *Pteranodon* and *Dsungaripterus*, and others again have high crests at the front end of the jaw, looking like a reversed ship's keel, like *Criorhynchus* from the Cambridge Greensand or *Tropeognathus* from the Santana Formation in Brazil.

The function of these crests can be explained in various ways. The evidence suggests that they were of different sizes in male and female animals. In this case they would be seen as a sexual characteristic, and were perhaps also used for display. But, as in the case of *Pteranodon*, they could have functioned as a stabilizer for the large, long head while flying. It is also possible that they served as a counterweight to the long beak, so the skull balanced more or less in equilibrium on the neck. Thus weight could be saved in the neck muscles.[19]

It has even been suggested that an elastic membrane was stretched between the long parietal crest of *Pteranodon* and its back, and that this functioned like the rudder of an aeroplane.[20] But this would have severely limited head movement, especially the extreme downward inclination needed for catching fish.

Crests at the front end of the skull and corresponding crests on the lower side of the lower jaw also played an important part when a pterosaur plunged its beak into the water in flight. In this event the head would swivel downwards and a long way backwards, so that its top side cleaved through the water. The high central crest over the front end of the upper jaw had exactly the same effect as the keel of a ship, automatically stabilizing the head and reducing its water resistance. Thus the crest fulfilled a hydrodynamic function.[21] This also

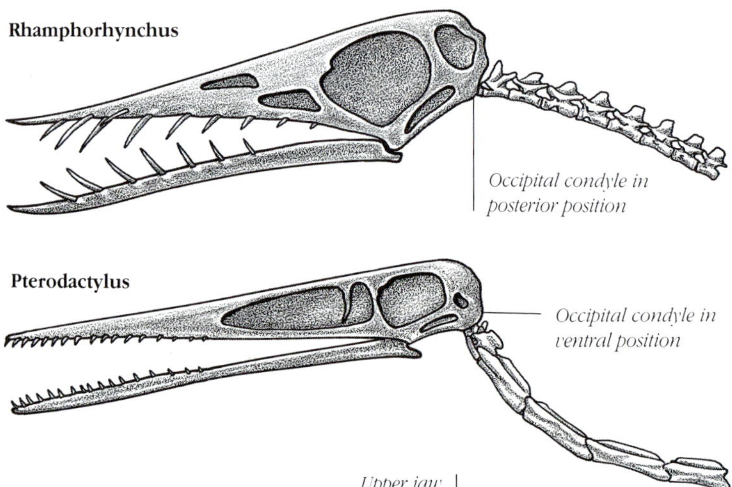

Head Postures (left)
Rhamphorhynchoid and pterodactyloid pterosaurs differed in the manner in which they carried their heads. Due to different positions of the occipital condyle, the head is in a straighter alignment with the neck in rhamphorhynchoids. In pterodactyloids a more angled arrangement is evident (see also photograph at top of page).

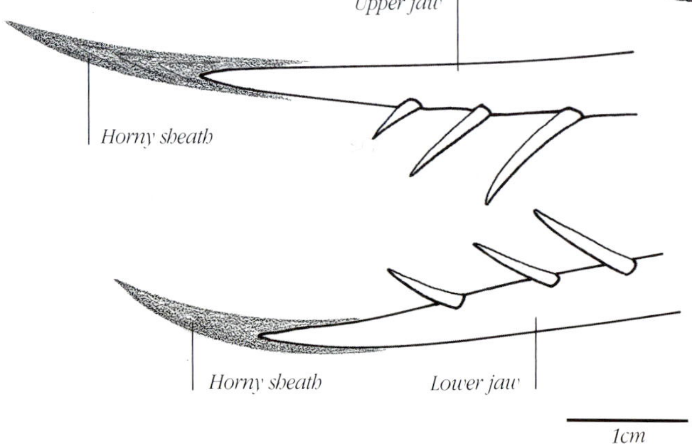

Beaks (below left)
Fossil preservation has indicated that in some cases the front ends of the jaws of *Rhamphorhynchus* were covered by long, pointed, horny beaks, attached like sheathes to the upper and lower jaws. They must have functioned like a pair of tweezers, enabling the pterosaur to winkle out and grasp its prey. They would certainly have been useful when fishing.

17 Wellnhofer, P., 1975. *Die Rhamphorhynchoidea (Pterosauria) der Oberjura-Plattenkalke Süddeutschlands*. Palaeontographica, 148: 1-33, 132-186; 149: 1-30, Stuttgart.

18 Zusi, R.L., 1962. *Structural adaptations of the head and neck in the Black Skimmer Rynchops nigra Linnaeus*. Publications of the Nuttal Ornithlogical Club, 3: 1-101, Cambridge, Mass.

19 Bramwell, C.D., and Whitfield, G.R., 1974. *Biomechanics of Pteranodon*. Philosophical Transactions of the Royal Society London, (B), 267: 503-581, London.

20 Stein, R.S., 1975. *Dynamic Analysis of Pteranodon ingens: a Reptilian Adaptation to Flight*. Journal of Paleontology, 49, 3: 534-548.

21 Wellnhofer, P., 1987. *New crested Pterosaurs from the Lower Cretaceous of Brazil*. Mitteilungen der Bayerischen Staatssammlung für Paläontologie und historische Geologie, 27: 175-186, Munich.

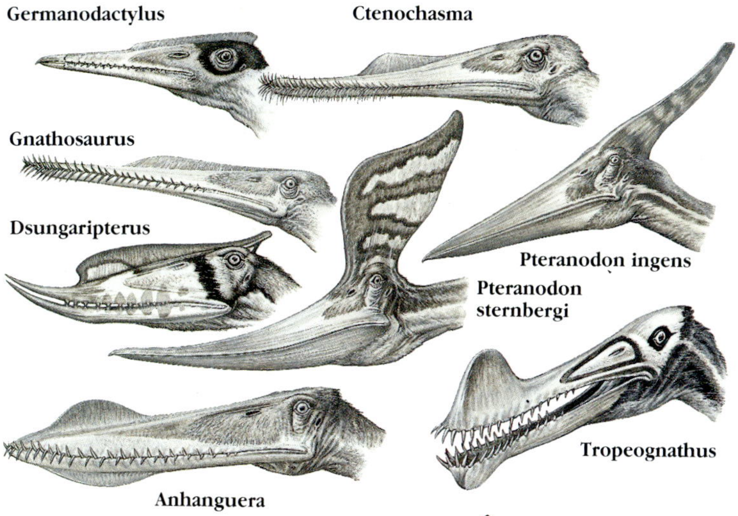

Crested Forms (left)
The drawing illustrates a variety of the forms of crest that we know pterosaurs exhibited. They vary from the low, median crests of *Germanodactylus*, to the large crests on the rear of the skull borne by *Pteranodon*, or the 'ship's keel'-like forms on the front of the jaws of *Tropeognathus*. Opinions differ as to their function. They may have served as an element in sexual display, or as counterbalances and stabilizers when flying, or even, in some cases, fishing on the wing.

meant that less muscle power was needed to raise the head out of the water again and restore it to a flying position.

Normally, the lower pterosaur jaw is as long and slim as the upper jaw. The hinge of the jaw has a particular structure. It works like a screw joint: when the beak was opened wide it splayed out the branches of the lower jaw in a similar way to the method used by the pelican, so that it was easier to stow fish caught in the throat pouch.[22] Perhaps this is also an indication that pterosaurs had a throat pouch in which food was collected and predigested.

22 Wellnhofer, P., 1980. *Flugsaurierreste aus der Gosau-Kreide von Muthmannsdorf (Niederösterreich) – ein Beitrag zur Kiefermechanik der Pterosaurier*. Mitteilungen der Bayerischen Staatssammlung für Paläontologie und historische Geologie, 20: 95-112, Munich.

Pteranodon's Crest (above)
Did *Pteranodon ingens*' crest serve as a mast on which a membrane that stretched to its back was mounted? Such a 'sail' might have served as a flight rudder, but it would have severely limited the mobility of its head in the vertical plane.

Jaw Articulation (right)
The articulations of the lower jaw are functionally mirror-image screw joints as is seen in this specimen of *Pteranodon*. Obliquely arranged ridges and grooves forced the lower jaw to widen as the beak was opened, probably in order to expand a throat pouch to hold fish. The photographs below show the position of the jaws when closed (lower right) and open (lower left). The posterior process of the lower jaw can swing upward and slide over the quadrate bone of the skull above.

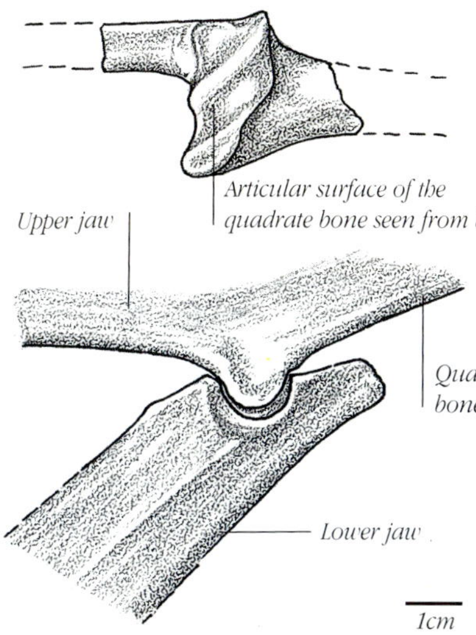

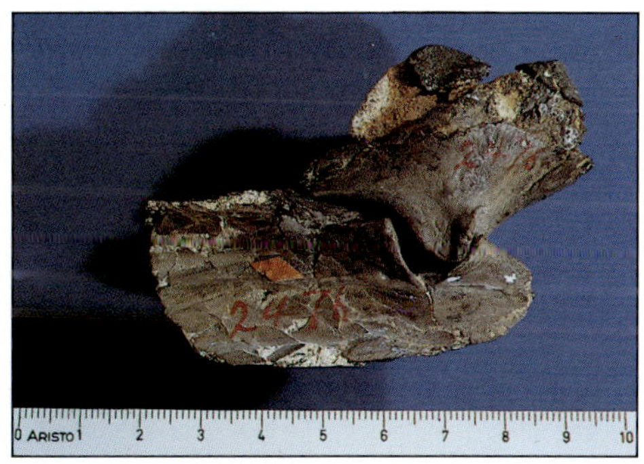

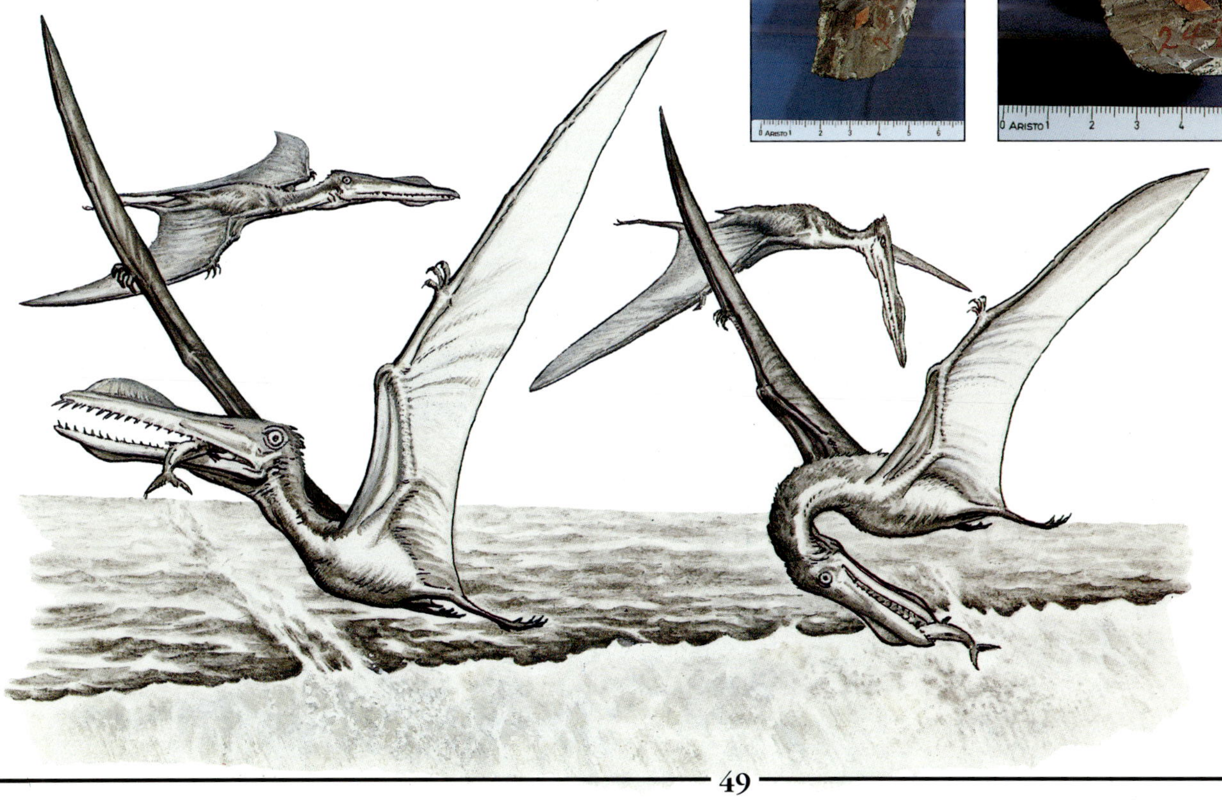

Fishing Techniques (left)
This sequence shows how the crest on the front of *Anhanguera*'s beak may have helped this pterosaur to fish when on the wing. As it dipped its jaws into the sea to snatch a fish, the crest would have ploughed through the water just below the surface, like the keel of a ship, reducing water resistance and stabilizing *Anhanguera*'s head at the critical moment of grasping its prey.

The Spinal Column

The spinal column is usually divided into cervical, dorsal, sacral and caudal vertebrae. There is a considerable difference in the number and length of the caudal (tail) vertebrae between the long-tailed pterosaurs (Rhamphorhynchoidea) and the short-tailed pterosaurs (Pterodactyloidea). The number of cervical, dorsal and sacral vertebrae is not uniform in all pterosaurs either.

There can be seven, eight or nine cervical (neck) vertebrae. It is often not clear whether the last two cervical vertebrae can be counted as dorsal vertebrae. In terms of definition, however, the first dorsal vertebra is the one on which ribs are the first to be connected to the sternum.

The number of dorsal (back) vertebrae varies between 11 and 16, according to whether one counts the last cervical vertebrae as dorsal or not, and whether the last dorsal vertebrae are already part of the sacrum. In the sacrum there are between three and five vertebrae in Rhamphorhynchoidea and up to ten in Pterodactyloidea. Long-tailed pterosaurs have about 40 caudal vertebrae with the exception of *Anurognathus*, whose short stubby tail has only 11 vertebrae. *Anurognathus* is the only 'long-tailed' pterosaur with a short tail.

The joints of all pterosaur vertebrae are concave at the front and convex at the back; such vertebrae are termed procoeleous. Only the caudal vertebrae are concave at both ends, and they are termed amphicoeleous.

The cervical vertebrae of pterosaurs are large and robustly built, in contrast with the rest of the vertebrae, though like the dorsal vertebrae they have lateral openings, so-called pneumatic foramina, which helped to save weight. In Rhamphorhynchoidea the neck is always shorter than, or at most the same length as, the vertebral column between it and the tail. In Pterodactyloidea the neck is always longer. This is not achieved by an increased number of vertebrae, but rather by extension of the cervical vertebrae themselves. The shorter cervical vertebrae of Rhamphorhynchoidea have high upper neural spines. They have short, thin cervical ribs close to the vertebrae. Pterodactyloidea have ribs only on the last two cervical vertebrae, and in their case the neural spines are mostly low and crest-shaped. The cervical vertebrae of *Pteranodon* and other Cretaceous pterosaurs also have a high neural spine however. In the gigantic *Quetzalcoatlus* the cervical vertebrae are also extremely long with very low, crest-shaped neural spines. Furthermore, in this genus and in other large pterosaurs, additional processes, so-called exapophyses, have formed, jutting out ventrally at both ends of the vertebra. This five-point joint in the cervical vertebrae prevented vertebral torsion and major sideways movement. But the cervical vertebral column could be considerably bent upwards and downwards, which was particularly important for taking in food, i.e. for catching fish.

In most Cretaceous pterosaurs some of the dorsal vertebrae are fused together. This is called a notarium, in which the vertebrae and to an extent the neural spines as well have fused. In *Pteranodon* for example these form a supraneural plate with a shallow articular facet on each side to accommodate the shoulder blades (scapulae). Functionally this notarium is a uniform complex for support and anchorage of the pectoral girdle. The supraneural plate did not only serve to articulate the free end of the scapula but was also the attachment

Rhamphorhynchus Vertebrae

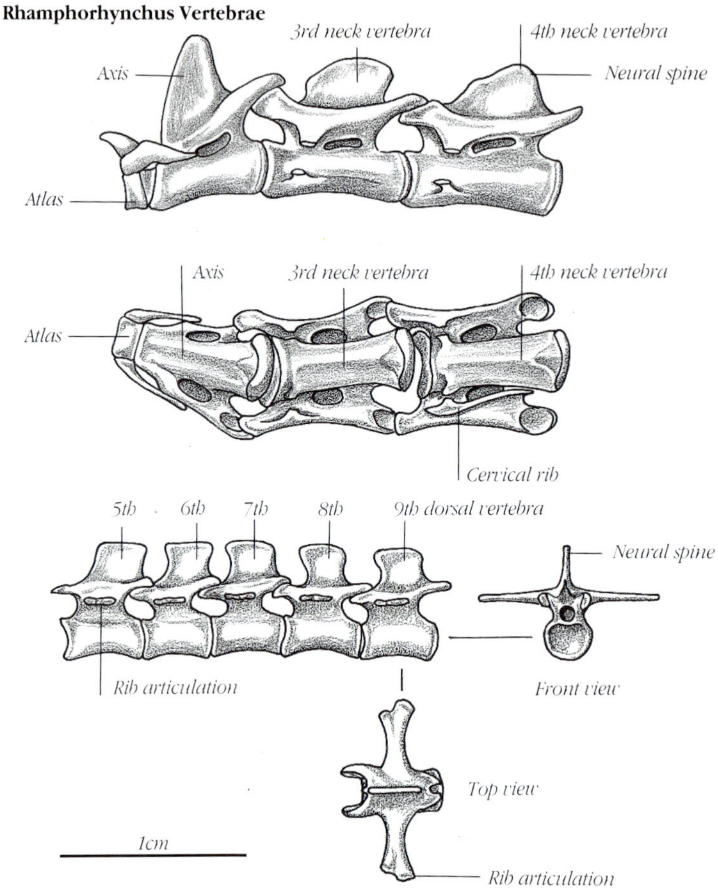

Pterodactylus

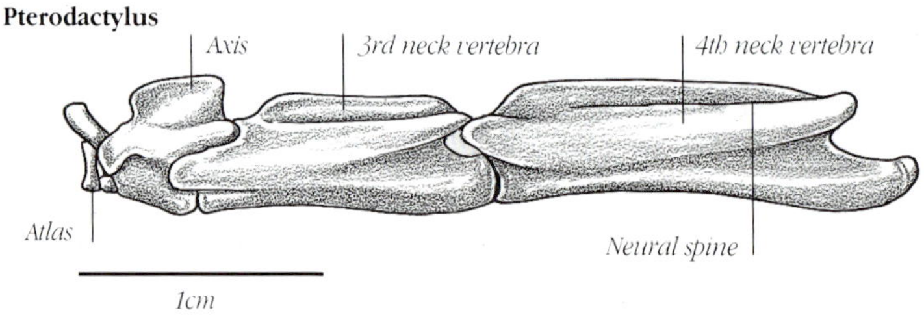

Pteranodon Notarium

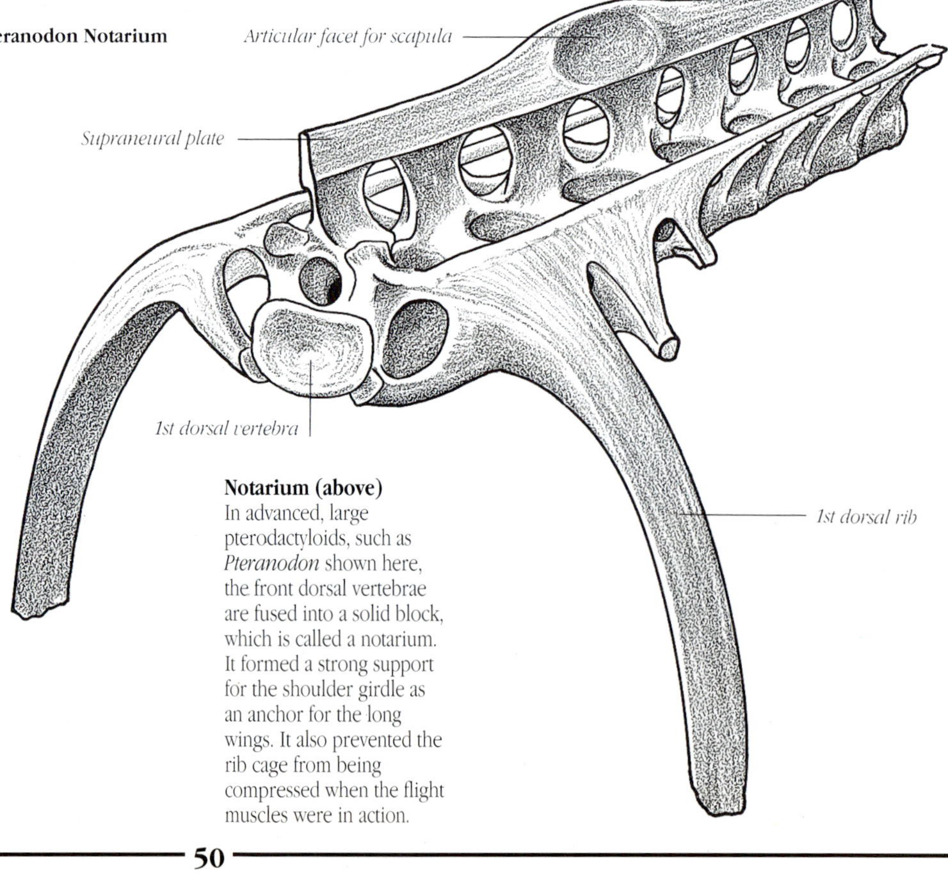

Notarium (above)
In advanced, large pterodactyloids, such as *Pteranodon* shown here, the front dorsal vertebrae are fused into a solid block, which is called a notarium. It formed a strong support for the shoulder girdle as an anchor for the long wings. It also prevented the rib cage from being compressed when the flight muscles were in action.

Rhamphorhynchus Vertebrae (left)
The general layout of the vertebrae of pterosaurs follows the archosaurian pattern. But compared to the trunk vertebrae (dorsals) the neck vertebrae (cervicals) are especially strong, providing large areas for the attachment of the neck musculature. Despite this, they are lightly built, partly hollow, and have holes in the bony walls, the pneumatic foramina, in order to save weight. The joints of all pterosaur vertebrae are generally concave at the front and convex at the back. Shown here are neck and trunk vertebrae in different aspects.

Pterodactylus Neck Vertebrae (below left)
The elongation of the neck in pterodactyloids is a consequence of the elongation of the individual neck vertebrae rather than of an increase in the number of neck vertebrae present. While the vertebrae grew longer, the neural spines on top of the vertebrae also became lower. As a result the head and neck could swing up and down quite freely, but sideways mobility was restricted.

point for an important flight muscle, the latissimus dorsi, which made it possible to raise the upper forelimb backwards. The chest ribs decrease in length and strength towards the rear of the creature. They are connected to the breastbone (sternum) by short intermediate pieces, the so-called sternal ribs. This produced a rigidly fixed rib cage, which could not be compressed by flight muscle activity.

The sacral vertebrae fuse to form the sacrum and are connected to the ilia of the pelvis by broad spines or sacral ribs. Often the last dorsal vertebrae are still included in the sacrum and in the case of *Pteranodon* for example they fuse to form a synsacrum, as in birds.

Rhamphorhynchoidea have a long tail consisting of up to 40 vertebrae. The first five to six caudal vertebrae are still normally developed and can move interdependently. The subsequent caudal vertebrae are increasingly longer and thinner. Their mobility is severely restricted by a system of elongated bone spines which surround and stiffen the tail. These thin rod-like extensions originate in the extended front and rear vertebral processes, the zygapophyses, then fork and reach up to six times the length of the vertebrae. The ventral vertebral connections, the so-called chevrons, also have thin spines of bone stretching forwards and backwards.

Right: The tail of the long-tailed pterosaurs, the rhamphorhynchoids, is composed of about 40 vertebrae which are elongated and stiffened by long, rod-like, bony extensions of the zygapophyseal processes (sometimes incorrectly called ossified tendons). This *Rhamphorhynchus* specimen from the Solnhofen limestone shows the structure of the tail and these stiffening extensions of the vertebrae clearly.

Below: This close-up of three *Rhamphorhynchus* tail vertebrae reveals that the bony rods originate as extensions from the vertebrae and from the ventral chevrons in between them (see also drawing below). The rods functioned as stiffening devices for the long, aerodynamic tail.

Rhamphorhynchus Vertebra

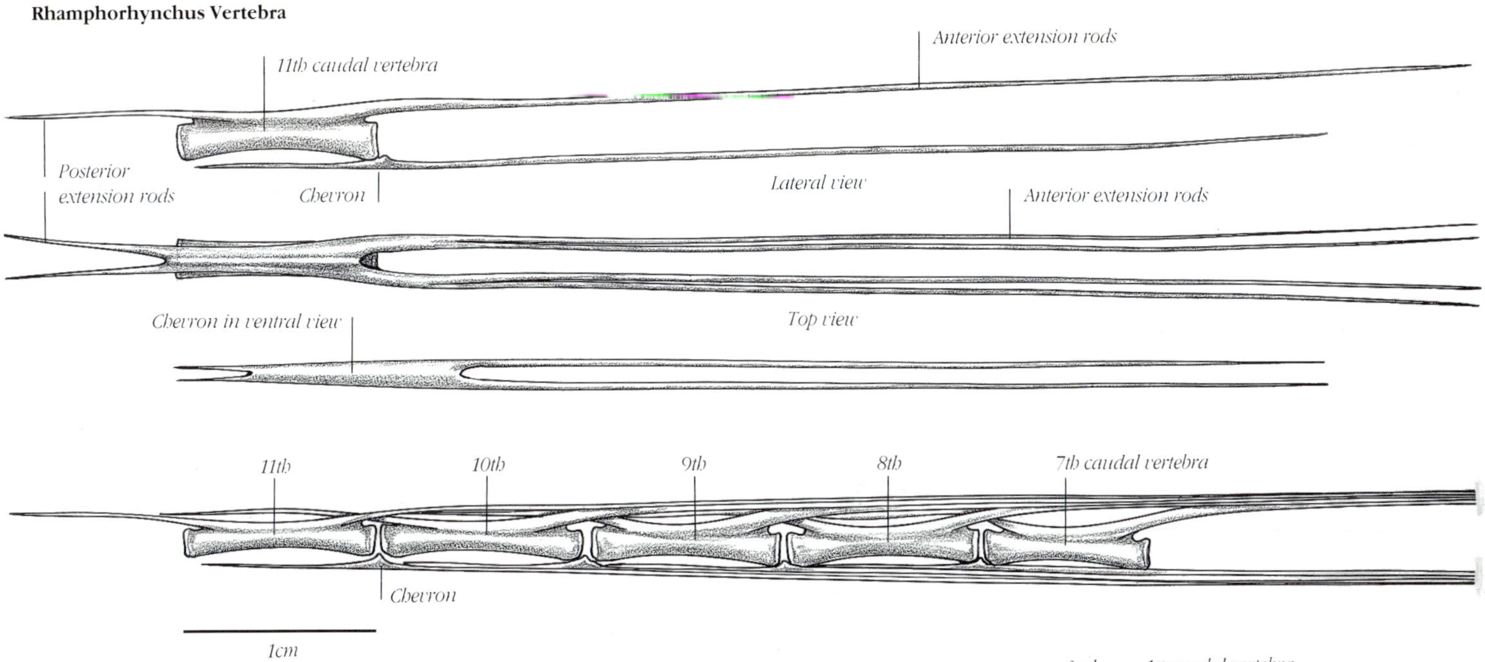

Tail Stiffeners (above)
This reconstruction shows the 11th tail vertebra of *Rhamphorhynchus* in isolation (in various views), and as part of a series of vertebrae. The intermediate vertebral elements, the chevrons, have also developed long, stiffening extensions. The tail was thus reinforced and stiffened both dorsally and ventrally.

Pterodactylus Tail

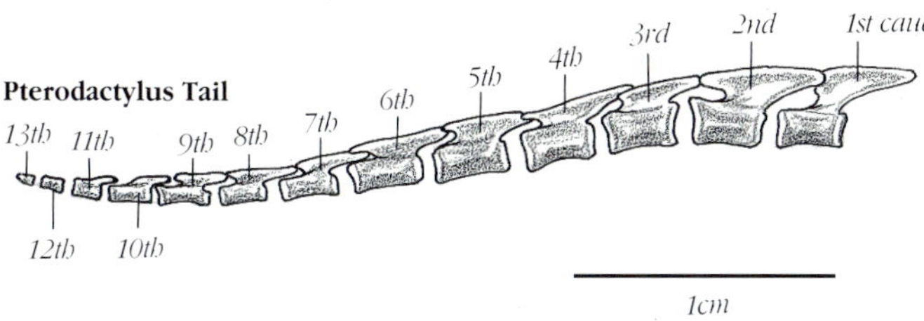

Pterodactylus Tail (left)
The tail of pterodactyloids is greatly reduced, and is composed of only a few small vertebrae which are not stiffened at all. This is *Pterodactylus'* tail.

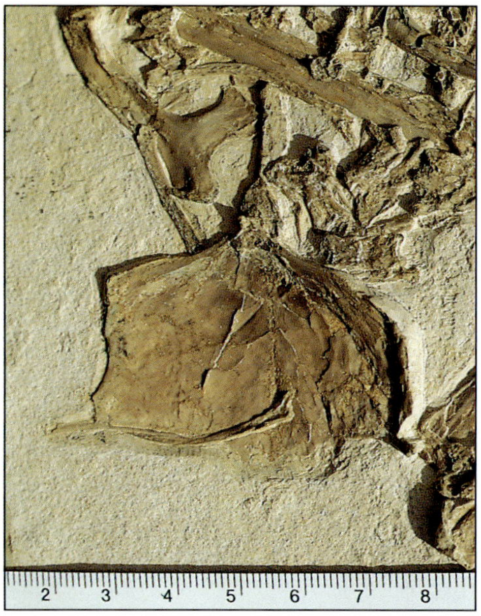

Breastbones (right) The pterosaur breastbone (seen here from below) is a wide plate with a crest (cristospina) protruding forward and ventrally. The large surface provided the area for the attachment of the large flight muscles.

Left: A *Rhamphorhynchus* breastbone (from below).

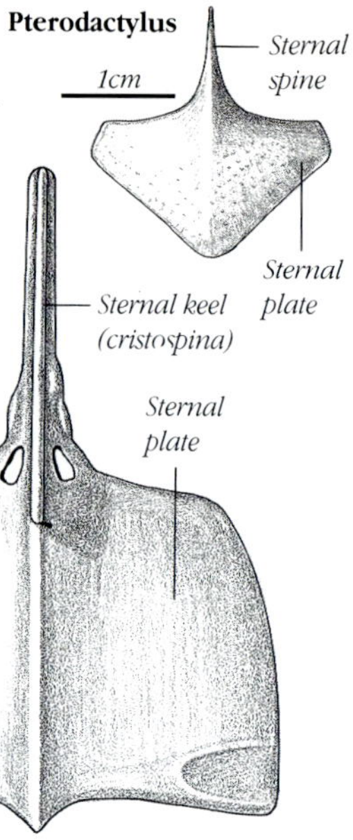

Thus long-tailed pterosaurs had a stiffened tail which moved principally upwards and downwards. A cross-section of the central tail area of *Rhamphorhynchus* reveals 26 dorsal and 12 ventral stiffening elements, which give the tail the stability of an elastic rod, similar to that of a carpet-beater. The function of a caudal vertebral column constructed in this way is clear: the tail was used for steering during sudden changes of direction in flight. This is also proved by the fact that at the end of this long tail there was a rhomboid tail membrane that must have had an aerodynamic function.[23]

While the posterior section of the tail of long-tailed pterosaurs was dominated by segmented axial musculature, the transverse processes of the front caudal vertebra still had epaxial muscles attached, which allowed the long tail to be moved sideways as well.

The tail of the short-tailed pterosaurs only contained a few short vertebrae, which rapidly taper, are not stiffened and apparently had no significance as far as flying was concerned. *Pterodactylus*' short tail had at most 16 vertebrae. The caudal vertebral column of *Pteranodon* was strange. It consisted of five very broad vertebrae with double joints. These were followed by six vertebrae which decreased in size, and at the end were long paired rods which were possibly integrated into the flight membrane. By means of an up-and-down movement, the tail of *Pteranodon* would have had a steering function in flight.[24]

The Sternum

Like birds, pterosaurs possessed a large ossified breastbone or sternum. It consisted of a widened bone plate and a forward-pointing keel or crest (carina). According to genus it was triangular, heart-shaped or rectangular in shape. The slightly bulging bone plate of the breastbone and its keel functioned as the attachment surface for the massive system of pectoral muscles that moved the wings. On the side edges of the sternal plate short sternal ribs were attached, and they formed a connection with the thoracic ribs.

At the point of transition from the sternal plate to the sternal keel (cristospina sterni) two articular facets were located dorsally to take the coracoids of the pectoral girdle.[25] This support function of the coracoids at the same time prevented compression of the rib-cage of the pterosaur when the flight muscles were contracted to flap its wings.

Gastralia

The gastralia or abdominal ribs are thin rods of bone set loosely in the skin of the abdomen. As a rule they are arranged in six rows one behind the other. Each row consists of a central piece angled in the middle with a slightly curved side-piece on either side. The gastralia are located immediately after the sternum, and bridge the space between sternum and pelvis. Together with this they served to protect the viscera. Gastralia are common in fossil and recent amphibians and reptiles, but are absent in modern birds.[26] It is assumed that they are skin ossifications of a former abdominal dermal armour that have been transferred inwards.

The Pectoral Girdle

Shoulder elements found in other reptiles, like the clavicle and interclavicle, are absent in pterosaurs. The pectoral or shoulder girdle consists of scapula and coracoid. They are both long and powerfully developed, and meet at a more or less acute angle in a V or U shape. They are usually fused to form a single bone, the scapulocoracoid. In Cretaceous pterosaurs which have a notarium the pectoral girdle is anchored to the vertebral column. The free end of the scapula is articulated here to a facet of the ossified neural spines. The ventral end of the coracoid is levelled off and in the form of a saddle-shaped articular facet, which meshes with the sternum in an appropriate articular socket. Both scapula and coracoid are involved in the shoulder joint which faces to the side and slightly upward. As a result, the humerus was particularly well suited to execute dorsally directed movements. Seen from the front, the two coracoids form a V shape with one another and in articulation with the sternum. In the

23 Similarly stiffened caudal vertebrae are found in the theropod dinosaur *Deinonychus* from the Lower Cretaceous of Montana, in which the tail also had a steering function, when the creature was running quickly on the ground.
Ostrom, J.H., 1969. *Osteology of Deinonychus antirrhopus an Unusual Theropod from the Lower Cretaceous of Montana*. Bulletin of the Yale Peabody Museum of Natural History, 30, 165 pp.; New Haven.
24 Bennett, S.C., 1987. *New Evidence on the tail of Pterosaur Pteranodon (Archosauria: Pterosauria)*. Fourth Symposium on Mesozoic Terrestrial Ecosystems, Short Papers (ed. P.M. Currie and E.H. Koster), pp.18-23; Drumheller.

25 Rhamphorhynchoidea and Pterodactyloidea differ principally in the nature of coracoid articulation at the sternum. In Rhamphorhynchoidea the articular facet for the left coracoid is behind the facet for the right coracoid. In Pterodactyloidea the coracoid articular facets are arranged in symmetrical juxtaposition on the sternum.

26 It is remarkable that the oldest known bird, *Archaeopteryx* from the Jurassic did not have an ossified breastbone, but typically reptilian gastralia.

Gastralia (right) Gastralia, sometimes called gastral ribs, are a typical structure in many amphibians and reptiles. They are dermal ossifications, and may have derived from the abdominal dermal armour protecting the underside of such creatures. In pterosaurs, such as in this *Rhamphorhynchus*, the gastralia are composed of three rod-like bones, an angled middle mediogastrale connected to two slightly curved lateral laterogastralia, arranged in six parallel rows one behind the other. They were originally embedded within the skin of the belly, between the sternal plate and the pelvis. Their function was probably to support and protect the intestines. Modern birds do not have gastralia.

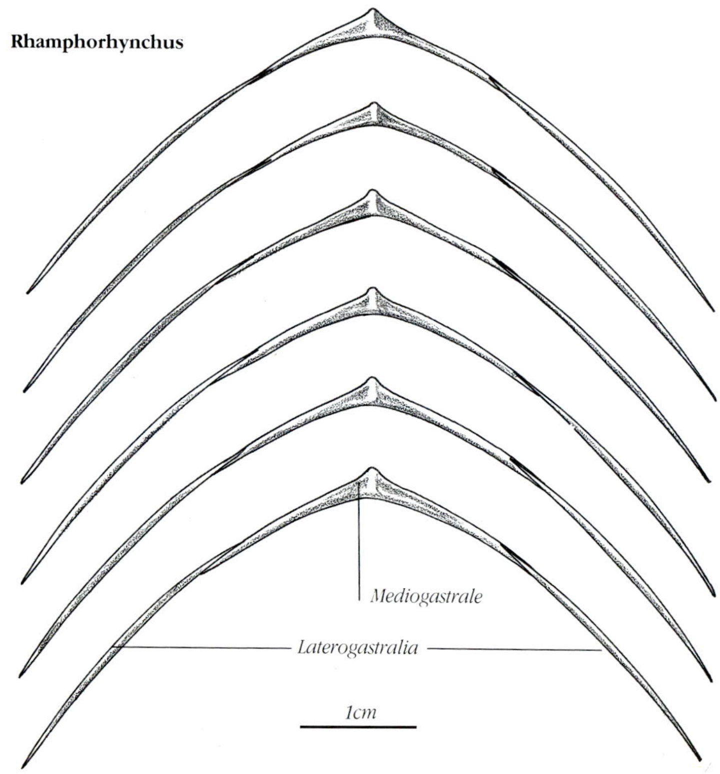

Rhamphorhynchus Shoulder Girdle

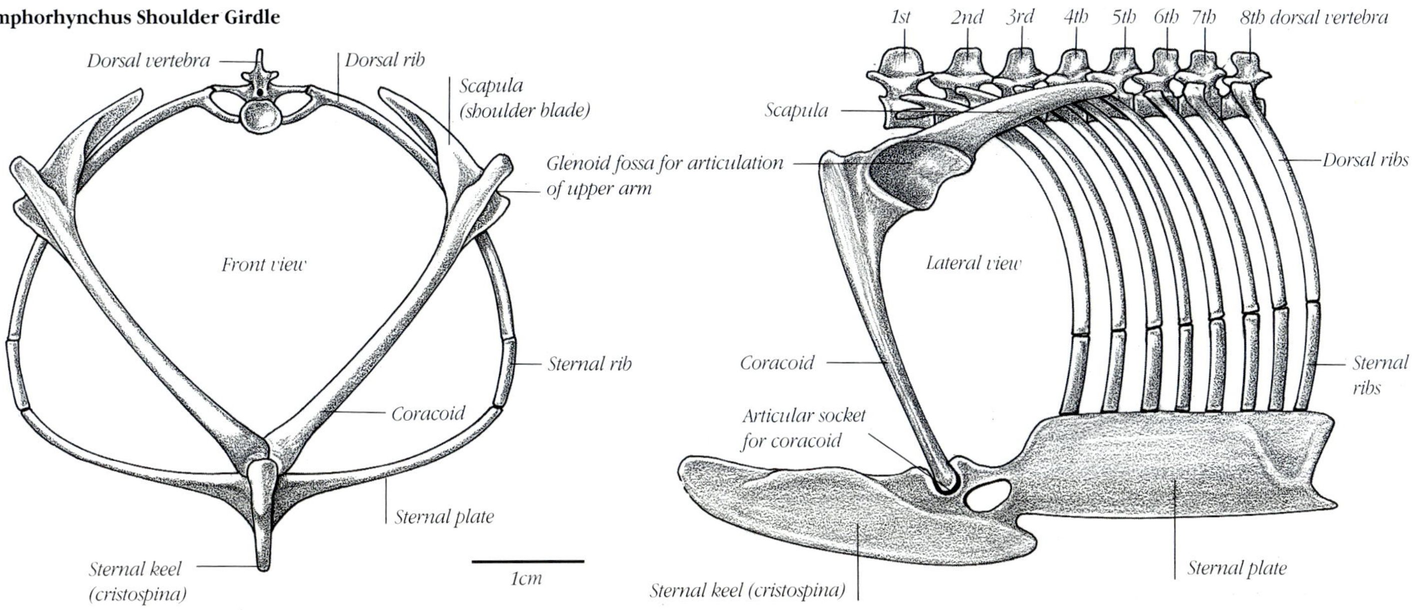

great Cretaceous pterosaurs both scapulacoracoids formed a closed ring with the notarium at the top and the sternum at the bottom, which created a stable part of the skeleton on which the long flight limbs could be mounted.

The Forelimbs

The forelimb of pterosaurs is developed into a wing. The fourth finger of the hand is greatly elongated and strengthened, while the first three fingers are normally developed and have claws. The fifth digit of the hand regressed in the course of evolution, and is absent.

In Pterodactyloidea a tendency to elongation of the wing is also reflected in the development of the metacarpals – the bones in the hand that form the palm. The bones of the flight limb are more powerfully developed in comparison with the bones of the hindlimbs. The long bones have very thin walls, and in some large Cretaceous pterosaurs, like *Pteranodon* for example, they are extremly thin, often only the thickness of a postcard.

A particular feature of pterosaur wings is the pteroid bone. It is connected by a joint to the distal, lateral carpal bone. Its function is disputed. Probably the pteroid reinforced the front edge of a small flight membrane between the upper and lower parts of the limb, the so-called propatagium.

The Humerus

The pterosaur humerus (upper arm) is relatively short and solid. At the upper end (proximal) and usually near the axis of the shaft a saddle-shaped articular head is developed; this was in contact with the glenoid fossa of the pectoral girdle. Laterally at the upper end there is always a broad protruding crest, the delto-pectoral process. The pectoral flight muscles were attached to this.

The shaft of the humerus is usually bent somewhat forwards. The distal articular facets to radius and ulna (the bones of the forearm) permit quite major bending of the elbow. The humerus is a hollow bone. Air openings (foramina pneumatica) have been found in the bones of large pterosaurs. Spongy bone tissue is found in joints and processes. In large Cretaceous pterosaurs the cavity of the shaft is stiffened with thin bony struts. This gives maximum bone strength combined with the greatest possible lightness. The internal support structures are effectively lines of force made material. Maximum bracing of the humerus was possible when it was approximately at right angles to the longitudinal axis of the body. It could swing up to 65° backwards. In *Pteranodon* vertical movement of the humerus was possible in a sweep from 25° below the horizontal to 70° above it. In the course of this motion it was directed 17° backwards at its lowest point.[27]

The Shoulder Girdle (above)
The shoulder girdle connected to the sternum, as shown here in *Rhamphorhynchus*. The glenoid fossa for the upper arm articulation faces slightly upward and backwards, indicating how the wing moved.

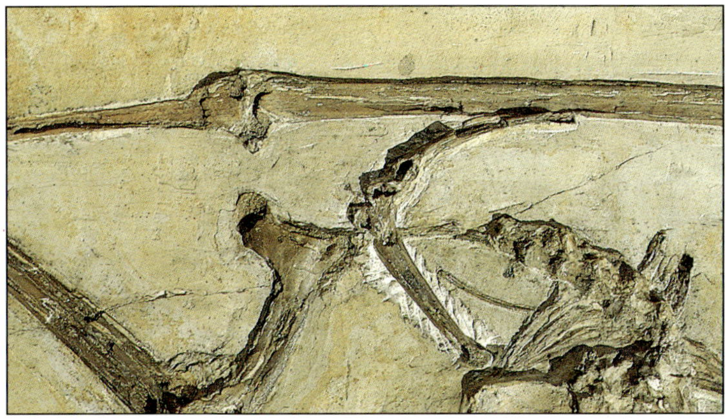

Right: The boomerang-shaped bone in the centre is the shoulder girdle of a *Rhamphorhynchus*.

27 Bramwell, C.D. and Whitfield, G.R., 1974, op. cit. (19).

The Radius and Ulna

The lower part of the limb is always longer than the humerus. The radius and ulna (the two bones of the forearm) are next to, and touch one another. The ulna is always the stronger bone. Movement at the elbow was only possible in one plane. The maximum extension between humerus and lower arm was about 150°. When it was bent, in other words when the wing was folded, the radius was pushed distally along the ulna. When the radius was moved in this way it pressed on the proximal carpal, which was twisted and slid away

The Upper Arm (right)
The upper arm (humerus) is relatively short in pterosaurs, but it is quite robust with large processes for the attachment of the flight muscles (the delto-pectoral process). Despite its strength, this bone is hollow, and, except for the articular extremities, it has very thin walls which makes it light. In large pterosaurs the shaft is internally strengthened by bony struts which stiffened the outer walls transversely. These drawings show the left humerus of *Rhamphorhynchus* in lateral (left), anterior (middle), and medial (right) views.

Rhamphorhynchus' Humerus

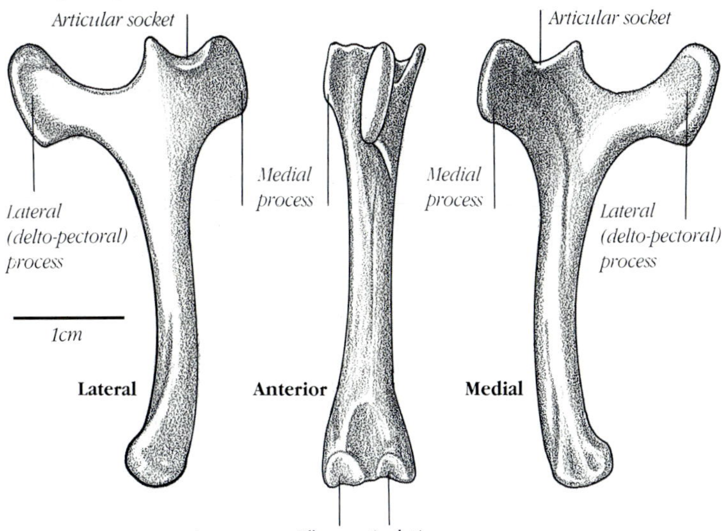

over a protrusion on the ulna. This caused the wing metacarpal to swing back, and with it the wing finger. Thus in pterosaurs there was automatic folding of the wing, a principle with which we are familiar in birds. It means that when the wing was folded the hand was automatically bent backwards in the wrist.

The Carpus

Originally the carpus (or wrist) was made up of five bones: two proximal and three distal carpals. There is a tendency for proximal carpals to fuse into one bone. This is the case with Rhamphorhynchoidea and Cretaceous Pterodactyloidea. In advanced forms there is a further reduction to two distal bones. The larger of the two supports the solid wing metacarpal. The smaller, laterally placed distal carpal has a forward oriented glenoid fossa to take the pteroid. This is a specific pterosaur bone, which does not occur in other vertebrates. It was probably connected to a tendon running along the front edge of a small flight membrane between the upper and lower parts of the limb, the propatagium. Presumably the function of the pteroid bone was to alter the angle of attack and shape of the propatagium, and hence to alter flight performance. It has been observed that the pteroid is always directed towards the body in undisturbed fossil skeletons, which supports this interpretation of its position.

However, another school of thought suggests that the pteroid was directed forwards and downwards, and that for this reason pterosaurs had a movable forewing, flexed and controlled by the pteroid. If this were the case, the angle of incidence of this forewing would have been altered by upward and downward movement and would have caused variable camber of the wing.[28] Another interpretation of the function of the pteroid postulates that it could have been directed both forwards and downwards – during slow flight, for example, and also towards the body in rapid flight. In this case a tendon would have been stretched over the free end of the pteroid bone, which would have caused the pteroid to click into one or the other position according to the position of the flight digit.[29] However, the fact that in many pterosaurs the pteroid was very long and thin – often up to two thirds of the length of the lower arm – tends to argue against this view.

The wrist could be moved in three ways: motion could occur between proximal carpal and the lower arm, between proximal and distal carpal in the form of a limited sliding movement, and thirdly as a rotation between distal carpal and the wing metacarpal.

The Metacarpus

The metacarpus (or palm of the hand) consists of four bones, of which the wing metacarpal is considerably stronger than the other three metacarpals. The wing metacarpal is jointed with the distal carpal in such a way as to allow it to twist via a tongue and socket joint. Distally it has a special joint formation to take the first joint of the flight digit. The three metacarpals of the three small digits are slender bone rods which lie one behind the other in close contact with the large metacarpal. The first two can also regress at their lower end, so that they no longer reach the carpus.

Long- and short-tailed pterosaurs also differ in the relative length of the metacarpal. In all Rhamphorhynchoidea it is shorter than half the length of the lower arm, in Pterodactyloidea it is longer than half the lower arm, and in Cretaceous pterosaurs it can often even be longer than the entire lower arm.

28 Frey, E. and Riess, J., 1981. *A new Reconstruction of the Pterosaur Wing*. Neues Jahrbuch für Geologie und Paläontologie, Abhandlungen, 161 (1): 1-27.
29 Pennycuick, C.J., 1988. *On the reconstruction of Pterosaurs and their manner of flight, with note on vortex wakes*. Biological Review, 63: 299-331.

Folding the Wing Finger (right)
The principal folding joint in the pterosaurian wing was the articulation between the wing metacarpal and the first phalanx of the wing finger. The drawings show this section of the hand of *Rhamphorhynchus* with the wing finger in maximum flexion (near left) and maximum extension (far right).

Rhamphorhynchus' Digits

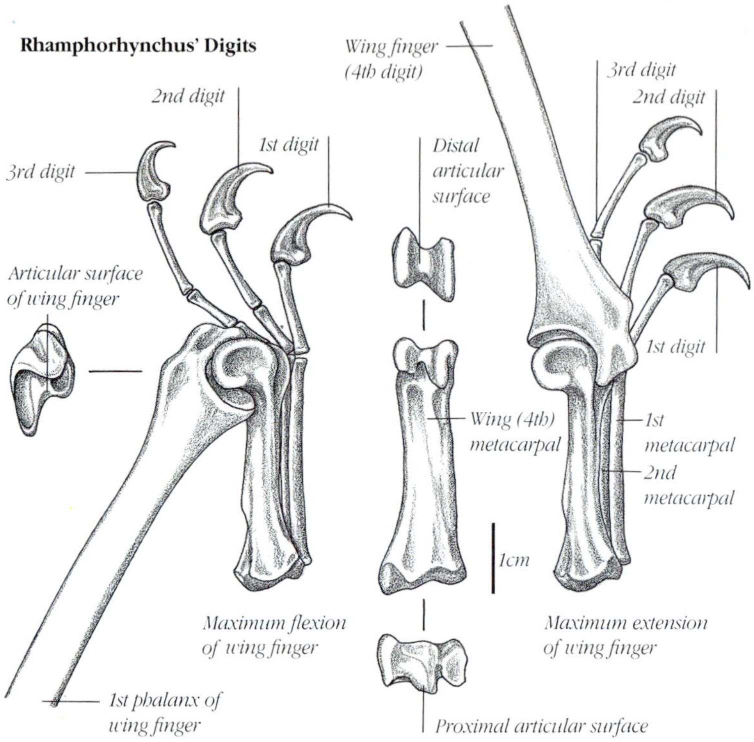

Below: A section of the wing skeleton of a *Pterodactylus* showing the principal folding joint of the pterosaurian wing. The first three fingers remained small, were equipped with sharp claws, and articulated with very slender metacarpals. The wing here appears folded to its maximum degree.

Pterosaur Wrists (right)
The wrist consists of two rows of small bones (carpals) which form the joint between the lower arm and metacarpus. They allowed twisting and sliding movements rather than much flexion in the wrist. A lateral carpal protruding to the front supported the slender pteroid bone, which is only known in pterosaurs.

Left: The wrists of a pterodactyloid from Brazil. The long, thin bone in the centre is the pteroid.

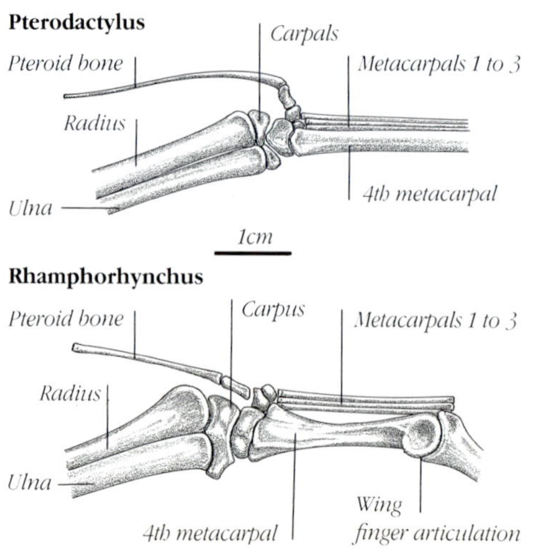

Appropriately to its function, the fourth metacarpal, the wing metacarpal, has a distinctive pulley on the distal end, into which the first flight digit joint was articulated. The axis of the joint is set somewhat at an angle, producing a twisting movement when the wing was folded, thus placing the flight digit against the body in such a way that its upper side pointed out sideways. To accommodate the double-bowled articular surface of the first flight digit phalanx there were corresponding depressions in the wing metacarpal, which indicate that the flight digit could be bent here to a very high degree, approximately to an angle of 25°-30°. At maximum extension the angle was 165°. Further extension was not possible, as a short process on the front side of the first flight phalanx, for the insertion of an extensor tendon, formed a stop and prevented forward hyperextension of the flight digit.

The Digits

The pterosaur hand had only four digits. The first three were normally developed and had powerful, sharp, curved claws which projected in front of the wing and were oriented towards the body. These small digits could be moved freely and, including the claws, had two, three and four phalanges on the first, second and third digit. The claws had well-developed so-called flexor tubercles for the insertion of strong flexor tendons. This indicates that pterosaurs could grip rock ledges or trees with their digit claws and presumably could also climb.

The fourth digit was extremely elongated to fulfil its function as bearer of the flight membrane. It was almost twenty times longer than the small digits, and also much stronger. It consisted of four long phalanges. The proximal joint of the first and thickest phalanx consists of two bowled glenoid fossae to accommodate the joint pulley of the wing metacarpal. The articular capsules of the four wing phalanges produce joints which are appropriately tightly articulated. The flight digit phalanges are slightly bent, resulting in what must have been a leading edge to the wing that bent backwards and downwards.

In Rhamphorhynchoidea all the flight digit phalanges have a longitudinal furrow on the flight membrane side, which obviously permitted secure attachment of the flight membrane. The fairly sharp edges of these ridges are drawn further back at the top than at the bottom. The leading edge of the flight digit phalanges is rounded, like the leading edge of an aeroplane wing. The flight digit phalanges of Pterodactyloidea exhibit different characteristics: they have no longitudinal furrow, and are triangular in cross-section.

Maximum wingspan was achieved with regard to the maximum angle of extension between the individual joints of the flight arm. In *Santanadactylus* for example the humeri were set at a backward angle of 15°, the elbow had an angle open to the front of 150°, and between lower arm and metacarpus and between metacarpus and flight digit there was an angle of 165° in each case.[30] Folding the flight arm was essentially achieved by pulling the humerus backwards and bending the elbow joint and the joint between the wing metacarpal and the wing finger.

30 Wellnhofer, P., 1985. *Neue Pterosaurier aus der Santana-Formation (Apt) der Chapada do Araripe, Brasilien.* Palaeontographica, A, 187: 105-182; Stuttgart.

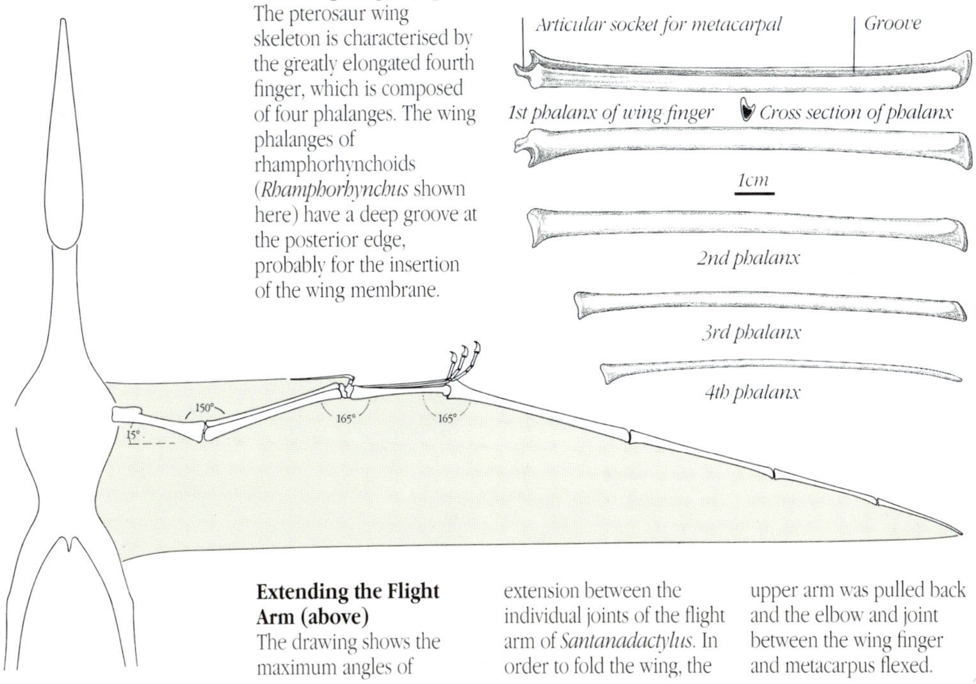

The Wing Finger (right)
The pterosaur wing skeleton is characterised by the greatly elongated fourth finger, which is composed of four phalanges. The wing phalanges of rhamphorhynchoids (*Rhamphorhynchus* shown here) have a deep groove at the posterior edge, probably for the insertion of the wing membrane.

Extending the Flight Arm (above)
The drawing shows the maximum angles of extension between the individual joints of the flight arm of *Santanadactylus*. In order to fold the wing, the upper arm was pulled back and the elbow and joint between the wing finger and metacarpus flexed.

The Pelvis

In principle the pterosaur pelvis is an archosaur pelvis. It is composed of ilium, ischium and pubis on each side. Another pelvic bone typical of pterosaurs is the prepubis, which is situated in front of the pubis. The structure of the pterosaur pelvis is markedly different from that of birds.

As a rule ilium, ischium and pubis are more or less fused and form a unit with the sacrum of the spinal column. The degree of ossification in the pelvis is obviously also dependent on the age of the individual creature. The ilium is extended forward to form a narrow bone, bent slightly upwards. In contrast the section of the ilium behind the socket of the hip joint is shorter. The socket of the hip joint (acetabulum) is oriented laterally and obliquely upwards to accommodate the femur. It is not perforated, as is the case with birds. All three pelvic bones, ilium, ischium and pubis, are involved in the formation of the hip socket. This orientation of the hip sockets prevented the femur from being placed vertically and makes it improbable that pterosaurs could walk on the ground like birds.[31]

31 Molnar, R.E., 1987. *A pterosaur pelvis from western Queensland, Australia.* Alcheringa, 11: 87-94.
Unwin, D.M., 1987. *Pterosaur locomotion – joggers or waddlers?* Nature, 327: 13-14.
Wellnhofer, P., 1988. *Terrestrial locomotion in pterosaurs.* Historical Biology, 1: 3-16.

Above: An isolated pelvis of *Campylognathoides*. It is seen from above and shows the hip sockets clearly. Due to compression their orientation is unnatural.

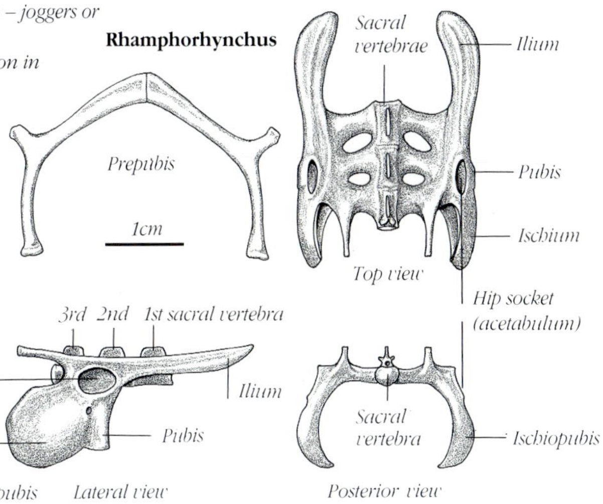

Rhamphorhynchus Pelvis (right)
The pelvis is seen here from various angles. The arrangement of the ilium, ischium and pubis follows the general reptilian pattern, although the proportions are unique to pterosaurs. A paired prepubic bone, originally connected to the pubis by cartilage, is also typical of pterosaurs.

The short, ventrally directed pubis may be fused with the ischium, broadening towards the rear to form a uniform bone plate, the ischiopubis, as in *Rhamphorhynchus* for example. The pubes are never fused along the midline below, but the lower ischia can meet in the middle and close the pelvis ventrally, as in *Pteranodon*, *Nyctosaurus*, *Dimorphodon* and possibly in *Campylognathoides*.[32] In other pterosaurs the pelvis is open ventrally.

The prepubes are not firmly connected to the pelvis and are very variable in shape, so that they can serve as diagnostic features. They are articulated with the pubes and in Pterodactyloidea are broadened into a scoop shape at their free ends. In Rhamphorhynchoidea they partially form long, slender bars of bone which meet on the midline and are held together either by cartilage or by bony fusion. The function of the prepubes was to support the viscera, and perhaps also as a 'sitting bone' when the creature was at rest.

There is no doubt that Rhamphorhynchoidea had strong dorsal and ventral tail muscles to control the long tail. It is highly probable that these muscles were attached to the rear and inner sides of the ischium and on the dorsal side of ilium and sacrum.

The Hindlimbs

By comparison with the forelimbs, the hindlimbs of pterosaurs are short, weak and slender, but relatively long when compared to the length of the trunk. In Triassic and Lower Jurassic pterosaurs they are relatively more strongly developed than in Upper Jurassic and Cretaceous genera. The femur is always shorter than the tibia.

32 Padian, K., 1983. *A functional analysis of flying and walking in pterosaurs*. Paleobiology, 9 (3): 218-239

Above: The pelvis of this *Pterodactylus*, seen here from the left, shows how it connected to the vertebral column and the tail. The prepubis is the spatulate bone below the pelvis. In life it was attached to the pubic bone by cartilage. This element is a typical pterosaurian bone, not found in other vertebrates. The prepubes probably supported the intestines.

The femur (upper leg bone) has a straight or slightly curved shaft with a circular cross-section. The head of the femur is more or less distinctly separated from the shaft by a narrow neck (collum). The ball of the joint (caput femoris) is directed obliquely upwards. The collum bends at an angle to the shaft, in most pterosaurs this is between 130° and 160°. If one considers the lateral orientation of the hip sockets mentioned earlier, then the femur could only have been directed obliquely downwards and outwards. Thus biped locomotion, like that of a bird, was not a possibility. In the case of optimal articulation in the hip sockets, the femur would be in the horizontal plane, in other words in the flight position. The distal joint of the femur, the trochlea, was broadened and orientated obliquely to the longitudinal axis of the shaft.

The tibia (the main shin bone) is a completely straight, slender hollow bone, which is always longer than the femur. At its lower end the tibia broadens towards the tarsus. Flexion of the knee joint permitted the tibia to be in an upright position when moving on the ground, despite the oblique position of the femur.

In early pterosaurs like *Dorygnathus* and *Campylognathoides*, the fibula (the smaller shin bone) is developed to full length. In the more advanced Jurassic forms the fibula no longer reaches the tarsus at its lower end, but ends above it as a wedge-shaped bone pressed closely to the tibia. In Cretaceous pterosaurs the fibula is markedly reduced or completely absent.

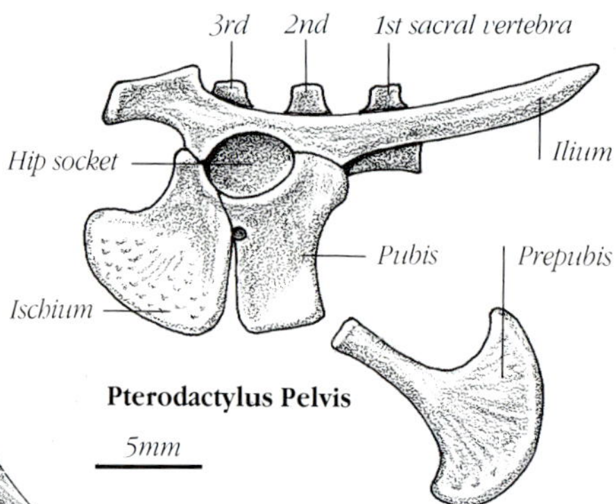

Pterodactylus Pelvis (above)
This pelvis is much like that of *Rhamphorhynchus*. Here, too, the hip socket is not perforated as in birds, and the ilium has a long extension to the front. However, the shape of the prepubis is different. It varies among all pterosaur genera, and can be interpreted as a diagnostic character.

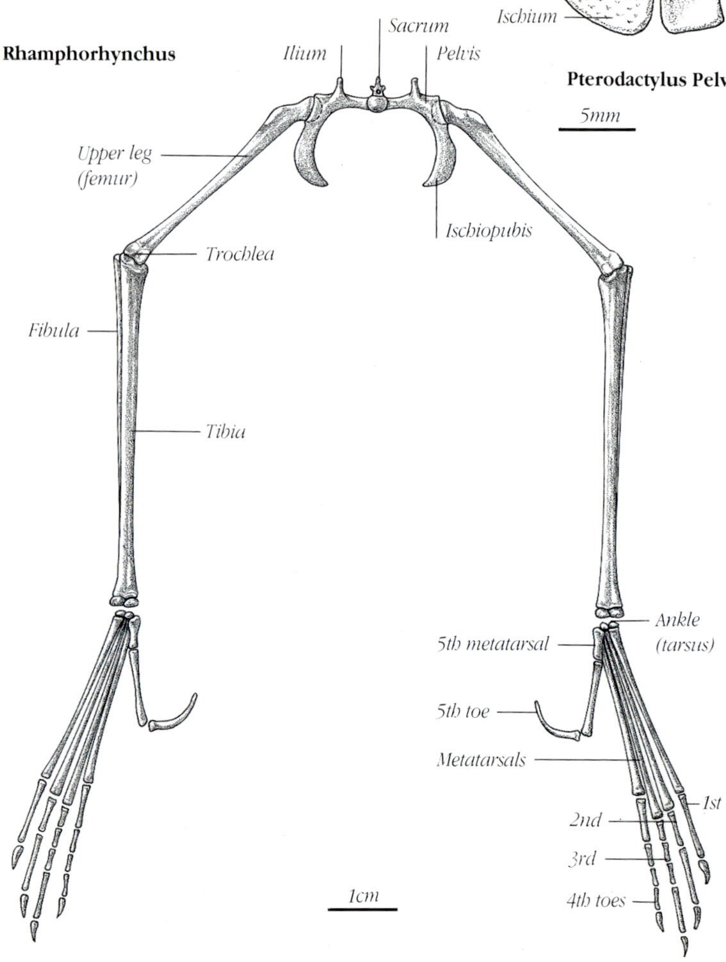

The Hind Legs (left)
This restoration shows how the hind legs of *Rhamphorhynchus* articulated in the pelvis. The hip socket pointed slightly upwards and backwards. The ball-like articular head of the femur (the caput) fitted into this socket in such a way that the femur pointed obliquely outwards rather than vertically downwards.

The Skeleton of the Foot

The foot of the pterosaur was generally narrow and long. Rhamphorhynchoidea differ from Pterodactyloidea in having a long fifth digit, which is flexed and has no claw. The phalangeal formula for the foot runs 2.3.4.5.2. in Rhamphorhynchoidea and 2.3.4.5.0. in Pterodactyloidea.

Pterosaurs have a typical mesotarsal ankle joint, i.e. the line of the joint runs between the proximal and distal bones of the tarsus. In early pterosaurs there were two proximal and three distal tarsals. In Cretaceous pterosaurs in particular the proximal tarsals fused with the tibia, while only two distal tarsals remain. The two proximal tarsals can also – particularly with age

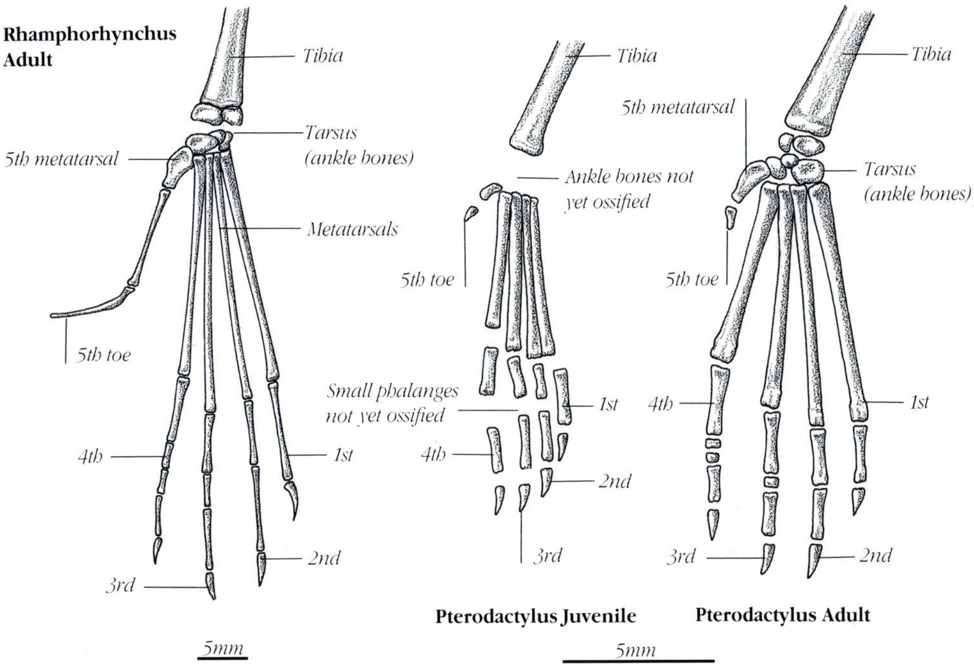

Left: The foot skeletons of *Rhamphorhynchus* (upper) and *Pterodactylus* (lower). In general the osteology of these feet suggests that pterosaurs were more likely to have walked on the soles of their feet, rather than on their toes like birds. Note that the fifth toe in pterodactyloids is reduced to a short stump.

Pterosaur Feet (above)
The drawings compare the feet of an adult *Rhamphorhynchus* (left), and a juvenile and adult *Pterodactylus* (middle and right). The tarsal bones and small phalanges in the third and fourth toes of the juvenile are missing; they were not yet ossified when this individual died.

– fuse to form a single bone, as in some *Rhamphorhynchus* species for example.

Undisturbed skeletons show that the ankle could allow right-angled articulation of the foot with the tibia. This is an indication of a plantigrade position of the foot (i.e. with the sole on the ground) when walking. Thus pterosaurs were not digitigrade like birds – walking on the toes – but plantigrade – walking on the soles of the feet – like man. So far there have been no confirmed finds of pterosaur footprints, so no conclusive proof of this is available.[33]

The metatarsals (upper foot bones) of the first four digits of the foot are long, and slender; they are staggered, and have somewhat broader, flattened joint ends. The fourth metatarsal is usually somewhat shorter than the first three. The fifth is extremely short, and projects into the ankle. The four long metatarsals and thus the digits could be spread, perhaps to stretch a web between the digits. The short fifth metatarsal could apparently also be moved in another plane.

The Foot Digits

All digits with the exception of the fifth had claws. But they were not as powerful as the claws on the digits of the hand, and also not as sharply curved. In Rhamphorhynchoidea the fifth digit has two phalanges, which are very long and meet at a marked angle. In Pterodactyloidea there is only a short phalanx on the fifth digit, and in *Pteranodon* and all other Cretaceous pterosaurs it has no phalanges.

In Pterodactyloidea the third and fourth toe has very short central phalanges, which only ossify in the adult growth phase. If these small intermediate phalanges are missing from a fossil specimen, then the creature is likely to be quite young. At that stage these small digital phalanges were still cartilaginous. As cartilage usually cannot be fossilized, there is a gap between the digital phalanges in such cases.

The long fifth digit on the foot of Rhamphorhynchoidea had more freedom of movement than the other digits and could apparently be moved in another plane as well. It is notable that in undisturbed skeletons the fifth digit is always flexed towards the tail. This position precludes stretching of the wing membrane by the fifth digit. There is also argument about whether this digit could have stretched a membrane on the inside of the hindlimbs, a so-called uropatagium. It must be assumed that the orientation of the other digits was vertical to this. Perhaps the feet of *Rhamphorhynchus* were held in flight in such a way that the soles pointed inwards, like those of modern flying foxes. The aerodynamic advantages of a uropatagium are not entirely clear. It was certainly a hindrance when swimming in water. But the long fifth digit could also have been used to stretch a web between the toes. However, the short-tailed pterosaurs no longer needed a long fifth toe of this kind.

Principal Osteological Differences between Rhamphorhynchoidea and Pterodactyloidea

Rhamphorhynchoidea	Pterodactyloidea
Jaws always completely filled with teeth	Jaws completely, partially or not at all filled with teeth
Nasal and preorbital opening separate	Nasal and preorbital opening not separate or confluent
Occipital joint (condylus occipitalis) directed backwards	Occipital joint (condylus occipitalis) directed downwards and backwards
Cervical ribs present	Cervical ribs only present on the last two cervical vertebrae
No notarium	Notarium possible
Articular facets for the coracoids on the sternum one behind the other	Articular facets for the coracoids side by side
Long tail made up of extended vertebrae	Short tail
Short metacarpus	Long metacarpus
Flight digit phalanges with longitudinal furrow on the rear side	Flight digit phalanges without longitudinal furrow, triangular in cross-section
Long fifth digit on foot	Short fifth digit on foot

33 The pterosaur tracks described by Stokes (1957) from the Jurassic Morrison Formation of Arizona clearly show the print of a plantigrade foot. However, in the meantime doubt has been expressed about whether these tracks were made by pterosaurs at all.
Stokes, W.L., 1957. *Pterodactyl tracks from the Morrison formation*. Journal of Paleontology, 31, 5: 952-954.
Padian, K. and Olsen, P.E., 1984. *The fossil trackway Pteraichnus: Not pterosaurian, but crocodilian*. Journal of Paleontology, 58, 1: 178-184.
Unwin, D.M., 1989. *A Predictive Method for the Identification of Vertebrate Ichnites and its Application to Pterosaur Tracks*. Dinosaur Tracks and Traces, edited by D.D. Gilette and M.G. Lockley, pp.259-274; Cambridge University Press.

PTEROSAURS OF THE TRIASSIC

The Triassic period at the beginning of the Mesozoic era was highly significant for reptilian evolution. This was the time at which such diverse groups as tortoises, crocodiles and dinosaurs appeared. Fossil finds of the oldest pterosaurs so far known came from the Upper Triassic. Fossil remains from the so-called Rhaetic in Baden, Württemberg and England were identified as pterosaur bones by various researchers as early as the nineteenth century, but these remains were far too sparse for a precise impression of these Triassic pterosaurs to be formed.[1] Recently, the discovery of bone fragments of Triassic pterosaurs has been reported from Gloucestershire in England.[2] They were collected from Upper Triassic fissure deposits at Cromhall Quarry among a few thousand vertebrate elements of different reptile groups, such as sphenodontids, terrestrial crocodiles, flying lizards and others. Although incomplete, the bones represent the first definite record of pterosaurs from the British Upper Triassic. Two small bones could be identified as wing metacarpals. However, they lack diagnostic characters and cannot be assigned to any known pterosaur, except to the Rhamphorhynchoidea. They may even represent a new genus of pterosaur.

In 1899, E.D. Cope in America also believed that bone fragments and isolated vertebrae from Triassic strata in Pennsylvania could be those of a long-necked, long-tailed pterosaur. He called it *Rhabdopelix* (=rod-basin, after a rod-like pelvic bone).[3] The Tübingen expert on fossil reptiles Friedrich von Huene also reported later on similar pterosaur bones from the Pennsylvanian Triassic.[4]

However, identification of these very incomplete fossil remains as pterosaurs was frequently questioned. One of the assumptions was that *Rhabdopelix* may have been related to lizards, similar to the gliding reptile *Icarosaurus*, known from Triassic deposits in New Jersey.[5] But for a long time a fossil reptile skeleton from the Middle Triassic in the southern Alps in Italy called *Tribelesodon* was considered to be the oldest known pterosaur.

Tribelesodon, a Triassic Pterosaur?

In 1886 Francesco Bassani (1853-1916), Professor of Geology in Milan, alluded to a pterosaur from Middle Triassic strata of Besano in Lombardy, northern Italy, in a monograph on fossils from the bituminous shales of Besano.[6] Milanese geologists had conducted systematic excavations there as early as 1863 and 1878, which had brought to light an interesting yield of hitherto largely unkown marine reptiles. Bassani identified a small fossil remnant of a skeleton as a pterosaur, which because of its tiny teeth with three cusps he called *Tribelesodon* (=three-cusped tooth).

Above: Franz Baron Nopcsa, Hungarian nobleman, geologist and acknowledged expert on fossil reptiles. He studied the specimen of *'Tribelesodon'* from Besano and interpreted it – in fact, erroneously – as the oldest known pterosaur.

Franz Baron Nopcsa (1877-1933), the Hungarian geologist and reptile expert, was the first to study these creatures from Besano in more depth. Baron Nopcsa was a many-sided, even enigmatic personality. He had extensive estates in Transylvania (modern Romania) and wrote numerous palaeontological treatises, even books. He was an internationally recognized specialist on fossil reptiles. In 1922 he produced a detailed description of the putative *Tribelesodon* pterosaur as well.[7] He even made a drawing of the reconstructed skeleton, showing it as a long-tailed pterosaur. The striking feature of this find, the only specimen of this creature, was a series of long bones, which both Bassani and Nopcsa took to be the elongated phalanges of a flight digit.

The Besano Triassic strata consists of black shale containing bitumen, a sort of mineral pitch, which was mined for commercial purposes. At the time it was distilled to produce raw materials for preparing 'Saurol', which like 'Ichthyol' (from the so-called fish-shale of Seefeld in the Tyrol) was used for pharmaceutical purposes, as an ointment. These were Middle Triassic marine deposits (Ladinian), about 235 million years old; they extend into the Swiss Canton Ticino and come to the surface at the southern end of Lake Lugano, on Monte San Giorgio. These strata, the 'Grenzbitumenzone', were also mined, and it was here that the Zürich palaeontologist Bernhard Peyer found reptile remains on a spoil heap in 1919. They sparked off very successful planned excavations, conducted by the University of Zürich on a regular basis in summer until the mid 1970s.

In September 1929 these produced an almost complete skeleton of a small reptile with extremely long cervical vertebrae. Vertebrae of this kind were already known from the Bayreuth Muschelkalk in northern Bavaria. In 1834 Georg Graf zu Münster had taken them to be limb bones, while in 1855 Hermann von Meyer in Frankfurt declared them to be the caudal vertebrae of a dinosaur, which he named *Tanystropheus* (=long vertebra).

1 Deffner, C. and Fraas, O., 1859. *Die Juraversenkung bei Langenbrücken*. Neues Jahrbuch für Mineralogie und Geologie: 1-38; Stuttgart. Flight digit phalanges from the Rhaet Bonebed are mentioned
Dawkins, W.B., 1864. *On the Rhaetic Beds and White Lias of Western and Central Somerset*. Quarterly Journal of the Geological Society, 20: 396-412; London. Dawkins discovered 'a crushed and hollow bone' in the Rhaetic (=Upper Triassic) of Somerset, England, and tentatively took it and two other bone fragments to be 'pterosaurian'.

2 Fraser, N.C. and Unwin, D.M., 1990. *Pterosaur remains from the Upper Triassic of Britain*. Neues Jahrbuch für Geologie und Paläontologie, Monatshefte, 1990 (5): 272-282; Stuttgart.

3 Cope, E.D., 1869. *Synopsis of the extinct Batrachia, Reptilia and Aves of North America*. Transactions of the American Philosophical Society Philadelphia, 14: 169-175; Philadelphia.

4 Huene, F.v., 1921. *Reptilian and Stegocephalian Remains from the Triassic of Pennsylvania in the Cope Collection*. Bulletin of the American Museum of Natural History, 44: 19, 561-574; New York.

5 Colbert, E.H., 1966. *A Gliding Reptile from the Triassic of New Jersey*. American Museum Novitates, 2246: 1-23; New York.

6 Bassani, F., 1886. *Sui fossili e sull'età degli schisti bituminosi triasici di Besano in Lombardia*. Atti Società Italiana Scienzi Naturali, 29: 15-72; Milan.

7 Nopcsa, F., 1922. *Neubeschreibung des Trias-Pterosauriers Tribelesodon*. Paläontologische Zeitschrift, 5: 161-181; Berlin.
Nopcsa is a colourful figure with an interesting history. He worked as a spy for the Austro-Hungarian Empire in the First World War. After that he directed the Hungarian Geological Survey in Budapest for three years. In 1913, at the end of the Balkan War, he tried to become King of Albania. To this end he asked Vienna for 500 soldiers, artillery and two fast steamers, in order to invade Albania and have himself proclaimed king. At the end of the First World War, Baron Nopcsa lost the greater part of his estates, and sometimes it seemed likely that he would lose his reason as well. He used the last of his money for an extended motorcycle tour of Europe with his secretary, who was also said to have been his homosexual lover. He shot him, and then himself, in 1933.

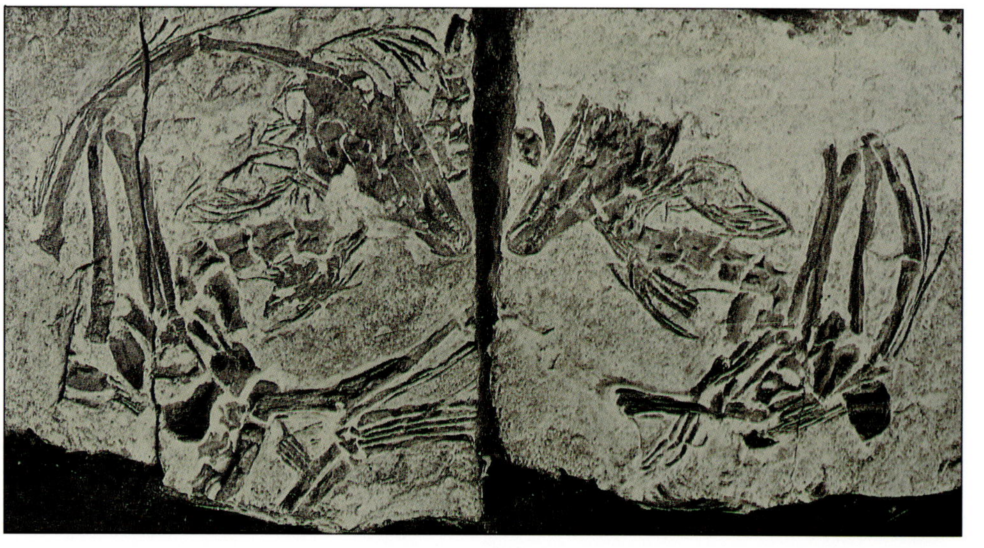

Emil Kuhn-Schnyder, a successor of Bernhard Peyer in Zürich, was present as his assistant at the time, and described those dramatic September days like this: 'As a stratum of slate about a centimetre thick was being split the small skull of a reptile and the long topmost cervical vertebra attached to it appeared. After searching for hours we succeeded in finding two pieces of shale which were the precise continuation of the first. They contained the rest of the skeleton, right down to the end of the tail . . . Thus we had a successful find that increased our knowledge of *Tanystropheus* decisively . . . A few days later Peyer travelled to Milan to study the collection of Triassic fossils in the Museo Civico di Storia Naturale there. He came back from Italy beaming, and I suspected at once what he had discovered: he established that *Tribelesodon* was not a pterosaur, but a small *Tanystropheus*, in poor condition. He was later able to prove this beyond doubt by a thorough examination.[8]

The Monte San Giorgio find solved two riddles at a stroke. Firstly, that the long *Tanystropheus* vertebrae were cervical vertebrae of a reptile with an extremely long neck. For this reason it is also known as 'giraffe-necked saurian'. Secondly it was clear immediately that the series of long bones that Bassani and Nopcsa had interpreted as long flight digit phalanges in *Tribelesodon* were nothing more than cervical vertebrae of *Tanystropheus*, the giraffe-necked saurian, a partially aquatic creature of the eosuchian group. It is a sign of Baron Nopcsa's stature as a human being and a scientist that he later acknowledged his error. He wrote in a letter to Bernhard Peyer in 1931: 'It is crystal clear that I made a mistake.'

So, the Triassic 'pterosaur' *Tribelesodon* stood discredited. Were the finds made in England a hundred years earlier, in 1829, by Mary Anning in the Jurassic cliffs of the Dorset coast in southern England to remain the oldest pterosaurs known to man? *Dimorphodon* from the Early Lias in Lyme Regis, Dorset, was already a perfectly developed pterosaur, which suggested that it must have had more primitive ancestors, even in the Triassic Period. So there must have been Triassic pterosaurs. But there had been no unquestionable fossil finds. Palaeontologists had to wait until 1973.

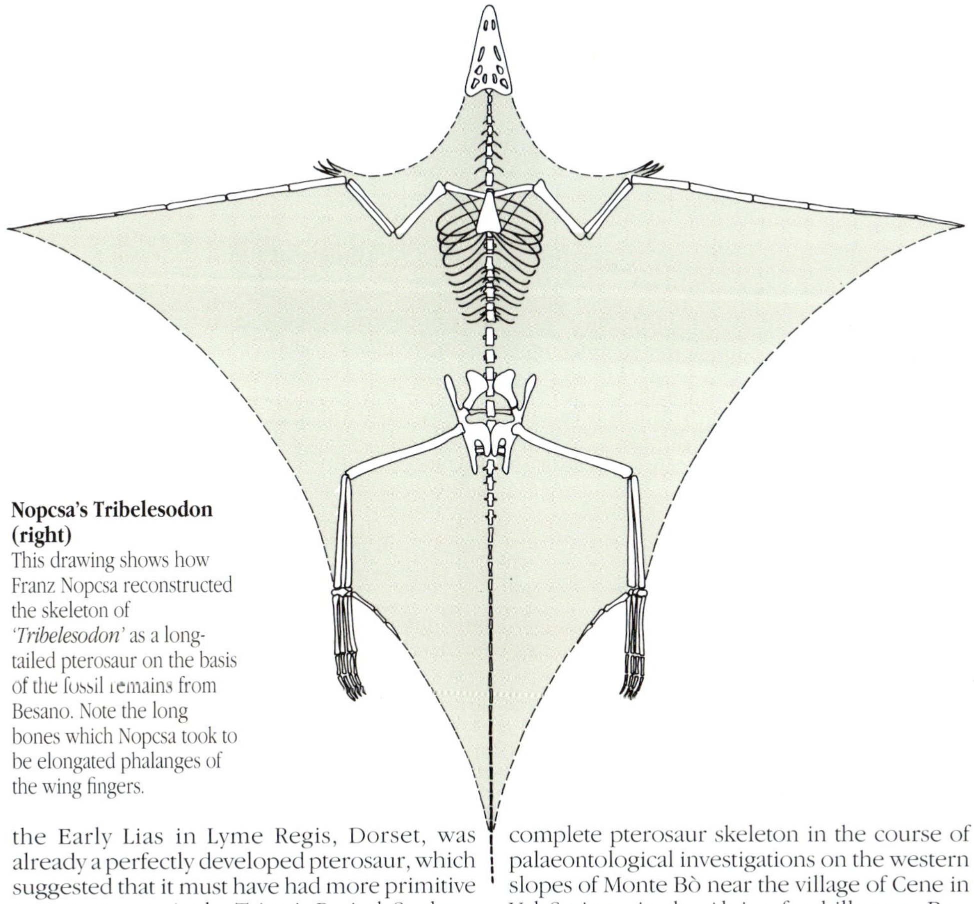

Nopcsa's Tribelesodon (right)
This drawing shows how Franz Nopcsa reconstructed the skeleton of *'Tribelesodon'* as a long-tailed pterosaur on the basis of the fossil remains from Besano. Note the long bones which Nopcsa took to be elongated phalanges of the wing fingers.

Eudimorphodon, the Oldest Pterosaur

In 1973 Mario Pandolfi of the Natural History Museum in Bergamo in northern Italy made a sensational find. He had unearthed an almost complete pterosaur skeleton in the course of palaeontological investigations on the western slopes of Monte Bò near the village of Cene in Val Seriana, in the Alpine foothills near Bergamo. He discovered it in a thin shale stratum from the rubble of a landslide in the former quarry of the village. The rock is one of the so-called Zorzino limestones, the geological age of which can be given as Upper Triassic (more precisely Middle to Upper Norian). These strata also correspond to the 'Hauptdolomit' of the southern Alps, with an absolute age of about 220 million years.

8 Kuhn-Schnyder, E., 1968. *Alles Lebendige meinet den Menschen*. Schweizer Spiegel, 12; Zürich Kuhn-Schnyder, E., 1974. *Die Triasfauna der Tessiner Kalkalpen*. Neujahrsblatt der Naturforschenden Gesellschaft Zürich, 1974: 119 p., Zürich.

Above: *Eudimorphodon ranzii* from the Upper Triassic limestone of Cene near Bergamo, northern Italy. It lived about 220 million years ago, and is thus one of the oldest known pterosaurs. The wing fingers and most of the hind legs are missing.

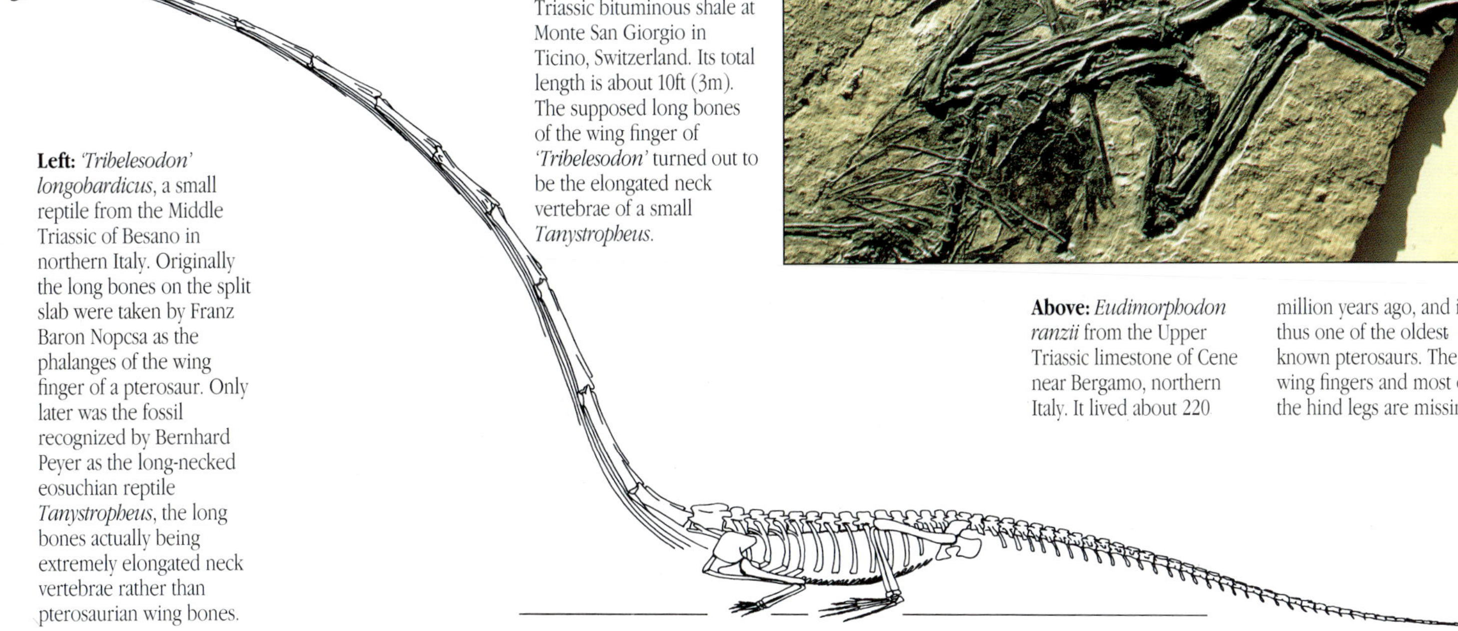

The Giraffe-Necked Saurian (below)
This skeletal restoration is of *Tanystropheus longobardicus*. Well preserved specimens of these long-necked eosuchian reptiles were discovered in Middle Triassic bituminous shale at Monte San Giorgio in Ticino, Switzerland. Its total length is about 10ft (3m). The supposed long bones of the wing finger of *'Tribelesodon'* turned out to be the elongated neck vertebrae of a small *Tanystropheus*.

Left: *'Tribelesodon' longobardicus*, a small reptile from the Middle Triassic of Besano in northern Italy. Originally the long bones on the split slab were taken by Franz Baron Nopcsa as the phalanges of the wing finger of a pterosaur. Only later was the fossil recognized by Bernhard Peyer as the long-necked eosuchian reptile *Tanystropheus*, the long bones actually being extremely elongated neck vertebrae rather than pterosaurian wing bones.

This meant that the oldest pterosaur now really had been found, the first that could be conclusively proved to date from the Triassic. Rocco Zambelli, the curator responsible in the Bergamo museum, called it *Eudimorphodon* (=true two-form tooth), following the next oldest *Dimorphodon* from the English early Jurassic. It was a pterosaur with a wing span of about 3·3ft (1m).[9]

It was expected that the oldest pterosaurs would have particularly primitive characteristics. It now emerged that this was only partially true. Original characters were the long vertebral tail and a short metacarpal, as was also typical of long-tailed Jurassic pterosaurs. The general structure of *Eudimorphodon*'s skeleton is entirely characteristic of this group,

9 Zambelli, R., 1973. *Eudimorphodon ranzii gen.nov., sp.nov., uno pterosauro triassico*. Rendiconti Scienc. Istituto Lombardo, B, 107: 27-32; Milan.

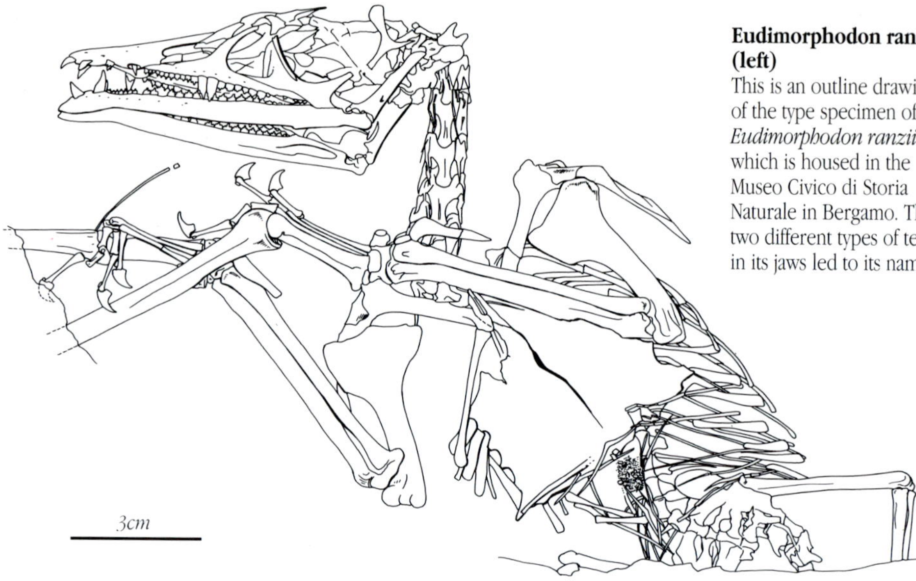

Eudimorphodon ranzii (left)
This is an outline drawing of the type specimen of *Eudimorphodon ranzii* which is housed in the Museo Civico di Storia Naturale in Bergamo. The two different types of teeth in its jaws led to its name.

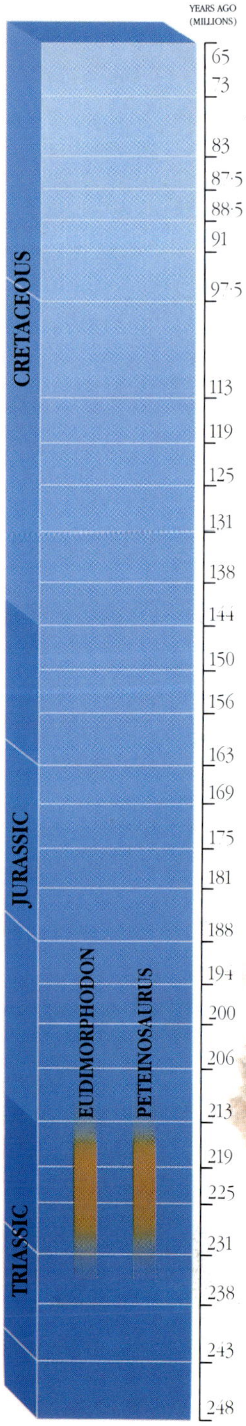

Time Chart (left)
The first pterosaurs appear in the fossil record in the Upper Triassic in marine limestone of Norian age. *Eudimorphodon* and *Peteinosaurus* already represent two distinct evolutionary lines, indicating that pterosaurs originated much earlier, perhaps in early Triassic or even late Permian times.

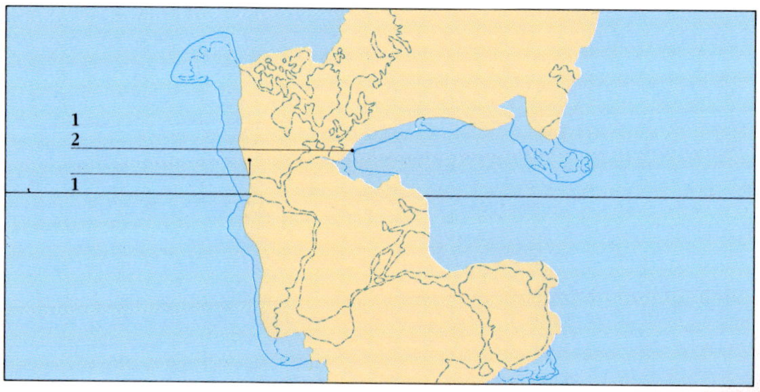

Map (left)
1 *Eudimorphodon*
2 *Peteinosaurus*

Peteinosaurus (right)
This long-tailed pterosaur, which lived on the shore of the late Triassic Tethys sea, was smaller than its contemporary *Eudimorphodon*. Its dentition consists of larger anterior fangs followed by a long series of small, pointed teeth which suggest that *Peteinosaurus* was insectivorous, perhaps catching insects on the wing. In the background we see the thecodont archosaur *Ticinosuchus*.

the Rhamphorhynchoidea, although the teeth are unique among pterosaurs.

The dentition does not consist, as is usual, of a row of teeth with a single cusp, but is divided into a few large front fangs and behind them a tight sequence of small teeth with three and five cusps. In the upper jaw there are two more large teeth with small additional cusps between the series with three and five cusps. In the upper jaw the dentition consists of 58 teeth, and in the lower jaw 56, thus a total of 114 teeth, and that in a jaw only 2·4in (6cm) long.

Dentition of this kind indicates a fish-catcher. Like other pterosaurs as well, *Eudimorphodon* must have lived by catching fish. And indeed hard scales of small ganoid fish, of the kind that frequently occur as fossils in Cene Triassic limestones were found in the region of this individual's stomach. On closer examination the Stuttgart museum curator Rupert Wild, an expert on Triassic reptiles, discovered wear facets in the teeth, which he attributed to

Eudimorphodon (below)
This Triassic pterosaur is one of the earliest known. Although it was a fully developed rhamphorhynchoid pterosaur with the characteristic long tail, it had unusual, unique dentition consisting of a series of teeth with single, three and five cusps. Such dentition would seem to belong to a fish eater, and indeed fossilized stomach contents have shown that *Eudimorphodon* preyed on small fishes such as *Parapholidophorus*, as shown here. Juvenile individuals had a different dentition and probably caught and ate insects such as the dragonfly seen at top right.

Comparative Sizes (above)
1 *Eudimorphodon:* wing span 3·3ft (1m)
2 *Peteinosaurus:* wing span 2ft (60cm)

Left: A close-up photograph of the skull of the type specimen of *Eudimorphodon ranzii*. The length of this skull is 3·4in (8·6cm). Rocco Zambelli, the curator of the Bergamo museum where the specimen is housed, named it *Eudimorphodon* which is Greek for 'true two-form tooth'.

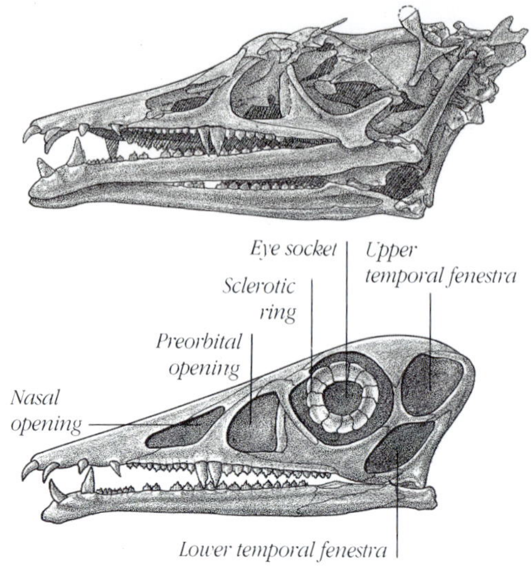

Eudimorphodon Skulls (right)
These drawings reveal the skull of *Eudimorphodon* as it was preserved (top), and as it was restored by Rupert Wild (below).

this pterosaur's fish diet. This is understandable, as the thick protective scales of these fish had to be crushed when the prey were caught.

The differentiated nature of the dentition and the fact that the teeth had more than one cusp are not primitive reptilian characters, but a specialization. The fact that no Jurassic pterosaur had multi-cusped teeth means that *Eudimorphodon* cannot be their direct ancestor, but represents a distinct line in the pterosaur family tree that became extinct in the Triassic.

Peteinosaurus

In subsequent intensive investigations five more sets of pterosaur skeletal remains were found in the Triassic limestone of Cene, and they were thorougly examined by Rupert Wild in 1978.[10] As well as *Eudimorphodon* he identified another genus, a smaller one, which he called *Peteinosaurus* (=winged reptile), from the Greek words *peteinos* (=winged) and *sauros* (=reptile). This genus turned out to be more primitive than *Eudimorphodon* in many ways. In particular *Peteinosaurus* had only single-cusped teeth, which were flattened, with sharp cutting edges at the front and back. At the front of the lower jaw were two large teeth. The upper skull remains unknown. Rupert Wild believes that *Peteinosaurus* was insectivorous. In any case this is the most primitive of the known pterosaurs. Its wings were still relatively short, only twice as long as the hind legs. In all other pterosaurs the wings were at least three times as long, or longer.

Besides the type specimen of *Peteinosaurus* only one other specimen is known. Both are housed in the Bergamo Natural History Museum. This second skeleton is also lacking the skull and neck completely, but a section of the dorsal vertebral column, of the stiff tail typically reinforced by rod-like bony extensions, and the bones of the wings, the hind legs and the feet are preserved. The skeleton is still in articulation but was washed together by currents before it was embedded. Both individuals of *Peteinosaurus* originate from the Zorzino limestone (Norian) of Cene, and are about the same size

Peteinosaurus has a wing span of only 24in (60cm) and can best be considered as a direct ancestor of the oldest Jurassic pterosaur *Dimorphodon*. Both can be placed within the

Eudimorphodon Life Restoration (left)
The drawing is a portrait illustrating how *Eudimorphodon* may have looked in real life. It is based on the type specimen which is pictured above left and on page 59.

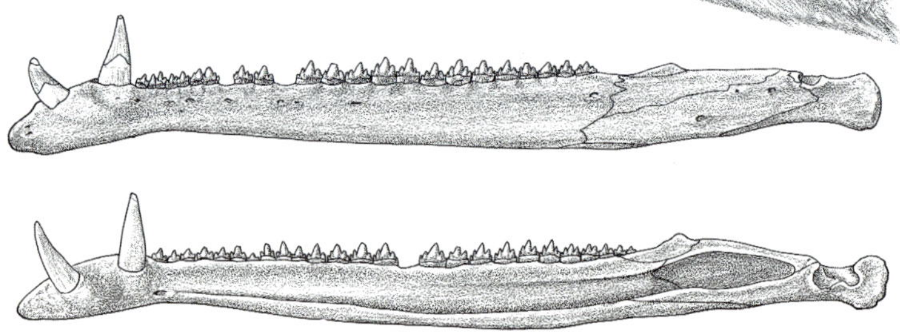

Eudimorphodon Jaw (left)
The lower jaw of *Eudimorphodon ranzii* is seen from the outer side (above) and from the inner side (below). The peculiar dentition consists of large fangs at the front, and small three-cusped and five-cusped teeth along the posterior section of the mandible.

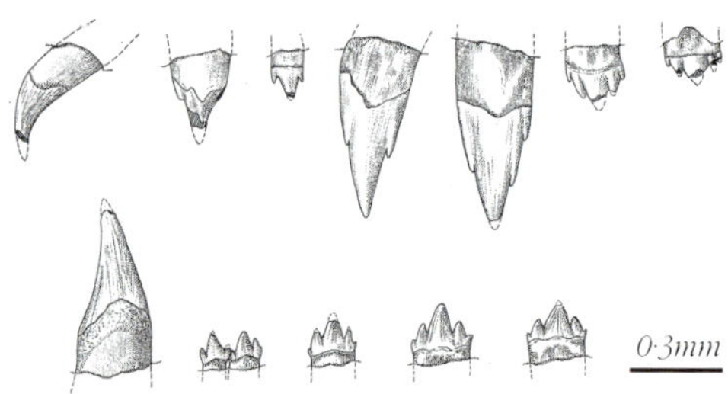

Eudimorphodon's Teeth (right)
In the upper drawing we see some typical teeth from the upper jaw (premaxilla and maxilla); in the drawing below are teeth of the lower jaw (mandible). The tips of the crowns are slightly worn from a diet of hard-scaled fish.

Below: The photograph shows in close-up detail the claws of the small fingers of *Eudimorphodon*. They are quite evidently robust and strong. They were probably used for climbing and for hanging on to tree trunks and steep cliff faces.

10 Wild, R., 1978. *Die Flugsaurier (Reptilia, Pterosauria) aus der Oberen Trias von Cene bei Bergamo, Italien.* Bolletino Società Paleontologica Italiana, 17 (2): 176-256; Modena.

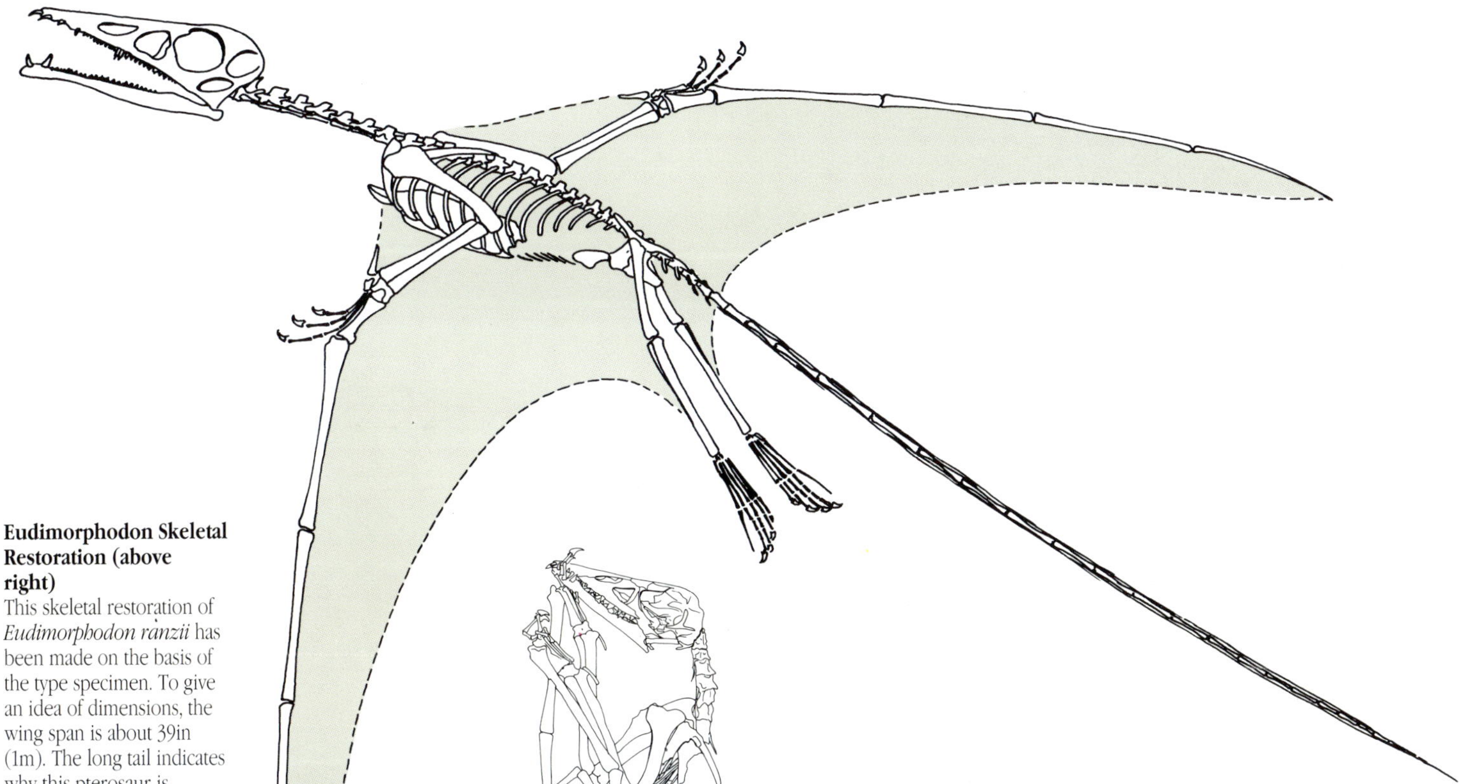

Eudimorphodon Skeletal Restoration (above right)
This skeletal restoration of *Eudimorphodon ranzii* has been made on the basis of the type specimen. To give an idea of dimensions, the wing span is about 39in (1m). The long tail indicates why this pterosaur is assigned to the Rhamphorhynchoidea. The general structure of *Eudimorphodon's* skeleton is entirely characteristic of this group of long-tailed pterosaurs, although its teeth are unique. The multi-cusped architecture of its teeth is not found in any other known pterosaur. The fact that no Jurassic pterosaur has multi-cusped teeth of this nature implies that *Eudimorphodon* cannot be their direct ancestor. Instead it seems to represent a distinct line in the family tree that became extinct in the late Triassic, relatively early in pterosaur history.

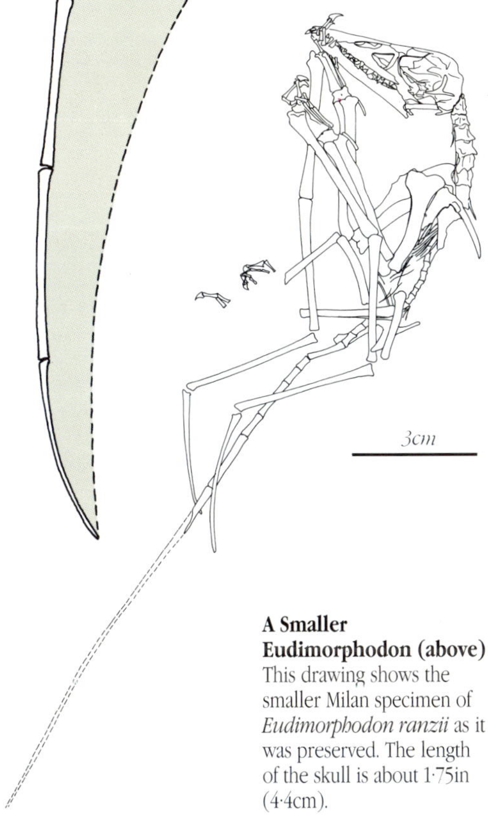

A Smaller Eudimorphodon (above)
This drawing shows the smaller Milan specimen of *Eudimorphodon ranzii* as it was preserved. The length of the skull is about 1·75in (4·4cm).

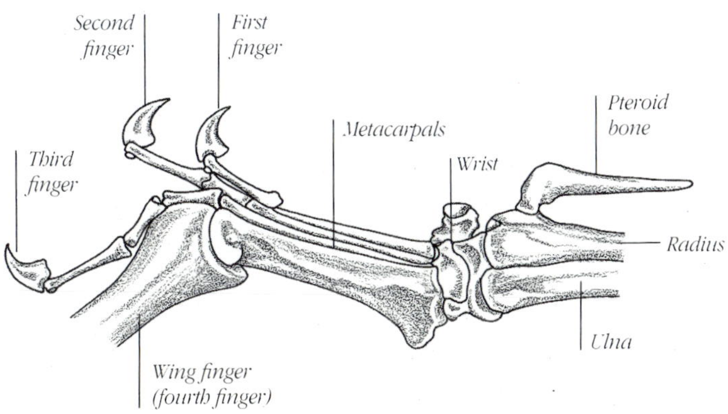

Eudimorphodon's Wrist and Fingers (above)
The wrist, the metacarpals, and the three small fingers (numbers one to three) of *Eudimorphodon* are seen here as preserved in the type specimen. The fourth finger, the wing finger, articulates with the wing metacarpal by means of a distinctive pulley-type joint on its distal end. It was this joint that enabled *Eudimorphodon* to fold its wing back. It is set at a slight angle which caused a twisting movement when the wing was folded, so positioning the flight digit against the body in such a way that the upper side of the wing faced outwards. Two of the claws are also to be seen in the photograph (bottom right of facing page).

Below: A second, but smaller and probably juvenile individual of *Eudimorphodon ranzii* was also found in the Upper Triassic Zorzino limestone of Cene near Bergamo. This specimen is now housed in the Museo Civico di Storia Naturale in Milan.

family Dimorphodontidae, while *Eudimorphodon* is so far the only representative of an independent family, the Eudimorphodontidae.

Preondactylus

Subsequently a third form of Triassic pterosaur came to light. In 1978 a fossil collector working in the Zorzino limestone of Endenna, only 6 miles (10km) from Cene, discovered three connected digital phalanges of a pterosaur wing, which Kevin Padian, a palaeontologist at the University of Berkeley, California traced back to a pterosaur with a wing span of about 4·9ft (1·5m).[11]

11 Padian, K., 1981. *Note of a new specimen of pterosaurs (Reptilia: Pterosauria) from the Norian (Upper Triassic) of Endenna, Italy*. Rivista Museo Civico Scienze Naturale 'E. Caffi', 2 (1980): 119-127; Bergamo.

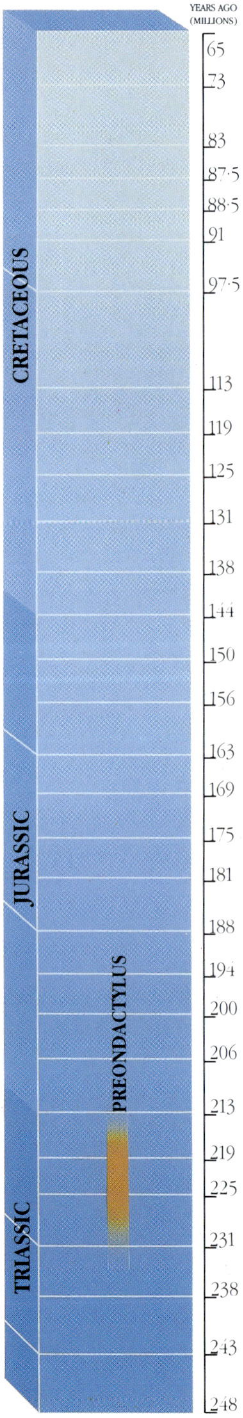

Time Chart (left)
Preondactylus is the third genus of Triassic pterosaur known to date. It is of late Triassic (Norian) age and represents a family, the rhamphorhynchids, otherwise known only from Jurassic strata.

Map (below)
1 *Preondactylus*

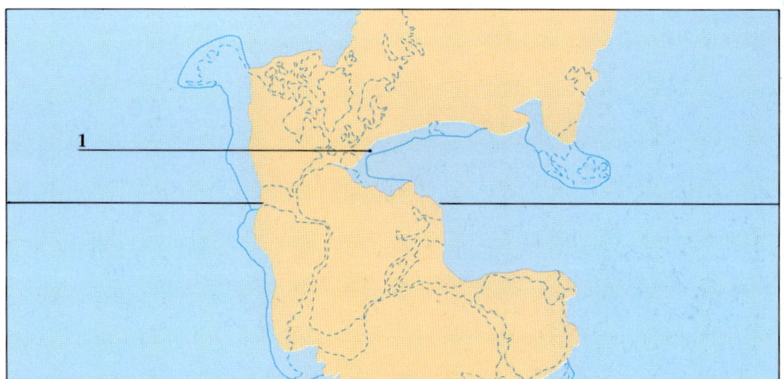

And in 1982 a collector in Udine in northern Italy found another pterosaur skeleton in bituminous, dolomitic limestone of the Upper Triassic (Early to Middle Norian) in the Preone valley in the Alps of the Veneto. Nando Buffarini, the finder, had a stroke of bad luck which is the nightmare of every fossil hunter. The slab of rock containing the valuable fossil shattered into several pieces as it was being extracted. The black bones were embedded in a marl stratum only fractions of an inch thick. When Signore Buffarini and his wife fitted the fragments of the slab of rock together again and washed them with water, the layer of marl was washed away with the bones, and lost. All that remained was the negative print of the skeleton on the surface of the rock. A cast of this negative relief was made with silicon rubber, and only then was the image of this pterosaur skeleton revealed in three dimensions, and in a form that allowed it to be studied.

Once more the investigation was conducted by Rupert Wild of the Museum für Naturkunde in Stuttgart.[12] His analysis enabled him to show that this was a specimen of an additional pterosaur family, hitherto known only in the Jurassic, the Rhamphorhynchidae. Rupert Wild gave the new genus from the Preone valley the name *Preondactylus* (=Preone finger) and established related characteristics with the Liassic pterosaur *Dorygnathus* (=spear-jaw). It is very probable that the three flight digit phalanges from Endenna described by Kevin Padian also belong to the genus *Preondactylus*. The dentition is differently developed from that of *Dorygnathus* from the Lias, consisting of single-cusped teeth. Skeletal proportions are also similar in the two genera. In 1984 one more pterosaur was discovered in the Preone valley of the Veneto Alps in northern Italy. This specimen too came from Upper Triassic dolomitic limestone like *Preondactylus*, but from a level that lay about 500-650ft (150-200m) deeper in the sequence of strata. Thus this fossil find can be dated as the oldest pterosaur yet known in geological history. The fossil consists of an accumulation of disarticulated skeletal bones, which are so tightly packed that they could be interpreted as the gastric pellet of a predatory fish.[13] This creature, probably

12 Wild, R., 1984. *A new pterosaur (Reptilia, Pterosauria) from the Upper Triassic (Norian) of Friuli, Italy*. Gortania, Atti Museo Friuliano di Storia Naturale, 5: 45-62; Udine.
Wild, R., 1984. *Flugsaurier aus der Obertrias von Italien*. Naturwissenschaften, 71: 1-11; Berlin, Heidelberg, New York (Springer).

13 Dalla Vecchia, F.M., Muscio, G. and Wild, R., 1989. *Pterosaur remains in a gastric pellet from the Upper Triassic (Norian) of Rio Seazza Valley (Udine, Italy)*. Gortania, Atti del Museo Friuliano di Storia Naturale, 10 (1988): 121-132; Udine.

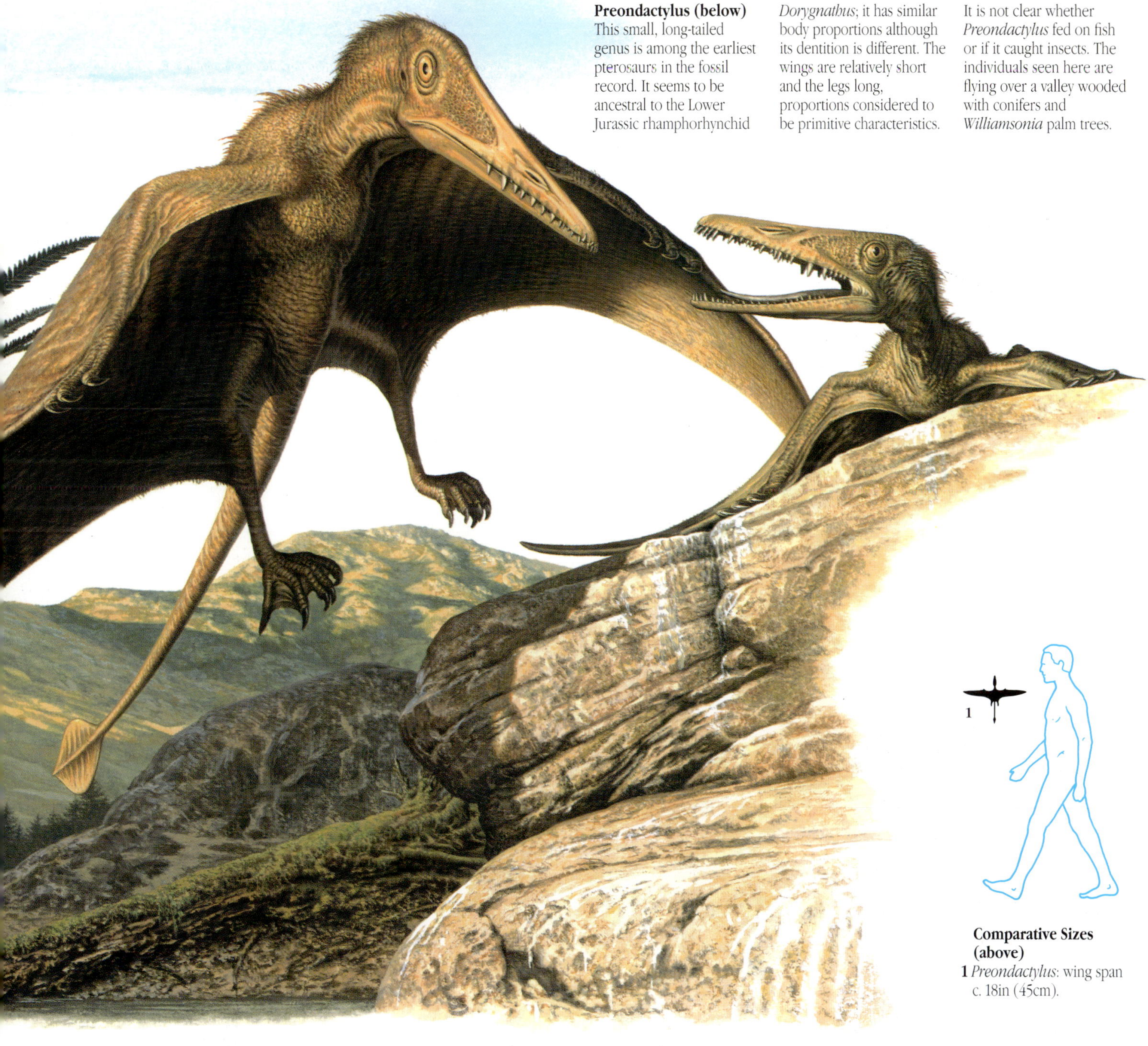

Preondactylus (below)
This small, long-tailed genus is among the earliest pterosaurs in the fossil record. It seems to be ancestral to the Lower Jurassic rhamphorhynchid *Dorygnathus*; it has similar body proportions although its dentition is different. The wings are relatively short and the legs long, proportions considered to be primitive characteristics. It is not clear whether *Preondactylus* fed on fish or if it caught insects. The individuals seen here are flying over a valley wooded with conifers and *Williamsonia* palm trees.

Comparative Sizes (above)
1 *Preondactylus*: wing span c. 18in (45cm).

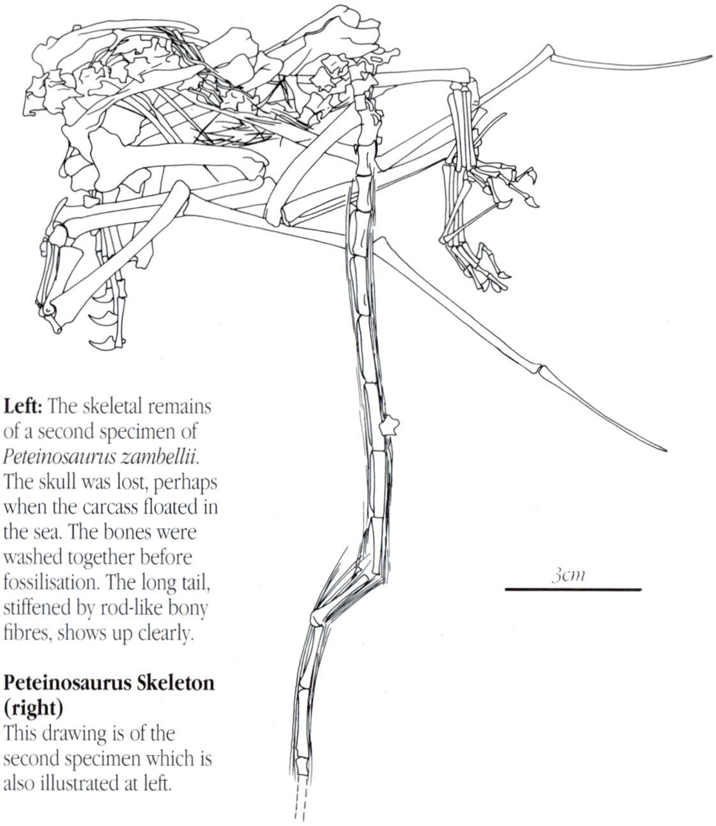

Left: The skeletal remains of a second specimen of *Peteinosaurus zambellii*. The skull was lost, perhaps when the carcass floated in the sea. The bones were washed together before fossilisation. The long tail, stiffened by rod-like bony fibres, shows up clearly.

Peteinosaurus Skeleton (right)
This drawing is of the second specimen which is also illustrated at left.

also a *Preondactylus*, fell prey to a predatory fish over 220 million years ago, and the fish then spewed up the indigestible remains of the skeleton. This gastric pellet sank to the bottom of the sea and was thus covered in ooze and fossilized.

Recently, reports of pterosaur remains from the late Triassic Dockum formation in West Texas suggest that the palaeogeographical distribution of Triassic pterosaurs was probably considerably more extended than hitherto assumed.[14] Although rather fragmentary, the teeth found in jaw fragments are multi-cusped and similar to *Eudimorphodon* from the Norian of Italy. If new finds confirm the pterosaurian nature of these remains, we may regard the pterosaurs as already well established and widespread in the Triassic.

Thus the oldest known pterosaurs occur in three distinct evolutionary lines: Eudimorphodontidae, Dimorphodontidae and Rhamphorhynchidae. Triassic pterosaurs are, as far as we know them, 'completely' developed and have all the typical skeletal characteristics of the order Pterosauria. They are not 'Propterosauria', or forerunners of the pterosaurs, and also not 'missing links' between pterosaurs and their ancestors. We must therefore accept that the evolutionary history of pterosaurs goes back much further into the past than was formerly believed. Perhaps they originated in the Lower Triassic, possibly even in the Permian, thus in the Palaeozoic Era. Certainly no fossils have so far been found that could be interpreted as ancestors of the pterosaurs or 'proto'-pterosaurs

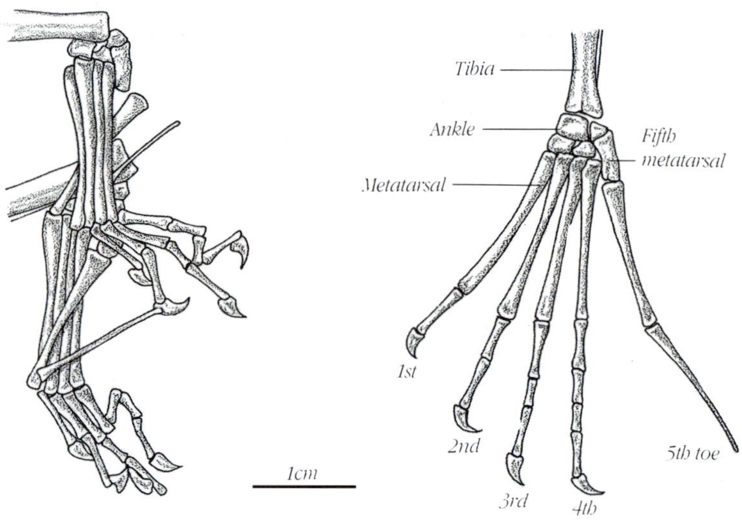

Peteinosaurus' Feet (right)
Seen here are the foot skeletons of *Peteinosaurus* as preserved in the second specimen (left) and as restored (right). The fifth toe is long, has no claw, and could be moved in a different plane from the other four toes. The function of this fifth toe is not clear. The specimens of *Peteinosaurus* are housed in the Museo Civico di Storia Naturale in Bergamo, northern Italy.

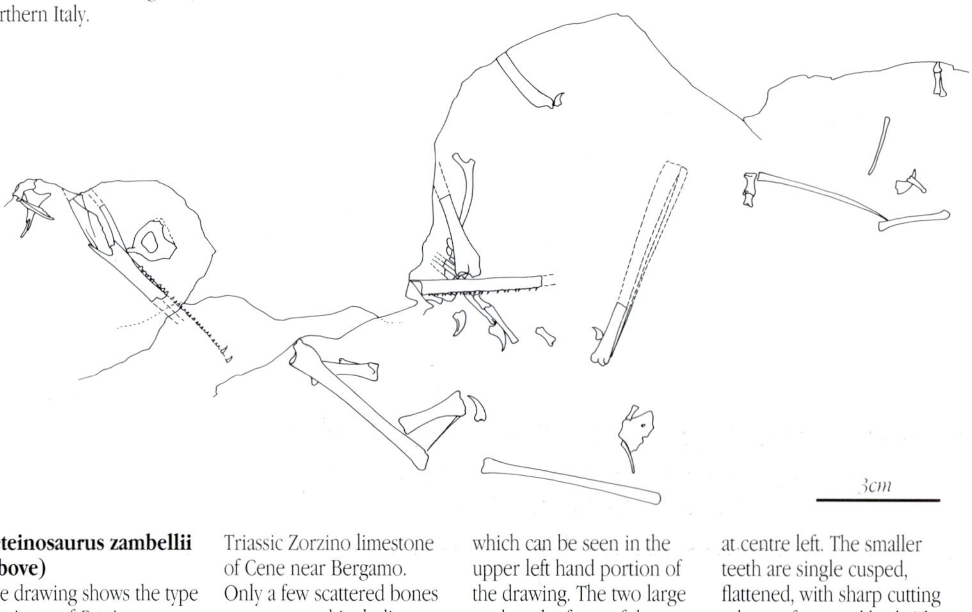

Peteinosaurus zambellii (above)
The drawing shows the type specimen of *Peteinosaurus zambellii* from the Upper Triassic Zorzino limestone of Cene near Bergamo. Only a few scattered bones are preserved including fragments of the lower jaw which can be seen in the upper left hand portion of the drawing. The two large teeth at the front of the lower jaw can be made out at centre left. The smaller teeth are single cusped, flattened, with sharp cutting edges at front and back. The upper skull is lost.

14 Murry, P.A., *Vertebrate paleontology of the Dockum Group, western Texas and eastern New Mexico*. The Beginning of the Age of Dinosaurs (K. Padian, ed.): 109-137; Cambridge University Press.

Left: The lower jaw of *Peteinosaurus* contains numerous small, single-cusped teeth. The jaw fragment is about 1·2in (3cm) long.

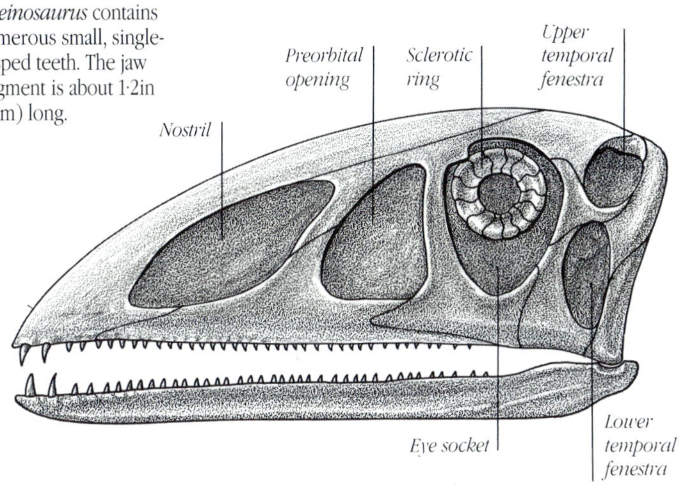

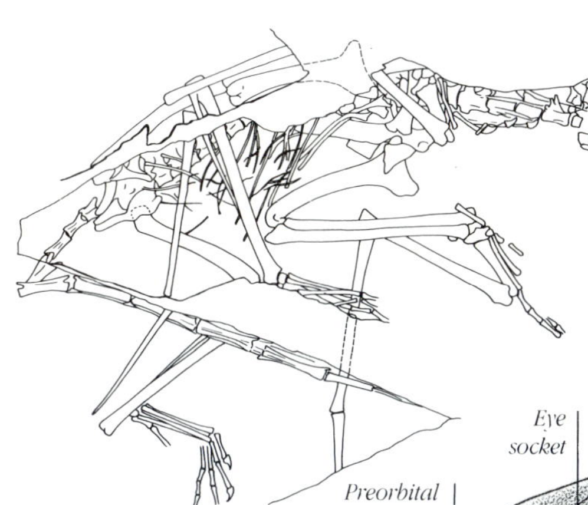

Preondactylus (left)
This is *Preondactylus buffarinii* from the Upper Triassic of the Preone valley near Udine. By misfortune only the impressions of the skeletal remains could be saved by the fossil collector, Nando Buffarini.

Peteinosaurus Skull (above)
This is a tentative restoration of the skull of *Peteinosaurus zambellii*. The shape of the skull is assumed to be similar to the Lower Jurassic genus *Dimorphodon*.

Untimely End (below)
This restoration shows a predatory fish, in this instance *Saurichthys*, about to consume an unwary *Preondactylus*. The indigestible bones of the pterosaur were vomited as a pellet by the predator.

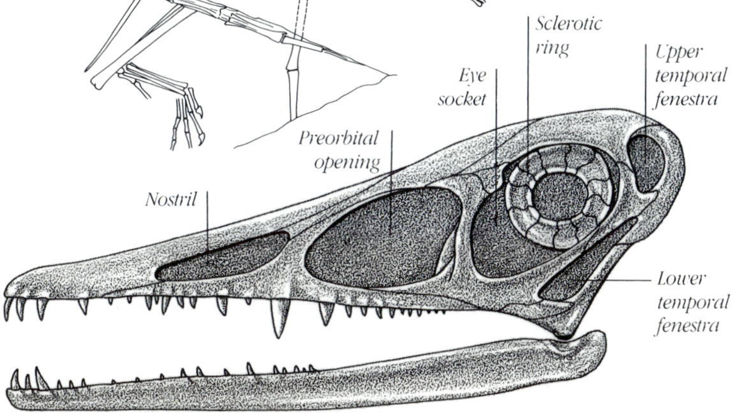

Preondactylus Skull (above)
This is a tentative restoration of the skull of *Preondactylus buffarinii*. The posterior section of the skull was not preserved and is assumed here to have been similar to the Jurassic *Dorygnathus*.

Gastric Pellet (below)
This accumulation of bones has been interpreted as a gastric pellet spewed up by a predatory fish. The pigeon-sized pterosaur, *Preondactylus buffarinii*, may have been caught when it was itself fishing, or after it had drowned.

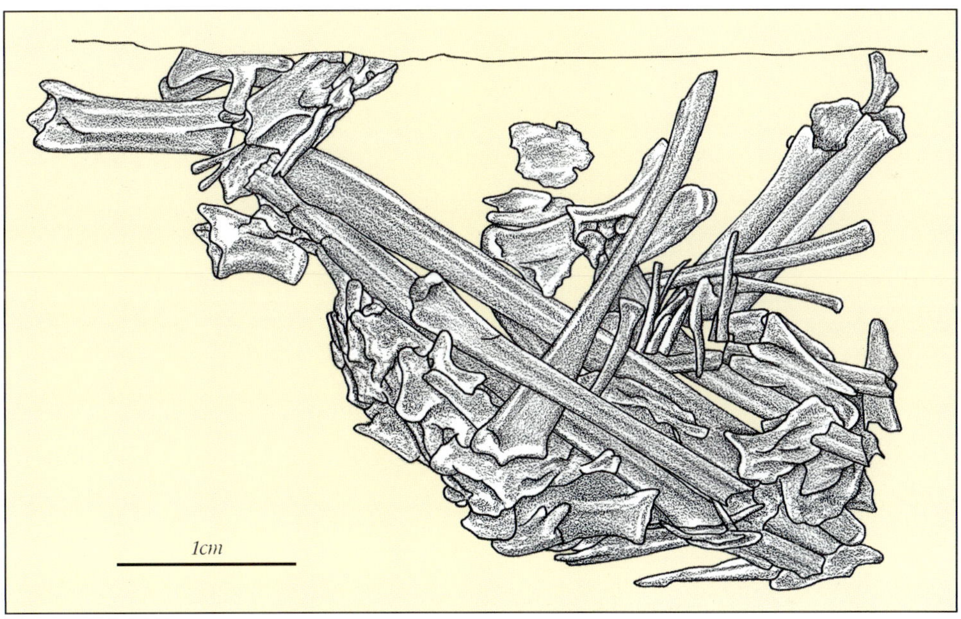

SUMMARY OF TRIASSIC PTEROSAURS

Rhamphorhynchoidea
Family Dimorphodontidae
Peteinosaurus zambellii
Upper Triassic, Middle to Upper Norian, Calcare di Zorzino, Cene near Bergamo, Italy.[10]
Family Eudimorphodontidae
Eudimorphodon ranzii
Upper Triassic, Middle to Upper Norian, Calcare di Zorzino, Cene near Bergamo, Italy.[10]
Eudimorphodon? sp.
Upper Triassic, Dockum Group, West Texas, USA.
In 1986, fragments of a lower jaw with two five-cusped teeth and of an upper jaw with multi-cusped teeth were reported from the Dockum Formation in West Texas.[14] The assignment of these jaw fragments must remain uncertain until more complete material is found
Family Rhamphorhynchidae
Preondactylus buffarinii
Upper Triassic, Early to Middle Norian, Dolomia di Forni, Madonna Peraries, Preone valley, province of Udine, Italy.[12]
?*Preondactylus* sp.
Upper Triassic, Middle to Upper Norian, Calcare di Zorzino, Endenna near Bergamo, Italy.[11,12]

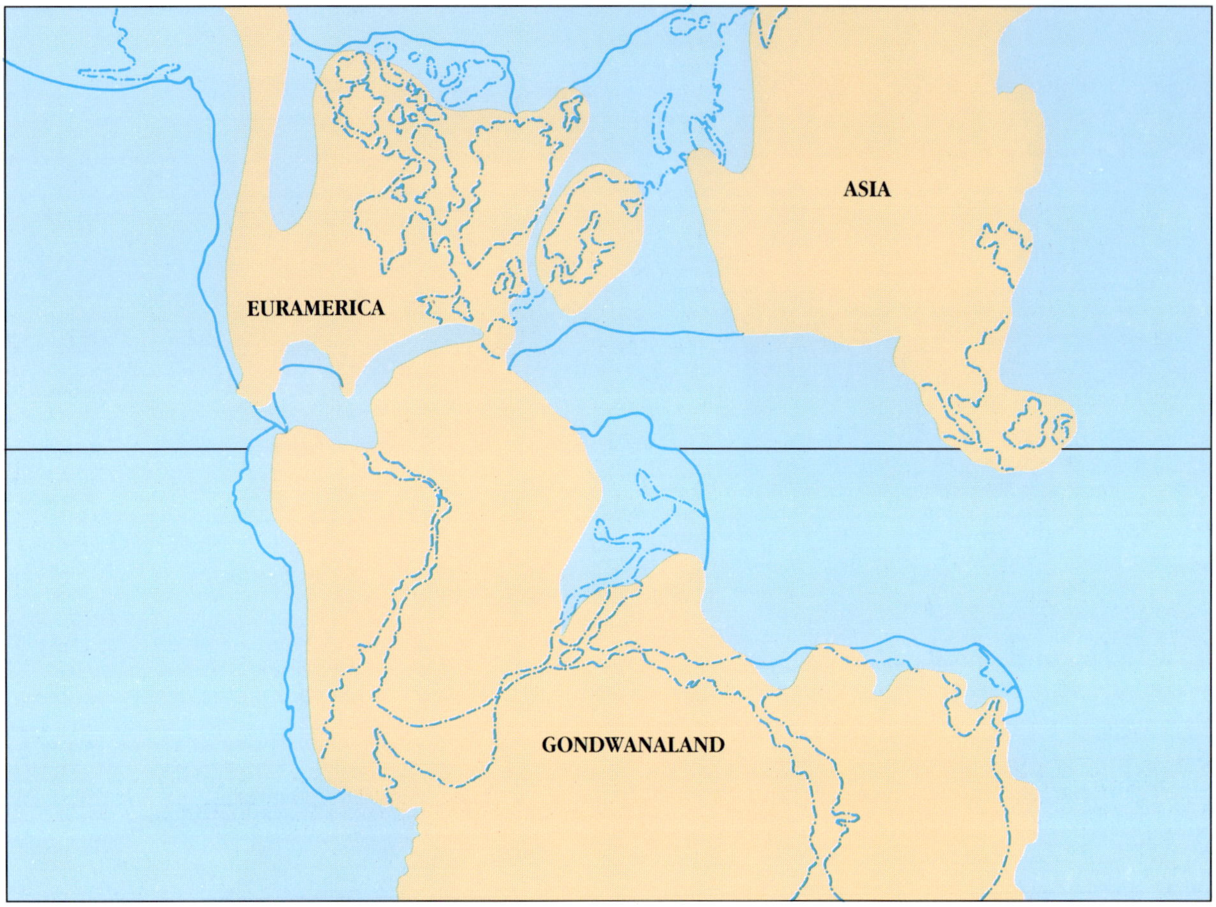

Palaeogeographic Map of the Jurassic Period (left) During the Triassic Period all the continents had been joined together to form the supercontinent of Pangaea. This began to split in the Jurassic Period as the sea flooded in and the Atlantic Ocean began to form. A sea also separated Europe from Asia. Despite this, as the map shows large land masses still existed in the form of Euramerica and Gondwanaland, which contained today's South America, Africa, India, Australia and the Antarctic. In the Jurassic Period fossil evidence shows that pterosaurs were distributed worldwide, and it is probable that they spread out from Europe where the oldest Triassic pterosaurs currently known have been found.

Below: A drawing (after Owen) of the skull of *Dimorphodon macronyx* as preserved from the Lower Lias of Lyme Regis, Dorset. The length of this skull is 8·5in (21·5cm).

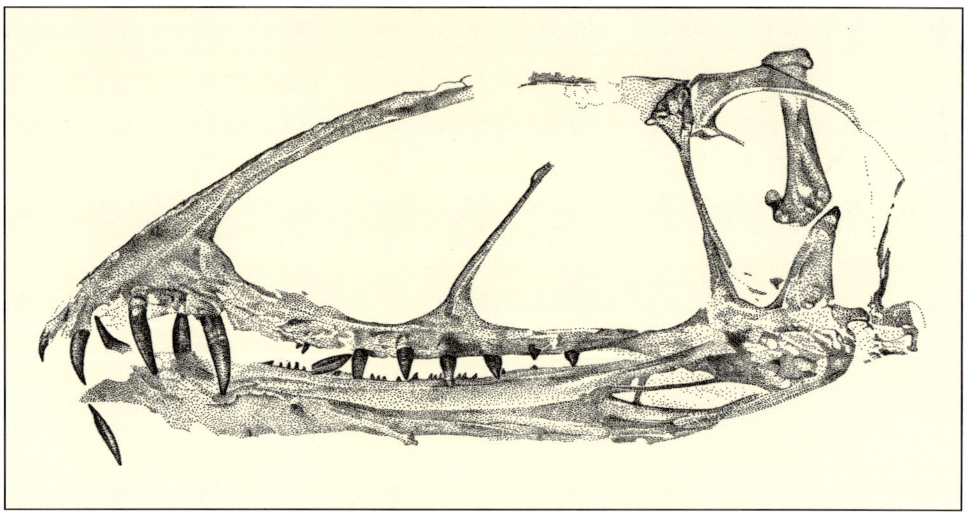

The middle division of the Mesozoic era is the Jurassic period, originally named after the Jura mountains in Switzerland. It followed the Triassic period and began about 213 million years ago. It ended 144 million years ago, and so lasted about 69 million years, or twice as long as the Triassic period.

The Jurassic is generally subdivided into Lias (Lower Jurassic), Dogger (Middle Jurassic) and Malm (Upper Jurassic). A major feature of the Jurassic period was extensive flooding of continents by the sea. For this reason finds of organisms that lived on land are particularly to be anticipated in near shore marine deposits. This is especially true of pterosaurs. The Atlantic Ocean began to open up in the Middle Jurassic. Despite this the map of the Earth still contained large, connected land masses, in the form of the supercontinents Euramerica and Gondwanaland, containing the modern continents of South America, Africa, India, Australia and the Antarctic. Asia was separated from Europe by a great seaway. The climate began to be differentiated in the course of the Jurassic. There was a wide tropical belt and only slightly cooler polar regions, but in comparison with today the Earth's climate was still warm and balanced. There were no polar ice caps as yet, and no distinct seasons.

So far Triassic pterosaurs have only been proved to exist in Europe, while it is probable that they already existed in North America. In the Jurassic, however, pterosaurs were distributed world-wide, and conditions were favourable for fossilization in numerous Jurassic 'fossil deposits' like Solnhofen in Bavaria, for example, or Holzmaden in Württemberg (southern Germany). We are familiar with pterosaur fossil finds from Jurassic deposits in Europe, Africa, Asia, North America, Central America and South America. The Lower Jurassic period saw pterosaurs spread all over the world, probably emanating from Europe, and also the development of various forms of pterosaur. Palaeontologists call this evolutionary process adaptive radiation: development in the course of the phylogeny showing adaptation to different life and environmental conditions.

In fact the first pterosaur finds came from Jurassic Solnhofen lithographic limestone: as we saw in Chapter 2, Cosimo Alessandro Collini described a fossil find from these strata as early as 1784.[1] However, the Solnhofen pterosaurs lived near the end of the Jurassic, about 150 million years ago. *Pterodactylus* appeared here as the first short-tailed pterosaur in the fossil record, while the last long-tailed pterosaurs on Earth existed at the same time in the same strata.

Pterosaurs of the Lias

The oldest Jurassic pterosaurs were discovered in the Lower Lias, in England. To the east of the little town of Lyme Regis in Dorset on the south coast of England are the Church Cliffs, a massive series of blue-grey limestone alternating with shale. English geologists call these marine strata, which are about 85ft (26m) thick, 'Blue Lias Limestone'. They also extend west of the town beyond Pinhay Bay, forming enormous cliffs.[2] W.J. Arkell describes it like this, 'Almost the entire succession of the Lower Lias in its typical development across England from the Dorset coast consists of clays and shales, in the lower part of which numerous bands of calcareous mudstone or clay-limestone have been formed by secondary chemi-

1 Collini, C.A., 1784. *Sur quelques Zoolithes du Cabinet d'Histoire naturelle de S.A.S.E. Palatine et de Bavière à Mannheim.* Acta Academiae Theodoro-Palatinae Mannheim, 5, pars physica: 58-103.

2 Arkell, W.J., 1933. *The Jurassic System in Great Britain.* Clarendon Press, Oxford.

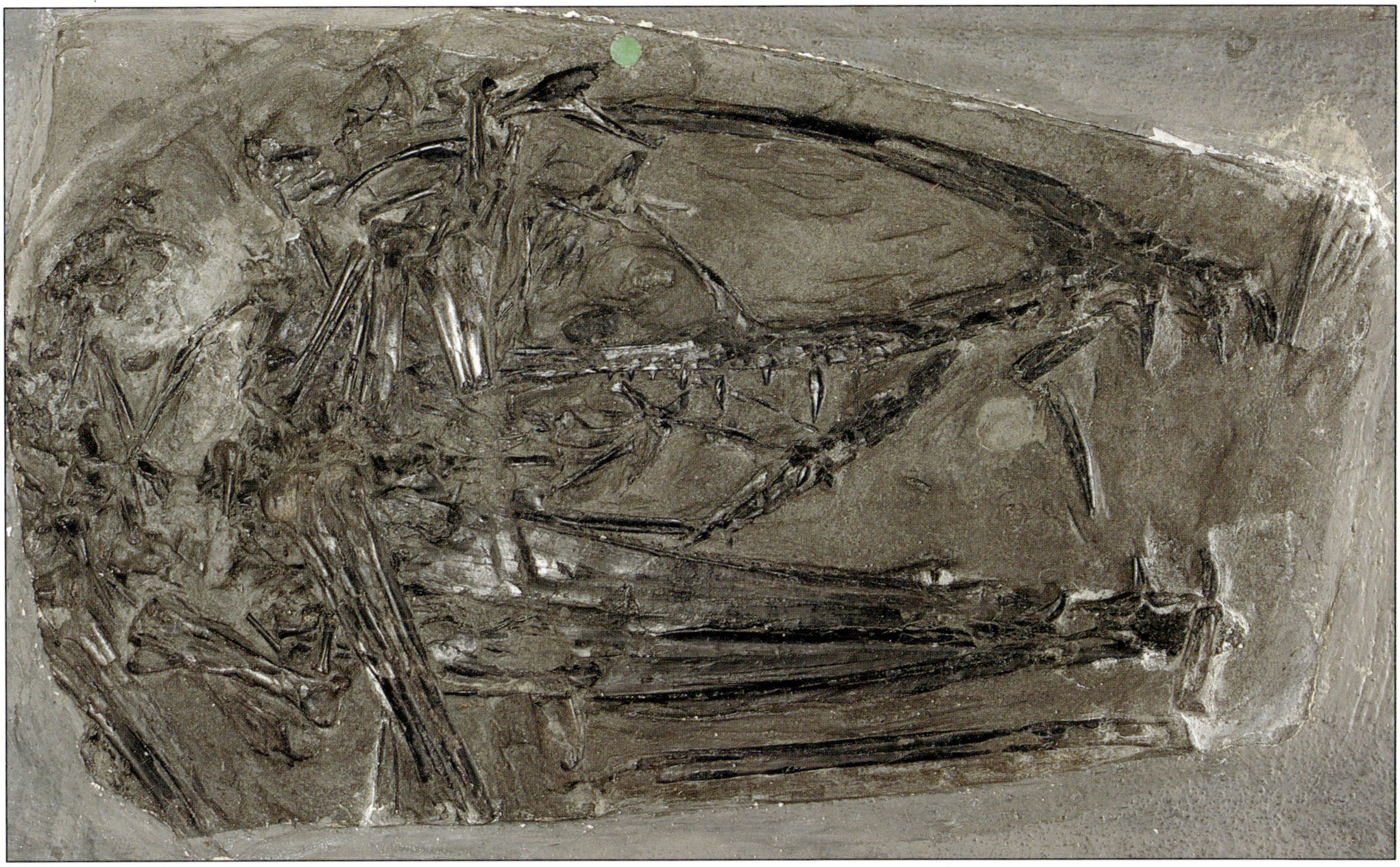

Above: *Dimorphodon macronyx*, a second specimen from Lyme Regis with the skull and lower jaw preserved. This specimen, which is now in the Natural History Museum in London, shows the large nasal and preorbital openings separated by thin bony struts which were a weight saving feature. The lower jaw is filled with a series of numerous tiny teeth, while the upper jaw contains larger teeth. The dual nature of the dentition in this genus gives rise to its name, *Dimorphodon* ('two form tooth').

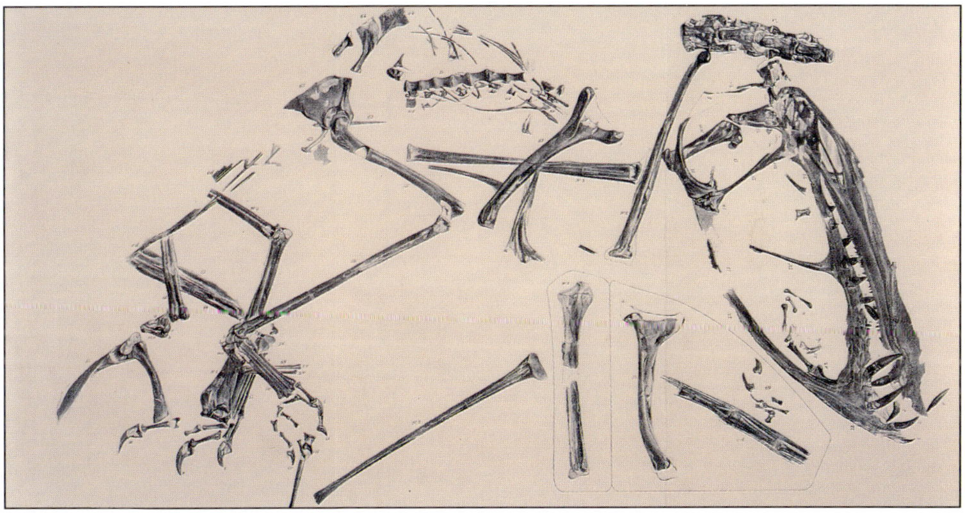

Right: These are the skeletal remains of *Dimorphodon macronyx* which Richard Owen received in 1858, as pictured in Owen's monograph of 1870. Again they are from the Lower Lias of Lyme Regis. Unlike previous finds, such as that acquired by William Buckland from Mary Anning, for the first time this skeleton had its skull preserved. It was so different from the late Jurassic *Pterodactylus* that Owen did not hesitate to establish a distinct genus for the Lyme Regis pterosaur.

cal processes . . . The most conspicuous remains are the skeletons of great marine reptiles. The presence of these, and also of fossil insects, points to deposition, not in deep water as frequently deduced from the clayey sediment, but in a comparatively shallow continental sea. The rivers flowing into the Liassic sea had reached maturity, for instead of bringing down detrital material from the surrounding land, they carried only fine mud. Mixed with the sand in suspension came a high proportion of iron, which, in the form of minutely disseminated pyrites, gave the usual dark grey to black colour to the Lias shales.' (p.120).

The first professional fossil collector, Mary Anning, worked intensively and successfully as a collector here in the first half of the nineteenth century.[3] Her most famous finds were skeletons of great marine reptiles, ichthyosaurs and plesiosaurs, then in 1828 came the first pterosaur of the Lias, acquired by William Buckland, an Oxford professor, who described it in 1829 as *Pterodactylus macronyx*.[4] The name refers to the great claws on the small digits of the hand. At first, however, the skull of this pterosaur was not known. It was not until 1858 that Richard Owen of the British Museum in London received more pterosaur material from the Lias of Lyme Regis, including remains of skeletons with skulls, belonging to the same species. Owen immediately established that the skull of the genus *Pterodactylus*, formerly known only from the Upper Jurassic of Solnhofen, was very different. He therefore called the Lyme Regis pterosaur *Dimorphodon* (=two-form tooth), a reference to the two kinds of teeth in the dentition of this genus.[5] As well as the very rare pterosaurs – so far only the species *Dimorphodon macronyx* is known – there are numerous fish and many ammonites in the Lyme Regis Blue Lias, organisms that show these strata to be former marine deposits. Geologists place them in the Sinemurian age, on the basis of the internationally valid time scale (a diagram explaining this time scale can be found on page 16 in Chapter 1). They have an absolute age of about 205 to 200 million years.

3 Mary Anning (1799-1847) from Lyme Regis earned her living by selling fossils she collected from Blue Lias strata on the coast near her home town. She found the first skeletons of ichthyosaurs and plesiosaurs in England. In 1828 she found the remains of a pterosaur skeleton in the Blue Lias: this was the first pterosaur to be found in England.

4 Buckland, W., 1829. Proceedings of the Geological Society, 1: 127.

5 Owen, R., 1865 and 1870. *A monograph on the fossil reptilia of the Liassic formations.* Palaeontographical Society London.
Owen, R., 1874. *Monograph of the fossil reptilia of the Mesozoic Formations. I. Pterosauria.* Palaeontographical Scoiety London.

Dimorphodon

Dimorphodon was a pterosaur about 3·3ft (1m) long overall with a maximum wing span of 4·6ft (1·4m). It had a relatively large, high skull with large side apertures. These 'windows', consisting of eye sockets, upper and lower temporal openings, preorbital openings and nostrils, were separated only by thin bars of bone. Thus despite its size the skull was very lightly built.

Dimorphodon had four large front teeth on each side of the upper jaw. Behind this was a row of smaller teeth. In the lower jaw as well four or five large front teeth were followed on each side by 30 to 40 tiny but pointed teeth. This specialization suggests that the creature was a fish eater. As a typical long-tailed pterosaur, *Dimorphodon* had a long vertebral tail made up of over 30 caudal vertebrae. The first five or six of these vertebrae were short, and could move against each other. Subsequent

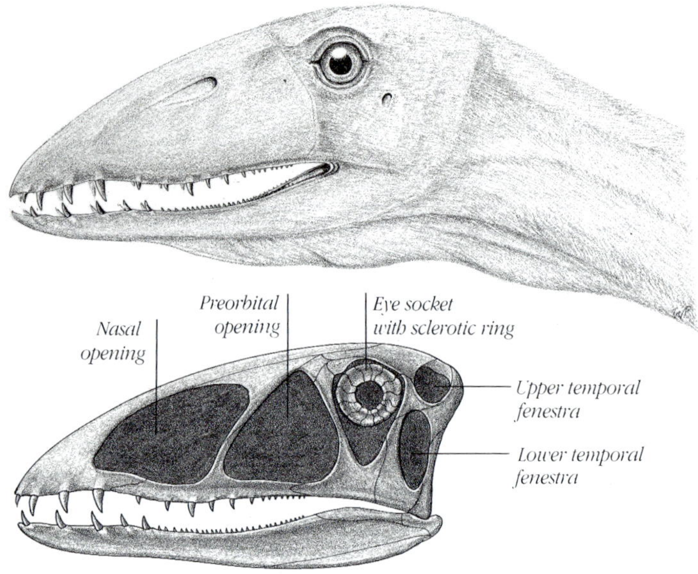

Dimorphodon's Head and Skull (left)
The life portrait of *Dimorphodon's* head seen at upper left is based on the fossil skulls from the Lower Lias of Lyme Regis. A restoration of one such skull is seen at lower left. The thin struts of bone separating the skull openings all helped to reduce the body weight that this creature needed to get airborne. The bony ring in the eye socket protected the eye against deformation. Note also the differently shaped teeth in the upper and lower jaws. These suggest that *Dimorphodon* was a fish eater.

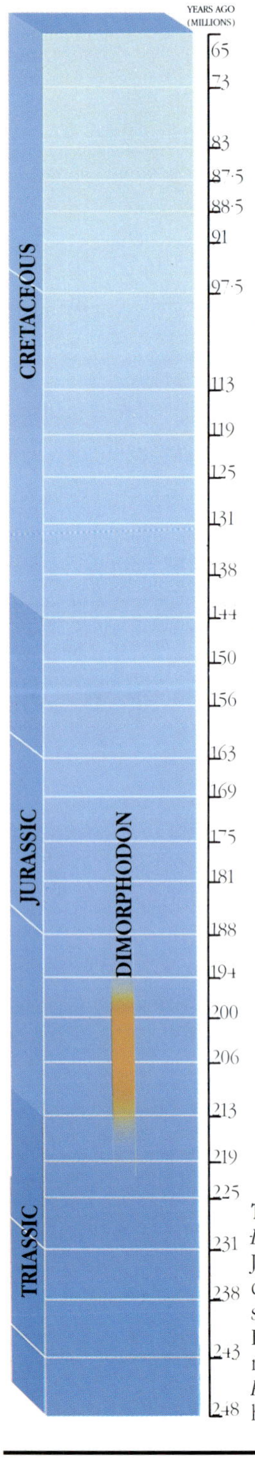

Time Chart (left)
Dimorphodon is the earliest Jurassic pterosaur, discovered in early Liassic strata in Southern England. It is a long-tailed form, related to the Triassic *Peteinosaurus* which may have been its ancestor.

Dimorphodon (below)
This early Jurassic pterosaur had a relatively large head and long hind legs, but short wings with a span of only about 4·6ft (1·4m). The finger claws were quite strong and could be used for climbing on rocks and cliffs. *Dimorphodon's* deep snout is similar to the high beak of a puffin, and the specialization of its teeth suggest that it also was a piscivore. The long tail was largely stiffened, as is characteristic of all rhamphorhynchoid pterosaurs. It was probably used as a drag rudder. The extent of the wing membranes and the shape of the terminal tail vane can only be tentatively reconstructed here, since no soft part impressions have been preserved with any of the fossil skeletons of this pterosaur so far discovered.

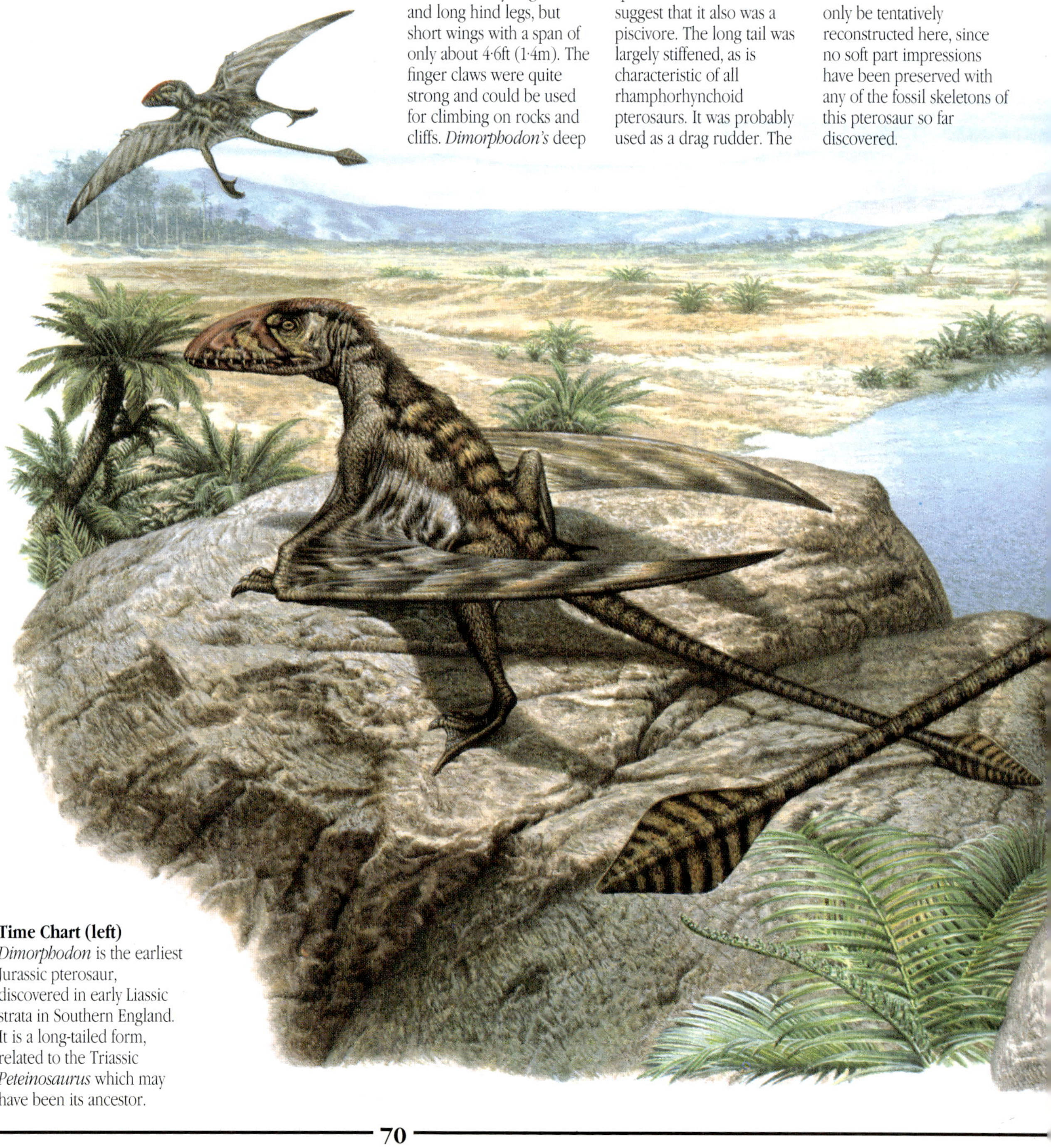

caudal vertebrae were increasingly elongated, and were stiffened against one another by long, thin vertebral processes. This structure of a long, stiff tail, only articulated at the beginning, was linked with its function as a means of stabilizing the pterosaur in flight, which was further emphasized by the fact that there was a small vane at the end that was presumably used as a drag rudder. Further discussion of the function of this vane will be found in the 'Life Style' chapter.

Dimorphodon's wings were still relatively short, a primitive characteristic. It is striking that the first of the four flight digit phalanges is only a little longer than the lower arm, and shorter than the second and third phalanges of the flight digit. The hind legs were extraordinarily powerfully developed and relatively long. The first four digits had claws; the fifth digit was fairly long and splayed sideways. A precise analysis of the structure of pelvis and hind leg led Kevin Padian to assume that *Dimorphodon* was well suited to biped, bird-like walking on the

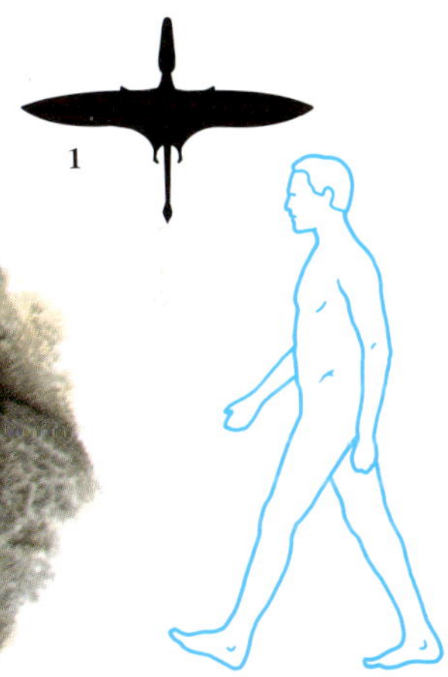

Comparative Sizes (above)
1 *Dimorphodon*: wing span 4·6ft (1·4m)

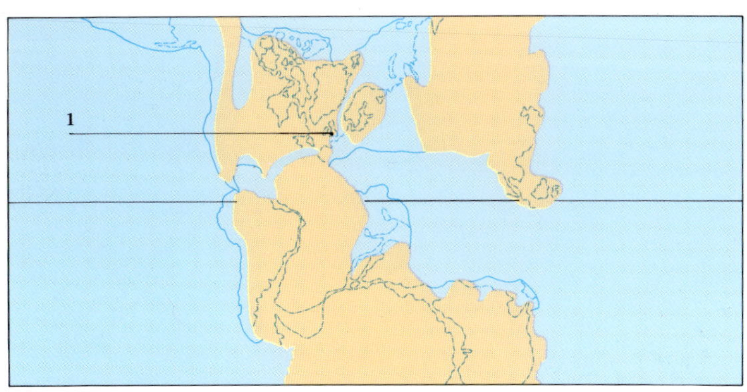

Map (right)
1 *Dimorphodon*

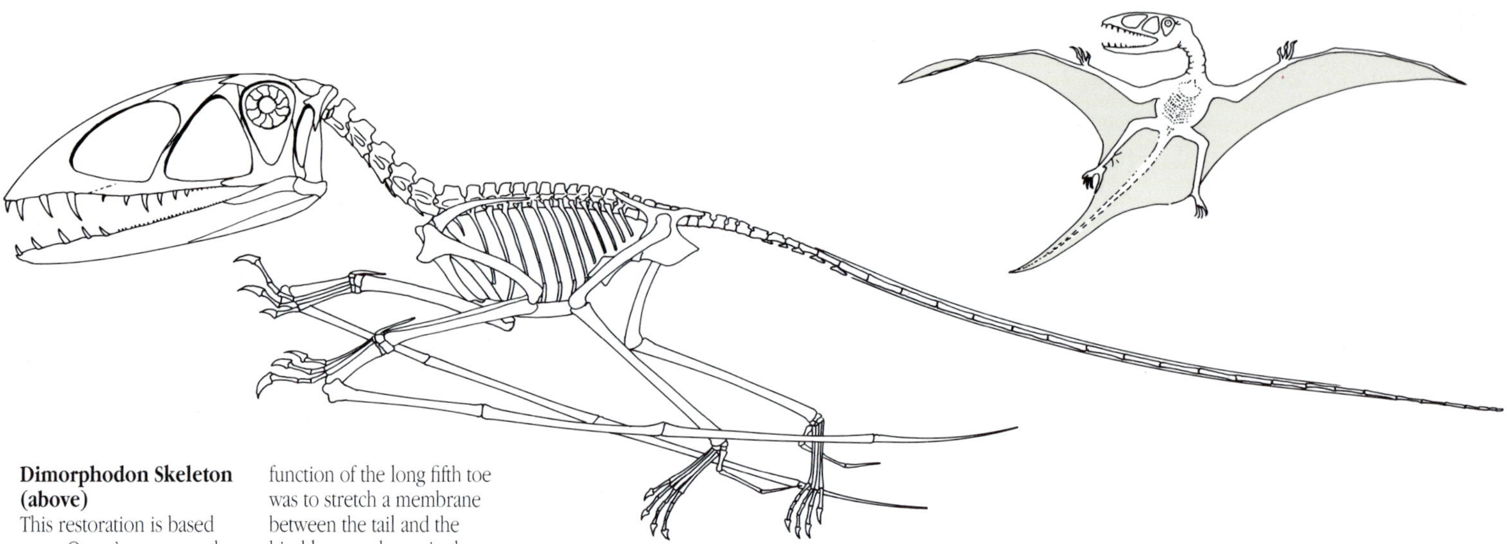

Dimorphodon Skeleton (above)
This restoration is based upon Owen's monograph. Owen thought that the function of the long fifth toe was to stretch a membrane between the tail and the hind legs, as shown in the drawing at top right.

ground, the long tail balancing the large head.[6] These problems of locomotion will be examined in greater detail in a later chapter.

Only a few specimens of *Dimorphodon* have been found, mainly incomplete skeletal remains. They all belong to the same species, *Dimorphodon macronyx*, and come from the Lower Lias of the Dorset coast in England. A single specimen, only a few bones of the flight arm and the hind legs, was found in the Lower Lias of Aust Cliff on the south bank of the Severn in Gloucestershire. With the exception of the Aust Cliff specimen, which was acquired for the Yale Peabody Museum, USA by O.C.Marsh from the London fossil dealer Bruce M. Wright in 1881, all the specimens are now in the Natural History Museum in London.

German Finds

The first Liassic pterosaur finds on the European mainland were made only a year after William Buckland had introduced the Lyme Regis pterosaur. This time finds were made in Upper Lias strata near the former Benedictine monastery of Banz, high above the valley of the Main near Staffelstein in Upper Franconia in Bavaria. The Lias rocks around the monastery were known for their skeletal remains of ichthyosaurs and sea crocodiles. Isolated pterosaur bones and jaw fragments were found by ducal chancellery councillor Carl Theodori in 1830.[7] He saw the fossil remains as a new species, which he first called *'Ornithocephalus' banthensis*. Later Andreas Wagner, Professor of Palaeontology in Munich, established that it must also be a new genus, which he called *Dorygnathus*.[8] The name comes from the Greek *dorys* (=spear) and *gnáthos* (=jaw), from the toothless, lance-shaped point of the jaw.

More complete *Dorygnathus* skeletons were later found, principally in Württemberg (southern Germany). As early as the late sixteenth century fossils had been found in the area around Bad Boll near Göppingen. In the mid eighteenth century, when the so-called Posidonian shale was mined, the first fossil vertebrates, ichthyosaurs and marine crocodiles, came to light. There were also quarries in the area near Holzmaden, Ohmden and Zell, small places in the northern foothills of the Schwäbische Alb, dominated by the towering Castle Teck.

Few of the quarries in this region are still worked today. Formerly roofing slate and flags for floors were mined here, but nowadays the decorative black slate slabs are only used for interior decoration. Here commercial mining is limited to a single stratum, the so-called Fleins, a layer 7in (18cm) thick, which gives three very firm slabs when split. This Fleins is up to 40ft (12m) below the surface of the site, so that some slate mines go down as deep as this. All the strata above this are technically unusable, although they still contain fossils, and find their way on to the spoil heaps.

Fossils are distributed throughout the profile of strata. Posidonian shale takes its name from a small frequently-occurring shell called *Posidonia bronni* (now called *Bositra parva*). Because of its bitumen content, many attempts have been made to extract oil from the rock, but the oil content is only 3-8 per cent, so the process is uneconomic.

Fossil content, especially ammonites as index fossils, mean that the Holzmaden Posidonian shale can be categorized as Lias epsilon, or the Lower Toarcian, according to the internationally accepted standard. The deposits came into being about 190 million years

A Bipedal Pterosaur? (above)
This drawing shows *Dimorphodon* running along on its two hind legs, with its long tail counter-balancing its head. Dr Kevin Padian's work on the pelvis and hindlimbs of this pterosaur has led him to argue that *Dimorphodon* was well suited to this form of locomotion, although his hypothesis is disputed.

6 Padian, K., 1983. *Osteology and Functional Morphology of Dimorphodon macronyx (Buckland) (Pterosauria: Rhamphorhynchoidea) Based on New Material in the Yale Peabody Museum.* Postilla, 189: 1-44; New Haven.
Padian, K., 1983. *A functional analysis of flying and walking in pterosaurs.* Palaeobiology, 9 (3): 218-239; Chicago.

7 Theodori, C., 1830. *Knochen vom Pterodactylus aus der Lias-Formation von Banz.* Froriep's Notizen für Natur und Heilkunde, 632: 101.
Theodori, C., 1831. *Über die Knochen vom Genus Pterodactylus aus der Lias-Formation von Banz.* Isis, p.277; Jena.
Theodori, C., 1852. *Über Pterodactylusknochen im Lias von Banz.* 1. Bericht des Naturforschenden Vereins Bamberg, 17-44; Bamberg.
On the subject of Carl Theodori and the Banz collection see also the chapter 'The History of Fossil Finds' and the accompanying footnote reference, no. 18.

8 Wagner, A., 1860. *Bemerkungen über die Arten von Fischen und Sauriern, welche im unteren wie im oberen Lias zugleich vorkommen sollen.* Sitzungsberichte der Bayerischen Akademie der Wissenschaften, math. – physikalische Classe, 36-52; Munich.

Above: *Dorygnathus banthensis* from the Upper Lias of Holzmaden in Württemberg, Germany. This complete specimen of the Naturkundemuseum in Stuttgart comes from the Posidonian shale (*Posidonienschiefer*) and nicely shows the long, stiff tail, and the slender bones of the wing finger. The length of the skull is 3·8in (9·8cm).

Above: Dr Bernhard Hauff (1866-1950), the pioneer of fossil excavations in the Upper Liassic Posidonian shales around Holzmaden in Württemberg. His skilful preparations of ichthyosaurs, plesiosaurs, fishes, and pterosaurs from the quarries nearby became world famous, and his fossil specimens are housed in many collections, as well as in the fine Museum Hauff in Holzmaden itself.

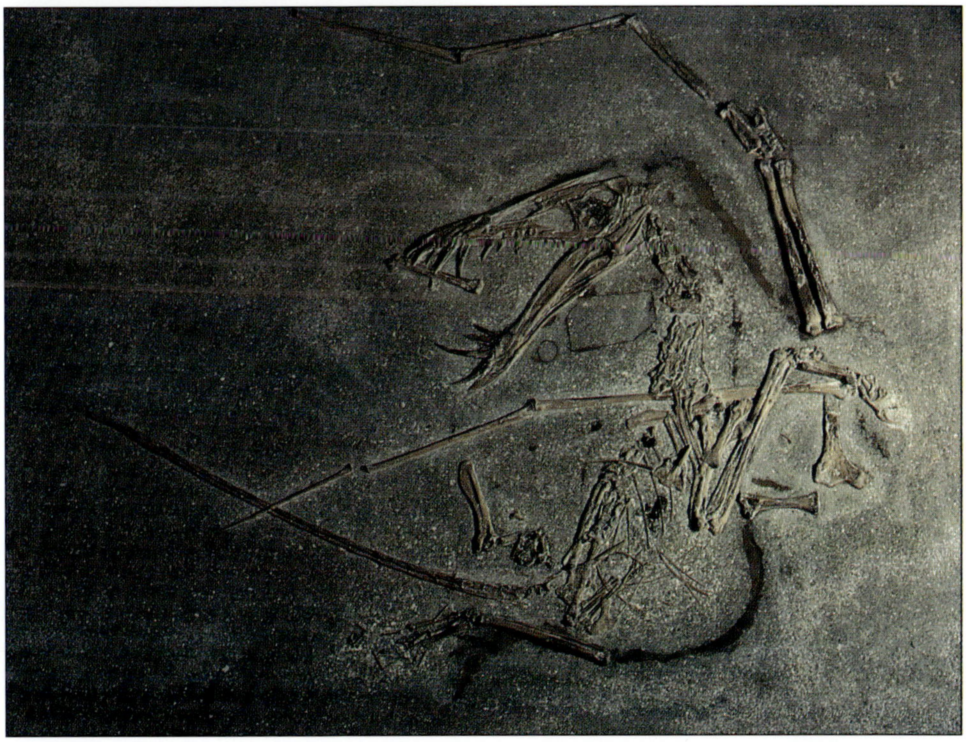

Below: This is another specimen of *Dorygnathus banthensis* from the Upper Lias of Holzmaden. The skull of this creature is particularly well preserved and shows clearly the formidable front teeth of this pterosaur which were used to grasp fish from the water. The skull length is 5in (12·8cm). This specimen is housed in the museum collection of Uppsala in Sweden.

ago in an extended shallow sea that flooded broad areas of central Europe in the Lias. The area of modern Holzmaden must at the time have lain about 60 miles (100km) north-west of a large island, the Vindelician Land. The eastern limit of the Liassic sea was the coast of the Bohemian mainland, in the centre was the Ardennes-Rhenish island, the east coast of which was roughly on the line of modern Bielefeld-Marburg-Frankfurt.

Reptile finds from Holzmaden have become famous, and are to be found in many of the world's museums today. Ichthyosaurs are particularly abundant, but plesiosaurs, marine crocodiles, and even the remains of a dinosaur have also been found. There are also many fish, ammonites, squids, bivalves, sea lilies and crabs. Most of the fossils are in excellent condition. The great expert on Holzmaden fossils was Dr Bernhard Hauff. He began his careful preparations around 1890, and opened a small museum of his finds in Holzmaden in 1937. This has been considerably extended and modernized, and still attracts a large number of visitors.[9]

The first pterosaur fossil finds from the Posidonian shale of Württemberg were described by Albert Oppel in 1856,[10] but more complete skeletons were not found until the late nineteenth century. Numerous outstanding pterosaurs from the slate quarries around Holzmaden and Ohmden are known today, representing two genera, *Dorygnathus* and *Campylognathoides*.[11] The area was declared a protected area for excavation by the state in

9 Hauff, B. and Hauff, R.B., 1981. *Das Holzmadenbuch.* 136 p.; Holzmaden/ Teck.
Urlichs, M., Wild, R. and Ziegler, B., 1979. *Fossilien aus Holzmaden.* Stuttgarter Beiträge zur Naturkunde, C, 11: 34 p.; Stuttgart.

10 Oppel, A., 1856. *Die Juraformation.* Jahreshefte des Vereins für Vaterländische Naturkunde in Württemberg; 12.

11 Plieninger, F., *Die Pterosaurier der Juraformation Schwabens.* Palaeontographica, 53: 209-313; Stuttgart.
Arthaber, G. von, 1919. *Studien über Flugsaurier auf Grund der Bearbeitung des Wiener Exemplares von Dorygnathus banthensis Theod.sp.*. Denkschriften der Akademie der Wissenschaften Wien, math.-naturwiss. Klasse, 97: 391-464; Vienna.

1979, and can be considered the principal area for Liassic pterosaur finds. Other sites where only a few fragmentary pterosaur remains were found are the area around Bayreuth in Upper Franconia (Bavaria)[12] and Braunschweig in Lower Saxony.[13]

Dorygnathus (=spear-jaw) is a long-tailed pterosaur with a wing span of about 3·3ft (1m). Its skull is elongated. The eye sockets are the largest apertures in the skull. The front teeth in the upper and lower jaw are long, powerful and curved, and meshed alternately when the beak was closed. The rear parts of the jaw had only very small teeth. Dentition of this kind was a very effective organ for seizing and holding slippery prey, in other words fish. *Dorygnathus* had a relatively small, triangular sternum. This served as an area of attachment for flight muscles. The flight digits and thus the wings were relatively short. The fifth digit of the foot was very long, however, and set at a lateral angle. Its function is not clear. Possibly it was used to spread a small web which permitted the animal to take off more easily if it had had to land on the surface of the sea.

12 Wild, R., 1971. *Dorygnathus mistelgauensis n.sp., ein neuer Flugsaurier aus dem Lias Epsilon von Mistelgau (Fränkischer Jura).* Geologische Blätter von Nordost-Bayern, 21(4): 178-195; Erlangen.

13 Stieler, C., 1922. *Neuer Rekonstruktionsversuch eines liassischen Flugsauriers.* Naturwissenschaftliche Wochenschrift, N.F. 21(20): 273-280; Jena.

Campylognathoides

The first pterosaur specimen from the Württemberg Lias was of the genus *Campylognathoides* (=curved jaw). However, the skeletal remains consisted only of a few bones of the flight arm, which had been described by Tübingen professor Friedrich August Quenstedt in 1858 under the name *Pterodactylus liasicus.*[14]

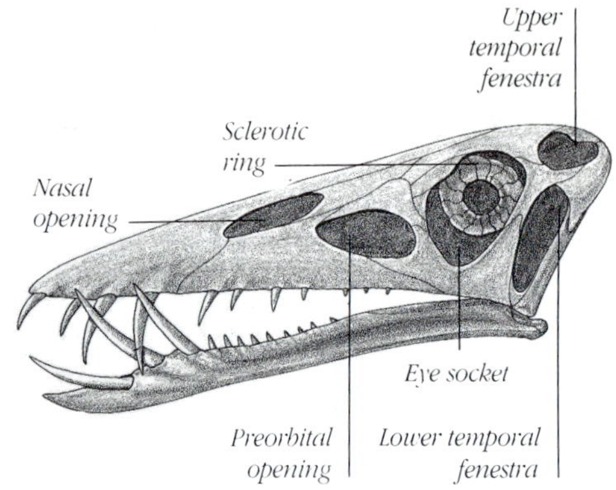

14 Quenstedt, F.A., 1858. *Über Pterodactylus liasicus.* Jahreshefte des Vereins für Vaterländische Naturkunde in Württemberg, 14: 299-336; Stuttgart.

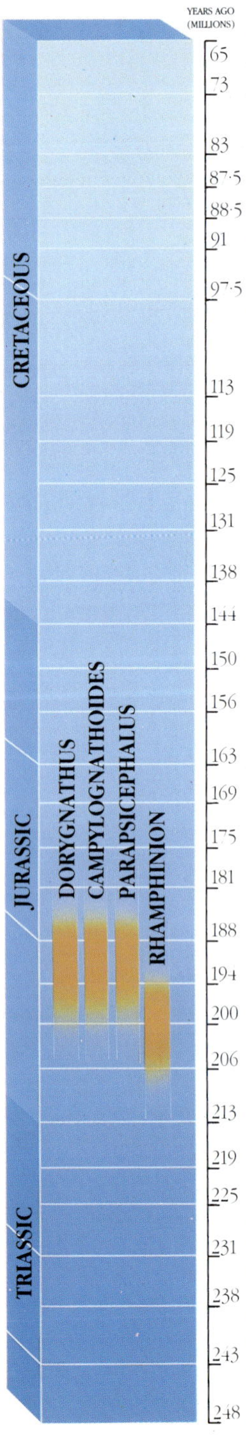

Time Chart (left)
Rhamphinion of the Lower Jurassic Kayenta formation of Arizona indicates a worldwide distribution of pterosaurs at that time. *Dorygnathus, Campylognathoides* and *Parapsicephalus* appear in the Upper Liassic (the Toarcian) of Germany and England.

Dorygnathus (above)
In this scene *Dorygnathus* is seen fishing on the wing a little offshore from the beach of the Upper Liassic Posidonian shale sea of Holzmaden, Germany. In the background, marine crocodilians, *Steneosaurus,* can be seen lumbering up the beach. *Dorygnathus* must have been abundant in the early Jurassic, particularly in this area. It was a long-tailed pterosaur with a relatively small wingspan. Its front teeth are long and point forward, thus forming a perfect gripping device for catching and holding slippery fish. It has long hind legs , and the fifth toe of the foot is also long with a bent phalanx. It may have been used to spread a small web between the toes.

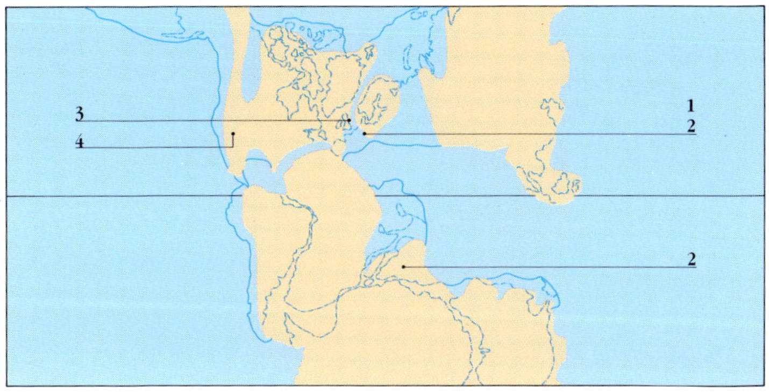

Map (above)
1 *Dorygnathus*
2 *Campylognathoides*
3 *Parapsicephalus*
4 *Rhamphinion*

Dorygnathus Heads (above)
The drawings show restorations of the skull of *Dorygnathus* and a life portrait based on such fossil skulls. Note the curving front teeth.

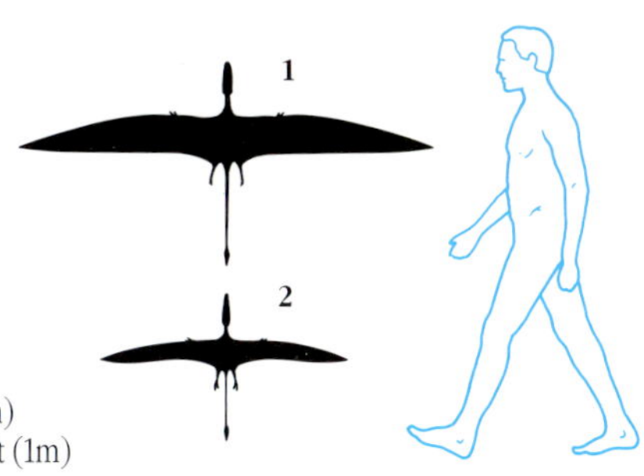

Comparative Sizes (right)
1 *Campylognathoides zitteli*: wing span 5·7ft (1·75m)
2 *Dorygnathus*: wing span 3·3ft (1m)

Campylognathoides (above)
This pterosaur lived in the same environment as *Dorygnathus* in the Holzmaden area of the Upper Liassic Sea, but it is also known from northern Germany, and possibly even India. It is also a long-tailed, rhamphorhynchoid pterosaur, but it has a shorter head and smaller teeth than its contemporary *Dorygnathus*. The skull is dominated by the large eye sockets. So far, two species are known, *Campylognathoides liasicus*, with a wing span of about 3·3ft (1m), and the larger *Campylognathoides zitteli* with a span of around 5·7ft (1·75m). In the background is seen *Ohmdenosaurus*, a middle-sized sauropod discovered in this area.

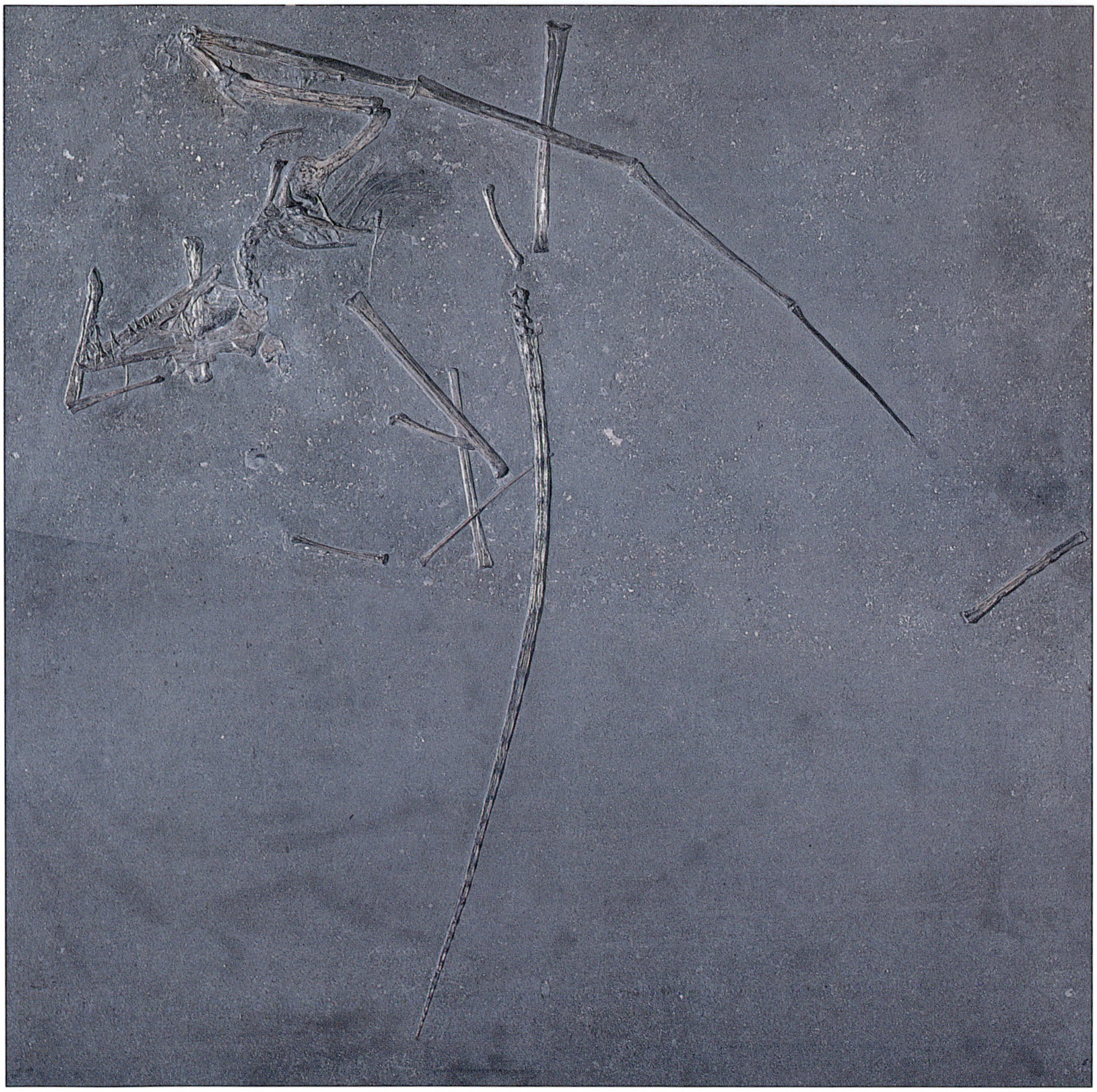

Left: These are the fossil remains of *Campylognathoides liasicus* which were excavated from the Upper Lias of Ohmden near Holzmaden in Württemberg, Germany. The skeleton is partly disarticulated, but it does show clearly the long tail characteristic of this genus and one wing. The tail is 10·2in (26cm) long, while the wing span is about 3·3ft (1m). This specimen is housed in the Staatliches Museum für Naturkunde, Stuttgart. It was as a result of the work of the Stuttgart palaeontologist Felix Plieninger that *Campylognathoides* was recognised as a genus in its own right, and not assigned to the genus *Pterodactylus* as Friedrich Quenstedt had originally suggested in 1858.

Getting Airborne (bottom of page)
While *Dorygnathus* had relatively short flight digits and wings, its feet had remarkably long fifth digits which were set at an angle which was laterally displaced from the remaining toes. The function of this fifth toe is not clear, but it may have served to spread a small web which could have helped the pterosaur to paddle to get airborne after it had alighted on water.

It was not until much more complete finds from the Holzmaden Posidionian shale had become known that the Stuttgart palaeontologist Felix Plieninger recognized that they were a genus in their own right.[15] The name *Campylognathoides* comes from the Greek *kampylós*, curved, and *gnáthos*, jaw, after the crooked ends to the jaw.

The genus has a characteristic relatively short skull, dominated by a large, circular eye socket. The end of the snout is pointed and toothed, with short, conical teeth set upright in the jaw. The sternum consists of a broad, rectangular plate of bone and has a short crest (cristopina) projecting forwards. In contrast with *Dorygnathus* the fifth toe is very short in this case. Two species of *Campylognathoides* are known from Holzmaden, *Campylognathoides liasicus* and *Campylognathoides zitteli*.[15,16] The former had a wing span of under 3·3ft (1m), and the latter had a span of about 5·7ft (1·75m).

In 1974 the Indian palaeontologist S.L. Jain described a fragment of a pterosaur skull with teeth as *Campylognathoides indicus*.[17] The fossil came from the Lower Jurassic of the Chanda district of India.

In 1986 a fossil collector found a small pterosaur pelvis that had survived in isolation in a Posidonian shale quarry in the area of Braunschweig in Lower Saxony, which could also be placed in the genus *Campylognathoides*.[18] In this case it was particularly significant that the hip sockets had survived in very good condition and their lateral and upward

15 Plieninger, F., 1895. *Campylognathus zitteli. Ein neuer Flugsaurier aus dem Oberen Lias Schwabens.* Palaeontographica, 41: 193-222; Stuttgart.
16 Wellnhofer, P., 1974. *Campylognathoides liasicus (Quenstedt), an Upper Liassic Pterosaur from Holzmaden. The Pittsburgh Specimen.* Annals of the Carnegie Museum, 45(2): 169-216; Pittsburgh.

17 Jain, S.L., 1974. *Jurassic Pterosaur from India.* Journal of the Geological Society of India, 15(3): 330-335.

18 Wellnhofer, P. and Vahldiek, B.W., 1986. *Ein Flugsaurier-Rest aus dem Posidonienschiefer (Unter-Toarcium) von Schandelah bei Braunschweig.* Paläontologische Zeitschrift, 60: 329-340; Stuttgart.

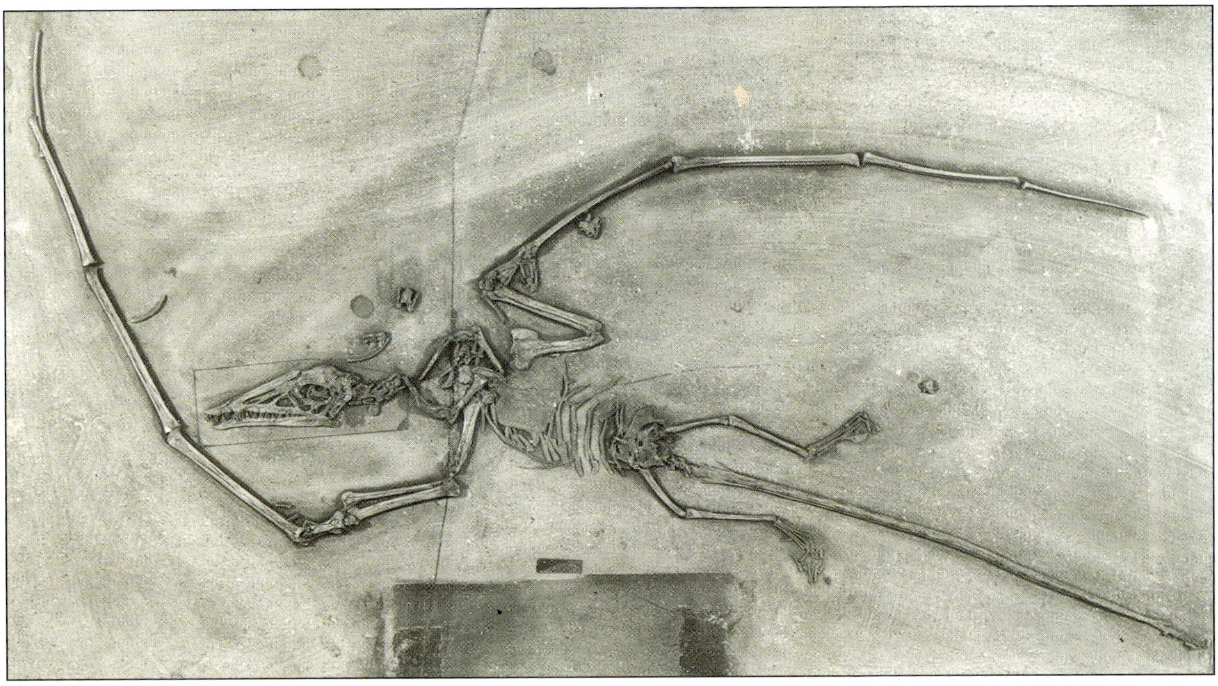

Left: This is one of the most complete specimens of *Campylognathoides liasicus* from the Upper Lias of Holzmaden. Called the Pittsburgh specimen, it is one of the prize fossils in the collection that the Carnegie Museum in Pittsburgh, Pennsylvania purchased from the Belgian Baron de Bayet in 1903. The Baron had bought this specimen from the noted Holzmaden palaeontologist Bernhard Hauff who had discovered it in 1897. Hauff had originally found the skeleton without the skull. It was not until a year later that he found the skull several yards away in the same part of the quarry, and was subsequently able to restore it to its rightful place. The length of the skull is 3·27in (8·3cm).

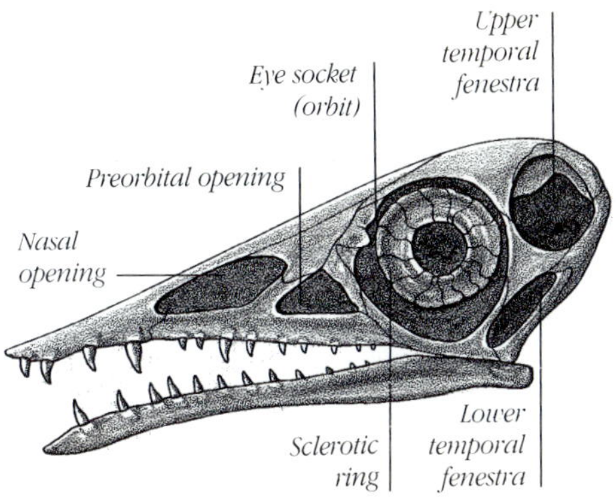

Campylognathoides Skull and Head (left and right)
Campylognathoides liasicus' eye sockets are very large, which suggests that it possessed acute eyesight. Some scientists argue that such large eyes indicate a nocturnal lifestyle. At right we see a life portrait based on the skull restoration shown left.

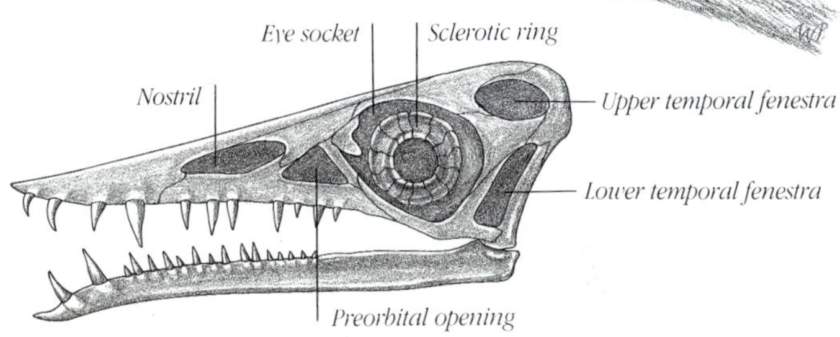

Campylognathoides zitteli Skull (right)
This second species of *Campylognathoides* was larger than *liasicus*. Its skull was 5in (13cm) in length.

Campylognathoides Pelvis (below)
The pelvis of a *Campylognathoides* specimen is seen here from the top (above) and the side. The hip sockets face laterally and slightly dorsally indicating that bipedal locomotion was unlikely.

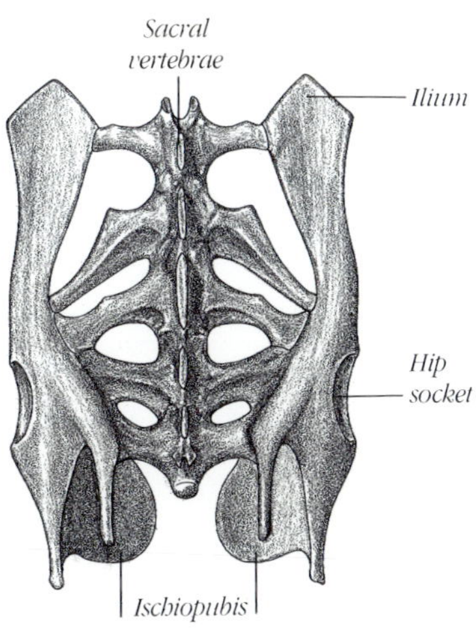

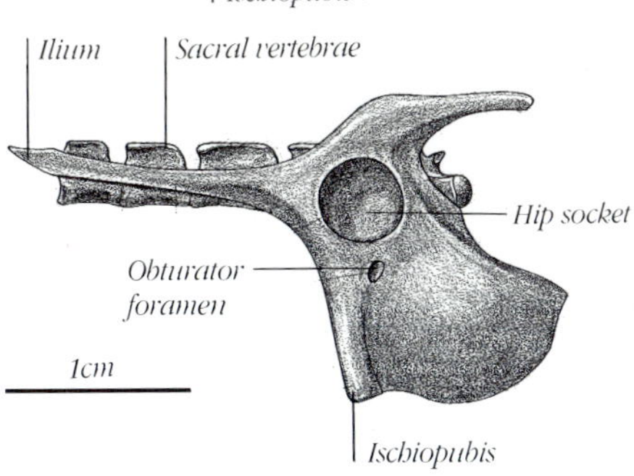

orientation indicated that pterosaurs were probably not in a position to walk on two legs like birds. Therefore the upper leg bones (femora) could not be oriented vertically, as would be needed to enable the legs to swing to and fro for walking and running on two legs.

A Whitby Discovery

A very important Upper Liassic pterosaur find came from England. The 'alum shales' of Whitby on the Yorkshire coast in northern England are grey, crumbly shale with a high pyrites content, in the higher strata of which limestone geodes (or concretions) occur. These were formerly exploited to produce cement. The alum shales in the Whitby area were mined in the eighteenth and nineteenth century in particular, on numerous sites. Alum is a potassium aluminium sulphate, used among other things in paper factories and dyeing works. The alum industry flourished in the eighteenth century, but declined when a cheaper production method was found.

The first fossil reptile found in the alum shales of the Yorkshire coast was a marine crocodile described by William Chapman in 1758.[19] Then in the 1880s the Reverend D.W. Purdon found a well-preserved pterosaur skull 'from the Alum Shale at Lofthouse' (modern Loftus). It was named *Scaphognathus purdoni* in honour of its finder by E.T. Newton of the Geological Survey in London in 1888.[20] However it later became clear that this pterosaur was different from *Scaphognathus* (=tub-jaw), which was only known from the Solnhofen flaggy limestones of Bavaria. The Whitby genus was thus given the name *Parapsicephalus* (=double-vaulted skull). The bulk of the fossils from the Upper Lias of Whitby are ammonites, however. These also permit precise dating of the alum shales. Like the Posidonian shales they belong to the Lower Toarcian.

19 Chapman, W., 1758. *An account of the fossile bones of an allegator, found on the sea-shore near Whitby in Yorkshire.* Philosophical Transactions of the Royal Society London, 50: 688-691.
William Chapman: 'The bones were covered five or six feet with water every full sea, and were about nine or ten yards from the cliff, which is nearly perpendicular, and about sixty yards high.'
See also Benton, M.J., and Taylor, M.A., 1984. *Marine reptiles from the Upper Lias (Lower Toarcian, Lower Jurassic) of the Yorkshire Coast.* Proceedings of the Yorkshire Geological Society, 44(4), no. 29: 399-429.
20 Newton, E.T., 1888. *On the Skull, Brain and Auditory Organ of a new species of Pterosaurian (Scaphognathus Purdoni) from the Upper Lias near Whitby, Yorkshire.* Philosophical Transactions of the Royal Society, 179: 503; London.

The only specimen of this genus is a fragment of a skull 5·5in (14cm) long which has survived three-dimensionally. The front end of the jaw and the teeth are missing. It was a medium-sized, long-tailed pterosaur with a wing span of over 3·3ft (1m). This is one of the rare pterosaur skulls in which it is possible to examine the structure of the brain. Of course the brain itself cannot survive, it decays immediately after the death of the animal. But the skull filled with ooze on the sea bed, this ooze finally became petrified and formed an inner cast of stone, or endocast. This endocast is a faithful, three-dimensional copy of the pterosaur-brain, the various structures of which can be analysed precisely, almost as though it were real brain.

Fragments from Arizona

In 1984 an interesting pterosaur fragment from the Kayenta formation in Arizona came to light. Some geologists date the Kayenta formation as Upper Triassic, others as Lower Jurassic. Kevin Padian of the University of California in Berkeley recognized that the fragmentary remains were a new species and genus, which he called *Rhamphinion jenkinsi*, named in honour of Dr Farish A. Jenkins Jr., who discovered the specimen.[21] Only four pieces were found. They come from the back of the head, the cheek-bone, the lower jaw and another indeterminate bone fragment. *Rhamphinion* (=beak and nape of the neck) is very probably a long-tailed pterosaur, a rhamphorhynchoid. Padian mentions another pterosaur bone from the Arizona Kayenta formation, a 1·65in (42mm) long flight digit metacarpal of a long-tailed pterosaur which must have had a wing span of about 4·9ft (1·5m).

Above: Cliffs of Lias rocks near Staithes north west of Whitby on the Yorkshire coast in England. Alum shales were mined in this area in the eighteenth and nineteenth centuries, and important fossil reptiles were discovered in the course of excavations.

Right: *Parapsicephalus purdoni*, a partial skull from the Upper Liassic alum shales of Whitby seen in top view. This important specimen is one of the few pterosaur skulls which have an endocast of the brain preserved within the brain cavity.

The Skull of Parapsicephalus purdoni (right)
The drawings show the skull fragment of *Parapsicephalus purdoni* that was found near Whitby in Yorkshire. It is seen from the left side (upper drawing), and from below (middle drawing). The lower drawing shows a restoration with the supposed front end of the jaw outlined in dotted lines. When this specimen was examined, its brain cavity was discovered to be filled with petrified sediment, which thus revealed a three-dimensional image of the pterosaur's brain. Such endocasts are rare. They help scientists determine the possible mental abilities and consequent behaviour of pterosaurs.

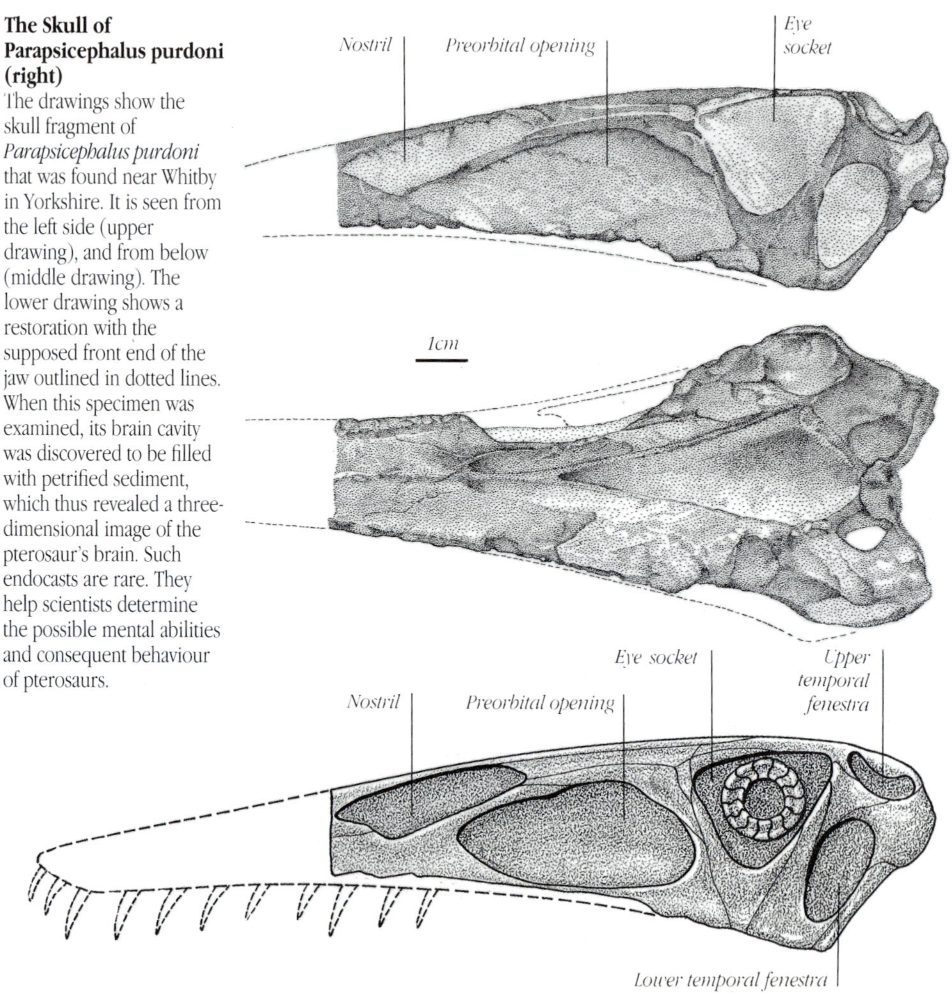

21 Padian, K., 1984. *Pterosaur remains from the Kayenta Formation (?Early Jurassic) of Arizona.* Palaeontology, 27(2): 407-413.

Middle Jurassic Pterosaurs

The Stonesfield Beds in Oxfordshire consist of sandstone and sandy limestone. They are a striking set of geological strata, extending from the Cotswolds south of Bath into Oxfordshire. In the early nineteenth century these slab-like rocks were even mined in subterranean quarries, for use as roofing slates.[22] Large fossil bones were found in these Stonesfield strata as early as the eighteenth century, and at the beginning of the nineteenth century Stonesfield slate became famous because of the first dinosaur to be scientifically described, *Megalosaurus*, as well as finds of fish, crocodiles and the jawbones of early, primitive mammals.

As well as vertebrate fossils the Stonesfield slate also contains fossils of invertebrates, namely brachiopods, bivalves and ammonites. These index fossils also permitted stratigraphic dating of the Beds to the Bathonian age, a subdivision of the Middle Jurassic or Dogger, for which an absolute age of 175 to 169 million years is accepted today.

In 1832 the Frankfurt vertebrate palaeontologist Hermann von Meyer described the first pterosaur remains from Stonesfield, which he called *'Pterodactylus' bucklandi*, in honour of William Buckland.[23] Subsequently further remains of pterosaur skeletons were found in various places in the Oxfordshire Stonesfield Slate and ascribed by T.H. Huxley to various species of the genus *Rhamphorhynchus*.[24] This was the same Thomas Huxley who later became known as Darwin's 'bulldog', a passionate defender of his theory of evolution. Thomas Henry Huxley (1825-1895) was one of the most brilliant scientists of his time. He had studied medicine, but concerned himself particularly with physiology and natural history. He was Lecturer in Palaeontology and Natural History at the School of Mines in London, and studied fossils, including the pterosaurs from the Stonesfield Slate and later the primordial

Right: For hundreds of years Stonesfield slate has been widely used as roofing tiles, as is seen here in the main street of the Cotswold village of Chipping Camden in Gloucestershire. The slate, which is quarried near the village of Stonesfield, is a grey, fissile, calcareous sandstone. Its greatest recorded thickness is 6ft (1·83m). The fossils found in these slate beds are renowned worldwide. For example, *Megalosaurus*, the first dinosaur to be scientifically described, was found in them during the last century.

22 Arkell, W.J., 1933 (see reference 2), p. 277: 'The Stonesfield Slate Beds consist typically of thin sands and sandy limestones, often in the form of spheroidal doggers called 'pot-lids' or 'burs', some of which split under the weather into fissile roofing tiles.'
p. 278: 'A few miles farther north-east, in the North Cotswolds, the Stonesfield Slate Beds attain their maximum development and are still worked in several places under the name of Cotswold Slates. The activity of the industry in the past may be judged by the fact that at Kineton Thorns alone no less than 120,000 slates were made in one season.'

23 Meyer, H. von, 1832. *Palaeologica zur Geschichte der Erde*. 560 pp.; Frankfurt.
24 Huxley, T.H., 1859. *On Rhamphorhynchus bucklandi, a Pterosaurian from the Stonesfield Slate.* Quarterly Journal of the Geological Society, 15: 658; London.

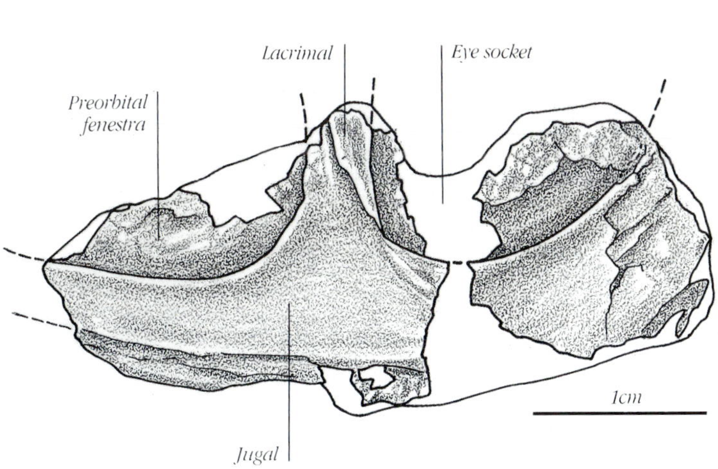

Rhamphinion Skull Fragment (left)
This is a fragment of the skull of *Rhamphinion jenkinsi* from the ?Lower Jurassic Kayenta formation of Arizona. We can see part of the jugal bone which forms the lower margin of the eye socket and the preorbital opening. This is one of a few, very scanty pterosaurian remains from these deposits which have also yielded dinosaurs. The drawing is based on an original by K. Padian who recognised *Rhamphinion* as a new genus.

SUMMARY OF LOWER JURASSIC PTEROSAURS

Rhamphorhynchoidea
Family Dimorphodontidae
Dimorphodon macronyx
Lower Lias, Blue Lias, Lyme Regis, Dorset; Aust Cliff, Gloucestershire, England.[5]

Family Rhamphorhynchidae
Dorygnathus banthensis
Upper Lias, Posidonian shale, Banz, Creez near Bayreuth, Bavaria; Holzmaden, Württemberg; Flechtorf, Lower Saxony, Germany.[11,13]
Dorygnathus mistelgauensis
Upper Lias, Mistelgau near Bayreuth, Bavaria, Germany.[12]
Campylognathoides liasicus
Upper Lias, Holzmaden, Württemberg, Germany.[15,16]
Campylognathoides zitteli
Upper Lias, Holzmaden, Württemberg, Germany.[15]
Campylognathoides sp.
Upper Lias, Schandelah near Braunschweig, Lower Saxony, Germany.[18]
Campylognathoides indicus
Upper Jurassic, Kota formation, Chanda District, Deccan, India.[17]
Parapsicephalus purdoni
Upper Lias, Alum shale, Whitby, Yorkshire, England.[20]

Undetermined family
Rhamphinion jenkinsi
Lower Jurassic, middle Kayenta Formation, Little Colorado River Valley, Arizona, USA.[21]

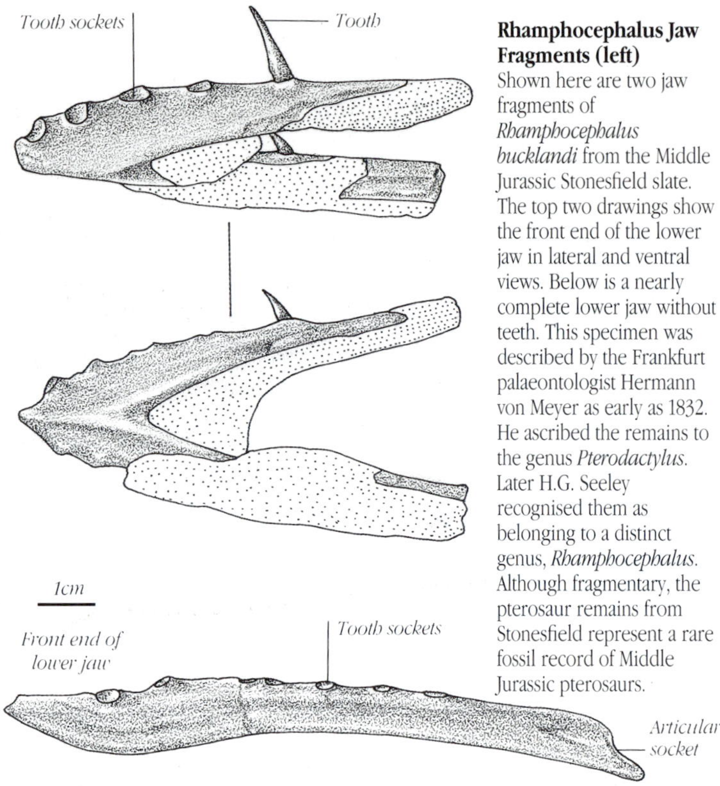

Rhamphocephalus Jaw Fragments (left)
Shown here are two jaw fragments of *Rhamphocephalus bucklandi* from the Middle Jurassic Stonesfield slate. The top two drawings show the front end of the lower jaw in lateral and ventral views. Below is a nearly complete lower jaw without teeth. This specimen was described by the Frankfurt palaeontologist Hermann von Meyer as early as 1832. He ascribed the remains to the genus *Pterodactylus*. Later H.G. Seeley recognised them as belonging to a distinct genus, *Rhamphocephalus*. Although fragmentary, the pterosaur remains from Stonesfield represent a rare fossil record of Middle Jurassic pterosaurs.

Above: This skull of the rhamphorhynchid pterosaur *Angustinaripterus* comes from the Middle Jurassic beds in Dashanpu Quarry, near Zigong in China. It is characterised by its large, interlocking teeth which are similar to those of *Dorygnathus*.

bird *Archaeopteryx*. This he interpreted by the application of Darwin's theory of evolution as a link between reptiles and birds, as the famous 'missing link'. As Darwin's 'bulldog' he came into conflict with Richard Owen of the British Museum (Natural History) in particular.

Later H.G. Seeley recognized a distinct genus in the Stonesfield slate pterosaurs, which he called *Rhamphocephalus* (=beak-head).[25] In no case were there complete skeletons, only fragments of lower jaw, skull and numerous individual bones like vertebrae, ribs, pectoral girdle and bones from the extremities. The dimensions of the fossil remains indicate that *Rhamphocephalus* must have reached a wing span of 3-4ft (0·9-1·2m).

The fragmentary state in which the remains of the skeletons have survived makes it difficult to ascribe Stonesfield slate pterosaurs to known species of the Jurassic. Skull and dentition of *Rhamphocephalus* indicate that they were long-tailed Rhamphorhynchoidea. On the other hand the wing phalanges have a triangular cross-section like the short-tailed pterodactylids, which do not occur until the Upper Jurassic. For this reason *Rhamphocephalus* is possibly a transitional form between the primitive Rhamphorhynchoidea and the more advanced Pterodactyloidea, as it seems to combine characteristics of both. Caudal vertebrae have not yet been discovered, however. More precise conclusions could only be drawn on the basis of more complete fossil finds, but such finds are hardly to be expected as the Stonesfield slate of Oxfordshire is now rarely quarried.

Pterosaurs from Middle Jurassic strata are found in very few places on Earth. Therefore *Rhamphocephalus* remains from the Stonesfield slate are of great significance despite their lack of completeness, as they form a link between the good fossil documentation in the Lower and Upper Jurassic.

Discoveries at Dashanpu

An enormous dinosaur cemetery was discovered and excavated a few years ago near the city of Zigong in the province of Sichuan, and a large museum has now been built above it. The Sichuan basin in central China is a gigantic sedimentation basin containing continental deposits from the Upper Triassic to the Jurassic-Cretaceous boundary. The strata are over-

25 Seeley, H.G., 1880. *On Rhamphorhynchus Prestwichi.* Quarterly Journal of the Geological Society, 36: 27; London.

Below: Excavations underway at Dashanpu Quarry. A vast dinosaur 'cemetery' was discovered here in the red beds of Sichuan. Amongst the many Jurassic dinosaur remains uncovered was the skull of *Angustinaripterus longicephalus*.

Angustinaripterus Skull (below)
The well preserved skull of *Angustinaripterus longicephalus* reveals dentition similar to that of *Dorygnathus* from Holzmaden in Germany. These teeth must have been perfect for gripping fish.

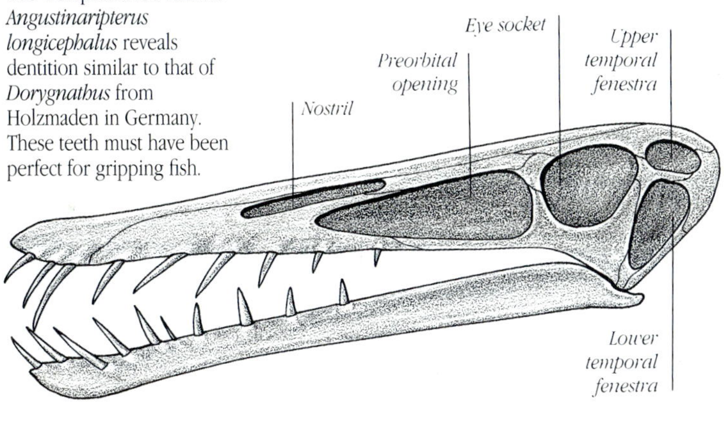

Below: The great limestone quarry of Solnhofen in Bavaria, Germany. This photograph was taken at the turn of the century when horses were still used to pull the carts laden with limestone slabs over the bridges to the workshops. The photograph shows nicely the typical layers of flaggy limestone which used to be used in this area for roofing. The series is composed of very fine grained, almost pure limestone slabs, deposited in the lagoons of the Jurassic Tethys Sea 150 million years ago.

10,000ft (3,000m) thick and consist mainly of reddish river and lake deposits, known overall as the 'Red Basin of Sichuan'.[26]

A number of different dinosaur faunas have been discovered throughout the strata of the Red Basin of Sichuan. They document the longest continuous sequence of Jurassic dinosaurs in the world. In one of these strata, the Lower Shaximiao formation of the Middle Jurassic, the skull of a pterosaur was found alongside the primitive sauropod dinosaur *Shunosaurus* and other vertebrate remains.[27] The rear end of the skull was missing, but nevertheless its shape and size can be reconstructed with some reliability. It is an elongated skull about 6·5in (16·5cm) long. The wing span must have been about 5ft (1·6m). The upper jaw had a total of 18 teeth and the lower jaw 18-20, which bit against each other alternately and must have been an excellent tool for catching fish. This dentition is similar to that of *Dorygnathus* from the Holzmaden Lias, but also shows similarities with *Rhamphocephalus* from the Stonesfield slate, to the extent that this can be judged.

The most striking characteristic of the Dashanpu pterosaur is the very narrow nostril. For this reason the genus was named *Angustinaripterus* (=wing with the narrow nostril), and because of its long head it was given the species name *longicephalus* (=long-skulled). Generally speaking the structure of the skull is the same as that of the long-tailed Rhamphorhynchoidea. This is recognizable not only from the dentition, but also from the fact that the pre-orbital opening and the nostril are divided by a bridge of bone. In short-tailed pterosaurs (Pterodactyloidea) these two skull openings are confluent and form a common nostril-preorbital opening in front of the eye socket.

Although *Angustinaripterus* from the Chinese Middle Jurassic shows similarities with pterosaurs from the European Lower and Middle Jurassic it can be considered as a representative of a distinct group of pterosaurs and thus is evidence of a wide distribution of flying reptiles to eastern Asia as early as the Middle Jurassic.

26 Dong Zhiming, 1988. *Dinosaurs of China.* English edition (text by A.C. Milner), 114 pp.; China Ocean Press, Beijing and British Museum (Natural History) London.

27 He Xinlu, Yan Daihan and Su Chunkang, 1983. *A new pterosaur from the Middle Jurassic of Dashanpu, Zigong, Sichuan.* Journal of the Chengdu College of Geology, supplement 1: 27-33; Chengdu, China.

Evidence from Neuquén

In 1975 R.M. Casamiquela described fragmentary skeletal remains of a small reptile from the Callovian formation of Neuquén in northern Patagonia as a coelurid dinosaur with the name *Herbstosaurus pigmaeus*, named after R. Herbst, who collected the specimen, and because of its small size (pygmy).[28] Pelvis, femur and other remains were found, and first related to *Compsognathus*, a small dinosaur from the Upper Jurassic of Solnhofen. However, it later became clear that *Herbstosaurus* must be a pterosaur.[29] The pelvis and the shape of the femur are similar to those of the Rhamphorhynchoidea, the long-tailed pterosaurs. More precise assignment is not possible, however.

Very few pterosaurs are known from Middle Jurassic strata. The lack of significant fossil sites is explicable on geological grounds. Nevertheless finds of *Rhamphocephalus* in Europe, *Angustinaripterus* in China and *Herbstosaurus* in Argentina show that pterosaurs had managed to spread all over the world in the course of the Middle Jurassic.

Upper Jurassic Pterosaurs

Solnhofen is a famous name for palaeontologists. It describes a classic fossil site, but is also used to identify an entire area in which fossils have been found in the Southern Franconian Alb in Bavaria. There the numerous quarries in which Solnhofen lithographic limestone is quarried extend over an area of about 50 miles (80km) in an east-westerly direction, essentially along the valley of the Altmühl, which flows into the Danube near Kelheim. For this reason one often finds the blanket term 'Solnhofen' in older museum collections, even when the fossil really comes from quarries in the region of Eichstätt, Pfalzpaint, Gungolding, Zandt, Painten or Kelheim.

28 Casamiquela, R.M., 1975. *Herbstosaurus pigmaeus (Coeluria, Compsognithidae) n.gen.n.sp. del Jurasico medio del Nequén (Patagonia septentrional). Uno de los más pequeños dinosaurios conocidos.* Acta primero Congreso Argentino Paleontologia et Bioestragrafia, 2: 87-102.

29 Ostrom, J.H., 1978. *The Osteology of Compsognathus longipes Wagner.* Zitteliana, 4: 73-118; Munich.

SUMMARY OF MIDDLE JURASSIC PTEROSAURS

Rhamphorhynchoidea
Family Rhamphorhynchidae
Rhamphocephalus bucklandi
Bathonian, Stonesfield Slate, Stonesfield, Oxfordshire, England.[23]
Rhamphocephalus depressirostris
Bathonian, Stonesfield Slate, Sarsden near Chipping Norton, Oxfordshire, England.[24]
Rhamphocephalus prestwichi Bathonian, Stonesfield Slate, Kineton near Stow-on-the-Wold, Gloucestershire, England.[25]
Angustinaripterus longicephalus
Middle Jurassic, Lower Shaximiao Formation, Dashanpu near Zigong, Sichuan, China.[27]

Undetermined family
Herbstosaurus pigmaeus
Callovian, Lotena Formation, Neuquén, northern Patagonia, Argentina.[28]

Above: *Rhamphorhynchus* is the most frequently occurring pterosaur to be found in the Solnhofen limestone of Bavaria. This individual is a *Rhamphorhynchus longicaudus* skeleton from Upper Jurassic Solnhofen deposits. The wing span of this long-tailed species was about 1·3ft (40cm).

We know from excavations that even ancient Roman builders used these flaggy limestones for building walls, as floor tiles and as slabs for inscriptions.[30] The Romans built a border wall, the so-called *limes*, a few miles north of the Solnhofen area. When the forts along the *limes* were being excavated, Solnhofen slabs were found that had been used by the Roman builders. The wide distribution and popularity of Solnhofen stone in the Middle Ages can be seen from the fact that even Hagia Sofia in Constantinople (now Istanbul, Turkey) was paved with it in the mid thirteenth century.

Alois Senefelder's invention of the lithographic printing process in 1796 gave a tremendous boost to the Solnhofen quarrying industry. It was not until lithographic printing was superseded by modern techniques that the light-coloured Solnhofen stone again became important as building material, especially for paving stones, wall tiles, steps and window sills, as well as stone for garden walls.

30 In the first three centuries AD this area was part of the Roman province of Raetia, the capital of which was Augusta Vindelicorum, the modern Augsburg.

Below: A disarticulated skeleton of *Rhamphorhynchus* from the Upper Jurassic Mörnsheim limestone. The skull, lower jaw, both wings, the long tail, and the large bony breastbone are preserved, as is the sclerotic ring of the eye (upper left corner).

When Solnhofen stone is being quarried, which is still done by hand today, the quarry workers continually find fossils, but systematic collecting and research on the Solnhofen fossils did not begin until the late eighteenth century. It was at that time that the first pterosaur find, the Eichstätt *Pterodactylus*, arrived in Elector Karl Theodor's natural history collection in Mannheim (see Chapter 2). Other fossil collections also played a significant role, like that of Duke Maximilian von Leuchtenberg, who became prince of Eichstätt in 1817, and that of the country doctor Carl Haeberlein in Pappenheim near Solnhofen. The most famous palaeontological finds passed through the hands of the elder and the younger Haeberlein, including the first two specimens of *Archaeopteryx lithographica*, the primeval bird. Carl F. Haeberlein (1787-1871) sold his collection of Solnhofen fossils with the first specimen of *Archaeopteryx* from Langenaltheim near Solnhofen for £700 to the British Museum (Natural History) in London in 1862. Ernst O. Haeberlein (1819-1896), his son, secured a price of 20,000 gold marks for the second *Archaeopteryx* specimen from Blumenberg near Eichstätt in 1881. Werner von Siemens, the industrialist, acquired the piece by means of interim financing for the Natural History Museum of the Humboldt-Universität in Berlin.[31]

As well as these famous and rare fossil specimens the Solnhofen strata have produced a mass of other fossil plants and animals for over 200 years. So far about 750 different species have been described. A striking feature of the fossil content is that it includes not only marine forms like ammonites, crabs, fish, and marine reptiles, but also terrestrial organisms like land plants, insects, tortoises and fresh-water crocodiles, sphenodontids, a small dinosaur, numerous pterosaurs and the six primeval birds so far known. This is a result of the way the Solnhofen flaggy limestones came into being: they were formed in a tropical lagoon about 150 million years ago (Lower Tithonian, early Upper Kimmeridgian).[32]

There are numerous theories about the origin of the Solnhofen limestone, and there is not space to discuss them all here. One theory proceeds on the assumption that in the warm climate of the Upper Jurassic the salt concentration of the sea water in the lagoon was raised by evaporation. This led to an oxygen deficit

31 Viohl, G., 1985. *Carl F. and Ernst O. Häberlein, the Sellers of the London and Berlin Specimens of Archaeopteryx.* The Beginnings of Birds, Proceedings of the International Archaeopteryx Conference Eichstätt 1984: 349-352; Eichstätt.

32 In the later Jurassic the area of Solnhofen and the Southern Franconian Alb was a lagoon preceding a continent to the north and protected from the open sea, the Tethys, to the south by a belt of coral reef. However, a connection with deep areas of the Tethys existed via shallows with algal and sponge reefs in the south-west. The Central German continent and its islands were covered with palm and seed ferns, gingko plants and coniferous trees, and populated by insects, land reptiles, pterosaurs and primordial birds. Tributary rivers washed dead plants and animals into the Solnhofen lagoon, lying between the reefs and islands. The most common fossils in the Solnhofen strata are marine animals and plants, however, especially crustaceans, ammonites, squids and fish. Reptile finds suggest that this was a flourishing animal group. There are some well-preserved and complete skeletons of turtles, lizards, rhynchocephalia, tuatara, crocodiles, ichthyosaurs and pterosaurs.

and at times to the poisoning of the water near the bottom of the lagoon by decaying matter. There are hardly any signs of organisms living on the sea floor in the Solnhofen strata. On the other hand these conditions explain the excellent state of fossil preservation, as carrion-eaters could not exist there either.[33]

Although they occur only rarely, pterosaurs are so well documented in Solnhofen that it must be considered the most important site in which they have been found in the world: the unique quality of fossilization has led to the preservation of complete skeletons and the documentation of soft parts like flight membrane prints. Long-tailed pterosaurs were still extant in Solnhofen in the Upper Jurassic, like *Rhamphorhynchus*, together with the first short-tailed pterosaurs like *Pterodactylus*. Together with primeval birds they dominated the air, but certainly had different life styles, or ecological niches, as biologists put it. Solnhofen pterosaurs showed a variety of diverse forms. In size they ranged from small flying saurians with a wing span of only 14in (36cm) to 8·2ft (2·5m).[34]

At the time of writing 17 different species of pterosaur from the Solnhofen limestone have been identified. They can be placed in eight different genera. The many Solnhofen fossils in excellent condition have made a major contribution to better understanding of this extinct reptile group. We have the finds in Bavaria to thank for the fact that pterosaurs are among the best-researched fossil reptiles today. Since the first discovery of a pterosaur skeleton in Eichstätt, described by C. Collini in 1784 and named *Pterodactylus* by G. Cuvier, two to three hundred specimens of pterosaur fossil remains have been unearthed in the Solnhofen lithographic limestone down to the present day. Many of them have found their way into museums and public collections and been investigated scientifically.

Above: This small *Rhamphorhynchus* skeleton was a very young individual when it died. Its total length from the end of its snout to the tip of its tail is only 7in (17·5cm). The large eye socket is a typical feature of immature vertebrates. The wings and legs were probably lost when the body was floating in the sea.

Right: A complete skeleton of *Rhamphorhynchus* from Solnhofen showing impressions of the wing membrane. Originally the right wing finger was folded beneath the body, but it has been removed and positioned symmetrically to the left wing. The traces where the bones were previously can still be made out. This masterpiece of the preparator of the Frankfurt Senckenberg-Museum is one of the most spectacular Solnhofen pterosaurs on display.

Rhamphorhynchus

The most frequently occurring genus is *Rhamphorhynchus* (=beak-snout), a long-tailed pterosaur after whom the whole suborder is named Rhamphorhynchoidea. In the Upper Jurassic, 150 million years ago, they already had a long period of evolution behind them then. The earliest known pterosaurs, those of the Upper Triassic, 220 million years ago, had been Rhamphorhynchoidea as well. They preserved the long vertebrate tail into the Jurassic, as a legacy of their reptilian ancestors. It is only on the basis of good fossil preservation in the rocks of Solnhofen that we know that there was a rhomboid membrane on the end of this long tail, which was certainly used as a rudder in flight.

Solnhofen long-tailed pterosaurs may have had a long past, but there was no future for them. They had reached the peak of their evolution, but it was also almost the end. Skeletons of *Rhamphorhynchus* have only been found subsequently in the strata above the Solnhofen limestone, the chronologically following geological unit, which is known as the Mörnsheim strata, or Malm Zeta 3. These strata were named

33 Barthel, K.W., 1978. *Solnhofen – Ein Blick in die Erdgeschichte.* 393 pp.; Ott-Verlag, Thun.
Malz, H., 1976. *Solnhofener Plattenkalk: Eine Welt in Stein.* (ed. Th. Kress). 109 pp.; Freunde des Museums beim Solenhofer Aktienverein, Maxberg.
Kuhn, O., 1977. *Die Tierwelt des Solnhofener Schiefers.* 5. Auflage, Neue Brehm-Bücherei, 318, 140 pp.; Ziemsen-Verlag, Wittenberg-Lutherstadt.
Viohl, G., 1985. *Geology of the Solnhofen Lithographic Limestone and the Habitat of Archaeopteryx.* The Beginnings of Birds, Proceedings of the International Archaeopteryx Conference Eichstätt 1984: 31-44; Eichstätt.

34 Wellnhofer, P., 1970. *Die Pterodactyloidea (Pterosauria) der Oberjura-Plattenkalke Süddeutschlands.* Abhandlung der Bayerischen Akademie der Wissenschaften, Neue Folge 141, 133 pp.; Munich.
Wellnhofer, P., 1975. *Die Rhamphorhynchoidea (Pterosauria) der Oberjura-Plattenkalke Süddeutschlands.* Palaeontographica (A), 148: 1-33; 132-186; 149: 1-30; Stuttgart.
Wellnhofer, P., 1983. *Solnhofener Plattenkalk: Urvögel und Flugsaurier.* (ed. Th. Kress). 64 pp.; Freunde des Museums beim Solenhofer Aktienverein, Maxberg.

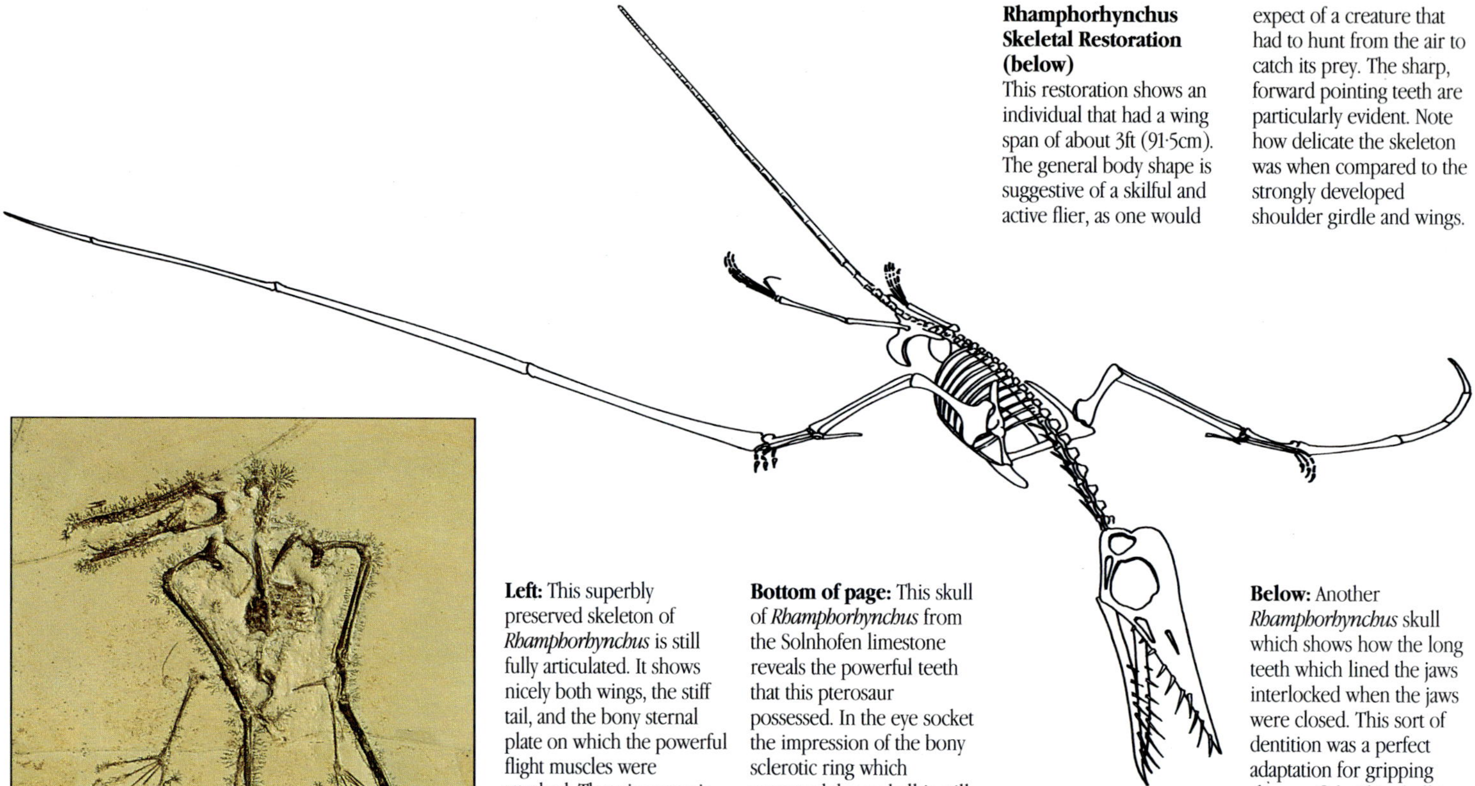

Rhamphorhynchus Skeletal Restoration (below)
This restoration shows an individual that had a wing span of about 3ft (91·5cm). The general body shape is suggestive of a skilful and active flier, as one would expect of a creature that had to hunt from the air to catch its prey. The sharp, forward pointing teeth are particularly evident. Note how delicate the skeleton was when compared to the strongly developed shoulder girdle and wings.

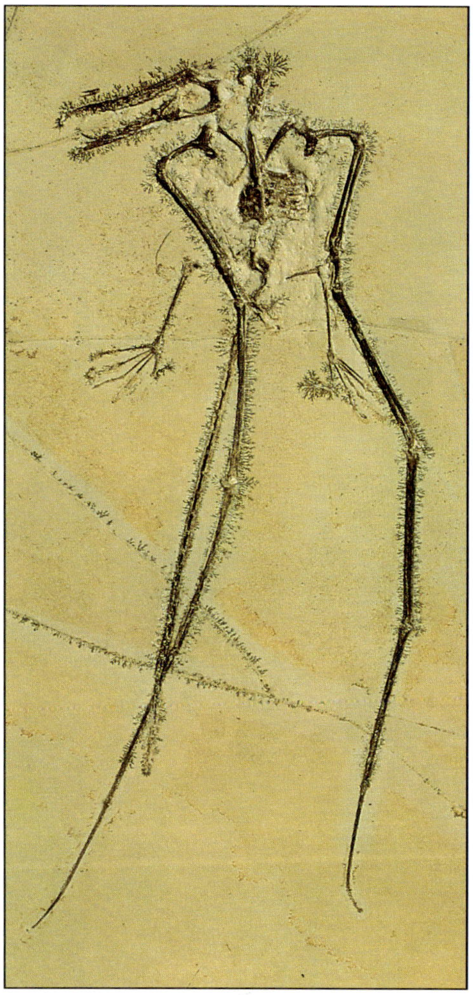

Left: This superbly preserved skeleton of *Rhamphorhynchus* is still fully articulated. It shows nicely both wings, the stiff tail, and the bony sternal plate on which the powerful flight muscles were attached. The wing span is c. 3·5ft (1·08m).

Bottom of page: This skull of *Rhamphorhynchus* from the Solnhofen limestone reveals the powerful teeth that this pterosaur possessed. In the eye socket the impression of the bony sclerotic ring which protected the eyeball is still visible.

Below: Another *Rhamphorhynchus* skull which shows how the long teeth which lined the jaws interlocked when the jaws were closed. This sort of dentition was a perfect adaptation for gripping slippery fish. This skull is 4·13in (10·5cm) long.

after the little village of Mörnsheim, which is near Solnhofen. They have also been quarried near Daiting, a few miles to the west.

Long-tailed pterosaurs do not occur in more recent Upper Jurassic strata or in the Lower Cretaceous. Thus they became extinct shortly after they had populated the coasts and islands of the Solnhofen lagoon.

Rhamphorhynchus was certainly a skilful flier. The sternum, where powerful muscles originated, was a broad plate of bone and had a forward-pointing crest, or cristopina. The neck is short, with compact, short vertebrae. The skull is always large, elongated and has a pointed front end. It has a large orbital opening and two smaller separate openings in the skull in front of the eyes, the nostril and the pre-orbital fenestra. The dentition is always powerfully adapted for catching prey, with long, pointed, slightly curved teeth directed forwards and outwards. When the jaw is closed they mesh alternately. There are 20 teeth in the upper jaw and 14 in the lower jaw. This denti-

Rhamphorhynchus Life Portrait (above)
This restoration shows the head of *Rhamphorhynchus muensteri*, the most common pterosaur found at Solnhofen. The top jaws contains 20 teeth, the lower jaws 14.

Rhamphorhynchus Skulls (right)
These drawings show the skulls of five different species of *Rhamphorhynchus* in ascending order of size. From the top they are: *longicaudus*, *intermedius*, *muensteri*, *gemmingi*, and *longiceps*. The lengths of the skulls in life range from 1·18in (3cm) up to 7·5in (19cm), and the corresponding wingspans from about 1·3ft (40cm) to 5·75ft (1·75m).

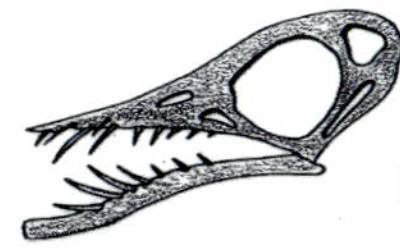

R. longicaudus

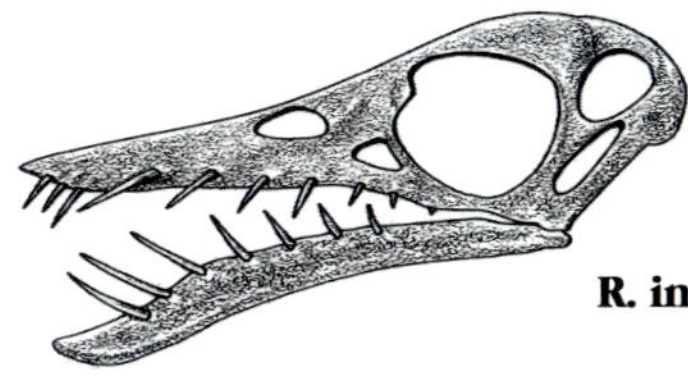

R. intermedius

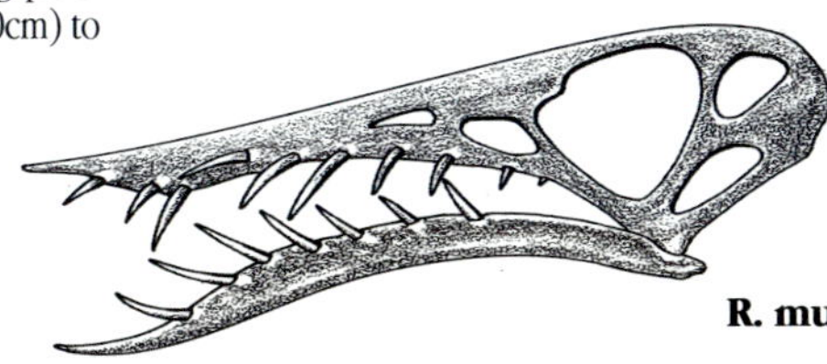

R. muensteri

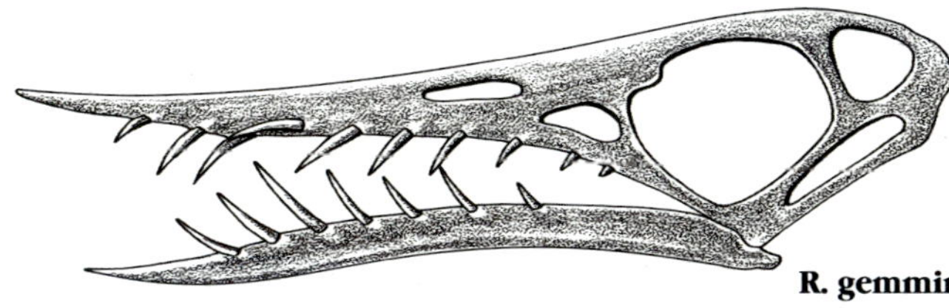

R. gemmingi

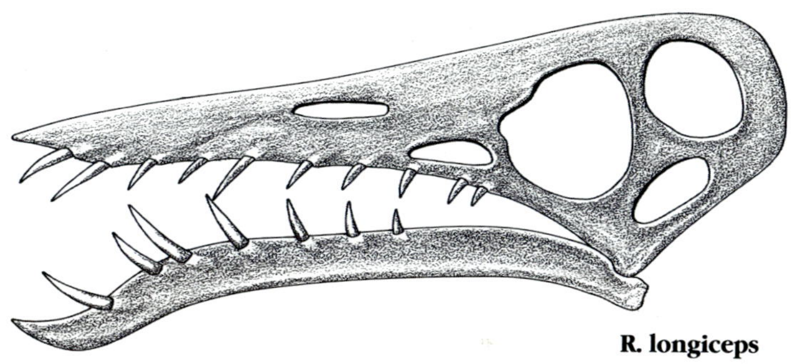

R. longiceps

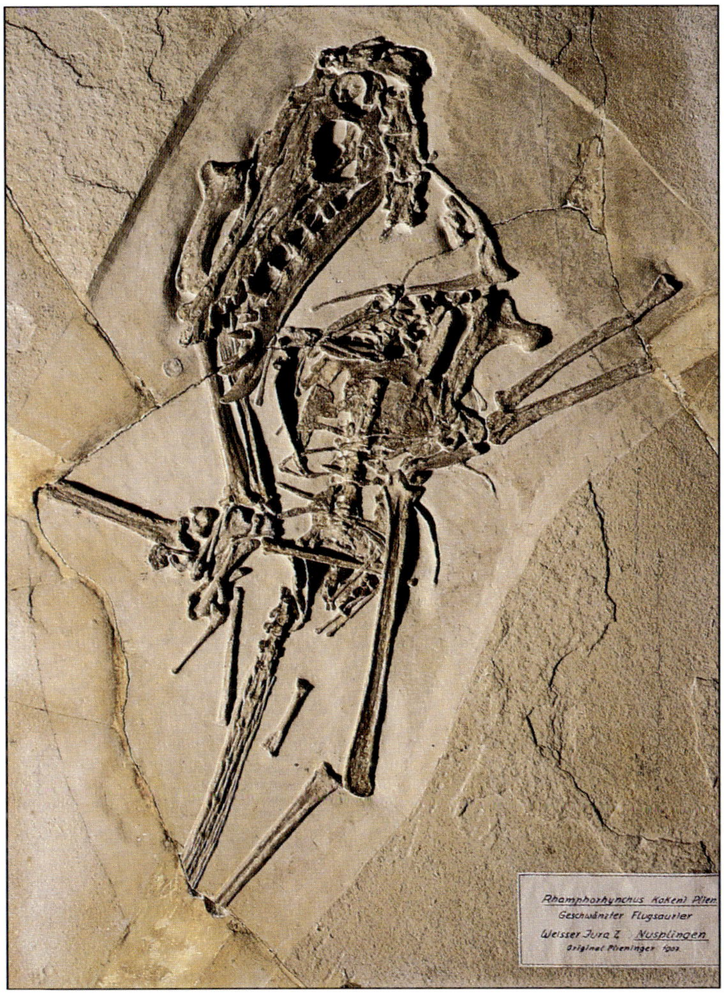

Left: *Rhamphorhynchus longiceps* from Nusplingen in Württemberg, Germany. This is the largest species of *Rhamphorhynchus* currently known. It has also been found in Solnhofen strata in Bavaria.

tion indicates that most species of *Rhamphorhynchus* were fish eaters. The remains of its last prey, a small fish, were in fact found in the stomach of a Solnhofen *Rhamphorhynchus*.

So far five different species of this genus have been found in Solnhofen. The smallest, *Rhamphorhynchus longicaudus* has a skull only 1·18in long (3cm) and a wing span of 15·75in (40cm). The largest species, *Rhamphorhynchus longiceps*, has a skull 7·5in (19cm) long and a wing span of 5·74ft (1·75m).

The most common species, *Rhamphorhynchus muensteri*, can be divided into two groups. One has a relatively long skull and a long flight digit, the other a relatively shorter skull and shorter flight digit. This suggests sexual dimorphism: the animals with larger skulls and longer wings could be assumed to be males, and the others females.[35] There are no distinctive characteristics on the skeleton to make it possible to distinguish male and female individuals in any other way.

Rhamphorhynchus also occurs in the Upper Jurassic flaggy limestone of Nusplingen in Württemberg (southern Germany).[36] This is the species *Rhamphorhynchus longiceps* (=long-skulled beak-snout), also known from Solnhofen in Bavaria. Remains of pterosaur skeletons from the somewhat older Oxford Clay (Upper Jurassic, Oxfordian) of Huntingdonshire in England were assigned to a new species, *Rhamphorhynchus jessoni*[37], and the same is true of individual bones from the Upper Jurassic Tendaguru strata of Tanzania, East Africa. These were discovered in the course of the famous German Tendaguru dinosaur expedition, 1909 to 1913.[38]

Finally individual teeth from the Middle Jurassic (Upper Callovian) and Upper Jurassic (Lower Kimmeridgian) of Portugal have also been assigned by R.A. Thulburn to the genus *Rhamphorhynchus*.[39]

35 Wellnhofer, P., 1975. Palaeontographica (A), 149, p. 3 (see reference 34).

36 Plieninger, F., 1907. (see reference 11).
At first Plieninger identified a large *Rhamphorhynchus* from the Upper Jurassic flaggy limestone of Nusplingen in Württemberg as a separate species, *Rhamphorhynchus kokeni*, but this species is identical with *Rhamphorhynchus longiceps* from the Solnhofen limestone of Eichstätt in Bavaria, described in 1902 by A. Smith-Woodward. With a skull 7·5in (19cm) long and a wing span of 5·74ft (1·75m) this is the largest known species of *Rhamphorhynchus*.
Smith-Woodward, A., 1902. *On two skulls of the Ornithosaurian Rhamphorhynchus.* Annals and Magazine of Natural History, (7), 9: 1-5; London.

37 Lydekker, R., 1890. *On Ornithosaurian Remains from the Oxford Clay of Huntingdonshire.* Quarterly Journal of the Geological Society, 46: 429; London. Description of pterosaur skeletal remains as *Rhamphorhynchus jessoni*.

38 Reck, H., 1931. *Die deutschostafrikanischen Flugsaurier.* Centralblatt für Mineralogie etc., B, 7: 321-336; Stuttgart.
Reck mentioned numerous individual bones and caudal vertebrae from the 'Obere Saurier-Mergel' of Tendaguru Hill in what was then German East Africa, modern Tanzania, and described them as a new species, *Rhamphorhynchus tendagurensis.* See also: Colbert, E.H., 1984. *The Great Dinosaur Hunters and Their Discoveries.* Dover Publications.

39 Thulburn, R.A., 1973. *Teeth of Ornithischian Dinosaurs from the Upper Jurassic of Portugal.* Servizio Geologico Portugal, Memoires, 22, N.S.: 89-134; Lisbon.

Pterodactylus

The earliest evidence of short-tailed pterosaurs, called Pterodactyloidea after the indicative genus *Pterodactylus* (=flight finger) also comes from Solnhofen limestone. They are about as rare as the Rhamphorhynchoidea, and their principal distinctive characteristic is that they had only a short tail consisting of a few small caudal vertebrae, which was probably meaningless in terms of flight. This new, more 'modern' type of pterosaur appears in several different forms in the Upper Jurassic. This indicates that the Solnhofen Pterodactyloidea had already gone through a long phylogeny, a period in which very different genera came into being, able to conquer differing habitats.

A gap in the fossil record means that we know nothing about the direct ancestors of short-tailed pterosaurs. But as the general structure of their skeleton corresponds basically with that of the long-tailed Rhamphorhynchoidea, we must work on the basis that short-tailed pterosaurs were descendants of this older group of pterosaurs, from which they originated in the Lower or Middle Jurassic.

In order to preserve and possibly improve their ability to fly, proportions were generally altered in the Pterodactyloidea, along with the loss of the tail. The beak became more markedly elongated, and the neck is clearly longer too, as a result of elongation of the individual cervical vertebrae. The head was to a certain extent balanced on the neck and held like that of a pelican in flight, in complete contrast with the Rhamphorhynchoidea, who predominantly held the head extended. The two apertures in the skull before the orbital opening, nostril and preorbital opening, are no longer separated by a bridge of bone, but confluent. The metacarpal bones are relatively long, often even longer than the lower arm. In contrast the flight digit is shorter and the fifth foot digit is reduced and only rudimentarily developed.

Pterodactylus Skulls (right)
Seen here are the skulls of four different species of *Pterodactylus* from the Upper Jurassic limestone of Solnhofen. The length of the skulls range from 1·65in (4·2cm) to 4·25in (10·8cm).

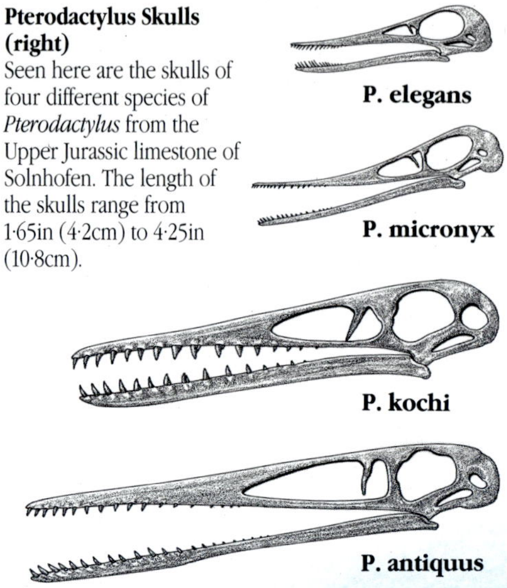

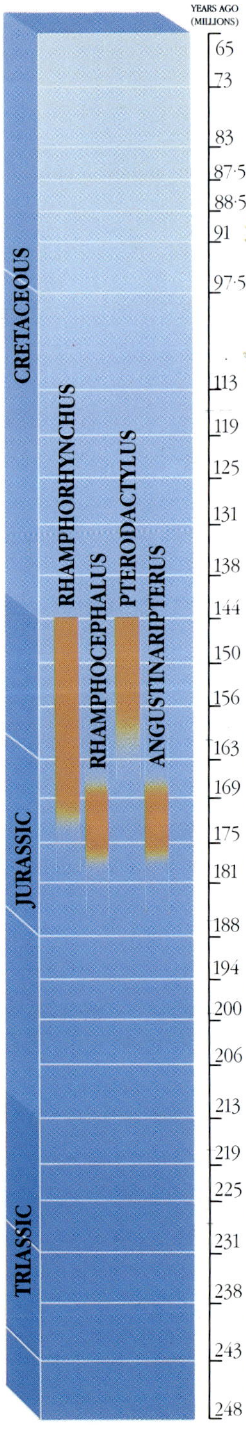

Pterodactylus (right)
Seen here in a landscape that is probably typical of the Solnhofen near-shore environment in the late Jurassic, *Pterodactylus* was an agile flyer that may have fed on insects or small fish. Strikingly, its tail is very much shorter than that of its contemporary *Rhamphorhynchus*. It was obviously no longer required to maintain flight stability. In the background can be seen the small dinosaur *Compsognathus*.

Time Chart (left)
Rhamphocephalus is known from the Bathonian of England, and *Angustinaripterus* from the Middle Jurassic of China. *Rhamphorhynchus* and *Pterodactylus* both peak in the Upper Jurassic of Germany, England, France and East Africa.

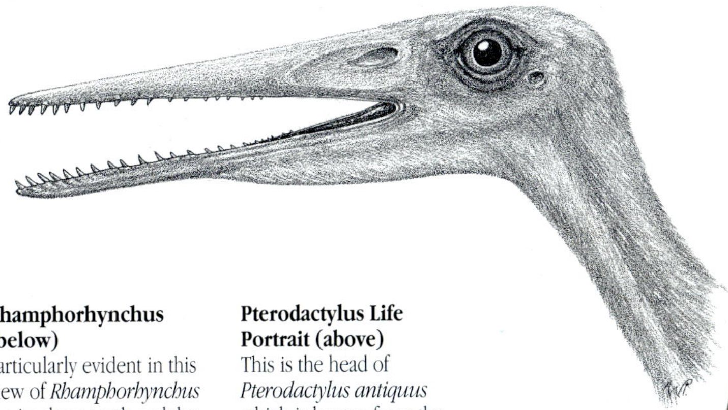

Rhamphorhynchus (below)
Particularly evident in this view of *Rhamphorhynchus* are its sharp teeth and the pointed horny beaks which formed the tips of the jaws.The shape of the wing membranes and of the vertical tail rudder are known from fossil evidence.

Pterodactylus Life Portrait (above)
This is the head of *Pterodactylus antiquus* which is known from the Solnhofen limestone. It is interesting to contrast the posture of this head with that of *Rhamphorhynchus* as shown on page 85. Note also the long beak.

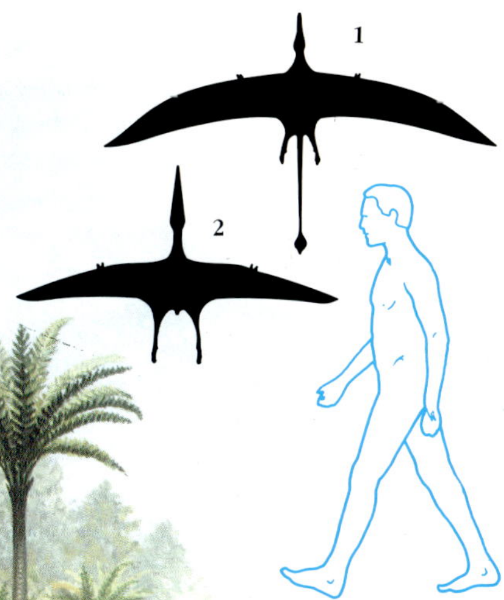

Comparative Sizes
1 *Rhamphorhynchus*: wing span 16-69in (40-175cm).
2 *Pterodactylus*: wing span 14-98in (36-250cm).

Map (below)
1 *Rhamphorhynchus*
2 *Pterodactylus*
3 *Rhamphocephalus*
4 *Angustinaripterus*

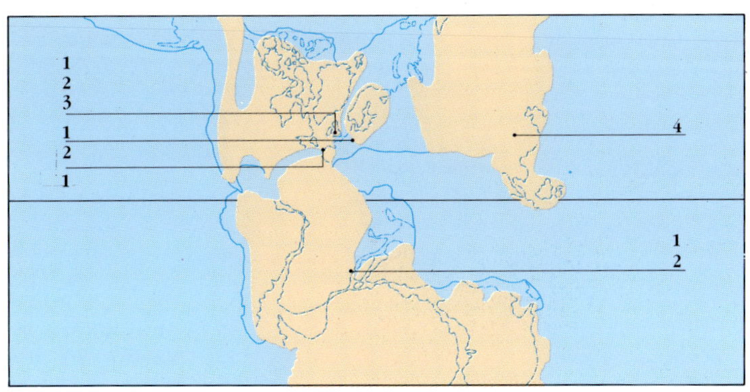

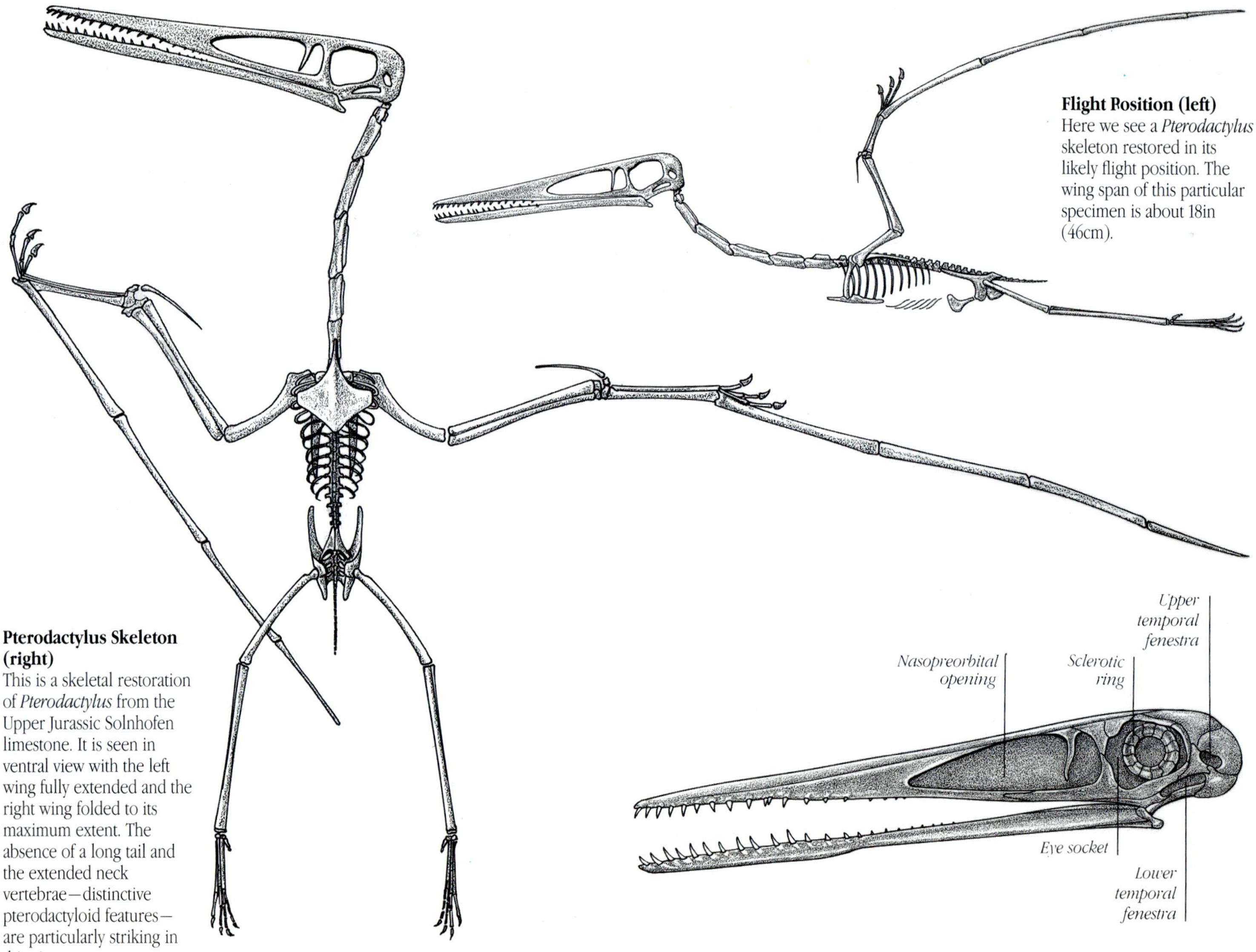

Flight Position (left)
Here we see a *Pterodactylus* skeleton restored in its likely flight position. The wing span of this particular specimen is about 18in (46cm).

Pterodactylus Skeleton (right)
This is a skeletal restoration of *Pterodactylus* from the Upper Jurassic Solnhofen limestone. It is seen in ventral view with the left wing fully extended and the right wing folded to its maximum extent. The absence of a long tail and the extended neck vertebrae—distinctive pterodactyloid features—are particularly striking in this view.

The short-tailed pterosaurs we encounter in the Solnhofen strata are still in the early stages of their evolution. They survived the change from the Jurassic to the Cretaceous, spread throughout the world even in the Lower Cretaceous and in their final form achieved bizarre shapes and gigantic dimensions. Five different genera of Pterodactyloidea have so far been found in the Solnhofen limestone, in which again a good ten species can be distinguished.[40] *Pterodactylus* specimens have also been found in the somewhat younger Mörnsheim strata as well (Malm Zeta 3).

Six species of this genus can be distinguished in the Solnhofen strata. The smallest is *Pterodactylus elegans*, diagnosed by its long, thin teeth, found only in the front part of the jaw. Its wing span is only 10in (25cm). The most common species is *Pterodactylus kochi*, of which there are fine, complete specimens in many museums and collections, some with surviving imprints of the flight membrane.[41] A series of smaller skeletons can be categorized as young animals of the species. When fewer finds were available, these small specimens were thought to be distinct species. However, it is possible to identify the younger animals quite definitively by the degree of ossification of the small toe phalanges. The smallest *Pterodactylus* and therefore probably the youngest ever found in the Solnhofen strata had a trunk length of only 0·75in (2cm) with a wing span of 7in (18cm). This baby pterosaur could only have been a few weeks old, but was certainly already able to fly properly.

Adult *Pterodactylus kochi* reach a wing span of about 20in (50cm). Thus they were about the size of a common moorhen. The larger Solnhofen *Pterodactylus* species like *Pterodactylus longicollum* (=long-necked flight finger), for example, reached wing spans of 4·75ft (1·45m), thus corresponding to the modern herring gull. We only know individual wing and leg bones of the largest species, *Pterodactylus grandis*. The bones suggest a wing span of about 8·2ft (2·5m), corresponding to the size of a bearded vulture. Thus this species was one of the largest of all the Jurassic pterosaurs.

Skeletal remains from Upper Jurassic strata in France, England and East Africa were also assigned to the genus *Pterodactylus*. Thus in 1859 H. von Meyer described a single bone from the upper arm (humerus) from the lithographic limestone of Cerin (Ain) in France as *Pterodactylus cerinensis*,[42] and a tibia was later assigned to the same genus.[43] In 1873 indivi-

dual skeletal remains of a large pterosaur from the Upper Jurassic of Boulogne-sur-Mer in northern France were described as *Pterodactylus suprajurensis*,[44] and again in 1874 R. Owen established two new species, *Pterodactylus manseli* and *Pterodactylus pleydelli* for indivi-

40 Wellnhofer, P., 1970. *Die Pterodactyloidea (Pterosauria) der Oberjura-Plattenkalke Süddeutschlands.* Abhandlungen der Bayerischen Akademie der Wissenschaften, N.F., 141 133 pp.; Munich.

41 Wellnhofer, P., 1987. *Die Flughaut von Pterodactylus (Reptilia, Pterosauria) am Beispiel des Wiener Exemplares von Pterodactylus kochi (Wagner).* Annalen des Naturhistorischen Museums Wien, 88 (A): 149-162; Vienna.

42 Meyer, H. von, 1859-1860. *Reptilien aus dem lithographischen Schiefer des Jura in Deutschland und Frankreich.* 144 pp.; Frankfurt.

43 Lortet, L., 1892. *Les Reptiles fossiles du Bassin du Rhône.* Archives du Musée d'Histoire Naturelle Lyon, 5: 1-139; Lyon.

44 Sauvage, H.E., 1873. *Note sur les reptiles fossiles.* Bulletin de la Société Géologique de France, ser. 3, 1: 365; Paris.

Right: *Pterodactylus kochi* from the Solnhofen limestone of Bavaria. *P. kochi* is the most common species of *Pterodactylus* to have been discovered. This Munich specimen is one of the most complete and best preserved pterodactyls in existence. Imprints of the wing membranes can be seen around the margin of the skeleton. The position of the head, rectangular on the neck, is typical of the short-tailed pterodactyloids. The wing span is 18in (46cm). Thus *Pterodactylus kochi* was about the size of a common moorhen. Larger species are known, the largest of all being *Pterodactylus grandis* which seems to have been about the size of a modern vulture.

Pterodactylus antiquus Skull (left)
This is a restoration of the skull of *Pterodactylus antiquus*, a species which is again known from the Solnhofen limestone of Bavaria. In contrast to *Rhamphorhynchus*, the nostril and the preorbital opening are not completely separated by a bridge of bone, but are confluent as shown here. The length of this particular skull is 4·25in (10·8cm). The drawing is based on the type specimen first described by Collini in 1784.

Left: The smallest pterodactyl ever discovered is a juvenile individual from the Solnhofen limestone. This baby pterosaur had a wing span of only 7in (18cm), and a body length of only 0·75in (2cm). It was probably just a few weeks old when it died. The indication that it is a juvenile, and not a distinct species in its own right, is the lack of complete ossification of the skeleton.

Left: *Pterodactylus elegans* from the Solnhofen limestone is one of the smallest pterosaur species. Its wing span was only about 10in (25cm). After death the head was pulled backwards by strong ligaments between the neck vertebrae, after the decay of the neck muscles. This bend of the neck may indicate that this animal died on dry land and was washed into the lagoon as a dried mummy.

dual pterosaur bones from the Kimmeridge Clay of Weymouth on the Dorset coast of south-east England.[5] Kimmeridge Clay, named after the small town of Kimmeridge east of Lyme Regis, corresponds in age in its higher layers to the Solnhofen Limestone of Bavaria. As well as large numbers of ammonite fauna it also contains fossil reptiles such as plesiosaurs, ichthyosaurs, crocodiles and, though very rarely, pterosaur bones as well. In the eighteenth century alum was mined from the clay and clay shales here. The famous, or rather infamous, 'Kimmeridge coal', a bituminous oil shale, had also been mined along the cliffs for centuries. The English geologist W.J.Arkell reported as late as 1933 that Kimmeridge oil shale was burned for heating purposes in the huts of the neighbourhood instead of coal, and that the inhabitants' noses had apparently become accustomed to the high sulphur content of the smoke produced.[2] In the second half of the nineteenth century oil was distilled from Kimmeridge Clay, but the high sulphur content and thinness of the strata that could be mined made the process uneconomic.

And finally, in the course of the dinosaur excavations on Tendaguru Hill in Tanzania, East Africa, *Pterodactylus* remains were found as well as *Rhamphorhynchus* fossils, and these were assigned to three different new species, *Pterodactylus arningi*, *Pterodactylus brancai* and *Pterodactylus maximus* by Reck in 1931,[38] though P.Galton of Bridgeport University believes that *Pterodactylus brancai* consisting of a bird-like tibiotarsus, corresponds better to the genus *Dsungaripterus* from the Lower Cretaceous of China than to the Solnhofen *Pterodactylus*.[45] There are also other species based on individual bones from France and England which cannot be assigned to the genus *Pterodactylus* with any certainty.

45 Galton, P., 1980. *Avian-like tibiotarsi of pterodactyloids (Reptilia: Pterosauria) from the Upper Jurassic of East Africa.* Paläontologische Zeitschrift, 54 (3/4): 331-342; Stuttgart.

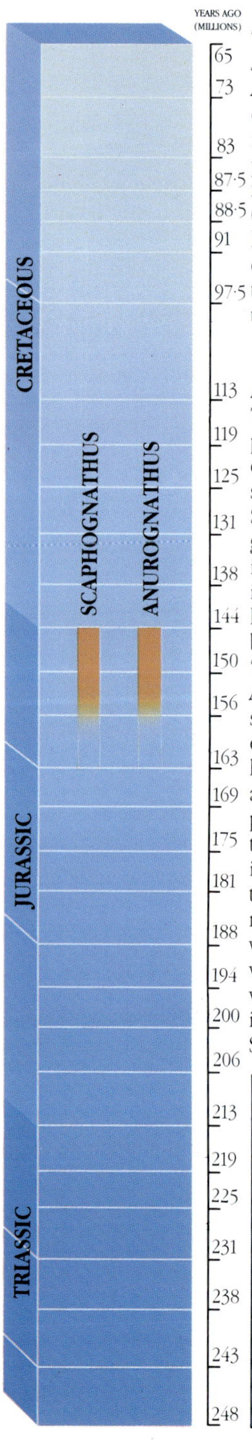

Time Chart (left)
Scaphognathus and *Anurognathus* were contemporary rhamphorhynchoid pterosaurs known only from the Upper Jurassic (Tithonian) Solnhofen limestone. Along with all other long-tailed pterosaurs, they both died out before the end of the Jurassic.

Anurognathus (right)
This is one of the strangest pterosaurs which is known only by a single fossil specimen from the Solnhofen limestone. In its general body proportions it is a rhamphorhynchoid, but its tail is greatly reduced in length and forms a kind of bird-like pygostyl or 'Parson's nose'. *Anurognathus* was a slender pterosaur with extremely long wings. It had a short, deep head with a broad mouth and small, peg-like teeth. This suggests that *Anurognathus* was insectivorous, and was therefore an agile, highly manoeuvrable flyer. In this view it is snapping at a woodwasp (*Pseudosirex*) which is also documented in the fossil record of Solnhofen.

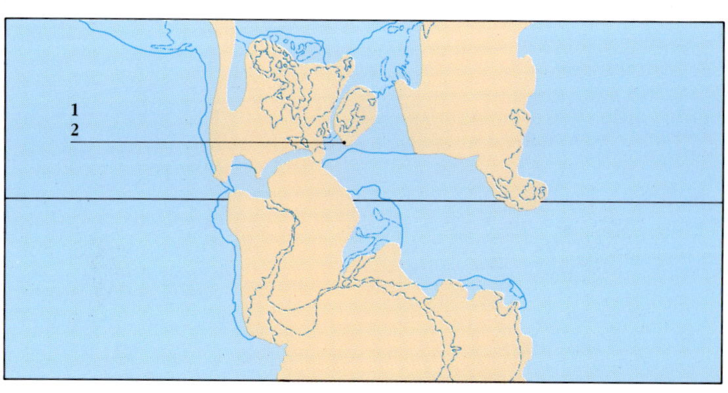

Map (left)
1 *Scaphognathus*
2 *Anurognathus*

Scaphognathus and Anurognathus

The genus *Scaphognathus* (=tub-jaw) is hitherto known only from two specimens from the Solnhofen limestone. The first was one of the earliest pterosaur finds of all and was described as early as 1831 by Bonn professor August Goldfuss.[46] As the tail region had not survived, Goldfuss thought he was dealing with a *Pterodactylus*, which he called *Pterodactylus crassirostris* (=thick-beaked flight finger). The find was made in Eichstätt and is in the collection of the University of Bonn. The second specimen of *Scaphognathus* came from the Solnhofen limestone of Mühlheim near Solnhofen. Here the long tail was preserved, mean-

46 Goldfuss, A., 1831. *Beiträge zur Kenntnis verschiedener Reptilien der Vorwelt.* Nova Acta Academiae Leopoldinae Carolinae, 15: 61-128; Breslau and Bonn. Wellnhofer, P., 1983. (see reference 34), p. 42-43.

Right: A young, half-grown *Scaphognathus* from the Solnhofen limestone. It is the second known specimen of this rare pterosaur genus. The immature skeleton was not fully ossified: the tail was still flexible, and some small foot bones had not yet developed.

Scaphognathus (below)
In its general body proportions and with its long tail, *Scaphognathus* resembles *Rhamphorhynchus*, but it does have some peculiarities. *Scaphognathus* had a somewhat shorter head and long teeth which were set in an upright position. The tips of the jaws are not as pointed as in *Rhamphorhynchus*, but rather blunt. It is not clear, however, whether *Scaphognathus* was an insect feeder or a fish catcher.

Comparative Sizes (above)
1 *Anurognathus*: wing span 20in (50cm).
2 *Scaphognathus*: wing span 3ft (90cm).

Right: *Scaphognathus crassirostris* is a long-tailed rhamphorhynchoid pterosaur with long, upright teeth. The skull length of this specimen is 4·5in (11·5cm). Its wing span is relatively small: just 3ft (90cm). This specimen was described by August Goldfuss as early as 1831.

An Immature Scaphognathus (below)
This is a drawing of the juvenile, half-grown individual of *Scaphognathus* that is pictured on the previous page. This fossil comes from the Solnhofen limestone of Mühlheim. The preservation of the long tail meant that this genus could be confidently classified with the Rhamphorhynchoidea. We can deduce that this individual died before it was fully grown because its skeleton was not fully ossified; note that there are some small bones missing in the foot.

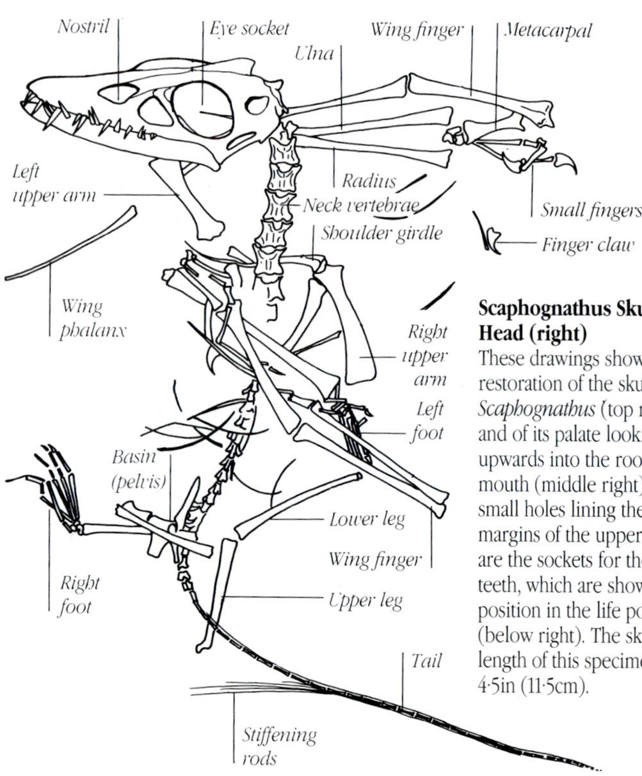

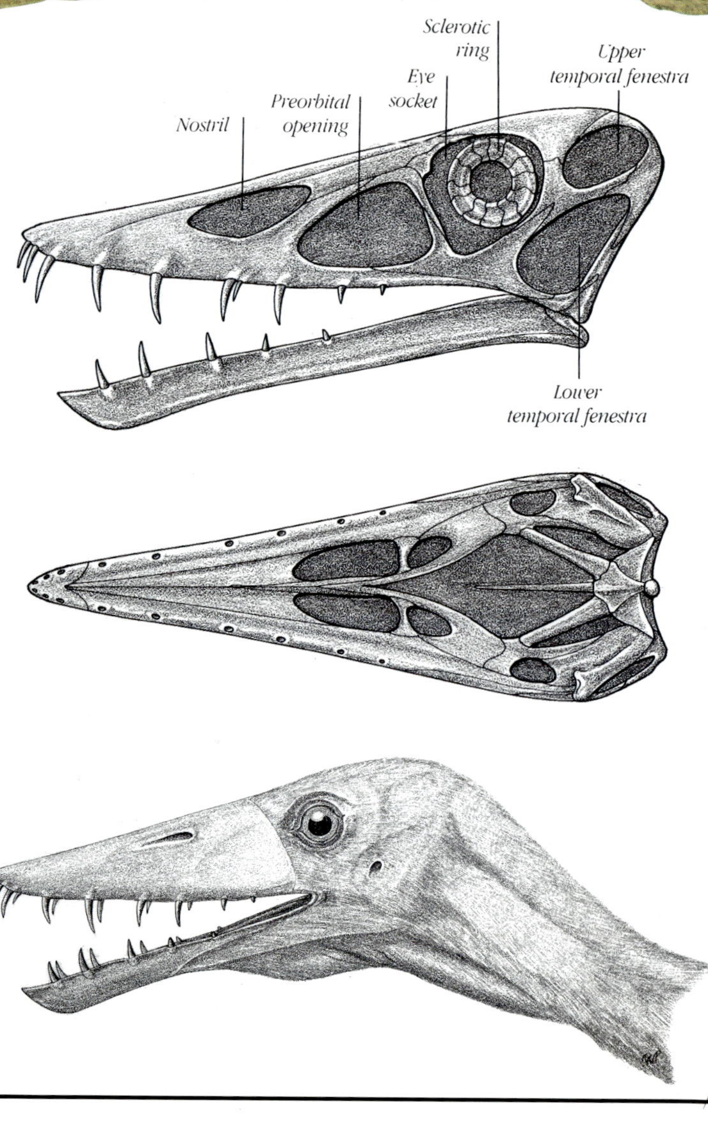

Scaphognathus Skull and Head (right)
These drawings show a restoration of the skull of *Scaphognathus* (top right), and of its palate looking upwards into the roof of the mouth (middle right). The small holes lining the margins of the upper jaw are the sockets for the 18 teeth, which are shown in position in the life portrait (below right). The skull length of this specimen is 4·5in (11·5cm).

Above: *Anurognathus ammoni* from the Solnhofen limestone of Bavaria. This is the only known specimen of this rhamphorhynchoid pterosaur which had a short tail.

Anurognathus ammoni (below)
This fragmentary skeleton is the unique specimen of *Anurognathus*. The actual fossil is shown in the photograph above. In fact, what we see are the impressions of bones rather than the bones themselves; it is like looking at a negative imprint of the skeleton. *Anurognathus* is an interesting genus because its tail is reduced to a short stump, similar to the Parson's nose or pygostyl of modern birds. Despite this feature, the general bauplan of the skeleton reveals that it must be grouped within the long-tailed pterosaurs; it is one of the rhamphorhynchoids.

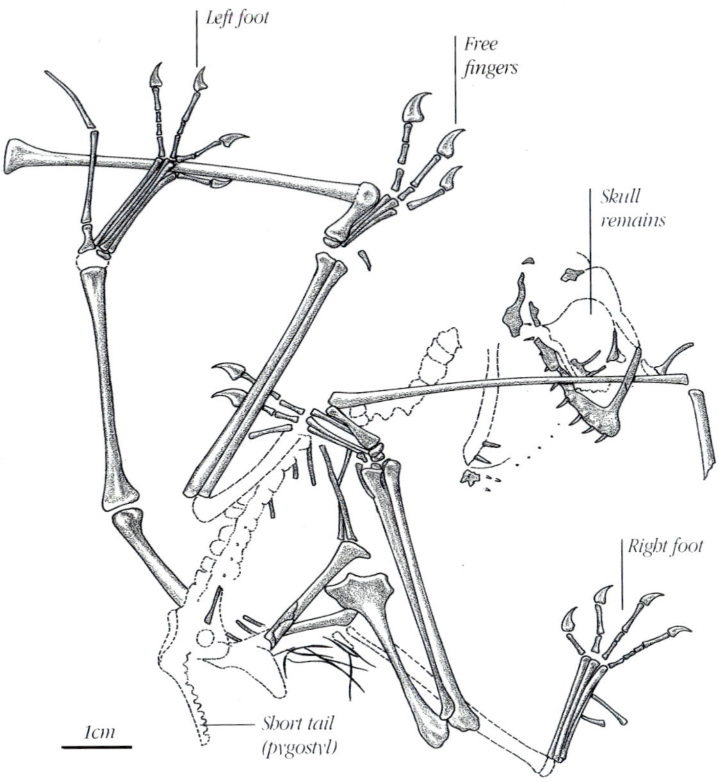

Anurognathus Skull and Head (below right)
The skull of *Anurognathus* is short and broad, with large openings. Is is only 1·2in (3cm) in length. Its broad, rounded jaws which are studded with short, peg-like teeth suggest that this pterosaur may have been an insect eater. If that were the case, it must have been a very agile flier to hunt down its prey on the wing, just as insectivorous birds today are quick and nimble in the air. The poor fossil preservation resulted in a rather tentative restoration.

ing that this genus could be classified with the long-tailed pterosaurs, the Rhamphorhynchoidea. The Mühlheim specimen is about half the size of the one from Eichstätt. From the small degree of ossification in the smaller specimen we must assume that this was a young animal that died before it was fully grown. But both specimens are of the same species. *Scaphognathus* had a shorter, more compact skull than *Rhamphorhynchus*. Its teeth, 18 in the upper jaw and 10 in the lower jaw, did not point forward, but were set upright in the jaw. The ends of the jaw do not meet at a point at the front, but are fairly blunt, like the bow of a boat, hence its name. Also the preorbital fenestrae are larger than in *Rhamphorhynchus*. The young *Scaphognathus* had a wing span of 20in (50cm), and the adult a span of 3ft (90cm).

So far only one specimen of *Anurognathus* (=tailless jaw) has been found, and this only as a negative imprint of skeletal remains on the surface of an Eichstätt limestone slab. It was described in 1923 as *Anurognathus ammoni*, after the Bavarian geologist Ludwig von Ammon, and is in the Paläontologische Staatssammlung in Munich.[47] This was a small and graceful pterosaur, one of the long-tailed group, the Rhamphorhynchoidea. It occupies a special place within this group, however. It has a short, reduced tail, a kind of 'Parson's nose' as in birds, but has to be classified as rhamphorhynchoid because of other skeletal characteristics. Thus it has a short metacarpus, a short neck and a long fifth digit on the foot. It also has a striking high, short skull, with small peg-like teeth. As *Anurognathus* apparently had a broad mouth it is to be assumed that it was insectivorous. Therefore it must have been an extraordinarily skilful flier if it caught its prey, dragonflies or wood-wasps, in flight. Its trunk was only 2in (5cm) long but it had extremely long wings with a span of 1·6ft (50cm).

47 Döderlein, L., 1923. *Anurognathus Ammoni, ein neuer Flugsaurier.* Sitzungsberichte der Bayerischen Akademie der Wissenschaften, math.-naturwiss. Klasse, 117-164; Munich.
Döderlein, L., 1929. *Über Anurognathus Ammoni Döderlein.* Sitzungsberichte der Bayerischen Akademie der Wissenschaften, math.-naturwiss. Klasse, 47-63; Munich.

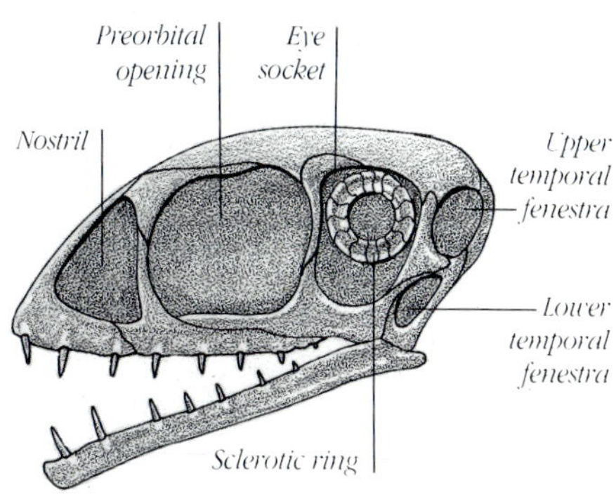

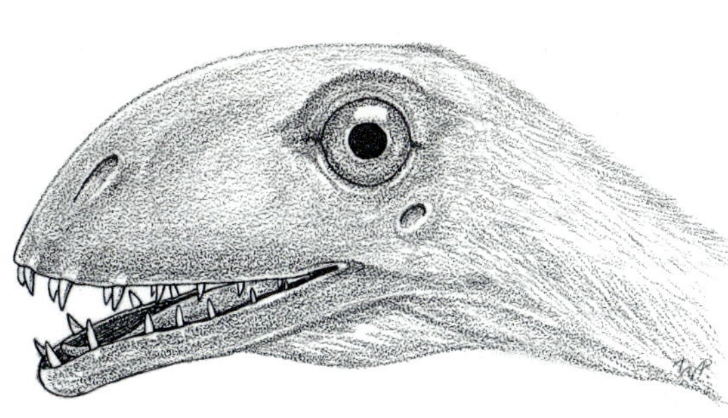

Germanodactylus (left)
Seen here in a late Jurassic landscape are two *Germanodactylus* short-tailed pterosaurs, one using its strong finger claws to climb up the trunk of a tree, and the other hanging head down from a branch by its feet in a bat-like fashion. Whether pterosaurs could actually rest in this position has been questioned by some scientists, however. The characteristic median crest on *Germanodactylus'* head is particularly evident in this view. It was probably covered with a horny carina. Two different species of this pterosaur have been discovered.

Gallodactylus (right)
In general appearance *Gallodactylus* resembled *Pterodactylus*. However, it had a number of distinct characteristics, such as the short medial crest at the rear of the head and teeth that were confined to the front ends of the long, slender jaws. The forwardly pointing teeth would have formed a very efficient gripping tool for grasping slippery fish from out of the water. *Gallodactylus* is known from late Jurassic strata in France and Germany.

Above: *Germanodactylus rhamphastinus*, the second species of *Germanodactylus* known from Bavaria. It is known only from the Mörnsheim limestone which is slightly younger than Solnhofen limestone.

Germanodactylus

We now turn our attention back to a pterodactlyoid pterosaur. The genus *Germanodactylus* (=German finger) was introduced in 1964 by the Chinese palaeontologist C.C.Young for a Solnhofen pterosaur which C.Wiman had already described in 1925 as *Pterodactylus cristatus* (=*Pterodactylus* with crest).[48] So far there have been very few finds of this genus, representing two different species, *Germanodactylus cristatus* from the limestone of Solnhofen and *Germanodactylus rhamphastinus* from Daiting. The Daiting specimen comes from the strata above the Solnhofen limestone, and is thus somewhat younger. It is therefore possible that in terms of phylogeny it should be seen as a descendant of the older Solnhofen species.

Germanodactylus' typical characteristic is a low bone crest on the mid-line of the skull, starting above the nostril and extending above the openings for the eye sockets. The dentition consists of a long row of powerful and relatively short teeth. The smaller *Germanodactylus cristatus* had a skull length of 5·1in (13cm) with a wing span of 3·2ft (98cm), the larger *Germanodactylus rhamphastinus* a skull length of 8·3in (21cm) and wing span of 3·5ft (1·08m).

David Unwin of Reading University also assigns vertebrae, radius, ulna, a first flight digit phalanx, and tibia and fibula of a pterosaur from the Kimmeridge Clay of the Dorset Coast of south-east England to the genus *Germanodactylus*.[49] This is not only the first *Germano-*

48 Wiman, C., 1925. *Über Pterodactylus Westmani und andere Flugsaurier.* Bulletin of the Geological Institute of the University of Uppsala, 20: 1-38; Uppsala.
Young, C.C., 1964. *On a new pterosaurian from Sinkiang, China.* Vertebrata Palasiatica, 8: 221-256; Beijing.

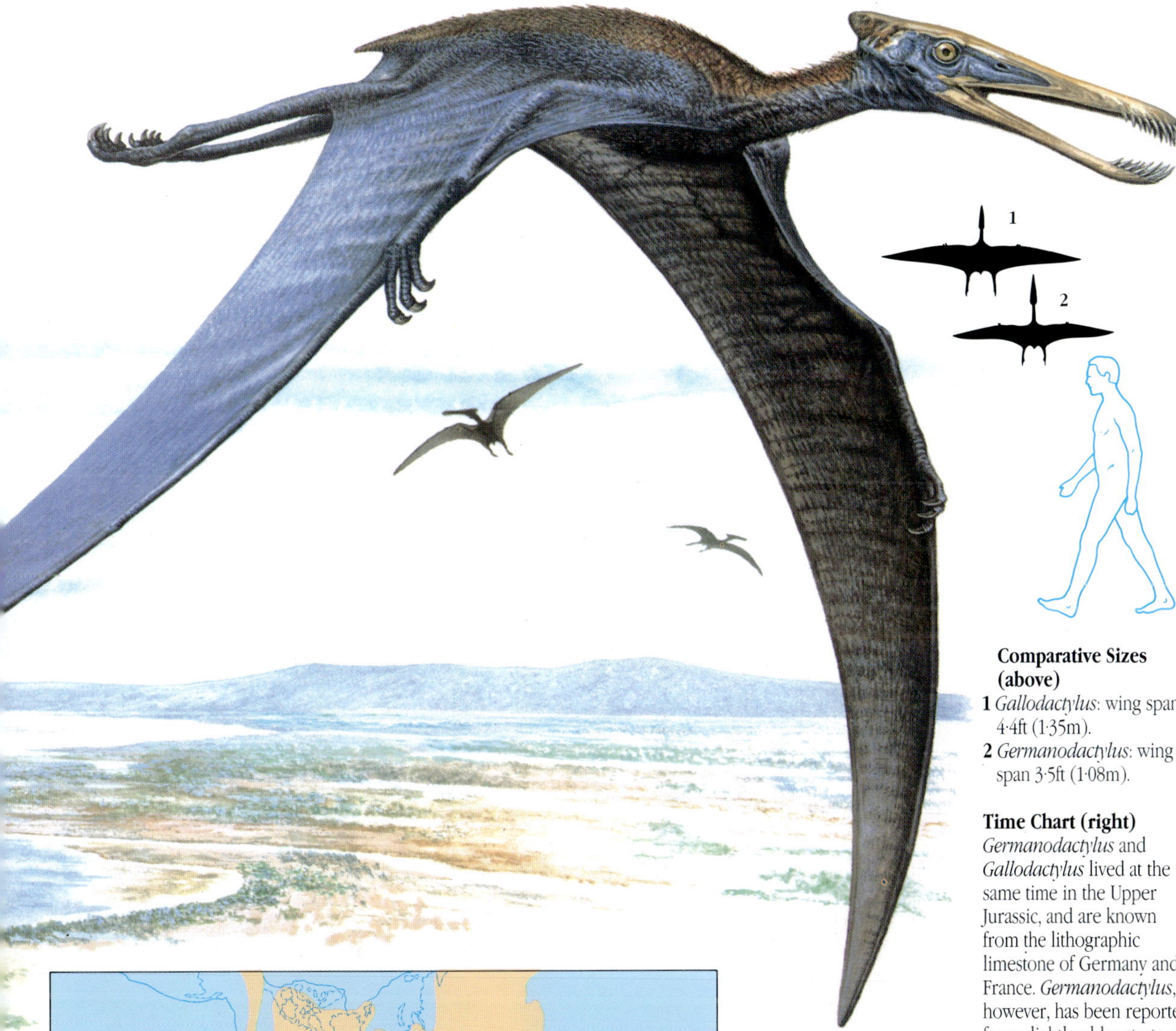

Comparative Sizes (above)
1 *Gallodactylus*: wing span 4·4ft (1·35m).
2 *Germanodactylus*: wing span 3·5ft (1·08m).

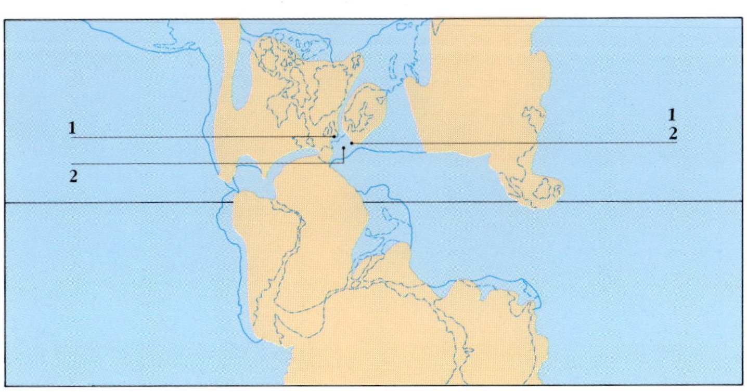

Map (left)
1 *Germanodactylus*
2 *Gallodactylus*

Time Chart (right)
Germanodactylus and *Gallodactylus* lived at the same time in the Upper Jurassic, and are known from the lithographic limestone of Germany and France. *Germanodactylus*, however, has been reported from slightly older strata of Kimmeridgian age in Dorset, southern England. Neither of these short-tailed pterosaurs survived into the Cretaceous.

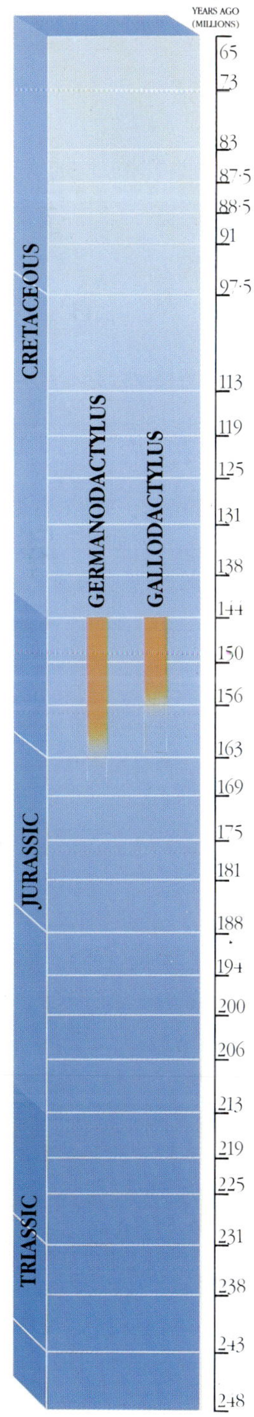

49 Unwin, D., 1988. *A new pterosaur from the Kimmeridge Clay of Kimmeridge, Dorset.* Proceedings of the Dorset Natural History and Archaeological Society, 109: 150-153.

Left and above: Shown here are the skeletal remains and skull of *Germanodactylus cristatus* from the Solnhofen limestone of Bavaria. The skeleton is disarticulated, with the trunk, leg and wing bones, and skull having become separated. This smaller species of *Germanodactylus* had a wing span of about 3·2ft (98cm). The sagittal crest is clearly visible.

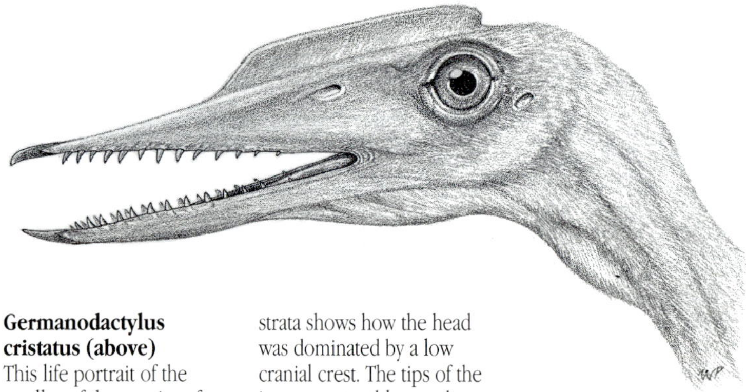

Germanodactylus cristatus (above)
This life portrait of the smaller of the species of *Germanodactylus* currently known from Upper Jurassic strata shows how the head was dominated by a low cranial crest. The tips of the jaws were toothless and were probably covered by pointed, horny beaks.

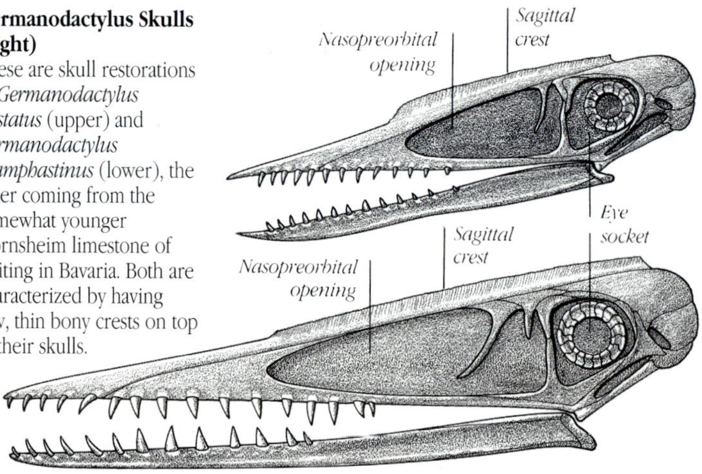

Germanodactylus Skulls (right)
These are skull restorations of *Germanodactylus cristatus* (upper) and *Germanodactylus rhamphastinus* (lower), the latter coming from the somewhat younger Mörnsheim limestone of Daiting in Bavaria. Both are characterized by having low, thin bony crests on top of their skulls.

dactylus to be found outside Bavaria, but also the oldest specimen of a pterodactyloid pterosaur, as the Lower Kimmeridge Clay is geologically somewhat older than the Solnhofen limestone.

Gallodactylus

The name *Gallodactylus* (=Gallic finger) was introduced in 1974 by the French palaeontologist J.Fabre for a pterosaur from the Upper Jurassic of Canjuers (Var) in southern France.[50] There were many features in common with pterosaurs from the Upper Jurassic limestones of Bavaria and Württemberg. Thus the first pterosaur which had been described as *Pterodactylus suevicus* from Nusplingen in Württemberg by Tübingen professor August Quenstedt as early as 1855[51] could now be assigned to the genus *Gallodactylus*. This species is also known from Solnhofen. There *Gallodactylus* is among the larger short-tailed pterosaurs. Its particular characters are an elongated beak with a small number of slender teeth limited to

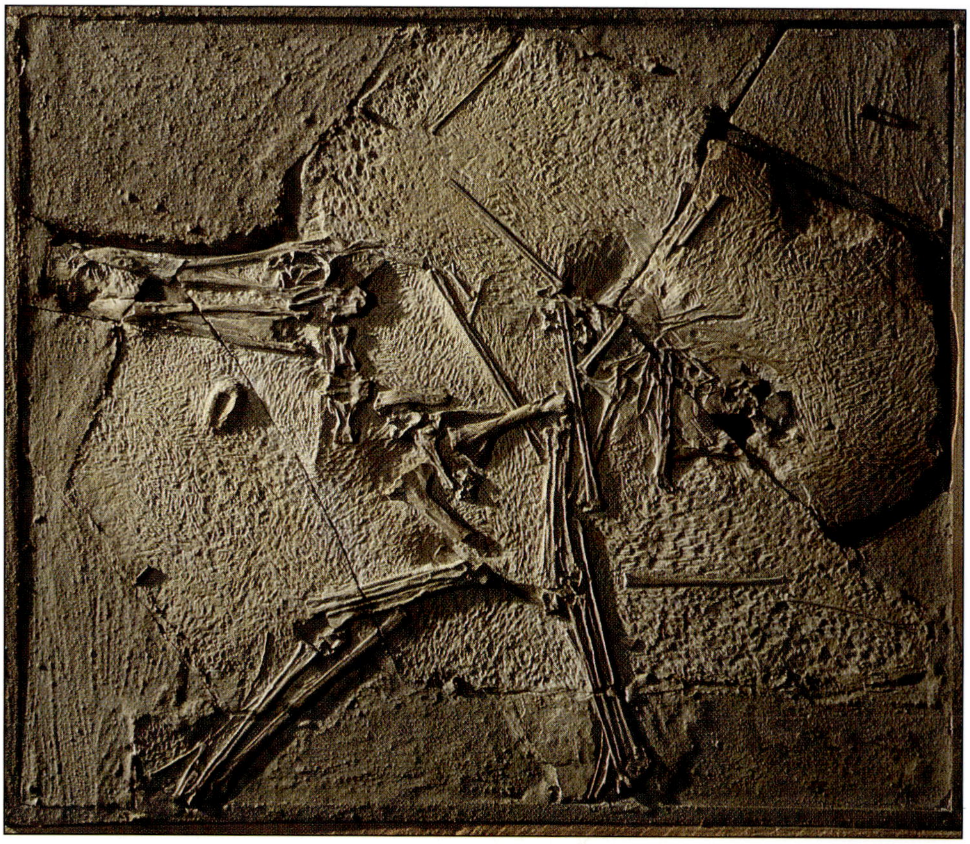

50 Fabre, J., 1974. *Un nouveau Pterodactylidae sur le gisement 'Portlandien' de Canjuers (Var): Gallodactylus canjuersensis nov.gen., nov.sp.* Annales de Paléontologie (Vertébrés), 62, fasc. 1: 35-70; Paris.
51 Quenstedt, F.A., 1855. *Über Pterodactylus suevicus im lithographischen Schiefer Württembergs.* 52 pp.; Tübingen.

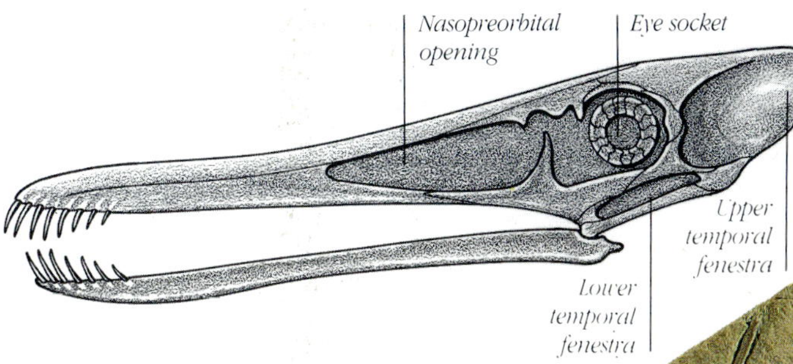

Gallodactylus Skull (above)
This is a restoration of the skull of *Gallodactylus suevicus*. The teeth are grouped together at the front of the jaws. The skull is 5·9in (15cm) long.

Below left: This is the only more or less complete skeleton of *Ctenochasma* of the six specimens known from the Solnhofen limestone. This filter-feeding pterosaur had a wing span of about 2·3ft (70cm). The specimen shows particularly well the very low skull with its long slender jaws which are slightly bent upwards and well adapted for filtering.

the front end of the jaw, and a short crest on the back of the head. *Gallodactylus* had a skull length of 5·9in (15cm) and a wing span of 4·4ft (1·35m).

Ctenochasma

This striking short-tailed pterosaur is one of the rarities of the Solnhofen limestone. It is also known from Upper Jurassic strata in France and northern Germany, however. The first *Ctenochasma* (=comb-jaw) was described as early as 1851 by H. von Meyer[52] although the specimen consisted only of the front section of a lower jaw with numerous long, tightly-packed and strong teeth. It came from Upper Jurassic marine limestone in the Hannover area (Lower Saxony) and was given the name *Ctenochasma roemeri*, after the palaeontologist F.A. Roemer.

The first specimen from the Solnhofen limestone was described in 1862 by Munich Professor of Palaeontology Albert Oppel.[53] It was only the fragment of an upper jaw from the Solnhofen region that had been acquired by King Max II for the Munich state collection in a fossil collection of the Pappenheim country doctor Carl Haeberlein in 1857. This was the same Dr Haeberlein who sold the first *Archaeopteryx* to the British Museum in London five year later. Later however complete skeletons of this species were found, and named *Ctenochasma gracile* by A.Oppel. It was not until 1981 that another species of this genus was described.[54] It was named *Ctenochasma porocristata* because of a porous bony crest on its skull. The specimen, an isolated skull with the front section of the

Right and below right: *Ctenochasma*, the 'comb jaw', is one of the most intriguing pterosaurs to have been found in the Solnhofen limestone. Its long jaws were equipped with hundreds of long, curved, slender teeth which made it a perfect filter feeder. It presumably waded in the water and swept its jaws from side to side in order to strain small marine organisms from the water. The specimens illustrated here show *Ctenochasma's* upper and lower jaw (top right) and the upper jaw with bristle-like teeth seen from above (below right).

Left: *Gallodactylus suevicus* from the Upper Jurassic limestone of Nusplingen in Württemberg. The Nusplingen quarry has yielded only a few pterosaur fossils; this is one of the best preserved examples. It was described by A. Quenstedt in 1855.

52 Meyer, H. von, 1851. *Ctenochasma Roemeri.* Palaeontographica, 2: 82; Stuttgart.
53 Oppel, A., 1862. *Über Fährten im lithographischen Schiefer.* Palaeontologische Mitteilungen des Museums des kgl. bayrischen Staates, 1: 121-125; Stuttgart.
54 Buisonjé, P.H. de, 1981. *Ctenochasma porocristata nov. sp. from the Solnhofen Limestone, with some remarks on other Ctenochasmatidae.* Proceedings of the Koninklijke Nederlandse Akademie van Wetenchappen, B, 84(4): 411-436; Amsterdam.

dentition came from the Eichstätt quarry district.

The apt name *Ctenochasma* was chosen because of this pterosaur's strange dentition. This consisted of a large number of long, thin, inward-bending teeth, arranged in a dense row like the teeth of a comb in the upper and lower jaw. They formed a regular straining apparatus, with which these creatures could filter their food out of the water. For this reason the expression 'comb dentition' is also used to describe its arangement of teeth. *Ctenochasma* had a total of 260 individual teeth in its jaws. The largest of the six specimens so far known had a skull 7·9in (20cm) long and a wing span of at least 3·9ft (1·2m). As it can hardly be supposed that *Ctenochasma* fed in flight, we can only imagine that these pterosaurs swam in the water or waded in it near the beach, as do most modern sea birds.

The genus *Ctenochasma* has also been proved to exist in the Upper Jurassic (Lower

Gnathosaurus (below)
This view shows clearly the specialized dentition of *Gnathosaurus*. The long, slender jaws are lined with a series of teeth which get increasingly longer as they near the front of the snout. This suggests that *Gnathosaurus*, seen here in late Jurassic Solnhofen near-shore environment, must have been a filter feeder. When closed, the beak could be used for catching and filtering small creatures from the water. In the background we can see *Archaeopteryx*, the primordial bird, and a small theropod dinosaur, *Compsognathus*.

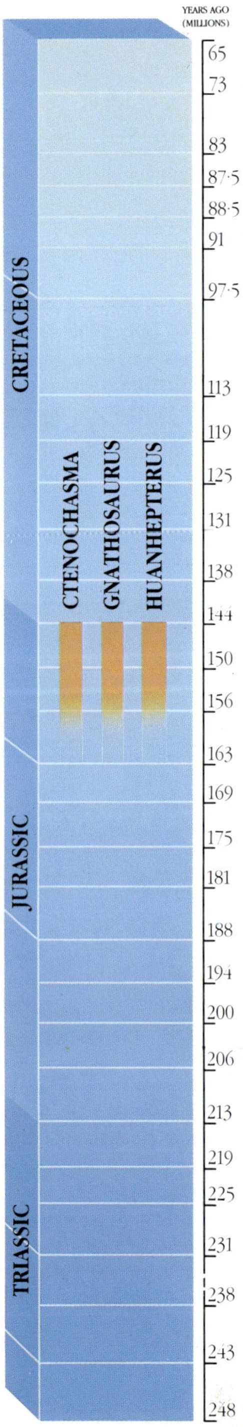

Time Chart (left)
The Upper Jurassic seems to have been the ideal time for pterodactyloid pterosaurs like *Ctenochasma*, *Gnathosaurus* and *Huanhepterus*. All were found in strata of about the same age.

Portlandian) of Haute Marne (France),[55] in the form of an isolated skull 9·5in (24cm) long, with dentition consisting of about 200 long, tightly-packed teeth, indicating a typical filter feeder.

Gnathosaurus

The last Solnhofen pterosaur genus to be discussed here is *Gnathosaurus* (=jaw reptile). This is another extremely rare pterosaur. The two specimens so far discovered are the only remains of this form which is only known from the Solnhofen limestone. The first find, an isolated remnant of a lower jaw from Soln-

55 Taquet, P., 1972. *Un crâne de Ctenochasma (Pterodactyloidea) du Portlandien inférieur de la Haute Marne, dans les collections du Musée de Saint Dizier.* Comptes Rendus de l'Académie des Sciences, 274: 362-364; Paris.

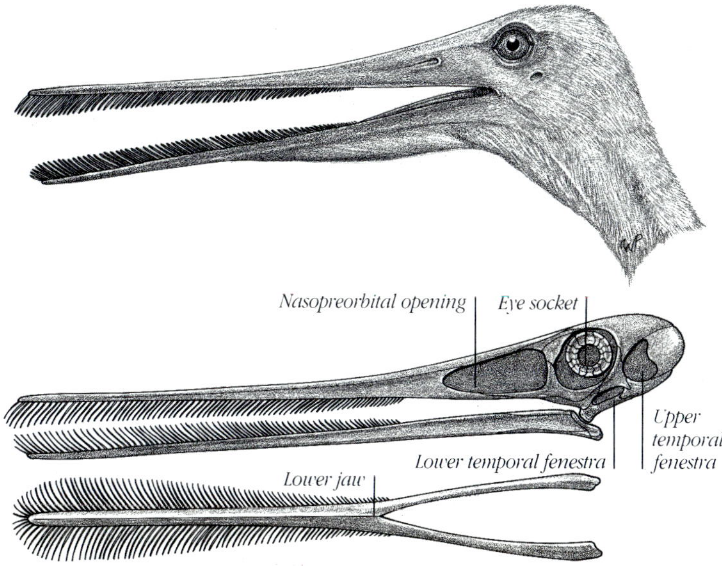

Ctenochasma Head and Skull (right)
These drawings show a life portrait of *Ctenochasma gracile* (above), and a restoration of the skull and lower jaw which is also shown from below in the bottom drawing. As can be seen, the jaws contain an extraordinary number of teeth; there are over 250 in an adult pterosaur. The skull length is 4in (10·4cm). This species had no crest on its skull. *Ctenochasma porocristata* by contrast had a long para-sagittal crest running along the top of its skull which, in life, was probably covered by a thin horny crest.

Ctenochasma (below)
Ctenochasma was even more extremely adapted as a filter feeder than *Gnathosaurus*. Its long jaws formed a filter basket for sifting out small aquatic planktonic organisms, such as crustaceans or the larvae of marine invertebrates, while standing in the shallows. Two distinct species are known, one (*porocristata*) with a bony crest along the midline of the skull.

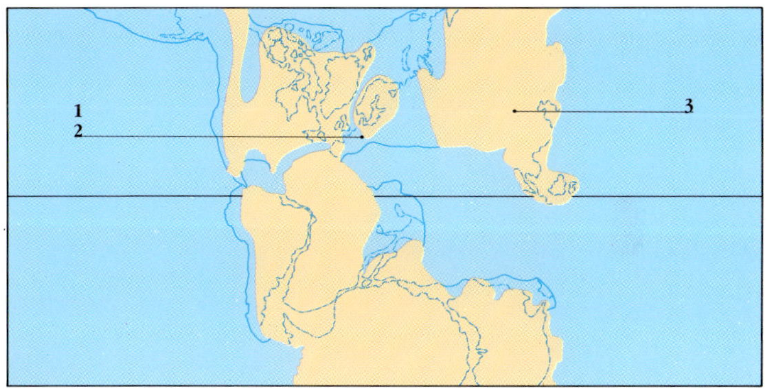

Map (right)
1 *Ctenochasma*
2 *Gnathosaurus*
3 *Huanhepterus*

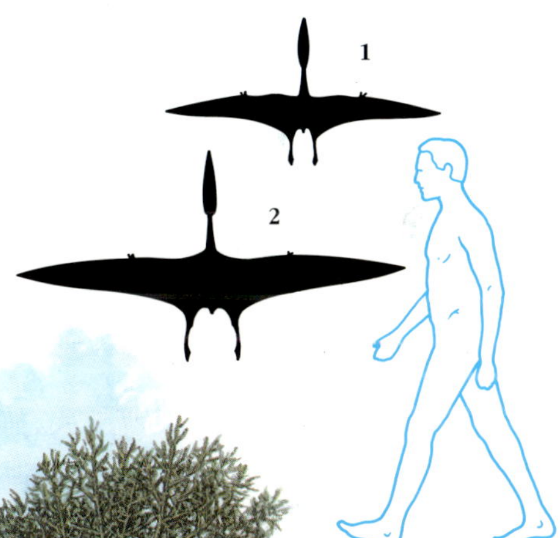

Comparative Sizes (above)
1 *Ctenochasma*: wing span 3·9ft (1·2m)
2 *Gnathosaurus*: wing span 5·6ft (1·7m)

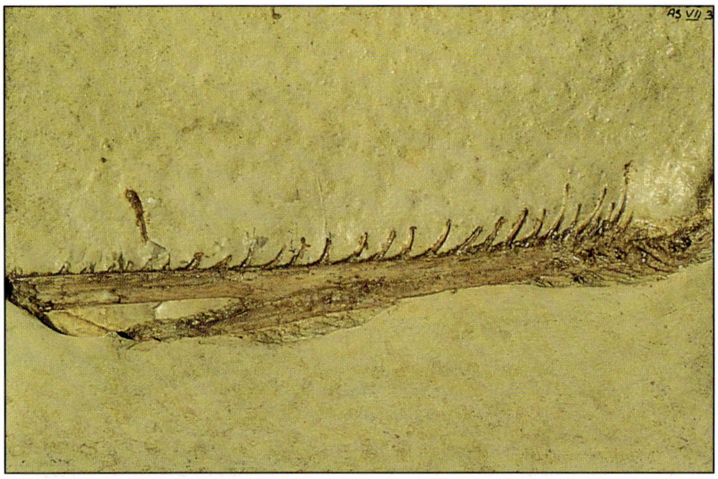

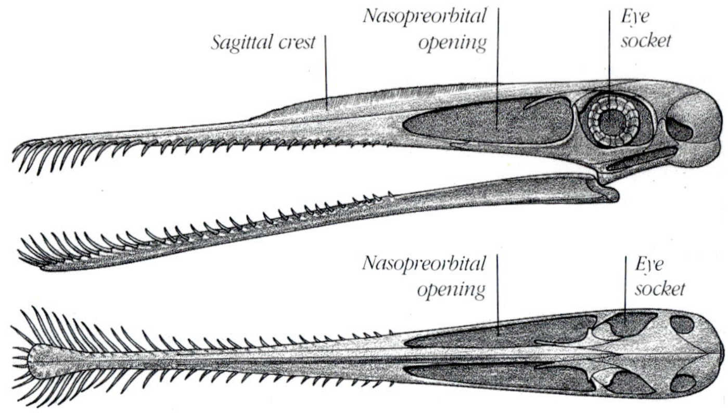

Above left: The first fossil of *Gnathosaurus* was recorded as early as 1832, but was mistaken for a crocodile. Only a partial lower jaw containing numerous teeth was known. It came from Solnhofen.

Gnathosaurus Skull (above)
This skull restoration of *Gnathosaurus* is seen in side and top view. A low, bony crest had developed along the midline of the skull roof. The dense arrangement of teeth points to a diet of small marine organisms. The skull length is 11in (28cm).

Left: A complete skull of *Gnathosaurus* discovered in 1951 in the Solnhofen limestone. The skull is exposed from the underside, and shows the complete dentition consisting of about 65 teeth.

hofen, was introduced as early as 1832 by the famous collector and palaeontologist Georg Graf zu Muenster in Bayreuth, and taken for a piece of crocodile jaw.[56] Later the Frankfurt palaeontologist Hermann von Meyer assigned the name *Gnathosaurus subulatus* (*subulatus* is the Latin for awl-like, after the shape of the teeth).[57]

It was not proved that the creature was a pterosaur until a second find was made in 1951, a skull 11in (28cm) long.[58] Nothing has so far been found of the rest of the skeleton, but it is not impossible that isolated bones from wings and hind legs of larger pterosaurs, as yet not classified with certainty, actually belong to the genus *Gnathosaurus*.

56 Muenster, G. zu, 1832. *Bemerkungen über eine neue Art Pterodactylus aus Solenhofen.* Neues Jahrbuch für Mineralogie, etc., 412-416; Stuttgart.

57 Meyer, H. von, 1834. *Gnathosaurus subulatus, ein Saurus aus dem lithographischen Schiefer von Solenhofen.* Museum Senckenbergianum, I: 3; Frankfurt.

58 Mayr, F.X., 1964. *Die Naturwissenschaftlichen Sammlungen der Philosophisch-Theologischen Hochschule Eichstätt.* Festschrift 400 Jahre Collegium Willibaldinum Eichstätt: 302-334.

Even though this was a large pterosaur with a wing span of about 5·6ft (1·7m) we must assume that it was a filter feeder, though the teeth are more powerful than in the case of *Ctenochasma*, and less densely arranged, there being only 130. But here again the jaws have teeth extending well towards the back, with the longest at the front and set around the spoon-shaped jaw end. Like *Ctenochasma porocristata*, *Gnathosaurus* also has a low bone crest on its skull.

Because of their strange filter dentition *Ctenochasma* and *Gnathosaurus* are significantly different from other pterodactyloids. For this reason they are placed together in a family of their own, which is known as the Ctenochasmatidae.

Russian Remains

Jurassic deposits in Kazakhstan in the southern Soviet Union have been noted for their rich fund of fossils for many years. Upper Jurassic limestone in the Karatau mountains, northwestern foothills of the Tien-shan, have provided fossil insects in excellent condition in particular, showing great similarity with insect fauna in the Solnhofen strata in Bavaria, which are approximately the same age.

A pterosaur from Karatau was first described in 1948.[59] These were the remains of a disarticulated and incomplete skeleton in which fragments of skull and jaw, vertebrae, ribs and bones from wings and hind legs can be recognized. In the jaws are peg-like teeth, with a total of 24 of them in the upper jaw. The shape of the jawbones suggests a high, short skull about 1·9in (48mm) long with a broad mouth, like a frog. For this reason the creature was given the name *Batrachognathus volans* (=flying frog-jaw). Very probably it was an insectivorous pterosaur that caught its prey in flight.

Thus this form is strikingly reminiscent of the somewhat smaller *Anurognathus* from the Solnhofen strata of Bavaria. It is not known whether *Batrachognathus* from Karatau had a reduced, partly fused tail, but the preserved characteristics and skeletal proportions in *Anurognathus* and *Batrachognathus* are very similar, and so they seem to be related, suggesting an assignment of both to the Anurognathidae family. It seems that only two speci-

59 Rjabinin, A.N., 1948. *Remarks on a Flying Reptile from the Jurassic of the Kara-Tau.* Akademia Nauk, Paleontological Institute, Trudy, 15, (1): 86-93; Moscow and Leningrad (in Russian).

mens of *Batrachognathus* have been found. It was a small rhamphorhynchoid pterosaur with a wing span of about 20in (50cm).

Sordes

In the sixties the Moscow zoologist A.G.Sharov discovered pterosaur remains while collecting fossil insects in the Upper Jurassic strata of the Karatau mountains, including an almost complete skeleton with imprints of soft parts of the body and flight membranes.[60] However, the sensational feature of this discovery was

60 Sharov, A.G., 1971. *New Flying Reptiles from the Mesozoic of Kazakhstan and Kirgizia.* Akademia Nauk, Paleontological Institute, Trudy, 130: 104-113; Moscow (in Russian).

Above and below: The skeletal remains of *Batrachognathus volans*, a rhamphorhynchoid pterosaur from the Upper Jurassic lake deposits of the Karatau mountains in Kazakhstan, USSR. The skeleton was disarticulated and washed together, so that many of the bones have got mixed up. Orientation is not easy, but the jaw bones of an obviously short skull are visible, indicating a broad, rounded mouth similar to that of *Anurognathus*. Such a mouth suggests a possible diet of flying insects.

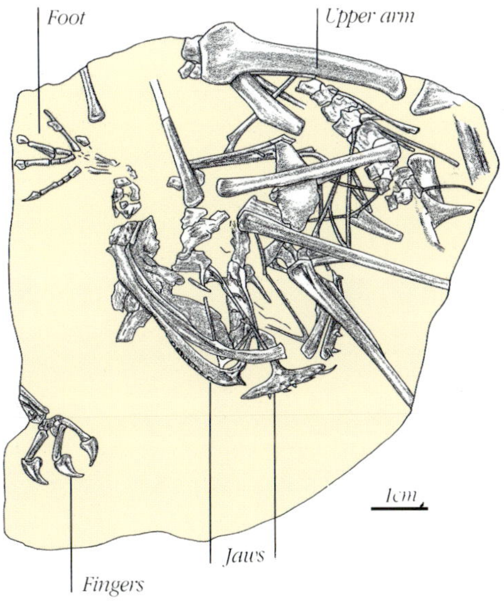

Right: Skeletal remains of *Sordes pilosus* from the Karatau mountains. *Sordes* was a rhamphorhynchid pterosaur with a wing span of only 2·1ft (63cm). The peculiarity of this specimen is the preservation of hairs covering the body, a sign of warm-bloodedness.

that because of the fineness of the sediment grain in which the Karatau fossils are contained, that even the hair that had covered the body of the creature in a thick fur was preserved. This was undeniable proof that pterosaurs were not naked or covered with reptilian scales, but that they were hairy, an indirect proof of their warmbloodedness. Traces of a hair covering had already been found in Holzmaden and Solnhofen pterosaurs,[61] but never the hairs themselves. Sharov called this new long-tailed pterosaur *Sordes pilosus* (=hairy evil spirit).

He writes in his examination, published in 1971, among other things: 'Long, dense and fairly thick hair covers the whole body, and the curvature of individual hairs suggests ample elasticity. There was also hair on the flight membrane, digits and the skin between the foot digits, although it was sparser here, and shorter. The root of the tail was also covered with hair, while the rest of the tail was apparently naked. The longest hairs on the body reached a length of 0·24in (6mm).'

The wing membrane outlines that have survived in *Sordes* show that the hind legs were integrated into the wing membrane. This left the long naked tail free to move, which was vital to its function as a rudder in flight. This all resulted in a fairly broad wing area.

Sordes was a rhamphorhynchoid pterosaur with the typical features of long-tailed pterosaurs, like a short metacarpal bone and long vertebrate tail. The end of the tail apparently flattened out slightly. There are no signs of the rhomboid terminal tail vane as in the Solnhofen *Rhamphorhynchus*. The fifth toe is very

61 Broili, F., 1927. *Ein Rhamphorhynchus mit Spuren von Haarbedeckung.* Sitzungsberichte der bayerischen Akademie der Wissenschaften, mathematisch-naturwissenschaftliche Abteilung: 49-67; Munich.

Above: A more complete specimen of *Sordes pilosus* from the Upper Jurassic of the Karatau mountains in Kazakhstan. The shape of the skull and the dentition resembles that of *Scaphognathus*.

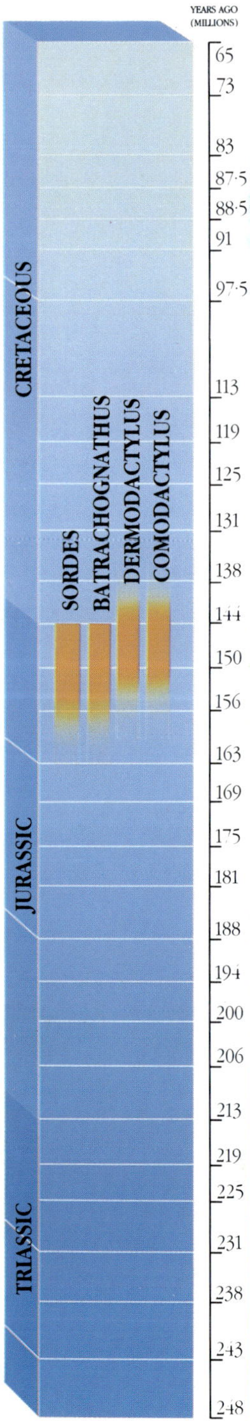

Time Chart (left)
Sordes and *Batrachognathus* are of about the same geological age as the Solnhofen pterosaurs. *Comodactylus* and *Dermodactylus* may be younger.

Map (below)
1 *Sordes*
2 *Batrachognathus*
3 *Dermodactylus*
4 *Comodactylus*
5 *Mesadactylus*

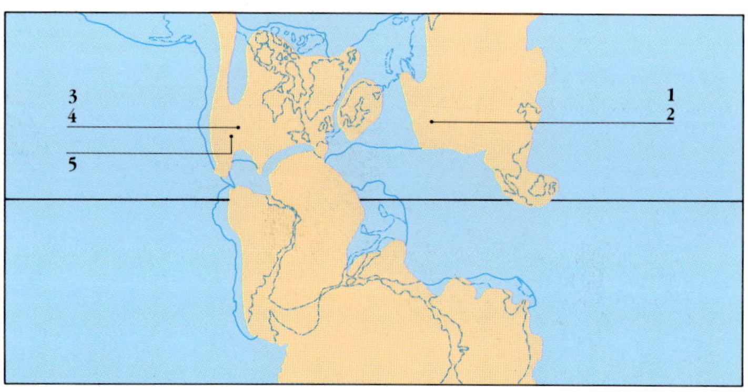

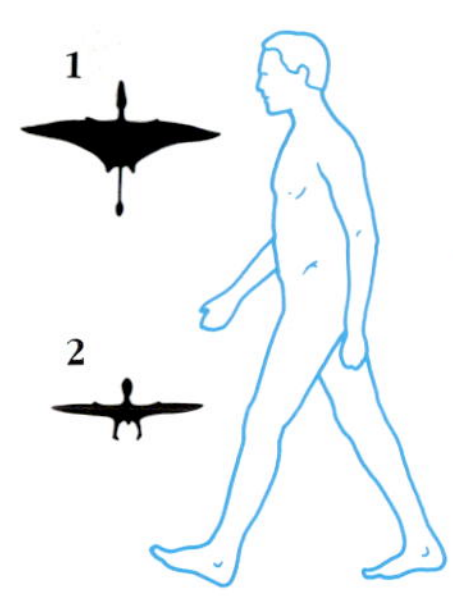

Comparative Sizes (left)
1 *Sordes*: wing span 24·8in (63cm).
2 *Batrachognathus*: wing span 20in (50cm).

Sordes (right)
Sordes is a small, long-tailed pterosaur that was probably a close relative of the Solnhofen *Scaphognathus*. The fossil preservation of its wing membranes shows that the wings were short and broad. A membrane also extended between the hind legs. *Sordes* is particularly fascinating because of the preservation of hairs which covered its body with a dense fur. This is evidence that *Sordes* and probably all pterosaurs were warm-blooded.

Batrachognathus (right)
Closely related to the Solnhofen *Anurognathus*, *Batrachognathus* was an insect eater as its name (='frog jaw') suggests. It had a short head and a wide mouth with small, peg-like teeth. It is known only by incomplete skeletal remains. Here it is shown with a short, reduced tail, by analogy to *Anurognathus* only.

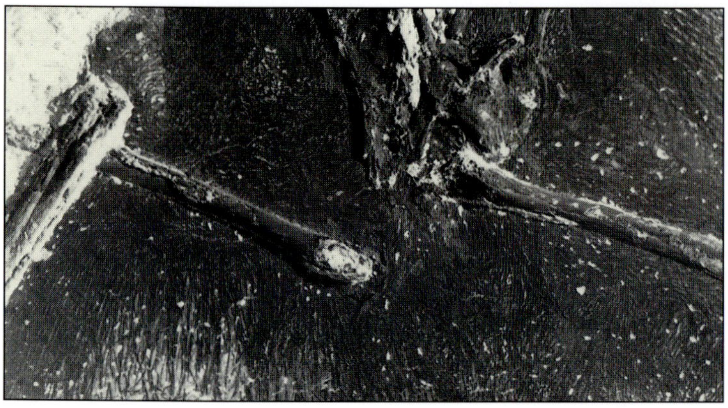

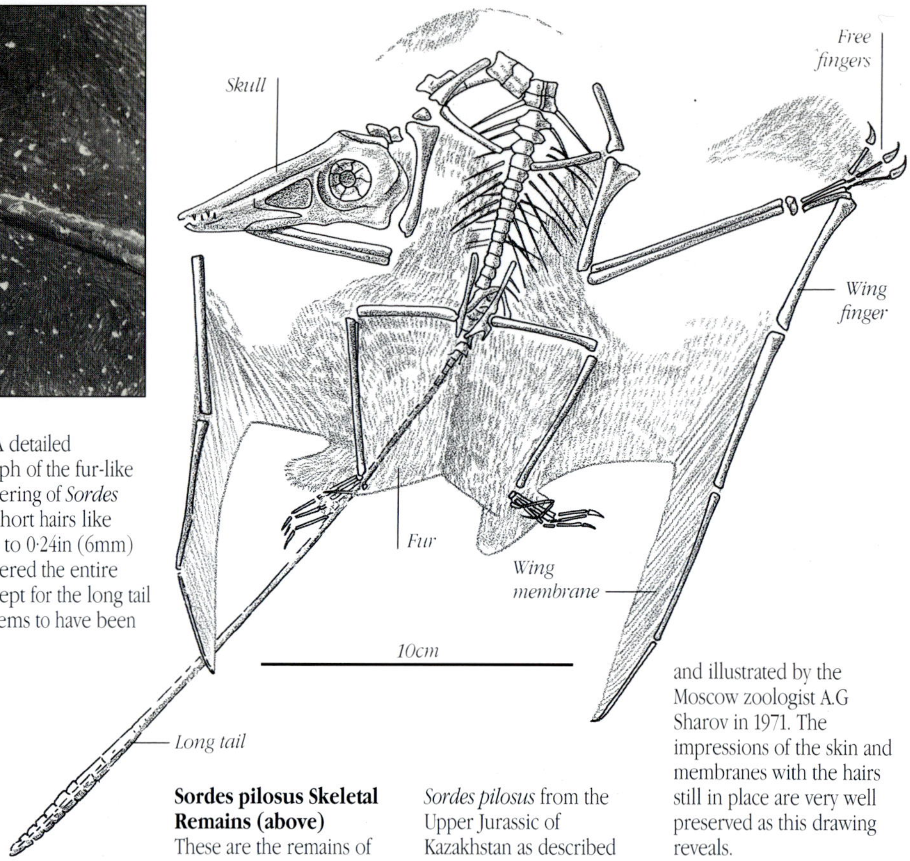

Above: A detailed photograph of the fur-like body covering of *Sordes pilosus.* Short hairs like these, up to 0·24in (6mm) long, covered the entire body except for the long tail which seems to have been naked.

Sordes pilosus Skeletal Remains (above)
These are the remains of *Sordes pilosus* from the Upper Jurassic of Kazakhstan as described and illustrated by the Moscow zoologist A.G Sharov in 1971. The impressions of the skin and membranes with the hairs still in place are very well preserved as this drawing reveals.

long and hook-shaped. Sharov assumed that it was used to spread the trailing edge of the flight membrane in flight.

Sordes' skull and dentition are similar to that of the Solnhofen *Scaphognathus*, and a close relationship can thus be assumed. Both have few teeth, set upright and far apart in the jaw, and relatively short wing digits. *Sordes* can therefore also be classified as a member of the Rhamphorhynchidae family. A total of three specimens of skeletons have been found. Its skull was 3·15in (8cm) in length, and its wing span 24·8in (63cm). It was probably piscivorous, but could also have fed on insects.

Excavation in China

The Ordos basin is in the central section of the Yellow River in North Central China. This area is also known as the Shanxi-Gansu-Ningxia Loess Plateau, after the extensive loess deposits (rock dust and silt blown together by ice age winds), which extend over parts of all three provinces. Red Mesozoic strata containing fossils, so-called Redbeds, form the stratum below the loess and have provided skeletons of dinosaurs and also part of a pterosaur skeleton.

The Chinese palaeontologist Dong Zhiming described this pterosaur in 1982, using the name *Huanhepterus* (=wing from the River Huanhe).[62] The first bone was a vertebra found by a farmer in a quarry on the bank of the Huanhe River in Upper Jurassic sediments of the Zhiden Group near Senshilipou, Qinyan County in Gansu Province in 1978. Only an incomplete skeleton of *Huanhepterus* has been found, of which only the front 13in (33cm) of the end of the snout, some cervical, dorsal and sacral vertebrae, the sternum, the skeleton of the left wing and the left hind leg survived. It was a short-tailed pterosaur whose dentition showed similarity with the Solnhofen *Gnathosaurus*. The teeth are very long, slender and pointed, and tightly packed in the front section of the jaw. Thus *Huanhepterus* was grouped in the Ctenochasmatidae family. On the mid-line of the skull is a thin, low bone crest similar to that of *Gnathosaurus* or *Ctenochasma poro-*

62 Dong Zhiming, 1982. *On a new Pterosauria (Huanhepterus quingyangensis gen. et sp. nov.) from Ordos, China.* Vertebrata Palasiatica, 20, (2): 115-121; Beijing (Chinese with English abstract).

Left: The so-called Redbeds of North Central China have proved a fertile hunting ground for palaeontologists. Red Mesozoic strata lie below the extensive loess deposits that were blown together by ice age winds, and the skeletons of dinosaurs and a pterosaur have been found in this region. We see here late Jurassic sediments of the Zhiden group that have been exposed at Senshilipou, Qinyan County, Gansu Province. It was here that the specimen of *Huanhepterus* was found. This specimen was described by Dr Dong Zhiming in 1982. Its name signifies 'wing from the River Huanhe'. So far, this find is the only record of late Jurassic pterosaurs in China and fills the gap between the Middle Jurassic pterosaurs at Dashanpu, and the Lower Cretaceous pterosaurs from the Junggar Basin in Sinkiang.

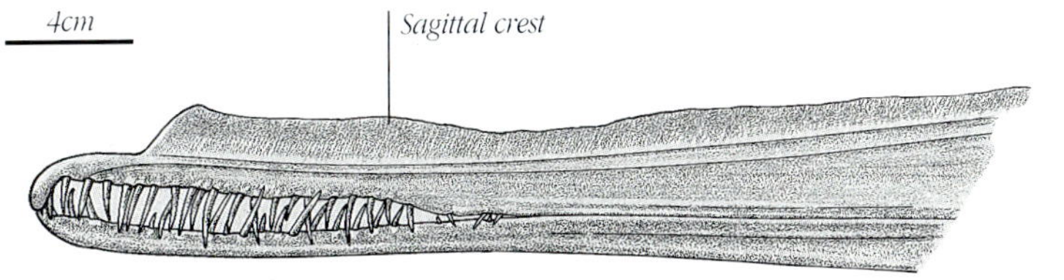

Above left: The front end of the jaws of *Huanhepterus quingyangensis* as preserved. As well as the jaws, part of a long, slender hindlimb has also been discovered in the late Jurassic sediments of the Ordos basin in China.

Huanhepterus quingyangensis (above) The drawing shows *Huanhepterus quingyangensis* from the Upper Jurassic of Gansu, China. The jaws are fringed with long, slender teeth. The teeth are tightly packed towards the front of the jaw, a characteristic that recalls *Ctenochasma*. On top of the skull there extended a low sagittal crest.

cristata from Solnhofen. The lower jaw had a total of 50 and the upper jaw about 52 teeth. The largest, 1in (25mm) long, is the eleventh tooth in the upper jaw. The end of the snout is somewhat broader towards the front.

A striking feature is the very long neck with markedly elongated cervical vertebrae. In this case there is no notarium, i.e. a fusion of the anterior dorsal vertebrae to anchor the pectoral girdle. The sacrum consists of seven vertebrae. Measured by the size of the humerus (5·7in; 14·5cm) and the first phalanx of the flight digit (14·2in; 36cm), *Huanhepterus* was a relatively large pterosaur, whose wing span can be estimated at about 8·2ft (2·5m). Its comb-like dentition indicates that it filtered out small organisms under water.

The Morrison Formation

The rarity of Jurassic pterosaur finds in North America is caused largely by the absence of fine-grained, Jurassic rocks like Solnhofen lithographic limestone or Holzmaden Posidonian shale, made up of fine, soft sediments in which the fragile and delicate pterosaur bones could survive as fossils. Skeletal remains of Jurassic pterosaurs have so far been found only in the Morrison formation of Wyoming and Colorado. It consists of a massive series of clay, silt and sandstones and also coarser conglomerates about 328ft (100m) thick. They originated from river and lake deposits about 135 million years ago, in the Jurassic, thus somewhat later than the Solnhofen strata in Bavaria.[63]

The Morrison Formation became famous because of the spectacular dinosaur finds made for the first time in Colorado in 1877, later continued very successfully in Utah and Wyoming. Here, near the place called Como Bluff, a station on the Union Pacific Railroad, already operating at the time, the principal explorers were the 'bone hunters' employed by O.C. Marsh, from Yale College, New Haven.[64] They excavated numerous skeletons of enormous dinosaurs, including *Apatosaurus* and *Diplodocus*, which can now be seen in the Yale Peabody Museum in New Haven.

Dermodactylus

In the very early stages of the Como Bluff excavations, in Reed's Quarry 5, Samuel Wendell Williston, then still one of Marsh's collectors, found, along with other small bones, the fragment of a pterosaur bone which Marsh identified as the distal section of the right wing metacarpal of a pterodactyloid pterosaur. He described it in 1878, using the name *Pterodactylus montanus*,[65] but in 1881 changed the name of the genus to *Dermodactylus* (=skin-finger).[66] The bone is only 1·26in (32mm) long. When complete it must have been at least twice as long. In comparison with the thickness of its shaft it was thus a relatively long and slender metacarpal bone and is thus typical of the Pterodactyloidea. The double pulley at the end to take the flight digit has survived. In his description, Marsh mentions that the shaft of the bone is thin-walled and hollow, and oval in cross-section. It is only possible to speculate about the size of the whole animal. *Dermodactylus* could have had a wing span of about 3·28ft (1m).

Fragmentary as this fossil is, it was still the first proof of the fact that pterosaurs existed in the New World as early as the Upper Jurassic. Only a very few other fossil proofs have appeared since.

63 Jensen, J.A. and Padian, K., 1989. *Small Pterosaurs and Dinosaurs from the Uncompahgre Fauna (Brushy Basin Member, Morrison Formation: ?Tithonian), Late Jurassic, Western Colorado.* Journal of Paleontology, 63, (3): 364-373.

64 Ostrom, J.H. and McIntosh, J.S., 1966. *Marsh's dinosaurs. The collections from Como Bluff.* Yale University Press, 388 pp.; New Haven.

65 Marsh, O.C., 1878. *New pterodactyl from the Jurassic of the Rocky Mountains.* American Journal of Science, ser. 3, 16: 233-234.

66 Marsh, O.C.: 1878. *Notes on American Pterodactyls.* American Journal of Science, ser. 3, 21: 342-343.

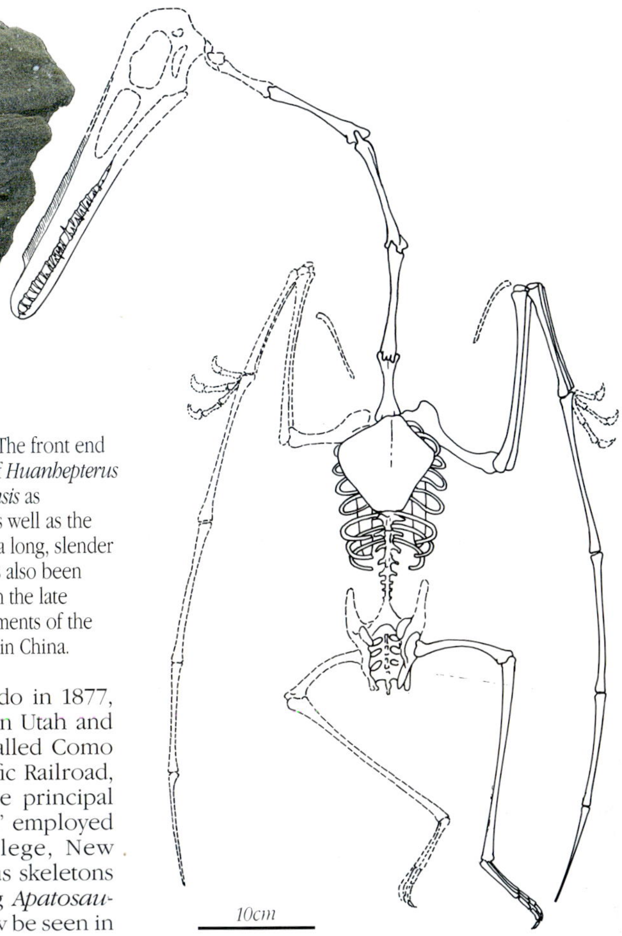

Huanhepterus Skeletal Restoration (above) The drawing shows the skeleton of *Huanhepterus quingyangensis* as restored by Dong Zhiming in 1982. The bones drawn in dotted lines were not preserved with the fossil specimen. *Huanhepterus* was a large pterodactyloid pterosaur with a wing span of about 8·2ft (2·5m). One of the most striking features of the skeleton is the length of its neck vertebrae.

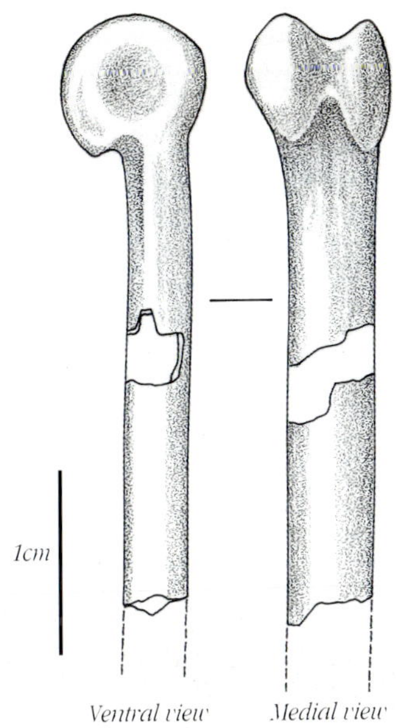

Dermodactylus Wing Fragment (above) *Dermodactylus montanus*, the fragment of a wing metacarpal bone of a pterodactyloid pterosaur from the late Jurassic Morrison Formation found by S.W. Williston at Como Bluff in Wyoming. Described by O.C. Marsh in 1878, this was the first record of a Jurassic pterosaur in the New World.

Comodactylus and Mesadactylus

Another wing metacarpal, which was only discovered 100 years later in the Yale Peabody Museum collection, came from the famous Quarry 9 on the Como Bluff in Wyoming, the so-called Mammal Quarry. Peter Galton of Bridgeport University identified this bone, which came to O.C. Marsh in New Haven along with fossil material found by W. Reed in 1879 in strata from the Morrison Formation in Quarry 9.[67] This metacarpal survived complete. Its short, compact shape shows that it is rhamphorhynchoid, and it is reminiscent of the same part of the skeleton in the Solnhofen *Rhamphorhynchus*. Galton called this second Jurassic pterosaur from Como Bluff *Comodactylus* (=Como-finger). With a length of 2·26in (57·5mm) it is bigger than comparable bones from all other long-tailed pterosaurs so far known. Its wing span is estimated at 8·2ft (2.5m), and so it must have been one of the biggest Jurassic pterosaurs of all.

The second fossil site in the Morrison Formation in which remains of pterosaurs were found was the 'Dry Mesa' Quarry in Montrose County in western Colorado. Fossil skeletal remains were found in the 1970s in particular in the course of several palaeontological expeditions mounted by Brigham Young University, Provo, Utah under the direction of James A. Jensen.[68] The most spectacular fossils from Dry Mesa Quarry were new large sauropod dinosaurs, including parts of the skeleton of the gigantic '*Supersaurus*' with an estimated height of 54ft (16·5m) and a body length of 82-98ft (25-30m). But there were also several skeletal remains of small vertebrates, including some from pterosaurs.[63] One bone was at first described as the shin bone of a bird, *Palaeopteryx thomsoni*,[69] but it later turned out to be a bone from the lower arm of a small dinosaur.

Finally Jensen and Padian identified numerous small bones as remains of a pterodactyloid pterosaur, including a synsacrum (fused sacral vertebrae in the pelvic region), vertebrae, pectoral girdle, wing bones and femurs. The new generic name *Mesadactylus* (=Mesa-finger) was introduced for them.[63]

The most remarkable find is the synsacrum, which consists of seven fused vertebrae, the first five forming a neural blade through fusion of the upper neural spines. It is this formation that is different from that of other pterosaurs and also that of many birds. The preserved upper arm bones (humeri), however, are more similar to pterodactylids from Solnhofen. The thickness of the walls of these bones is between 0·01-0·02in (0·25-0·5mm).

Finally a flight digit phalanx from the Dry Mesa Quarry, described by Jensen and Ostrom as an indeterminate pterodactyloid pterosaur in 1977,[70] also belongs to *Mesadactylus*.

The bones from the Morrison Formation are evidence for a diverse pterosaur fauna in North America during the late Jurassic.

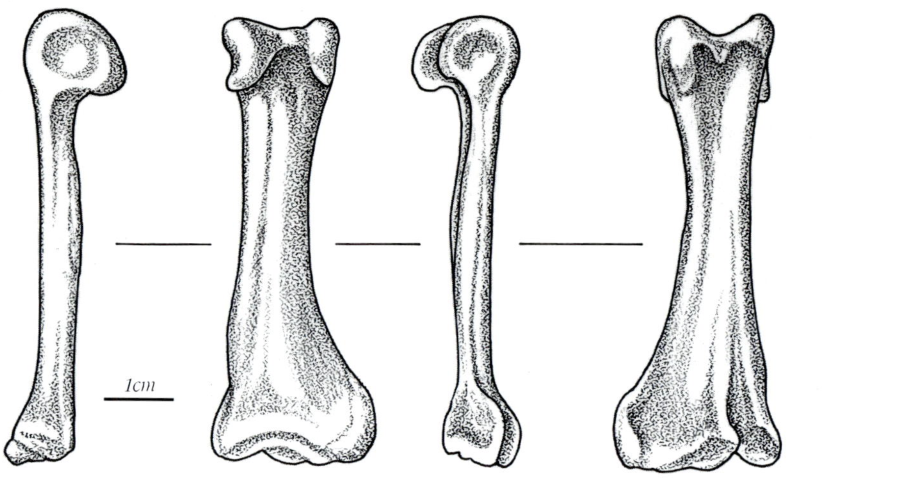

Dorsal view *Medial view* *Ventral view* *Lateral view*

Comodactylus ostromi (above)
This is a metacarpal bone from the right wing of *Comodactylus ostromi* which was discovered at Como Bluff in Wyoming in 1879. However, it was not until 100 years later that Peter Galton identified it. For the first time it showed that rhamphorhynchoid pterosaurs lived in North America contemporary with the giant dinosaurs from the same area.

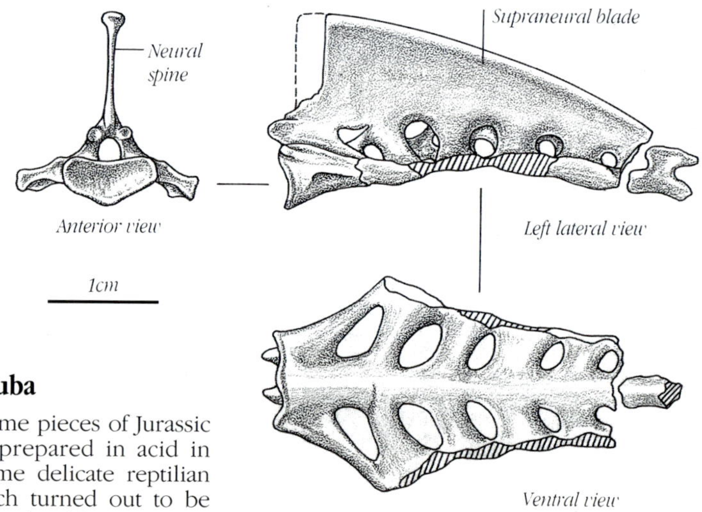

Mesadactylus ornithosphyos (above)
This specimen was discovered by Jim Jensen in the Dry Mesa quarry in Colorado. Of the several small bones, this synsacrum is the most diagnostic element. It was first taken to originate from a prehistoric bird, but was recently identified as pterosaurian.

A Pterosaur from Cuba

In the 1960s, when some pieces of Jurassic limestone were being prepared in acid in search of fossil fish some delicate reptilian bones were found which turned out to be those of a pterosaur. The fossil-bearing rock had been collected half a century earlier by Barnum Brown, for many years Curator of Fossil Reptiles at the New York American Museum of Natural History. It came from a site near Viñales in the province of Pinar del Rio in western Cuba, and was known for its marine fauna, ammonites, bivalves, fish and marine reptiles. The geological age of this black limestone is thus clear, and can be given as Upper Jurassic, Oxfordian to be more precise.

When Edwin H. Colbert, then Curator of Fossil Reptiles at the American Museum in New York examined this specimen scientifically in 1969[71] only one Jurassic pterosaur, indeed literally only a single bone, was known in the western hemisphere, and that was *Dermodactylus* from Como Bluff in Wyoming. This was a short-tailed pterosaur, but Colbert saw in this new and much more complete set of skeletal remains from Cuba the typical characteristics of a rhamphorhynchoid pterosaur, i.e. a long tail with long caudal vertebrae, surrounded by ossified 'tendons' (actually long processes of the vertebrae to stiffen the tail) and also the typical short metacarpal bone.

The specimen consists of skeletal remains of an individual preserved in seven small pieces of rock that had been treated with acid. What has survived are fragments of skull, isolated vertebrae, the two pectoral girdles, a sternum

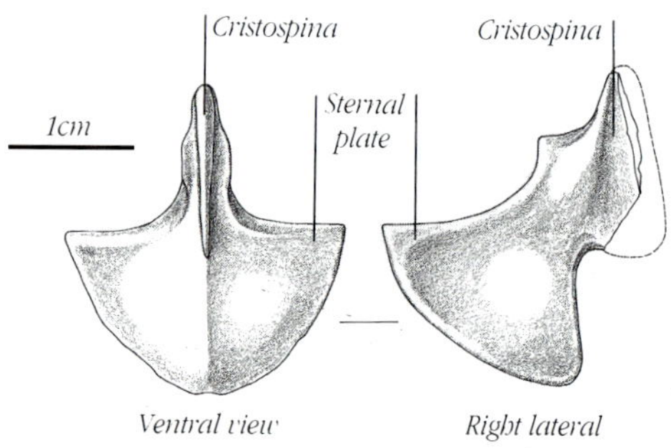

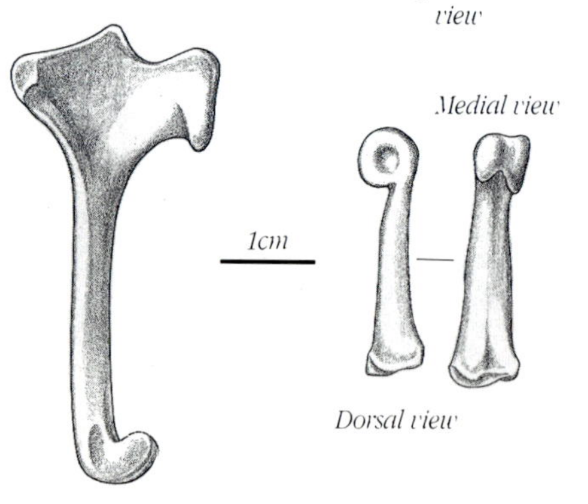

67 Galton, P.M., 1981. *A Rhamphorhynchoid Pterosaur from the Upper Jurassic of North America.* Journal of Paleontology, 55, (5): 1117-1122.

68 Jensen, J.A., 1975. *Continuing study of new Jurassic/Cretaceous vertebrate faunas from Colorado and Utah.* National Geographic Society Research Reports, 1975: 373-381.

69 Jensen, J.A., 1981. *Another look at Archaeopteryx as the world's oldest bird.* Encyclia, 58: 109-128.

70 Jensen, J.A. and Ostrom, J.H., 1977. *A Second Jurassic Pterosaur from North America.* Journal of Paleontology, 51, (4): 867-870.

71 Colbert, E.H., 1969. *A Jurassic Pterosaur from Cuba.* American Museum Novitates, 2370: 26 pp.; New York.

SUMMARY OF PTEROSAURS OF THE UPPER JURASSIC

Rhamphorhynchoidea
Family Rhamphorhynchidae
Rhamphorhynchus longicaudus
Rhamphorhynchus intermedius
Rhamphorhynchus muensteri
Rhamphorhynchus gemmingi
Rhamphorhynchus longiceps
Lower Tithonian, Solnhofen lithographic limestone, Bavaria; in part Nusplingen lithographic limestone, Württemberg; Germany.[34]
Rhamphorhynchus jessoni
Oxfordian, Oxford Clay, Huntingdonshire, England.[37]
Rhamphorhynchus tendagurensis
Upper Jurassic, Obere Saurier-Mergel, Tendaguru, Tanzania.[38]
Rhamphorhynchus sp.
Upper Callovian, Pedróga; Lower Kimmeridgian, Guimarota near Leiria, Portugal.[39]
Odontorhynchus aculeatus
Lower Tithonian, Solnhofen lithographic limestone, Bavaria, Germany.[34]
Scaphognathus crassirostris
Lower Tithonian, Solnhofen lithographic limestone, Bavaria, Germany.[34,40]
Sordes pilosus
Upper Jurassic, Karatau, Kazakhstan, USSR.[60]

Family Anurognathidae
Anurognathus ammoni
Lower Tithonian, Solnhofen lithographic limestone, Bavaria, Germany.[34,41]
Batrachognathus volans
Upper Jurassic, Karatau, Kazakhstan, USSR.[59]

Undetermined family
Nesodactylus hesperius
Oxfordian, Viñales, Pinar del Rio, Cuba.[71]
Comodactylus ostromi
Tithonian, Morrison Formation, Como Bluff, Wyoming, USA.[67]

Pterodactyloidea
Family Pterodactylidae
Pterodactylus antiquus
Pterodactylus kochi
Pterodactylus micronyx
Pterodactylus elegans
Pterodactylus longicollum
Pterodactylus grandis
Pterodactylus grandipelvis
Lower Tithonian, Solnhofen lithographic limestone; in part Mörnsheim strata; Bavaria, Germany.[34]
Pterodactylus cerinensis
Kimmeridgian, lithographic limestone, Cerin, Dept. Ain, France.[44]
Pterodactylus suprajurensis
Purbeckian, Boulogne-sur-Mer, France.[46]
Pterodactylus manseli
Pterodactylus pleydelli
Kimmeridgian, Kimmeridge Clay, Weymouth, Dorset, England.[5]
Pterodactylus arningi
Pterodactylus maximus
Upper Jurassic, Obere Saurier-Mergel, Tendaguru, Tanzania.[38]

Family Gallodactylidae
Gallodactylus suevicus
Lower Tithonian, Solnhofen lithographic limestone, Bavaria; Nusplingen lithographic limestone, Württemberg; Germany.[51]
Gallodactylus canjuersensis
Portlandian, Gisemens des Bessons, Canjuers, Dept. Var, France.[50]

Family Germanodactylidae
Germanodactylus cristatus
Lower Tithonian, Solnhofen lithographic limestone, Eichstätt, Bavaria, Germany.
Germanodactylus rhamphastinus
Lower Tithonian, Mörnsheim strata, Daiting, Bavaria, Germany.[34]
Germanodactylus sp.
Lower Kimmeridgian, Kimmeridge Clay, Kimmeridge Bay, Dorset, England.[49]

Family Ctenochasmatidae
Ctenochasma roemeri
Upper Jurassic, Deister near Hannover, Lower Saxony, Bavaria.[52]
Ctenochasma gracile
Ctenochasma porocristata
Lower Tithonian, Solnhofen lithographic limestone, Bavaria, Germany.[34,53,54]
Ctenochasma sp.
Lower Portlandian, Haute Marne, France.[55]
Gnathosaurus subulatus
Lower Tithonian, Solnhofen lithographic limestone, Bavaria, Germany.[34,57,58]
Huanhepterus quingyangensis
Upper Jurassic, Ordos, Gansu Prov., China.[62]

Family Dsungaripteridae
Dsungaripterus brancai
Upper Jurassic, Obere Saurier-Mergel, Tendaguru, Tanzania.[47]

Undetermined family
Dermodactylus montanus
Tithonian, Morrison Formation, Como Bluff, Wyoming, USA.[65,66]
Mesadactylus ornithosphyos
Tithonian, Morrison Formation, Dry Mesa Quarry, Colorado, USA.[63,68,69,70]
Pterodactyloidea indet.
Tithonian, Vacca Muerta Formation, Neuquén, Argentina.[72,73]

with an unusually large and deep keel, bones from the wing skeleton, pelvic bones, fragments of the femur, metatarsal bones and various ribs. The skeleton was no longer articulated naturally, but had fallen apart. Colbert called this Jurassic pterosaur from Cuba *Nesodactylus* (=island-finger) and placed it in the Rhamphorhynchidae family. *Nesodactylus* seems to have had somewhat longer and more heavily built wings and hindlimbs than the Solnhofen *Rhamphorhynchus*. Jaws and teeth are not preserved.

Nesodactylus hesperius (left)
Of the incomplete skeleton of this rhamphorhynchoid pterosaur from Cuba, we see here the breastbone (upper drawings), the upper arm bone (lower left) and the metacarpal bone of the left wing finger (lower right).

Right: The fossil remains of *Nesodactylus hesperius* from the late Jurassic of Pinar del Rio were preserved in several pieces of rock that were treated with acetic acid. In this way the bones could be exposed three-dimensionally.

Neuquén Fossils

Fossils that have survived in a very similar way to the Solnhofen fossils have been found in the Upper Jurassic flaggy limestones of the province of Neuquén in the eastern foothills of the Andes in central Argentina. These flaggy limestones can also be called lithographic limestone. They are part of the Vacca Muerta formation and because of the ammonite fauna they are placed in the Middle to Upper Tithonian age of the Upper Jurassic. Thus they are younger than the Solnhofen strata (which are Lower Tithonian) in Bavaria, and although they look very similar to them they were not formed in a lagoon protected by reefs but in an open shallow sea. It is estimated that the coast of the Jurassic sea at the time was about 62 miles (100km) away.[72]

The pterosaur specimens consist of one disarticulated and incomplete skeleton which palaeontologists are still studying and an isolated tibia 3·75in (95mm) long from an as yet undetermined pterodactyloid pterosaur, possibly a new species.[73]

But the find does show the important fact that short-tailed pterosaurs, the Pterodactyloidea, were distributed all over the world right at the beginning of their evolution in the Upper Jurassic. Fossil finds establish that at this time they lived in Europe, Africa, Asia, North and South America.

72 Cione, A., Gasparini, Z., Leanza, H. and Zeiss, A., 1987. *Marine oberjurassische Plattenkalke in Argentinien.* Archaeopteryx, 5: 13-22; Eichstätt.
73 Gasparini, Z., Leanza, H. and Garata Zubilliga, J., 1987. *Un pterosauria de las Calizas Litograficas Tithonianas de Area de los Catutos, Neuquén, Argentina.* Ameghiniana, 24, (1-2): 141-143; Buenos Aires.
Leanza, H. and Zeiss, A., 1990. *Upper Jurassic Lithographic Limestones from Argentina (Neuquén Basin): Stratigraphy and Fossils.* Facies, 22: 169-186; Erlangen.

Pterosaurs of the Cretaceous

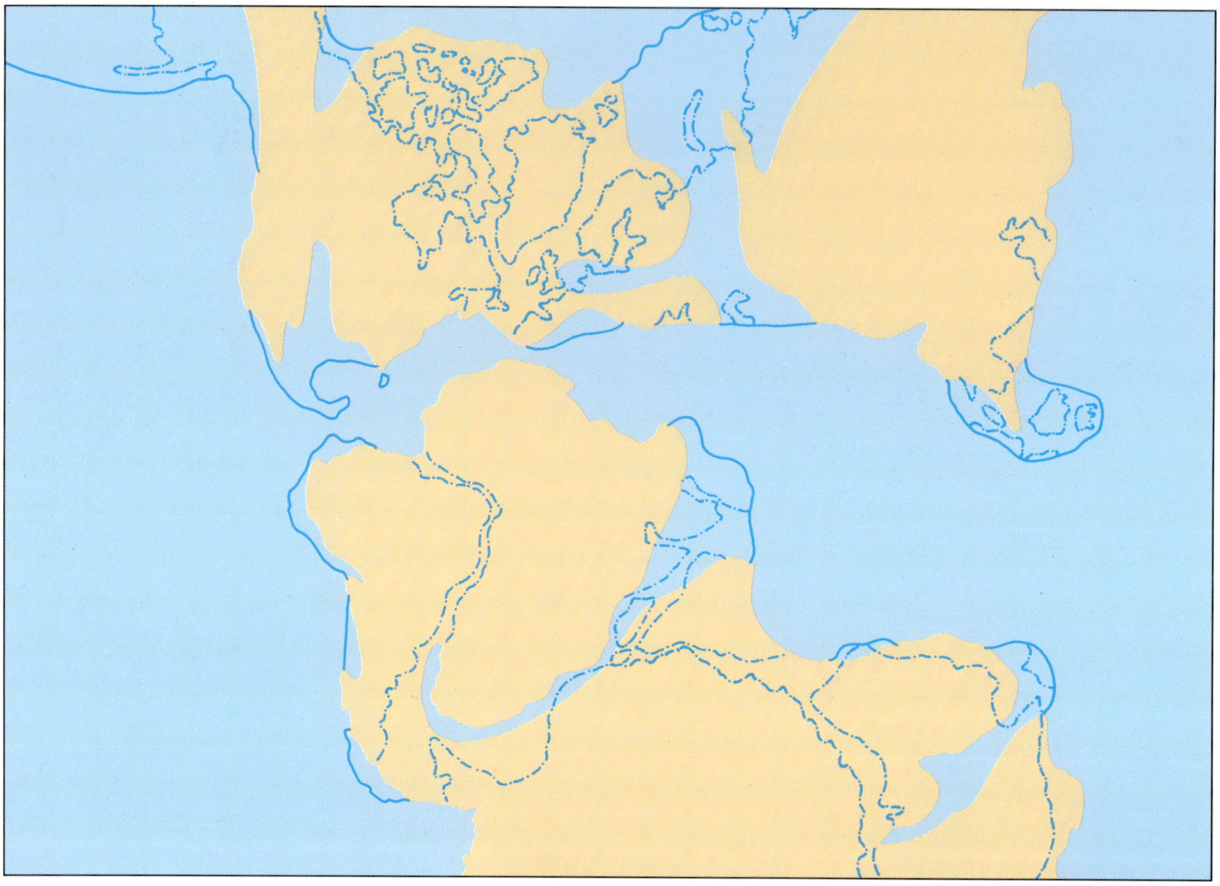

Palaeogeographic Map of the Lower Cretaceous (left)
At the beginning of the early Cretaceous, shallow seas had started to divide the southern continents although South America and Africa were still connected to one another. The Atlantic Ocean had started to develop, and the northern and southern continents were completely separated. The sea dividing Europe from Asia was growing.

Palaeogeographic Map of the Upper Cretaceous (right)
By late Cretaceous times the continents had moved to more familiar positions. Africa and South America were beginning to drift apart, and India was moving eastwards across the Indian Ocean. Australia and Antarctica had also detached themselves from what had been Gondwana. Extensive marine flooding in the north meant that seaways divided western from eastern North America.

The Jurassic was followed by the Cretaceous Period, which began about 144 million years ago and ended 65 million years ago. This final period of the Mesozoic Era was the longest, lasting for 79 million years. It is usually subdivided into Lower and Upper Cretaceous. Many of the geological and biological events of the time were highly significant for Earth as it is today.

In the Upper Cretaceous in particular there was extensive marine flooding all over the world, continents drifted apart and most of Earth's present high mountain ranges were formed. The two supercontinents, Laurasia in the north and Gondwana in the south, separated. At the beginning of the Cretaceous, South America and Africa were still connected, but towards the end of the period they had moved over 1,250 miles (2,000 km) apart: the South Atlantic had come into being. North America and Eurasia were still connected. In the south, Australia, Antarctica and India detached themselves from the Gondwana land mass and drifted apart.

These changes also affected the climate and further development of the plant and animal kingdoms. Thus this period also saw the first appearance of flowering plants and deciduous trees. Dinosaurs were still the dominant land animals in the Cretaceous as well, and some of their evolutionary lines reached their peak. Although birds were becoming more numerous and had conquered various habitats, pterosaurs continued to dominate the Earth's air space.

Cretaceous pterosaurs were found on all continents with the exception of the Antarctic, again overwhelmingly in marine deposits, although only short-tailed species, the Pterodactyloidea, lived in the Cretaceous. Long-tailed pterosaurs, the Rhamphorhynchoidea, did not survive the transition from the Jurassic to the Cretaceous. The major feature of the period was enormous increase in size. The largest Jurassic pterosaurs had a maximum wing span of 8·2ft (2·5m), but in the Lower Cretaceous pterosaurs with a 20ft (6m) wing span began to emerge, and in the Upper Cretaceous wing spans of nearly 40ft (12m) were found as well. These pterosaurs were the largest flying creatures that have ever lived on Earth.

Towards the end of the Cretaceous the oceans withdrew from large continental areas. Subsequently global cooling occurred, and distinct climatic zones came into being on Earth. Then, 65 million years ago, at the end of the Cretaceous, it was all over. This amazing happening went into the annals of the history of the Earth as the 'mass extinction at the Cretaceous/Tertiary boundary'. Dinosaurs, ichthyosaurs, plesiosaurs, mosasaurs, and many fish and marine invertebrates disappeared from the Earth at that time, and the pterosaurs went as well. The reason for their extinction will be discussed in a separate chapter.

English Finds

The first pterosaur bones in England were found by Gideon Mantell, a country doctor, in about 1827, in Tilgate Forest near Cuckfield in Sussex. A few years before he had discovered the first skeletal remains of the dinosaur *Iguanodon* there as well. The fossiliferous strata are sandy and clayey delta deposits from rivers and seas, and were formed in the south of England in early Cretaceous times. They are known as Wealden, a geological formation also found in Belgium, Northern France, and northwestern Germany. At first Mantell thought the delicate bones he had found in Tilgate Forest were those of birds. He described them as 'Bones of Birds' in his 'Illustrations of the Geology of Sussex' in 1827. He later gave them the new generic name *Palaeornis* (=old bird), but he finally diagnosed them as pterosaur bones and they were later placed in the genus *Ornithocheirus* (=bird-hand) by the English palaeontologist Harry Govier Seeley.

The first pterosaur remains from the Upper Cretaceous of England were described by James Scott Bowerbank in 1845. They were also the first pterosaurs so far known that were larger than the largest Jurassic pterosaurs. The material consisted of the front end of a snout with typical teeth, part of a pectoral girdle and a few other bones, and came from the Chalk (Turonian) of Burham in Kent.[1] Bowerbank estimated the wing span of this species at 8 or 9ft (2·75m). Richard Owen, who had hitherto thought it impossible that a cold-blooded reptile with a higher body weight than a warm-blooded mammal like the flying fox could lift itself into the air, had to revise his opinion. In an 1851 monograph (p.80),[2] he concedes: 'Of

1 Bowerbank, J.S., 1845. Proceedings of the Geological Society of London.

2 Owen, R., 1851. *Monograph of the fossil Reptilia of the Cretaceous Formations.* Palaeontographical Society London.

Left: The upper jaw of *Ornithocheirus cuvieri* from the Burham chalk pit in Kent. The fossil of this large Upper Cretaceous pterosaur was described by J.S. Bowerbank in 1845. This jaw measures 7in (18cm) in length.

Below: 'Mantell's Quarry' in Tilgate Forest near Cuckfield. It was in Wealden strata such as this that Gideon Mantell discovered the first pterosaur bones to be found in England. They were from the genus *Ornithocheirus*.

the remarkable Reptiles now extinct, which, like the Bats, had their anterior members modified for plying a broad membranous wing, no species had been discovered prior to 1840, which surpassed the largest of the *Pteropi*, or 'Flying-foxes', in the spread of those wings, and there was a *priori* a physiological improbability that the cold-blooded organization of a reptile should, by any secondary modification, be made to affect more in the way of flight, or be able to raise a larger mass into the air, than could be done by the warm-blooded mammal under an analogous special adaptation . . . The subsequent discovery of portions of the skull of the Pterodactyls shows that the manifestations of creative power in past time surpass the calculations that are founded upon actual nature.'

Ornithocheirus

This genus was described by Harry Govier Seeley from fossil material found at English Cretaceous localities.[3] Despite the fact that no more complete skeletons were found, but essentially only fragments of jaws, individual bones or vertebrae, a total of 36 species were distinguished. Most of them are based on specimens from the Cambridge Greensand, a sandy marl full of phosphatic nodules. Numerous pits in the Cambridge area exploited the valuable phosphate of lime.

The Cambridge Greensand is a marine sediment deposited when the Cenomanian Sea expanded at the beginning of the Upper Cretaceous. As the fossils of pterosaur bones often seem to have been worn down, frequently as a result of rolling, it is suspected that they were washed out of older strata on the seashore and redeposited on a 'secondary deposit', this time in younger Cenomanian strata. Pterosaurs of the Cambridge Greensand could thus be considerably older, and possibly came from the late Lower Cretaceous, the Albian.[4]

3 Seeley, H.G., 1869. *Index to the fossil remains of Aves, Ornithosauria and Reptilia in the Woodwardian Museum Cambridge.* Proceedings of the Cambridge Philosophical Society, 3: 169.

4 Rawson, P.F. et al., 1978. *A correlation of Cretaceous rocks in the British Isles.* Geological Society London, Special Reports, 9: 70pp.
Seeley H.G., 1870. *The Ornithosauria. An elementary study of the bones of Pterodactyls.* 130 pp., Cambridge University Press. p.2: '. . . perfect bones are almost unknown. Even those bones like the carpals, which almost retain their entirety, invariably show indications of having been rolled on the sea-shore among the nodules of phosphate of lime with which they now occur, in their angular margins being rounded . . .'

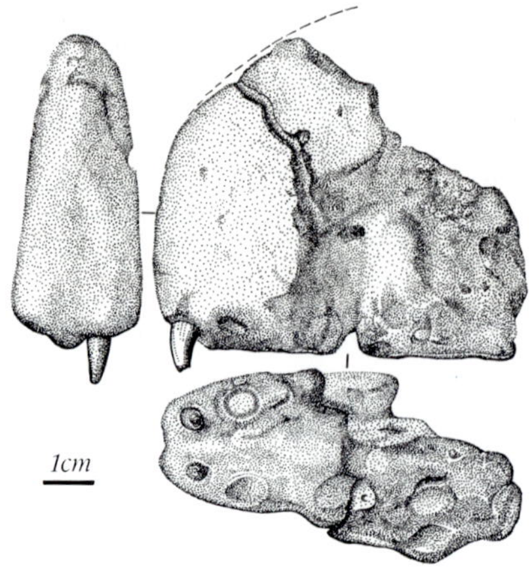

Criorhynchus Jaw Fragments (above)
This jaw fragment is seen from the front (top left), from the left side (middle top) and from below (lower middle). Enlarged is a single incomplete tooth.

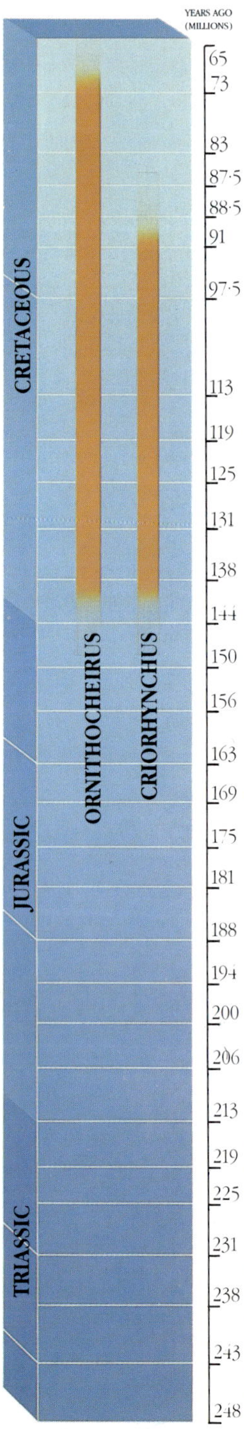

Time Chart (left)
Ornithocheirus is something of a 'waste bin' generic name which may include several distinct genera. Most of the indisputable fossil material comes from the Upper Cretaceous (Cenomanian) of England. *Criorhynchus* is probably related. It is known from early Lower to early Upper Cretaceous.

Map (right)
1 *Ornithocheirus*
2 *Criorhynchus*

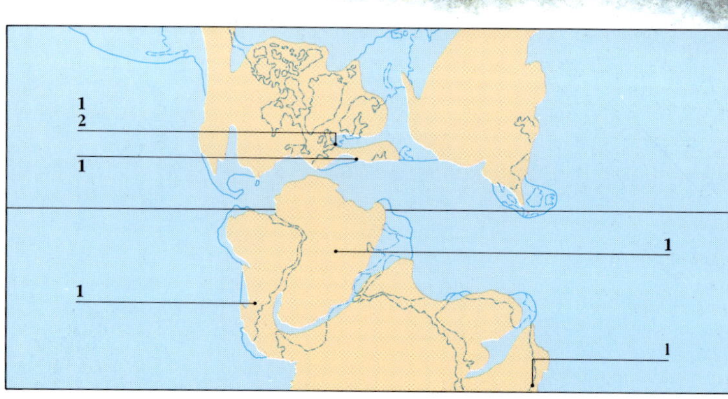

Criorhynchus (below)
Formerly known only by fragments of the snout, *Criorhynchus* has turned out to be a close relative of *Tropeognathus* discovered more recently in early Cretaceous strata of Brazil. *Criorhynchus*, known only from England, must have looked similar, although the cranial crest at the front end of its snout was blunter and more robust. Its dentition of few, quite strong teeth suggests that this pterosaur was probably a fish eater.

In 1869 Seeley was commissioned to arrange and catalogue the pterosaur collection of the Woodwardian Museum (now Sedgwick Museum) of the University of Cambridge. Almost all the specimens came from the Cambridge Greensand, over 1,000 bones, all 'more or less broken and battered'. The name *Ornithocheirus* (=bird-hand) occurs for the first time in his 'Index to the fossil remains of Aves, *Ornithosauria* and Reptilia'. Because of the fragmentary and incomplete condition of the skeletal remains, controversy persists about what belongs to this genus and the family Ornithocheiridae named after it. In any case they were mainly large or very large pterosaurs with long, slender skulls, some probably with a bony crest on the snout, all armed with powerful teeth, coming right to the foremost point of the jaw. The fact that this genus was not too clearly defined meant that as time passed *Ornithocheirus* and the Ornithocheiridae became something of a 'waste bin' for many finds

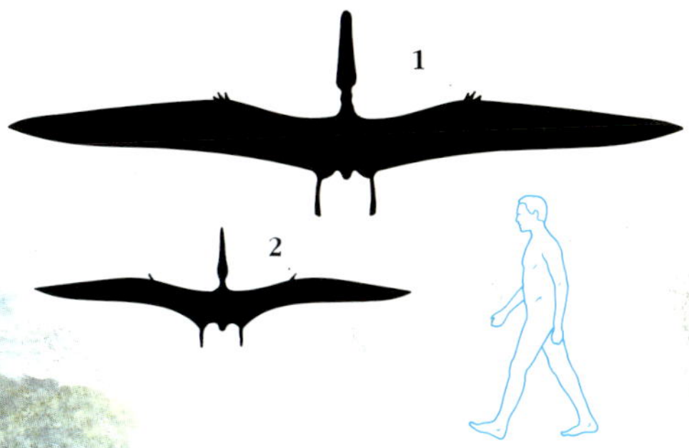

Comparative Sizes (above)
1 *Criorhynchus*: wing span c16·4ft (5m).
2 *Ornithocheirus*: wing span c8·2ft (2·5m).

Ornithocheirus (above right)
Orntibocheirus was one of the most abundant pterosaurs in Cretaceous times. Known mainly from the Cambridge Greensand, it was a middle-sized, short-tailed pterosaur which was distributed worldwide. Its strong teeth, set in long, tapering jaws, suggest that it was piscivorous. In this scene a parent is shown feeding its young which is picking stored fish prey out of the parental throat pouch.

in England, France, Germany, Bohemia, Austria, Africa, South America and Australia.

In 1914 R.W. Hooley tried to bring an element of order into the English *Ornithocheirus* material. He distinguished between five different groups, to which he gave names of their own, particularly on the basis of jaw fragments.[5]

As well as the pterosaur bones G. Mantell had found in the Sussex Wealden, Richard Owen also described part of a tibia from there as *Ornithocheirus curtus* (=short *Ornithocheirus*), and an incomplete lower jaw of a large species as *Ornithocheirus sagittirostris* (=arrow-beaked *Ornithocheirus*),[6] from the somewhat older 'Hastings Beds' of St. Leonards-on-Sea.

More remains of *Ornithocheirus* skeletons came from the late Lower Cretaceous, the so-called Albian of Folkestone in Kent. The latest English Cretaceous pterosaurs come from the Chalk (Turonian). They too are known only from fragments of bones, and were also assigned to *Ornithocheirus*.

5 Hooley, R.W., 1914. *On the Ornithosaurian genus Ornithocheirus with a review of the specimens from the Cambridge Greensand in the Sedgwick Museum, Cambridge.* Annals and Magazine of Natural History, Ser. 8, 13: 529-557.

6 Owen, R., 1874. *Monograph of the fossil Reptilia of the Mesozoic Formations. I. Pterosauria.* Palaeontographical Society, 27: 1-14; London.

Ornithocheirus Skull (below right)
This is a restoration of the skull of *Ornithocheirus* made on the basis of several fragmentary remains from the English Cretaceous. The skull is long and slender and the jaws are armed with numerous short, sharp teeth which extend to the very front of the upper and lower jaws. It was probably a fish eater.

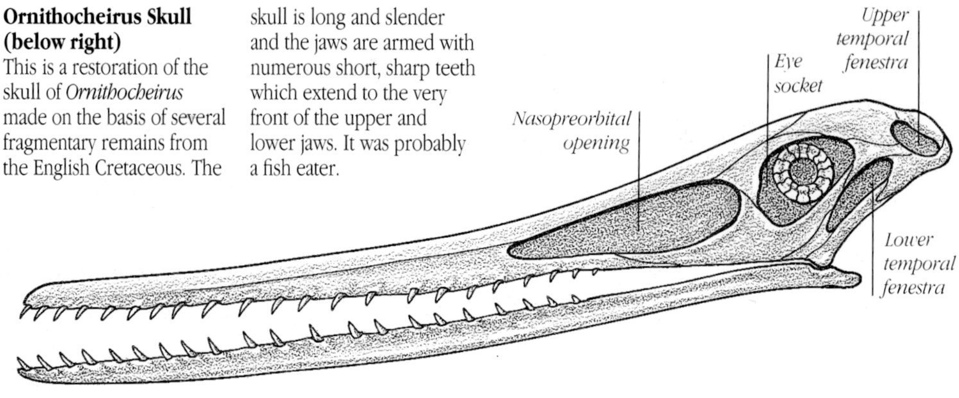

Criorhynchus

Richard Owen identified a second type of large pterosaur in the copious pterosaur material from the Cambridge Greensand, and called it *Criorhynchus* (=ram-snout). Only fragments of bones have been found for this species as well. The most marked characteristic is the front end of the jaw. Unlike *Ornithocheirus* it is not slender and pointed, but blunt and solid, though compressed laterally. The front end of the upper jaw is slightly flattened and dented at the front. The jaws had powerful teeth, set upright, and curved slightly backwards. These typical features of the genus *Criorhynchus* were based only on a fragment of a front end of a snout. Owen did place other bones from the Cambridge Greensand in this genus, but definite classification was impossible.

Palaeontologists had always been puzzled about what *Criorhynchus*, a pterosaur with a wing span of about 16·4ft (5m), might have

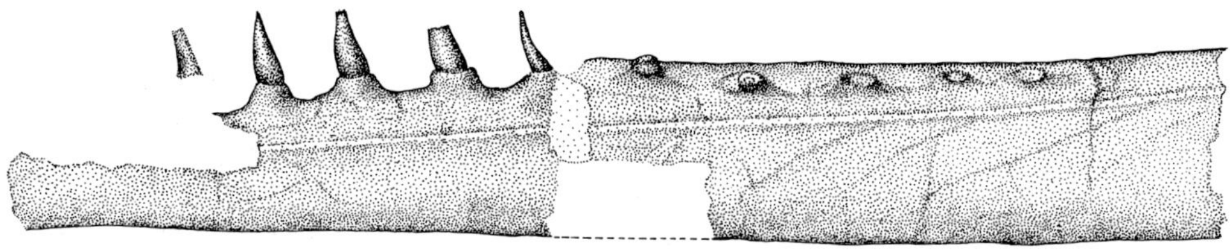

Ornithocheirus sagittirostris

1cm

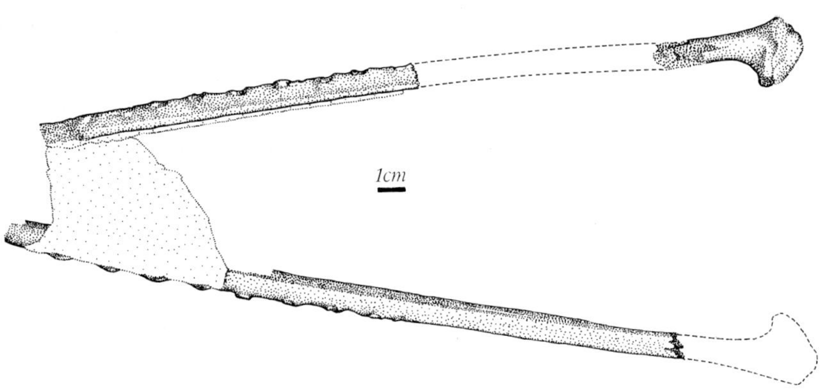

Ornithocheirus Jaw Fragments (left, below and right)
The drawings on this page and at near left on the opposite page show jaw fragments of different species of *Ornithocheirus* from various English Cretaceous localities. Immediately to the left is a lower jaw of *Ornithocheirus sagittirostris* seen in lateral view and, below that, from above. *O. sagittirostris* is from the Wealden of Sussex. At the foot of this page we see the upper jaw of *Ornithocheirus cuvieri* in lateral view and from below. *O. cuvieri* is from the Burham chalk pit in Kent, as is *O. giganteus* which is illustrated on the opposite page (middle left). The remaining specimen on this page is a jaw fragment of *O. daviesi* from the late Lower Cretaceous (Albian) of Folkestone in Kent. The final specimen illustrated at the foot of the opposite page is *O. sedgwicki* from the Cambridge Greensand. Here we see the tip of the upper jaw in lateral and dorsal views, and below that the tip of the lower jaw in lateral and dorsal views. No complete skeletons of *Ornithocheirus* are known; we only have jaw fragments such as this, individual bones or vertebrae. Despite this, 36 species have been distinguished, most of them based on specimens from the Cambridge Greensand. Many fragmentary remains from Europe, Africa, South America, Australia and New Zealand have also been assigned to *Ornithocheirus* on the basic of insufficient diagnostic characters. As a result, the generic name has become something of a 'waste bin' which needs clearer definition.

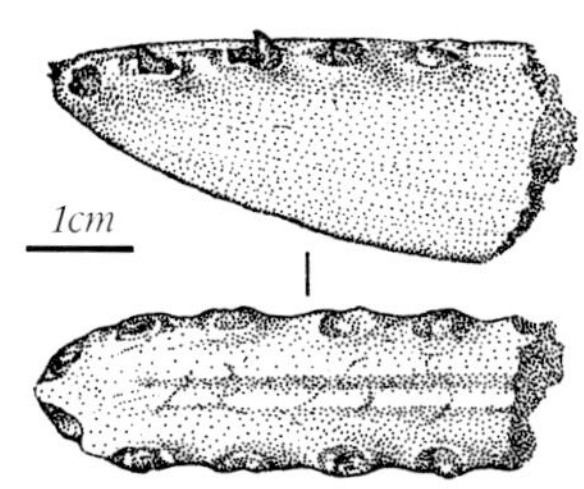

Ornithocheirus daviesi

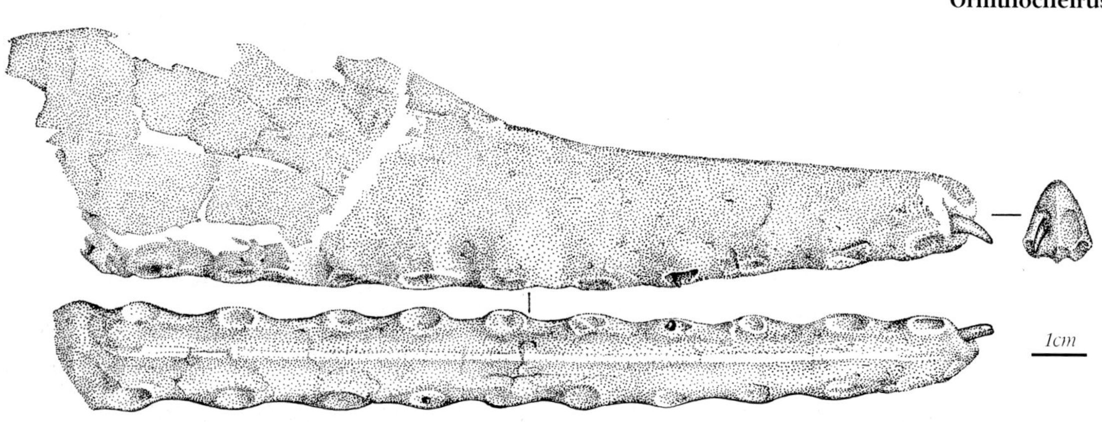

Ornithocheirus cuvieri

Above: Richard Owen, the British comparative anatomist who became the first superintendent of the British Museum (Natural History). It was Owen who first identified specimens from the Cambridge Greensand as belonging to a genus distinct from *Ornithocheirus*. He named it *Criorhynchus*. Owen conducted a fierce debate with Harry Govier Seeley on whether pterosaurs were warm-blooded or not.

looked like. The Viennese palaeontologist G. von Arthaber attempted a reconstruction of the skull in 1919. The result was a fairly short, tall head about 7in (18cm) long, which looked different from all the Cretaceous pterosaurs so far known.[7] The solution came from a South American find.

A few years ago a complete pterosaur skull in good condition from the Santana Formation (Aptian) in Brazil was discovered and described as *Tropeognathus* (=keel-jaw).[8] The front end of the jaw was startlingly similar to that of *Criorhynchus* from the Cambridge Greensand. It sloped sharply at the front and extended backwards in the form of a bony crest tapering towards the top. This crest is almost semicircular at the top and limited to the front end of the upper jaw. It was presumably drawn through the water like the keel of a ship while fish were being caught in flight, and helped to stabilize the head in this phase.

7 Arthaber, G. von, 1919. *Studien über Flugsaurier auf Grund der Bearbeitung des Wiener Exemplares von Dorygnathus banthensis Theod. sp.*. Denkschriften der Akademie der Wissenschaften, 97: 391-464; Vienna.
8 Wellnhofer, P., 1987. *New Crested Pterosaurs from the Lower Cretaceous of Brazil.* Mitteilungen der Bayerischen Staatssammlung für Paläontologie und historische Geologie, 27: 175-186; Munich.

It now became clear that the high snout end in *Criorhynchus* was the front part of a bone crest that may have looked similar to that of *Tropeognathus*. *Criorhynchus* must thus have had a much longer skull than Arthaber thought. The common features of these two genera meant that they were closely related and placed together in the family Criorhynchidae.

Doratorhynchus

In 1875 H.G.Seeley reported that at Christmas 1868, while staying in Swanage on the Dorset coast, he had been given a long vertebra and part of a large lower jaw found by a quarry worker in the Purbeck Limestone of Langton Matravers. The geological age of Purbeck Limestone is late Upper Jurassic to early Cretaceous (Tithonian-Berriasian). Seeley took the slender vertebra, about 5·2in (12 cm) long, to be the caudal (or tail) vertebra of a large long-tailed pterosaur to which he gave the name *Doratorhynchus* (=spear-snout).[9] He also included other bones in this, among them a wing phalanx 12in (30cm) long, described by Owen in 1870 as *Ornithocheirus validus*. Seeley had certainly allowed for the possibility that the vertebra could actually come from the neck, an assumption that later turned out to be correct. Further investigation of the cervical vertebrae of Cretaceous pterosaurs led to emphasis of the great similarity of this *Doratorhynchus* vertebra with the fifth cervical vertebra of the giant pterosaur *Quetzalcoatlus* from the Late Cretaceous of Texas.[10] This meant that many long vertebrae from the Cambridge Greensand that had formerly been taken for caudal vertebrae must be cervical vertebrae of pterosaurs. This shows that *Doratorhynchus* from the early Cretaceous in England was a large pterosaur with an extremely long neck, possibly an ancestor of the giant pterosaurs of Texas.

9 Seeley, H.G., 1875. *On an Ornithosaurian (Doratorhynchus validus) from the Purbeck Limestone of Langton near Swanage.* Quarterly Journal of the Geological Society London, 31: 465-468.
10 Howse, S.C.B., 1986. *On the cervical vertebrae of the Pterodactyloidea (Reptilia, Archosauria).* Zoological Journal of the Linnean Society, 88: 307-328; London.

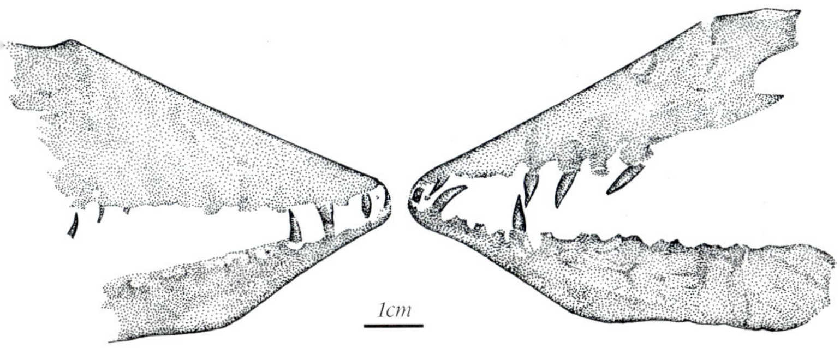

Ornithocheirus giganteus

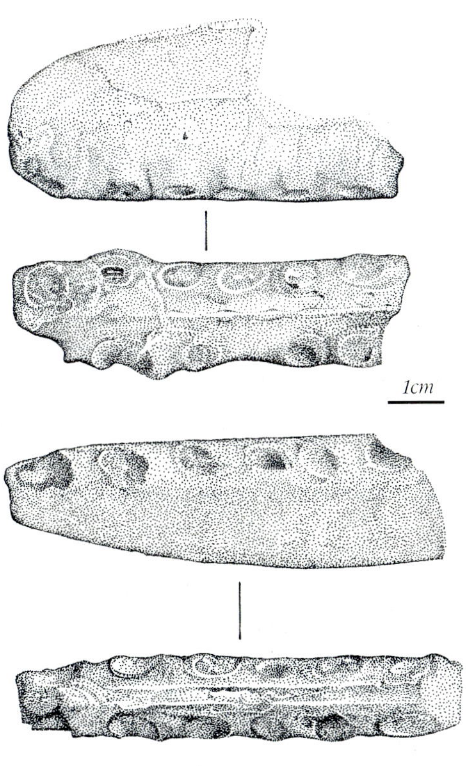

Ornithocheirus sedgwicki

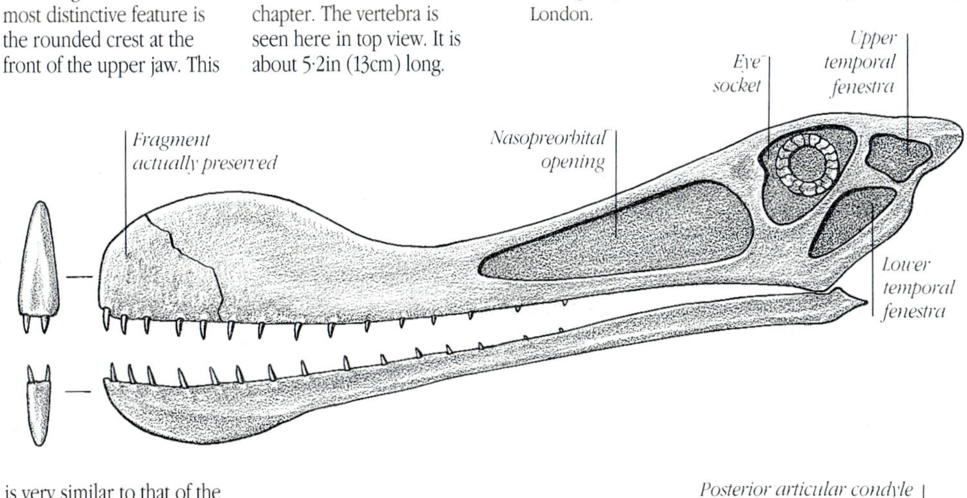

Criorhynchus Skull (below right)
This is a tentative restoration of the skull of *Criorhynchus* from the Cambridge Greensand. The most distinctive feature is the rounded crest at the front of the upper jaw. This is very similar to that of the Brazilian *Tropeognathus*, and as a result the two genera have been grouped in the family Criorhynchidae. We may deduce that this crest stabilized *Criorhynchus*' head while fishing. The skull length is about 20in (50cm).

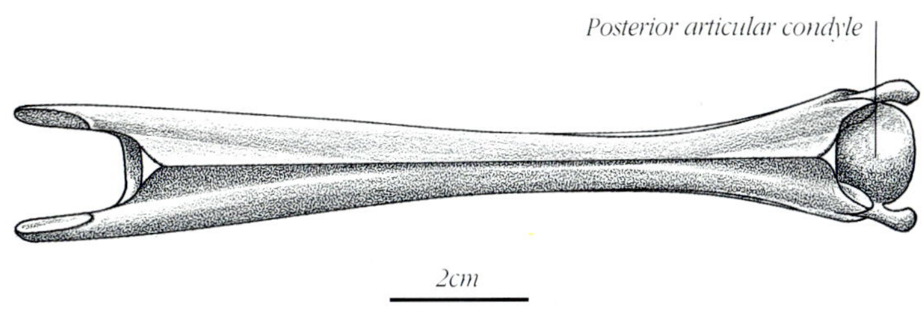

Doratorhynchus Vertebra (bottom of page)
This long vertebra was discovered in the Purbeck limestone near Swanage on the Dorset coast in 1875. It was thought to have come from the tail of a long-tailed pterosaur, named as *Doratorhynchus* by H.G. Seeley. Later it was discovered that it is in fact the neck vertebra of a long-necked, short-tailed pterosaur similar to the giant Texan pterosaur *Quetzalcoatlus*, which is described at the end of this chapter. The vertebra is seen here in top view. It is about 5·2in (13cm) long.

Ornithodesmus

A fourth genus of Cretaceous pterosaur was discovered in England, from the Wealden of the Isle of Wight. It was again Seeley who in 1887 described a sacrum from the Lower Cretaceous of Atherfield (Aptian) which he thought came from a bird, and which he named *Ornithodesmus* (=bird-ribbon).[11] Later he diagnosed it as a pterosaur, and also placed in this genus parts of a skull, neck and dorsal vertebrae, the breastbone and parts of the wing skeleton of an individual which also came from Atherfield in the Isle of Wight. Because of the broad teeth of this individual he called the species *Ornithodesmus latidens*.[12]

11 Seeley, H.G., 1887. *On a sacrum apparently indicating a new type of Bird, Ornithodesmus cluniculus Seeley.* Quarterly Journal of the Geological Society London, 43: 206.

Ornithodesmus was a large pterosaur with a wing span of 16·4ft (5m). The front ends of the jaw were similarly formed to those of a duck. Despite this it was not a 'duck-billed pterosaur', as the jaws were equipped from back to front with a series of strong, lancet-shaped, laterally compressed teeth. The eye socket is small and placed fairly far back in the skull. The first six dorsal vertebrae are fused to form a notarium to which the pectoral girdle attached. The strong, alternately-meshing dentition and

12 It has recently become clear that the sacrum of *Ornithodesmus cluniculus* that Seeley at first thought came from a bird could not come from a pterosaur either, but belonged to a small theropod dinosaur. Thus the generic name *Ornithodesmus* designates neither a bird, nor a pterosaur, but a dinosaur. Therefore a new generic name will have to be given to the pterosaur species *'Ornithodesmus' latidens* (from a letter from Dr Andrew Milner, London).

Ornithodesmus (below) *Ornithodesmus* was quite a large pterosaur; its wing span measured about 16·4ft (5m) while its skull length was about 22in (56cm). It is known from the Wealden (Lower Cretaceous) of England, and must have been a contemporary of the *Iguanodon* dinosaurs that can also be seen browsing on the vegetation in this Wealden landscape. To give an idea of comparative size, these *Iguanodon* were about 23ft (7m) long. *Ornithodesmus* is distinguished from all other Cretaceous pterosaurs by the peculiar broad and rounded front end of its beak This characteristic has given rise to its popular name of 'the duck-billed pterosaur'. However, unlike a duck, the front of its beak was lined with short, robust, alternately meshing teeth, a dentition that is suggestive of a diet of fish.

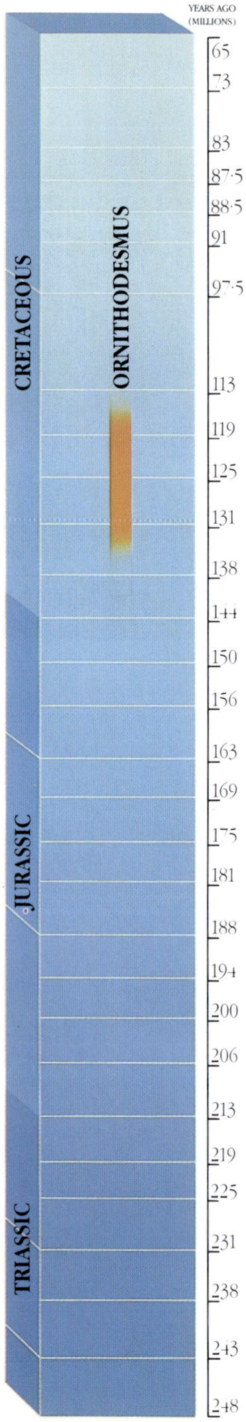

Time Chart (left) *Ornithodesmus* is known only from the Lower Cretaceous (the Wealden) of the Isle of Wight, England, and has so far not been found anywhere else. The genus represents a family of its own. It was the fourth genus of Cretaceous pterosaur to be discovered in England.

Map (below)
1 *Ornithodesmus*

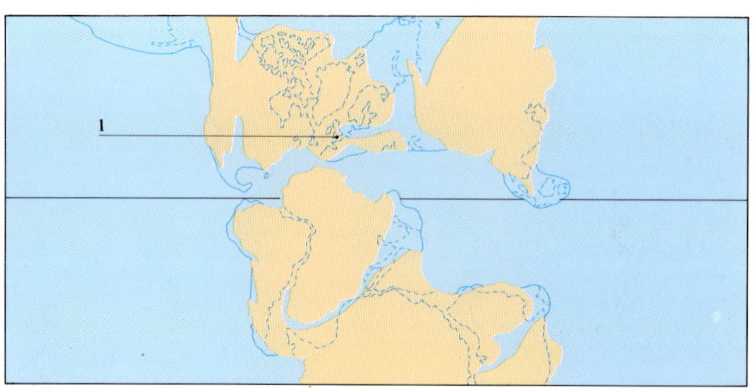

Above: This specimen from the Natural History Museum in London consists of skull fragments of *Ornithodesmus*, a large, short-tailed pterosaur from the early Cretaceous of the Isle of Wight. At the left we can see the front end of the jaws. They are broad, in the manner of a duckbill, but equipped with short teeth. A reconstruction of the skull is shown on the following page.

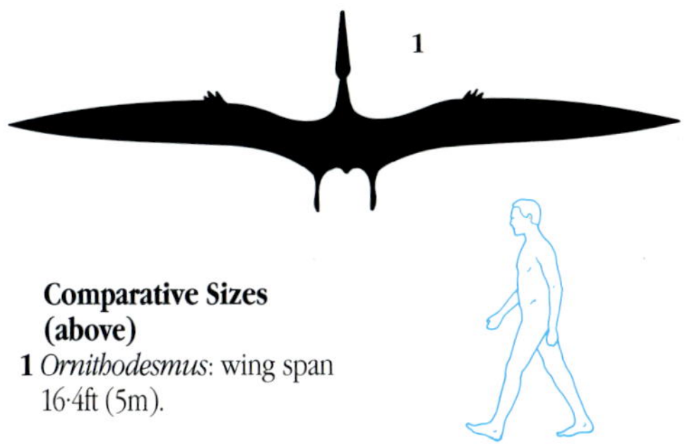

Comparative Sizes (above)
1 *Ornithodesmus*: wing span 16·4ft (5m).

broad snout suggest that *Ornithodesmus* was a fish eater, and that it possibly used a different catching technique from its contemporaries, the Ornithocheirids, with their pointed snouts.

European Finds

So far few Cretaceous pterosaurs have been found on the continent of Europe. Here too we are dealing with individual bones and fragments, fossils which were usually assigned to the English *Ornithocheirus*.

From the early nineteenth century onwards, hard Cretaceous coal has been mined in the so-called 'New World' near Grünbach in Lower Austria. It was formed in an estuary delta region about 70 to 80 million years ago. The deposits in which the coal seams are found are part of the Gosau Formation (Campanian) which occurs widely in the Northern Calcareous Alps. With the coal, which originated from ferns, conifers and palms, fossil remains of fauna have been found that suggest a tropical climate: dinosaurs, crocodiles, turtles and lizards. H.G. Seeley turned his attention to

Ornithodesmus Skull (below)
This drawing is a restoration of the skull of *Ornithodesmus latidens*, the species name *latidens* referring to the broad teeth which can be seen at the front of the upper and lower jaws. The drawing has been made on the basis of incomplete material from the early Cretaceous of the Isle of Wight. The eye socket is situated quite a long way back in the skull, and it extends downwards into a slot-like opening. The length of this skull is about 22in (56cm). When seen from above, the ends of the beak are rounded and broad, which gives the bill a characteristic rather duck-like appearance.

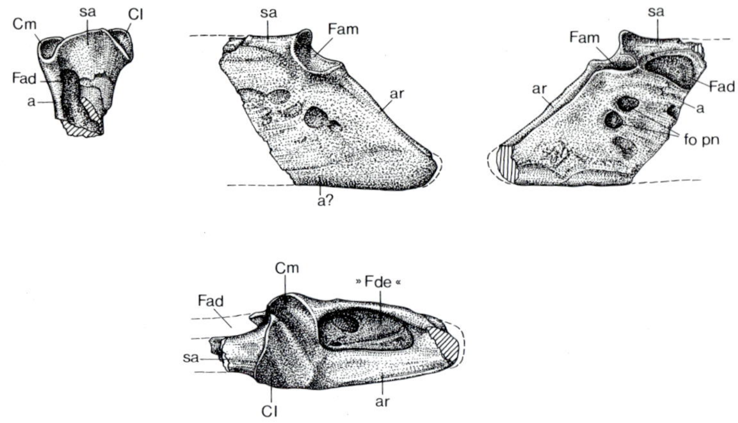

Ornithocheirus bunzeli (above)
Seen here are remains of the pterosaur *Ornithocheirus bunzeli* from late Cretaceous coal deposits that were mined near Grünbach in Lower Austria. The drawings show the articular end of the lower jaw in different aspects. These bones were described by H.G. Seeley in 1881. They showed that this pterosaur could open its jaws very wide indeed, presumably to grasp fish that it had caught.

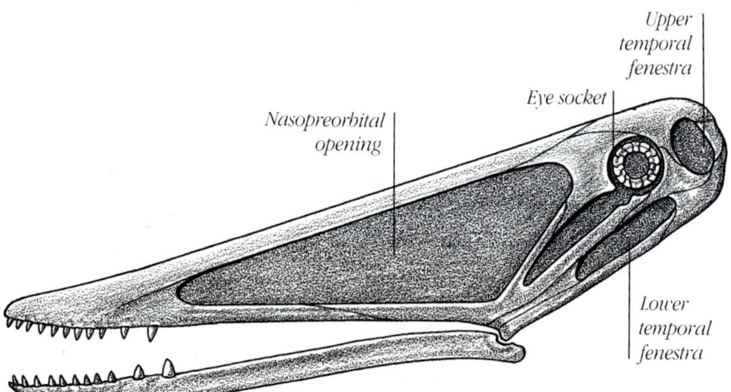

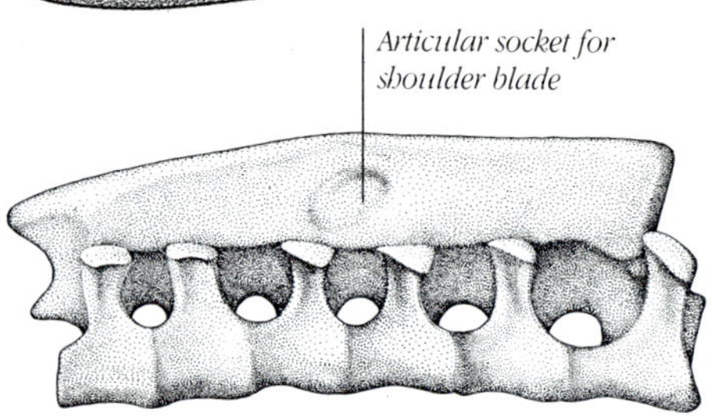

Ornithodesmus Trunk Vertebrae (left)
The first six trunk vertebrae of *Ornithodesmus* arc fused to form a notarium. The shoulder blade was supported in a shallow depression in the fused neural spines. The length of this notarium is 6·8in (17·25cm).

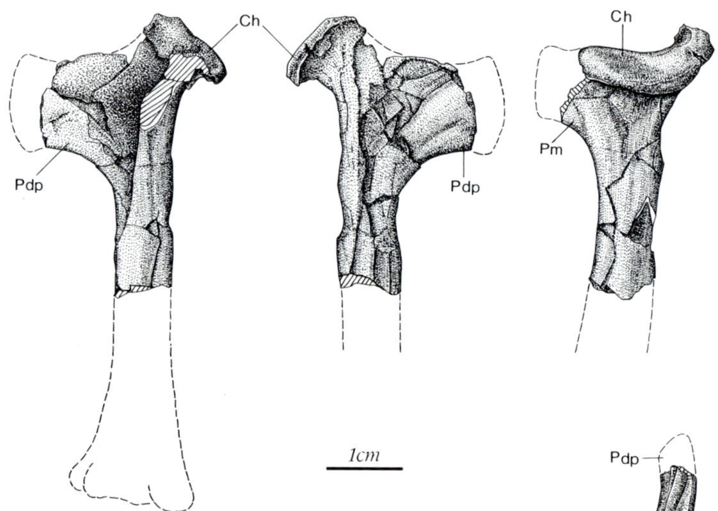

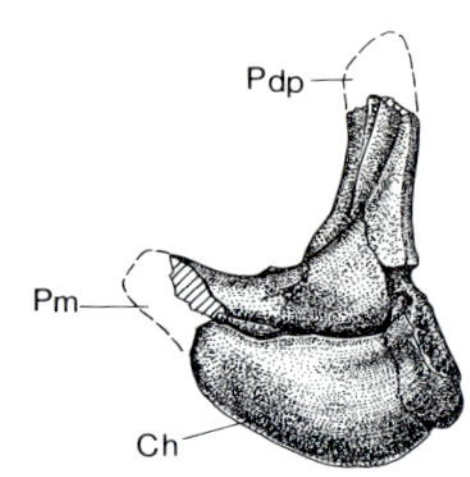

Ornithocheirus Arm Fragments (above)
These drawings show the upper arm of *Ornithocheirus* in various aspects. Although very incomplete, the bones of *Ornithocheirus* are the only fossil record of Cretaceous pterosaurs in the Alps.

A Cretaceous Landscape (left)
This drawing recreates the sort of landscape in which scientists believe *Ornithocheirus* flourished during the late Cretaceous, in the part of the world that is now Grünbach in Austria. It shows the mouth of a river with tropical vegetation: screw trees, reeds, willows and palms. A predatory *Ornithocheirus* is seen soaring above the water, while turtles and crocodiles bask on sand bars by the river. A lizard is perched on a creeper, while in the foreground and background hadrosaurs can be seen foraging for food. Such fauna have been preserved in the deposits of coal which originated from the lush plant material of this locality.

these reptiles in 1881, and described the articular bone of a lower jaw as *Ornithocheirus bunzeli*, because the bone had been described as a lizard by E. Bunzel in 1871.[13] Other pterosaur remains from this site were a humerus fragment and fragments of wing phalanges suggesting a wing span of 4·92-5·74ft (1·5-1·75m).

This lower jaw articular bone showed in particular that these pterosaurs could open their mouths very wide. At the same time the branches of the lower jaw widened, to create a larger opening for the throat pouch, in which fish that had been caught were stored.[14] As far as it is possible to determine, pterosaurs from the Gosau Cretaceous of Lower Austria were very similar to some *Ornithocheirus* species from the Cambridge Greensand.

In the Turonian strata of Bohemia (Czechoslovakia) A. Fritsch found fossil bones that he thought to be those of a bird, which he called *Cretornis* (=chalk-bird). Later the relatively small, well-preserved wing bones were recognized as those of a pterosaur, which was given the name *Ornithocheirus hlavatschi*, named after the collector of the fossils Hlaváč, a pharmacist in the town of Chotzen.[15]

In 1885 E. Koken described the metacarpal bone of a very large pterosaur with an estimated wing span of 28ft (8·5m) from the Lower Cretaceous of the Hannover region as *Ornithocheirus hilsenis* (from the Hils mountains).[16] However, other researchers doubted the pterosaurian nature of this bone, and thought it was the foot digit phalanx of a carnivorous dinosaur. Unfortunately this dispute cannot be cleared up, as the specimen has disappeared.

13 Seeley, H.G., 1881. *The Reptile Fauna of the Gosau Formation preserved in the Geological Museum of the University of Vienna.* Quarterly Journal of the Geological Society London, 37: 620-704.

14 Wellnhofer P., 1980. *Flugsaurierreste aus der Gosau-Kreide von Muthmannsdorf (Niederösterreich) – ein Beitrag zur Kiefermechanik der Pterosaurier.* Mitteilungen der Bayerischen Staatssammlung für Paläontologie und historische Geologie, 20: 95-112; Munich.

15 Fritsch, A., 1881. *Über die Entdeckung von Vogelresten in der böhmischen Kreideformation.* Sitzungsberichte der königlich-böhmischen Gesellschaft der Wissenschaften, 1880: 85; Prague.

16 Koken, E., 1885. *Über Ornithocheirus hilsensis Koken.* Zeitschrift der deutschen Geologischen Gesellschaft, 37: 214; Berlin.

Above: Exposures of the Tugulu Group, of early Cretaceous age, in the Wuerho region of the Junggar Basin in China. These deposits have yielded remains of *Dsungaripterus* and *Noripterus*.

Finally, fossil remains of Cretaceous pterosaurs were discovered also in France. As early as 1882 a dubious cervical vertebra and teeth were found in the Lower Cretaceous strata (Gault) of the Paris basin, and compared with *Ornithocheirus* species from the Cambridge Greensand by H.E. Sauvage. In 1983 the upper end of an ulna from the Lower Cretaceous (Hauterivian) of the Haute-Marne was discovered. This bone has also been placed in the genus *Ornithocheirus*. The size of the bone suggests a wing span of 12ft (about 3·7m).[17] And last of all a cervical vertebra that also came from the Lower Cretaceous (Aptian) of eastern France (Aube) has been described.[18]

17 Buffetaut, E. and Wellnhofer, P., 1983. *Un reste de Ptérosaurien dans l'Hauterivien (Crétacé Inférieur) de la Haute-Marne.* Bulletin de la Société Géologique de France, 1983 (7), 25(1): 111-115, Paris.

However, it cannot be proved with certainty that all these finds belong to the genus *Ornithocheirus* or the family Ornithocheiridae.

Discoveries in Asia

The Junggar basin is in north-western China, between the Altai and Tianshan mountain ranges, in the province of Xinjiang. Its sequence of geological strata includes a long series of continental sediments, sandstones, slates and shales, deposited from the late Permian to the late Cretaceous. Some Upper Jurassic and Cretaceous horizons have produced dinosaurs. Pterosaurs have also been found in one of these formations, the Tugulu group. Their age can be given as late Lower Cretaceous.

Dsungaripterus and Phobetor

It was Professor Young Chung-chien (C.C Young for western palaeontologists), the Grand Old Man of Chinese vertebrate palaeontology, who discovered the first Chinese pterosaur: *Dsungaripterus weii* (=Junggar-wing).[19] The fossil material consisted of the front sections of skull and lower jaw, and a large part of the rest of the skeleton, preserved in excellent three-dimensional condition. In 1973 a palaeontological expedition excavated more *Dsungaripterus* skeletal material on the same site near Wuerho, in the north west of the Junggar basin, including complete skulls, a sternum, a sacrum and pelvic bones.[20]

18 Buffetaut, E., Dubus, B. and Mazin, J.-M., 1989. *Une vertèbre de ptérosaure (Reptilia: Archosauria) dans l'Aptien de l'Aube.* Bulletin annual de l'Association Géologique Auboise, 11: 3-8.

19 Young, C.C., 1964. *On a new pterosaurian from Sinkiang, China.* Vertebrata Palasiatica, 8: 221-256; Beijing.

20 Young C.C., 1973. *Reports of Paleontological Expedition to Sinkiang (II). Pterosaurian Fauna from Wuerho, Sinkiang.* Memoirs of the Institute of Vertebrate Palaeontology and Palaeoanthropology, Academia Sinica, 11: 18-35; Beijing (in Chinese).

Right: The skull of *Dsungaripterus weii* from the Lower Cretaceous of Xinjiang Province, China, seen from the side (upper photograph) and from the top (lower). These views show particularly clearly the extremely pointed, toothless tips of the jaws which are bent upwards at the front. It seems possible that *Dsungaripterus* used these like a pair of forceps to winkle out small crabs and shellfish on the shore. Once caught, such creatures could have been broken open by means of the flattened bony knobs that are situated further back in the jaws. The crests running along the midline of the skull are also readily apparent in this view. The length of this skull is 16in (41cm).

Dsungaripterus weii was a fairly large pterosaur with a wing span of 9·8-11·5ft (3-3·5m), with a skull up to 1·6ft (50cm) long. The toothless tips of its jaws, slightly bent upwards, are a striking feature. They seem to have worked like a pair of tweezers. Further back, both upper and lower jaw have blunt knobs on their margins, which look like breaking tools. Perhaps *Dsungaripterus* used its jaw like the beak of a shore bird, in order to find and crack open bivalves, snails and crabs.

Other particular features of *Dsungaripterus* are cranial crests on the skull: an elongated crest on the snout along the mid-line extending over the eyes, and a short crest rising above the back of the head. The eye socket is quite high in the skull, and relatively small. The largest aperture in the skull is the nasopreorbital opening. *Dsungaripterus* had a series of fused front dorsal vertebrae, a so-called notarium, and fused sacral vertebrae, a synsacrum, similarly to birds.

Phobetor (right)
Phobetor was a close relative of *Dsungaripterus*, although only about half its size. It had similar bony crests adorning its head. Its pointed jaws are straighter than those of *Dsungaripterus* and they contain real, conical teeth rather than tooth-like bony knobs. *Phobetor* is seen here skimming over the water and feeding in the fast-moving river shallows.

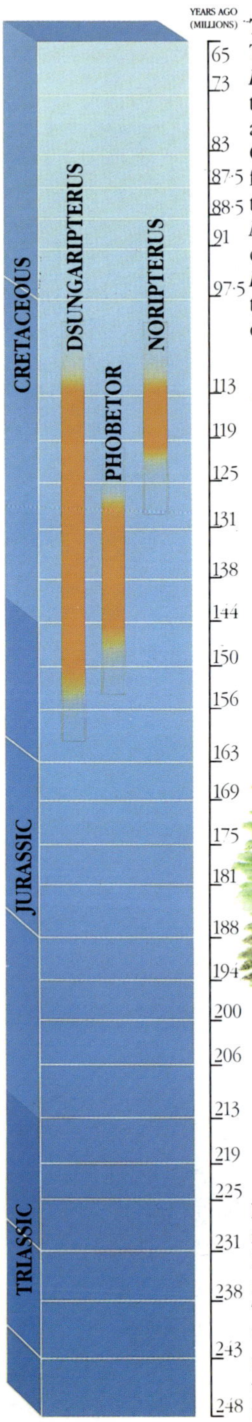

Time Chart (left)
There is evidence that *Dsungaripterus* existed in the late Jurassic of Africa, and survived into the early Cretaceous where it was found in China along with the related genus *Noripterus*. Another dsungaripterid pterosaur, *Phobetor*, is recorded from the early Lower Cretaceous of West Mongolia.

Dsungaripterus (above right)
In this scene we see a small river flowing into a lake. The stegosaur *Wuerhosaurus* is entering the shallows with one of its young. It is a landscape typical of the early Cretaceous environment of Xinjiang, China, where *Dsungaripterus* lived. It probably fed on shellfish, or perhaps even fish, from such a habitat. *Dsungaripterus* was a moderately large pterodactyloid pterosaur with distinctive pointed and curved jaws which may have been used like tweezers to probe for small aquatic creatures. The blunt, bony knobs further back in the jaws could have been used to crack open the shells of such organisms.

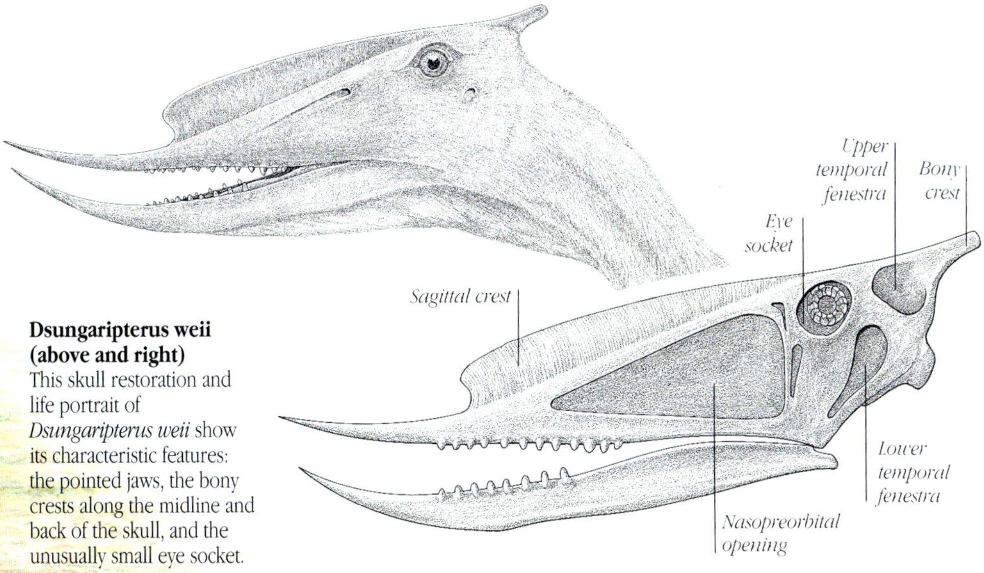

Dsungaripterus weii (above and right)
This skull restoration and life portrait of *Dsungaripterus weii* show its characteristic features: the pointed jaws, the bony crests along the midline and back of the skull, and the unusually small eye socket.

In 1982 N. Bakhurina of the Moscow Palaeontological Institute described skeletal remains of a small *Dsungaripterus (D. parvus)* from the early Lower Cretaceous, Zagan Zabsk Formation, of western Mongolia. At first there were only bones from the wing skeleton and the hind legs. Later skulls were also discovered, and there were such clear distinctions between these bones and the Chinese *Dsungaripterus* specimens that a new genus, *Phobetor* (=the frightening one) was suggested for the Mongolian form. *Phobetor* did have *Dsungaripterus*' toothless jaw points and a cranial crest, but also had genuine pointed teeth. Its wing span was about 4·9ft (1·5m).[21]

21 Bakhurina, N.N., 1982. *Pterodactyl from the Lower Cretaceous of Mongolia.* Palaeontological Journal, 4: 104-108; Moscow (in Russian).
Bakhurina, N.N., 1986. Priroda, Akademia Nauk SSR; Moscow (in Russian).

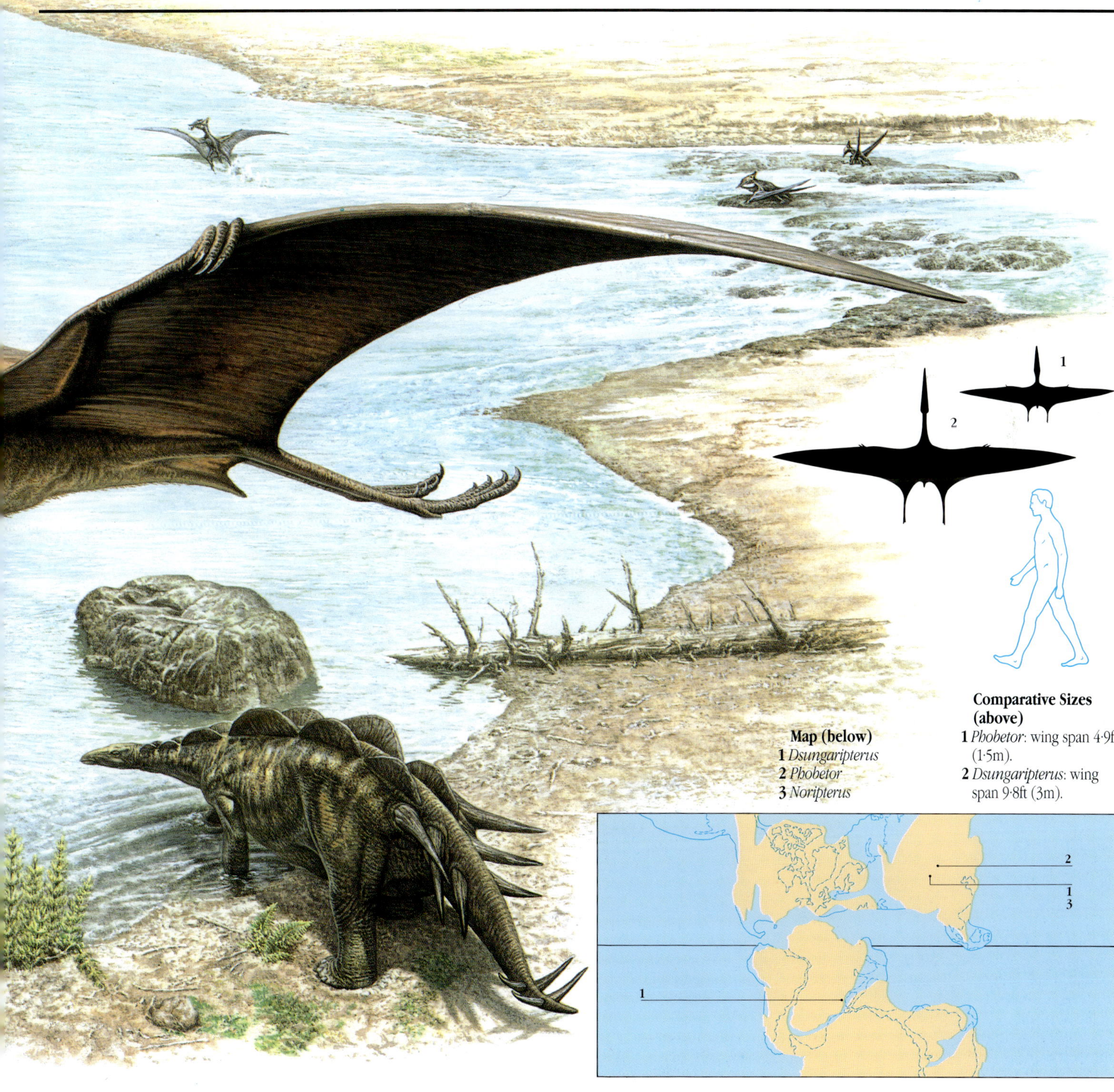

Comparative Sizes (above)
1 *Phobetor*: wing span 4·9ft (1·5m).
2 *Dsungaripterus*: wing span 9·8ft (3m).

Map (below)
1 *Dsungaripterus*
2 *Phobetor*
3 *Noripterus*

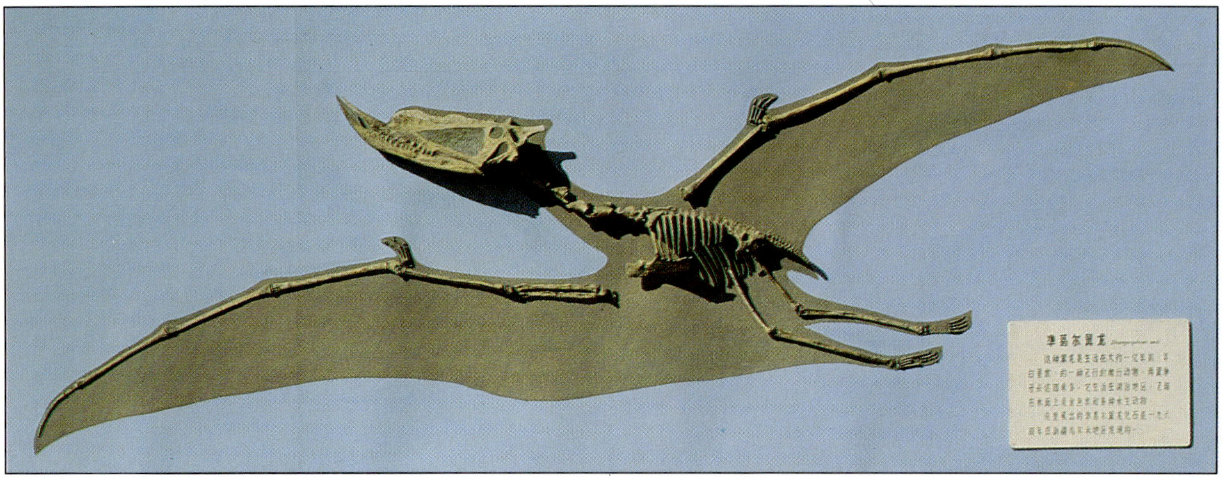

Left: A wall mount of the skeleton of *Dsungaripterus weii* from early Cretaceous sediments in Xinjiang, China which is on display in the Natural History Museum, part of the Institute of Vertebrate Palaeontology and Palaeoanthropology in Beijing. Missing bones have been restored and put in their correct places. The outlines of the body and the wing membranes are indicated against the blue background. The wing span of this mounted skeleton is nearly 10ft (3m).

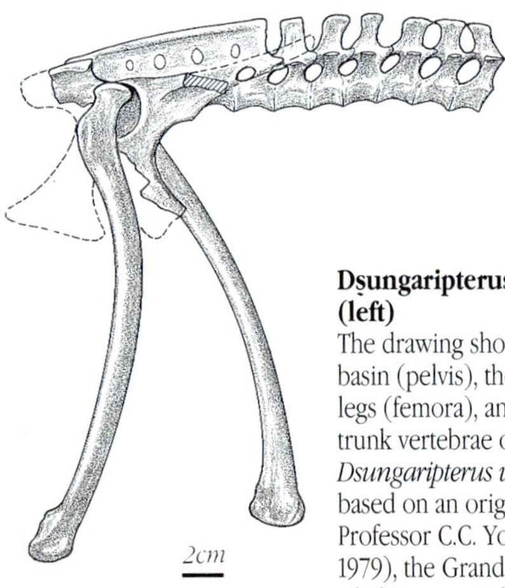

Dsungaripterus Pelvis (left)
The drawing shows the basin (pelvis), the upper legs (femora), and the last trunk vertebrae of *Dsungaripterus weii*. It is based on an original by Professor C.C. Young (1896-1979), the Grand Old Man of Chinese vertebrate palaeontology who discovered and described this, the first Chinese pterosaur, in 1964.

Noripterus

During the second palaeontological expedition to the Junggar basin of Xinjiang another pterosaur was discovered, which was a third smaller than *Dsungaripterus*. Professor Young gave it the name *Noripterus* (=lake-wing).[20] The skeletal remains were of various individuals, including the front end of an upper jaw and bones from the wing skeleton and hind legs. Like that of *Dsungaripterus*, the lower jaw was toothless at the front. The rest of the teeth were strong, and set fairly far apart. *Noripterus*' neck vertebrae were long and narrow. Both Lower Cretaceous pterosaurs from Xinjiang are closely related and have therefore been placed in the family Dsungaripteridae.

Professor Young tentatively placed pterosaur skeletal remains from the Lower Cretaceous of Shantung and Inner Mongolia in this category.[19] Dsungaripterid pterosaurs apparently occurred all over the world. They lived in East Africa in the Upper Jurassic,[22] and in South America in the Lower Jurassic,[23] as well as in China and Mongolia.

The first mention of a Cretaceous pterosaur in Russia was by N.N.A. Bogolubov, who described a fragment of a neck vertebra of considerable size in 1914.[24] It was found in Upper Cretaceous deposits in Saratov in the Petrovsk district and given the name *Ornithostoma* (=bird-mouth) *orientalis*. The genus *Ornithostoma* had already been suggested by H.G. Seeley in 1871 for the jaws of toothless pterosaurs from the Cambridge Greensand, but later synonymized with the North American *Pteranodon* by S.W. Williston. The incompleteness of the English finds has not so far made it possible to confirm this.

Long-necked Giants

No more Cretaceous pterosaurs were found in the Soviet Union until 1984. L.A. Nessov of the University of Leningrad found skeletal remains of a large pterosaur in Upper Cretaceous strata of Uzbekistan in that year, and gave them the name *Azhdarcho lancicollis*, from the Uzbek name for a mythical dragon.[25] The species name *lancicollis* refers to the species' long neck. The neck vertebrae are very long and slender, and also very similar to the long cervical vertebrae of the giant pterosaur *Quetzalcoatlus* from Texas. For this reason they were classified in a common family which was named Azhdarchidae.[26]

A very long neck vertebra described by the French palaeontologist C. Arambourg in 1959 is also assigned to the Azhdarchidae family.[27] The fossil came from Cretaceous strata in the Amman region of Jordan, and was allotted the

Above and below: This is the skull of *Phobetor parvus* from early Lower Cretaceous rocks of Western Mongolia. This dsungaripterid pterosaur had toothless pointed tips to its jaws, and crests along the top and at the rear of its skull.

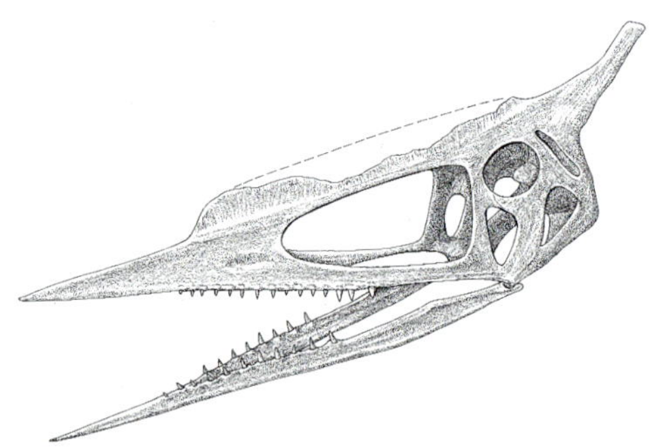

22 Galton, P., 1980. *Avian-like tibiotarsi of the pterodactyloids (Reptilia: Pterosauria) from the Upper Jurassic of East Africa.* Paläontologische Zeitschift, 54: 331-342; Stuttgart.
23 Bennet, S.C., 1989. *A Pteranodontid Pterosaur from the early Cretaceous of Peru, with comments on the Relationships of Cretaceous Pterosaurs.* Journal of Paleontology, 63(5): 669-677.
24 Bogolubov, N.N.A., 1914. *A propos' d'une vertèbre de Ptérodactyle des dépots crétacés supérieurs du gouvernement de Saratoff.* Annales de géologie et minéralogie de la Russie, 16 (1): 1-7.
25 Nessov, L.A., 1984. *Pterosaurs and Birds from the Upper Cretaceous of Middle Asia.* Paleontological Journal, Academy of Sciences SSSR, 1984 (1): 47-57; Moscow (in Russian).
26 Padian, K., 1986. *A taxonomic note on two pterodactyloid families.* Journal of Vertebrate Palaeontology, 6 (3): 289.
27 Arambourg, C., 1959. *Titanopteryx philadelphiae nov. gen., nov. sp., ptérosaurien géant.* Notes et Mémoires du Moyen Orient, 7: 229-234.

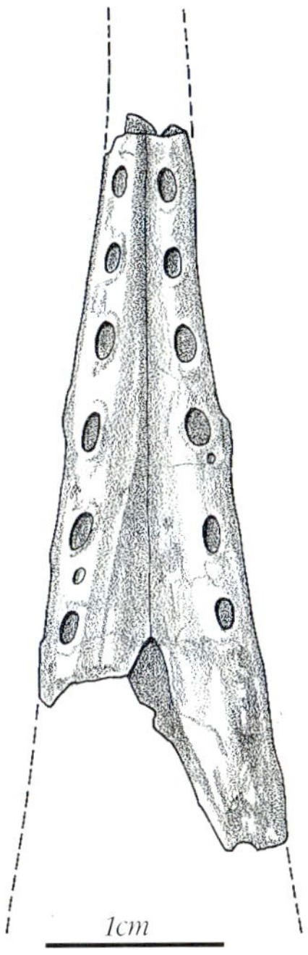

Noripterus Jaw (above)
We see here, from below, a fragment of the upper jaw of *Noripterus*. The holes are empty tooth sockets.

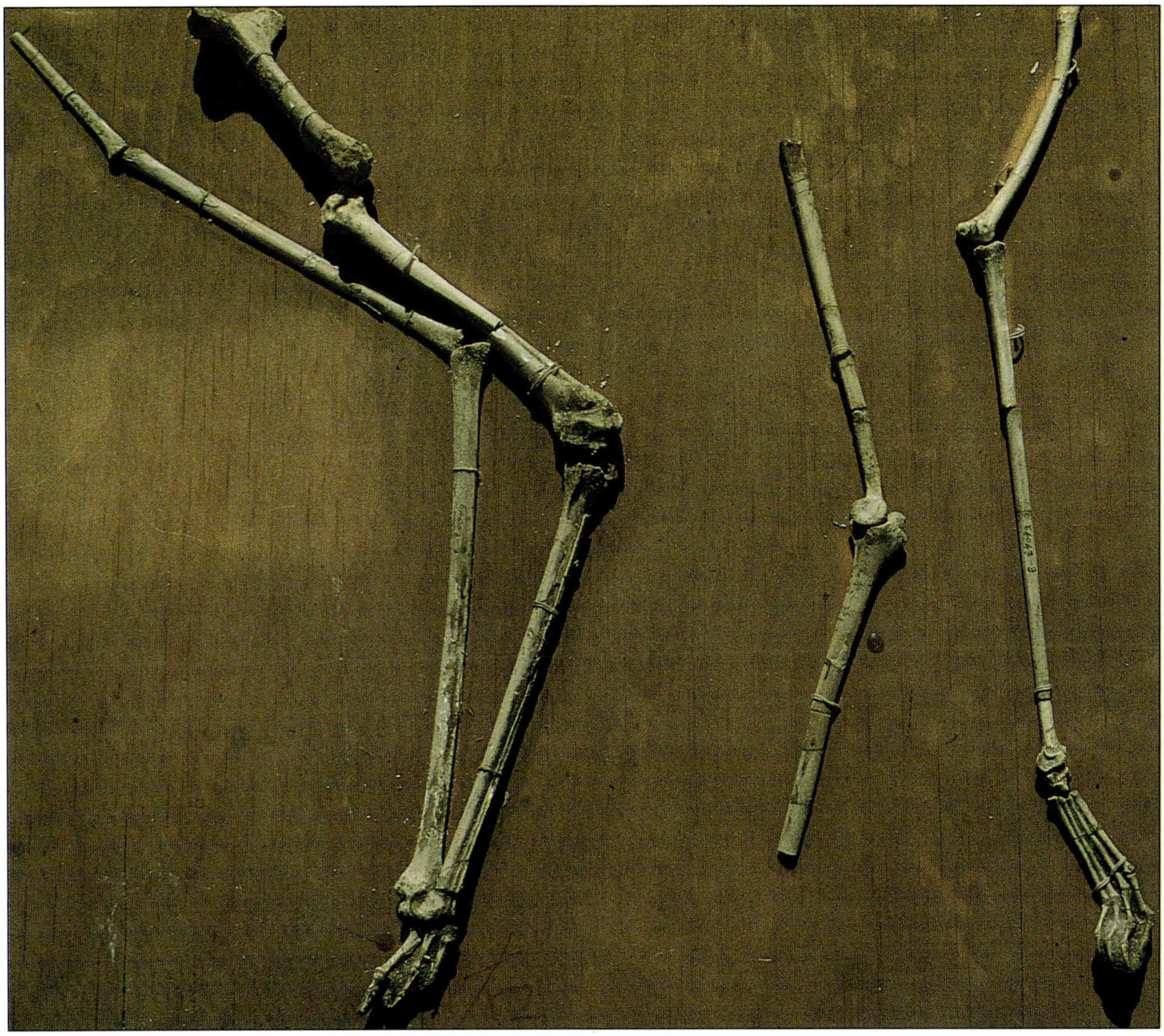

Right: These are remains of *Noripterus*. On the left is a nearly complete folded wing, in the centre the main folding joint of the wing which connected the wing finger to the metacarpal bone, and on the right a complete hind leg.

name *Titanopteryx* (=titan-wing) as a new genus. In fact Arambourg believed that the 2ft (60cm) long bone was the wing metacarpal of a giant pterosaur. It was not until the extremely long neck vertebrae of *Quetzalcoatlus* were found in Texas that it became clear that the *Titanopteryx* bone was also a neck vertebra. It thus achieved comparable size with the giant pterosaurs of Texas.

Long-necked giant pterosaurs occurred all over the world in the Cretaceous, as is shown by the discovery of fossil remains of *Quetzalcoatlus* in Texas and possibly also in Alberta, *Azhdarcho* in Uzbekistan, *Titanopteryx* in Jordan, a long vertebra in Senegal and perhaps *Doratorhynchus* in England. They must have been soaring fliers of great stamina, able to cover great distances.

The find of a 'limb bone' in the Upper Cretaceous of Hokkaido in Japan, classified in the genus *Pteranodon*[28] and a fragment of a pterosaur jaw with three teeth from Rajasthan in India[29] should also be mentioned. Both finds are too incomplete to be determined more accurately. They are simply proof of the palaeogeographic distribution of pterosaurs in the Cretaceous.

28 Ikuwo, O., Hasegawa, Y, and Otsuka, H., 1972. *Preliminary Report on the Cretaceous Reptile Fossils from Hokkaido.* Memoirs of the National Science Museum, 5: 213-222; Tokyo.

29 Dubey, V.S. and Narain, K., 1946. *A Note on the occurrence of Pterosauria in India.* Current Science, 15 (10): 287-288; Bangalore.

Out of Africa

Proof that pterosaurs also lived on the African continent in the Cretaceous rests upon two examples only. The first is an incomplete wing metacarpal of a large pterosaur from Cretaceous (Cenomanian-Turonian) deposits in the former Belgian Congo, modern Zaïre. W.E. Swinton described it in 1948 and established similarities with *Ornithocheirus* species from the English Cambridge Greensand.[30] The total length of this metacarpal was about 14in (36cm), and thus the wing span of this creature can be estimated at 13-16ft (4-5m) at least.

30 Swinton, W.E., 1948. *A Cretaceous Pterosaur from the Belgian Congo.* Bulletin de la Société Belge de Géologie, Paléontologie et Hydrologie, 47: 234-238; Brussels.

Azhdarcho Neck Vertebra (below right)
Azhdarcho from the Upper Cretaceous of Uzbekistan, was a large pterosaur with an extremely long neck. The long neck vertebrae, one of which is seen here in different aspects, are of the same type as those of the giant Texas pterosaur, *Quetzalcoatlus*. They have been assigned to the same family, Azhdarchidae.

Below: This 2ft (60cm) long neck vertebra (first taken to be a wing bone) is the only record of the giant pterosaur *Titanopteryx* which was discovered in Cretaceous rocks in Jordan. With its extremely long neck, *Titanopteryx* can also be assigned to the Azhdarchidae.

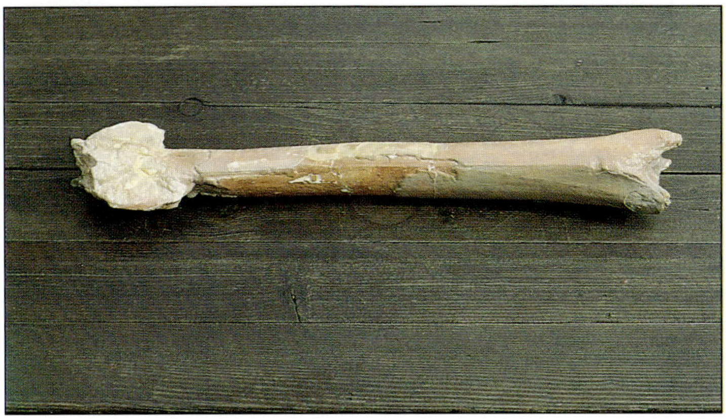

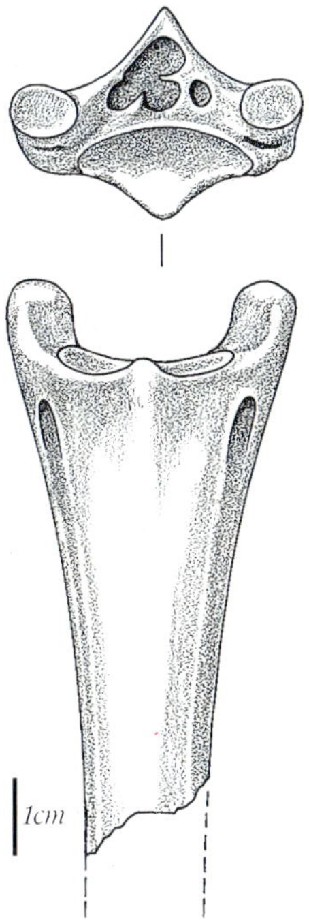

The other specimen of an African Cretaceous pterosaur is an elongated neck vertebra 9·6in (24·5cm) long, from the Upper Cretaceous (Campanian-Maastrichtian) of Paki in Senegal.[31] The vertebra is similar in form to those of the giant pterosaurs *Quetzalcoatlus* of Texas and *Titanopteryx* of Jordan, and shows that Azhdarchidae occurred in Africa as well.

This sparse fossil record of Cretaceous pterosaurs in Africa is surprising as pterosaurs were amply documented in South America in this period, particularly in the Lower Cretaceous, thus at a time when the two continents were not yet separated by the South Atlantic, so that unrestricted interchange of fauna must have been possible. Other reptiles, like crocodiles and turtles, occur correspondingly in Lower Cretaceous strata of equal age, in both Brazil and Niger. Brazilian Cretaceous pterosaurs must have extended their habitat to the African section of the Gondwana continent 110 million years ago. Clearly there were less favourable conditions for fossil preservation in suitable sediments on the African side. Perhaps in future more intensive searches for fossils, and geological exploration, will also lead to the discovery of more, and more complete, Cretaceous pterosaurs in Africa.

Australian Excavations

It was a long time before pterosaur fossils were discovered in Australia as well: the first confirmed find was reported by R.E. Molnar and R.A. Thulborn in 1980.[32] The fossil material consisted merely of the front end of a lower jaw with empty tooth sockets, a single vertebra and an isolated scapulocoracoid. All these skeletal remains were found in marine Lower Cretaceous sediments in the Eromanga Basin in western Queensland. The fossil strata are part of the Toolebuc Formation and dated to the late Lower Cretaceous (Albian). Thus they are about 100 million years old and were deposited alongside plentiful marine fauna, especially bivalves, snails, belemnites, ammonites, fish and marine reptiles.

The characteristics of the lower jaw of this Toolebuc pterosaur are reminiscent of parts of the jaw of *Ornithocheirus* in England. The vertebra, also incomplete, is similar to those of English and Chinese pterosaurs of the Lower Cretaceous, and the pectoral girdle like that of *Pteranodon* and *Nyctosaurus* from the Upper Cretaceous of North America.

It seems that in any case the early Cretaceous Australian pterosaurs belonged to various genera, but that there were clearly close links with English, American and Chinese forms.

In 1987 R.A. Molnar identified another pterosaur fragment from the Toolebuc Formation in Queensland.[33] It was part of a pelvis, successfully freed from the rock, complete and in three dimensions, with acid. It corresponded in many ways with *Pteranodon*. The hip socket still showed its natural orientation, in other words sideways, upwards and backwards. This leads one to assume that the hind legs could not have functioned like those of birds, an argument against bipedal locomotion.

31 Monteillet, J., Lappartient, J.R. and Taquet, P., 1982. *Un Ptérosaurien géant dans le Crétacé supérieur de Paki (Sénégal).* Comptes rendus Académie des Sciences Paris, 295, série II: 409-414; Paris.

32 Molnar, R.E. and Thulborn, R.A., 1980. *First pterosaur from Australia.* Nature, 288 (5789): 361-363.

33 Molnar, R.E., 1987. *A pterosaur pelvis from Western Queensland, Australia.* Alcheringa, 11: 87-94.

Above: The front end of a lower jaw from the Lower Cretaceous Toolebuc Formation of Western Queensland collected by Ralph Molnar in 1979. The fossil is similar to some species of *Ornithocheirus*.

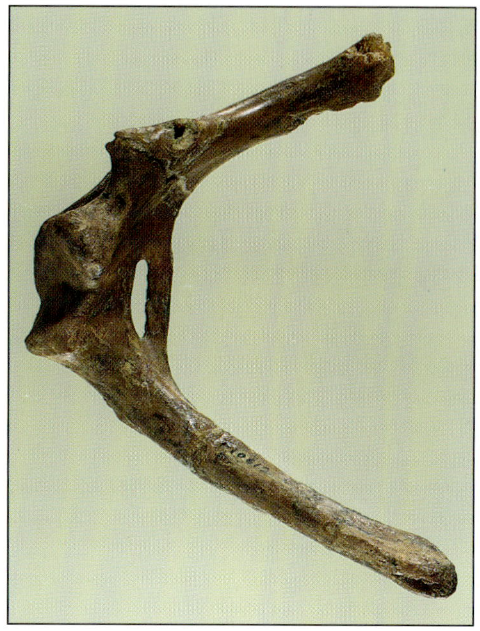

Above: A shoulder girdle of the 'Toolebuc pterosaur' also discovered by Ralph Molnar. This boomerang-shaped bone shows close similarities to the Upper Cretaceous *Pteranodon* from North America.

However sparse and fragmentary the Australian pterosaur material may be, so far only four fragments of bones, it still shows that pterosaurs had penetrated a considerable distance into southern latitudes in the Lower Cretaceous, that they probably belonged to various groups of Pterodactyloidea, and that there were close correspondences with pterosaur fauna in America, Europe and possibly China as well.

The most southerly occurrence of pterosaurs is documented by a single fragment of bone from Upper Cretaceous strata in New Zealand. It is the distal section of an ulna from a middle-sized pterodactyloid coming from marine sandstone in the Mangahouanga Stream on the North Island of New Zealand, described by J. Wiffen and R.E. Molnar in 1988.[34] A single pterosaur tooth was found with the ulna, proving that toothed pterosaurs still lived at this late stage of the Cretaceous (Campanian-Maastrichtian).

The bone shows certain similarities with English and Brazilian genera, but cannot be classified with certainty. The special feature of this find, however, is its palaeogeographic situation. It is not only the first New Zealand pterosaur, but also the most southerly. In the Upper Cretaceous this area of New Zealand was at a latitude of 60°S. That proves that even shortly before they became extinct pterosaurs were in a position to live in extremely high latitudes and in a cool to cold-temperate climate with clear seasonal variations. This was only possible if these large, actively flying reptiles were warm-blooded.

South America's Nyctosaurus

It was Arthur Conan Doyle who told of the exciting discovery of pterosaurs that were still alive in South America, in his Professor Challenger story *The Lost World*. Of course this notion sprang from the imagination of a brilliant writer. In 1912, when the book appeared, the inventor of the legendary Sherlock Holmes could certainly not have suspected that some of the most significant fossil pterosaur finds were to be made in South America. It was not until 1953 that the Brazilian L.I. Price described the first South American pterosaur find.[35]

The fossil was the upper part of a humerus from the Upper Cretaceous (Gramame Formation) of Paraiba in Brazil. The complete bone would have been about 6·5in (16·5cm) long. This suggests a larger pterosaur with a wing span of about 11·5ft (3·5m).

The form of this humerus corresponds very well with specimens from the Chalk of the Niobrara Formation in Kansas, already named as *Nyctosaurus* (=naked reptile) by O.C. Marsh as early as 1876. For this reason Price called this first South American pterosaur *Nyctosaurus*

34 Wiffen, J. and Molnar, R.E., 1988. *First pterosaur from New Zealand.* Alcheringa, 12: 53-59.

35 Price, L.I., 1953. *A presenca de Pterosauria no Cretáceo superior do Estado da Paraiba.* Divisão Geologia Mineralogia, Notas preliminares, Estud., 71: 1-10; Rio de Janeiro.

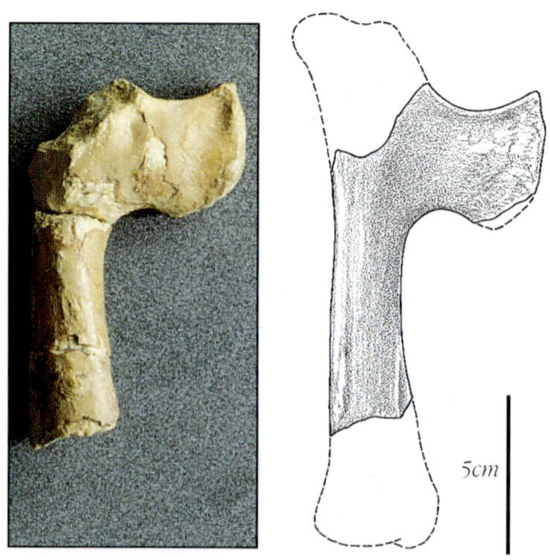

Above: This upper arm from late Cretaceous rocks in Paraiba, Brazil was the first discovery of pterosaur remains in South America. It was described as *Nyctosaurus lamegoi* in 1953. The wing span of this animal may have reached 11·5ft (3·5m).

Above: The city of Juazeiro do Norte in Ceará in north-eastern Brazil with the Araripe Plateau in the distance. The area has become famous for the fossils occurring in the early Cretaceous Santana Formation exposed along the edge of the vast plateau.

lamegoi (named after A.R. Lamego, then director of the geology and mineralogy division of the Department of Mineral Production in Rio de Janeiro), a species which was clearly larger than *Nyctosaurus gracilis* from Kansas. *Nyctosaurus* was a toothless pterodactyloid, which differed from its giant contemporary *Pteranodon* not only by being smaller, but also by having no crest at the back of its head. Of course a single arm bone cannot tell us anything about such particular features of the skull; we can only make these deductions by assuming that it belongs to the genus *Nyctosaurus*. If more complete skeletal material of *Nyctosaurus lamegoi* should ever be found, it could very well turn out that it is different from the Kansas forms and perhaps represents a new genus.

It was again L.I. Price who announced the first pterosaur skeletal remains from the Santana Formation in north-eastern Brazil, from a stratum that has produced finds among the most significant and productive in the world. Probably the best known fossils of the Santana Formation are the fish, which are preserved typically in limestone concretions. Local fossil collectors split these nodules, also known as geodes, and thus reveal the fossil. This kind of fossil preservation is particularly favourable for the delicate, thin-walled pterosaur bones, as the calcareous concretions form a protective case for the bones and skeletal elements, in which they are mostly preserved in three-dimensional form, and have thus survived the passage of millions of years as if they had been kept in a safe.

Preparation of the fossils is a lengthy process, using hammer and chisel or a chemical method with acid, but it produced individual bones with which it is often possible to test the possibilities of articular movement, just as though these were the bones of living creatures. These unique fossil preservations and the range of new pterosaurs, some of them looking very strange indeed, all contributed to the fame of the Santana Formation in Brazil.

The site is on the slopes of the Araripe Plateau (Chapada do Araripe), at the border of the states of Piauí, Ceará and Pernambuco in north-eastern Brazil. The Santana Formation reaches a thickness of about 656ft (200m). But it is only in the uppermost stratum, the Romualdo Member, that the fossil concretions that were formed about 115 million years ago, in the late Lower Cretaceous (Aptian), in a marine environment near the coast, are to be found. Even though it is only in recent decades that increasing attention has been paid to pterosaur remains, the fossil manifestations of the Araripe Plateau have been known for a long time. In fact it was two Bavarian naturalists, zoologist J.B. Spix and botanist C.F.P. Martius, who undertook an expedition to Brazil from 1817 to 1820, into country that at the time was completely unexplored. In 1819 they reached the province of Pernambuco and discovered the fossil sites of what is called the Santana Formation today, with their fish fossils.[36] Zoologist Dr Johann Babtist Spix and botanist Dr Carl Friedrich Philipp Martius undertook an extended scientific and anthropological expedition to north-eastern Brazil from 1817 to 1820 on behalf of the Bayerische Akademie der Wissenschaften in Munich. At the time an ocean voyage was accompanied by privations, hardship and danger. The Munich scientists found this out all too quickly. On the very second night, after they had sailed from Trieste, the Bora, the cold north wind of the Adriatic, struck their frigate *Austria* with full force, so that they were nearly shipwrecked.

Three months later, on 15 July 1817, they reached Rio de Janeiro at last and began their journey of exploration, which was to last over three years, into the north-east of the country and the Amazon region. They covered about 6,200 miles (10,000km), mostly on foot or by boat. At that time the Brazilian interior was still wild, completely unexplored territory. Their journey was undertaken under the most primitive conditions, and they suffered unbelievable hardship, hunger, thirst and severe illness.

In the three-volume description of their journey, published after their return to Munich, they also report on the discovery of sites in the Araripe region where fossils were found: 'Almost on the south-eastern border of the province, near the little Villa do Bom Jardim, in the Cayriris Novos district, there is a fairly extensive marly lime formation, in which there are numerous fossils of fish. The same appear both in the bedded strata of rock and in the segregated and rolled pieces.' They also provide an illustration of a fossil fish in their work, the front part of a *Rhacolepis*, a marine fish similar to the tarpon.

Santana Specimens

Price called the first pterosaur from the Araripe Plateau *Araripesaurus*, (=Araripe reptile) and classified it as an ornithocheirid.[37] The specimen consisted of skeletal remains of an individual preserved in a calcareous concretion typical of the Santana Formation. They in-

36 Spix, J.B. and Martius, C.F.P., 1828. *Reise in Brasilien.* 3 volumes; Munich.

37 Price, L.I., 1971. *A Presença de Pterosauria no Cretáceo Inferior da Chapada do Araripe, Brasil.* Anais Academia Brasileira Ciencias, 43 (suppl.): 451-461.

Right: These wing bones encased in a limestone nodule were the first pterosaurian fossils discovered in the Lower Cretaceous Santana Formation of the Araripe Plateau. They were recorded in 1971, and described as *Araripesaurus castilhoi* by the Brazilian palaeontologist Llewellyn Price. Not only was the concretion split in half, the individual bones were split and present mirror images. The specimen consists of a section of the lower arm, wrist and metacarpals, and the upper end of the wing finger. The wing span of this particular individual can be estimated at 7·2ft (2·2m).

Above: The lower jaw of *Santanadactylus araripensis* from the Santana Formation of the Araripe Plateau in north-eastern Brazil. We are here looking down on the lower jaw from above.

Above right: The incomplete skull of *Santanadactylus araripensis* shown after preparation. It was found in a limestone nodule.

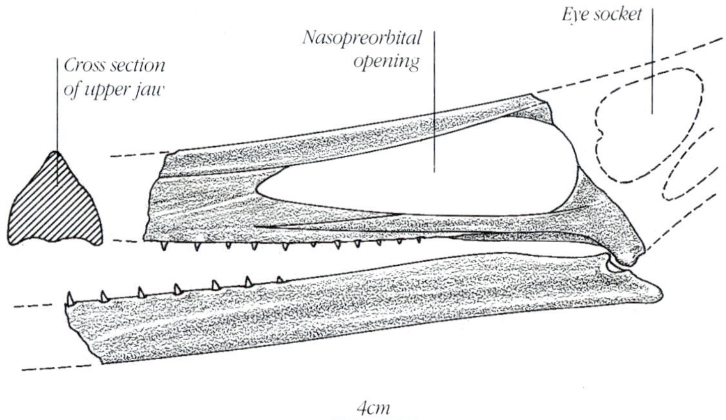

Santanadactylus araripensis Skull Restoration (right)
This restoration shows the skull and lower jaw of *Santanadactylus araripensis* as far as it is preserved. *Santanadactylus* is classified as an ornithocheirid pterosaur with teeth. Its wing span was over 10ft (3m).

cluded parts of the right-hand wing skeleton of a short-tailed pterosaur. The only full-length bone to have survived is the fourth metacarpal, which supported the flight digit. On the basis of its size (4·6in; 11·8cm) the wing span of this individual can be estimated at 7·2ft (2·2m). It is true that the skull is missing, and this is the most important part of the skeleton for diagnostic purposes. For this reason the classification of *Araripesaurus* as an ornithocheirid must remain dubious.

The second pterosaur find from the Araripe Formation also belongs to an undetermined family. It is a single flight digit phalanx, discovered in 1977 and named *Araripedactylus* (=Araripe finger).[38] This long, robust bone 1·8ft (55cm) long was the first proof that large pterosaurs also occur in the Santana Formation; their wing span was at least 16ft (4·8m). Here too more complete skeletal material is needed before the relationships of *Araripedactylus* to other genera can be cleared up.

In 1980 another pterosaur genus from the Santana strata was described, and named *Santanadactylus* (=Santana-finger).[39] But again there were only isolated bones, fragments of a humerus and shoulder articulation, and of two neck vertebrae; they came from different individuals, however. The neck vertebrae are elongated, and suggest a long-necked form. The characteristics of the humerus make it possible to classify *Santanadactylus* as an ornithocheirid. However, these have short neck vertebrae, which means it is doubtful whether the long neck vertebrae belong to *Santanadactylus* at all, or to a hitherto unknown genus.

Later further and more complete examples of *Santanadactylus* skeletal remains were discovered.[40] They included a spinal column that was still connected, with dorsal vertebrae fused to form a notarium, skull, bones of the wing skeleton and even the two almost complete wings of an individual.[41] So far four distinct species of *Santanadactylus* have been identified. Most of them were relatively large ornithocheirid pterosaurs with teeth, with a wing span of 9·5-18·7ft (2·9-5·7m). One species, *'Santanadactylus' spixi*, was probably a dsungaripterid, and represents a new genus.

Above: An upper arm of *Santanadactylus* from the Santana Formation of the Araripe Plateau in north-eastern Brazil. The length of this humerus is 6·7in (17cm). It was the characteristics of this bone that enabled scientists eventually to classify *Santanadactylus* as an orntithocheirid.

The genus *Brasileodactylus* (=Brazil-finger) was based on the front end of a lower jaw from the Santana Formation in Brazil.[42] The surviving section is 4·4in (11·2cm) long, slightly bent upwards and triangular in cross-section. The jaw was toothed, but the teeth had fallen out, so that only the empty alveoli could be seen. These are rounded-elliptical, and more widely spaced towards the back. Characteristics suggest that it belongs to the Ornithocheiridae. Only more complete finds would help to determine whether *Brasileodactylus* is identical to *Santanadactylus*.

The 'Old Devil'

D.A. Campos and A.W.A. Kellner described a new pterosaur genus from the Santana Forma-

38 Wellnhofer, P., 1977. *Araripedactylus dehmi nov. gen., nov. sp., ein neuer Flugsaurier aus der Unterkreide von Brasilien.* Mitteilungen der Bayerischen Staatssammlung für Paläontologie und historische Geologie, 17: 157-167; Munich.

39 Buisonjé, P.H. de, 1980. *Santanadactylus brasilensis nov. gen., nov. sp., a longnecked, large pterosaur from the Aptian of Brazil.* Proceedings of the Koninklijke Nederlandse Akademie van Wetenschappen, B, 83 (2): 145-172; Amsterdam.

40 Wellnhofer, P., Buffetaut, E. and Gigase, P., 1983. *A pterosaurian notarium from the Lower Cretaceous of Brazil.* Paläontologische Zeitschrift, 57: 147-157; Stuttgart.
Wellnhofer. P., 1985. *Neue Pterosaurier aus der Santana-Formation (Apt) der Chapada do Araripe, Brasilien.* Palaeontographica, A, 187: 105-182; Stuttgart.

41 Leonardi, G. and Borgomanero, G., 1987. *The skeleton of a pair of wings of a pterosaur (Pterodactyloidea, ?Ornithocheiridae, cfr. Santanadactylus) from the Santana Formation of the Araripe Plateau, Ceará, Brazil.* Anais do X Congresso Brasileiro de Paleontologia, 1987: 123-129; Rio de Janeiro.

42 Kellner, A.W.A., 1984. *Ocorrencia de uma mandíbula de pterosauria (Brasileodactylus araripensis, nov. gen., nov. sp.) na Formaçao Santana, Cretáceo da Chapada do Araripe, Ceará, Brasil.* Anais XXXIII Congresso Brasileiro de Geologia, 1984: 578-590; Rio de Janeiro.

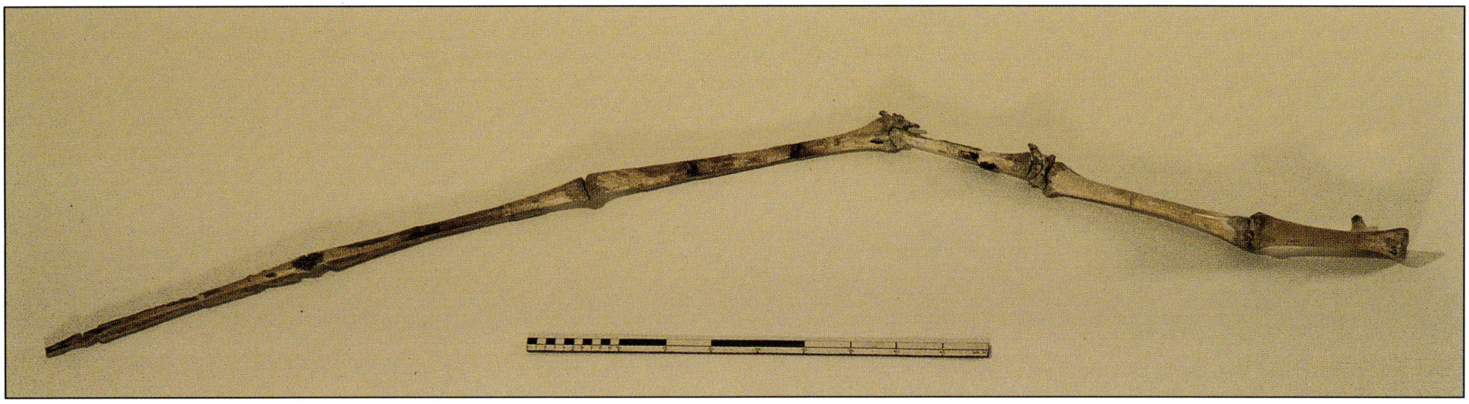

Above: A complete wing of *Santanadactylus* after the wing bones had been isolated from the large limestone nodule in which they were preserved. The length of the wing is 5ft (1·52m).

Above: The wrist of *Santanadactylus* with the pteroid bone pointing towards the body. The bones were isolated from the limestone with acid.

Below: This front end of the lower jaw is the only known skeletal fragment of the genus *Brasileodactylus*. It measures 4·4in (11·2cm) in length.

tion, which they called *Anhanguera* (=old devil) after a name from the Indian Tupi culture.[43] The type specimen is a slender skull 1·6ft (50cm) long with a medial crest on the snout. This crest is an outgrowth of the upper jaw bone. A similar crest may also have developed on the lower side of the lower jaw. *Anhanguera* had teeth, and was probably a fish eater. The crest on the snout stabilized the head when the tip of the snout was drawn through the water while fishing in full flight.

Anhanguera is now one of the best-known pterosaurs from the Santana Formation. Two more skeletal remains of this genus were found, including a fairly complete specimen, in which the skull and a large proportion of the post-cranial elements had survived, like spinal column, ribs, pectoral girdle, pelvis and parts of the wings and hind legs. Some of the bones were found still articulated naturally in a large

43 Campos, D.A. and Kellner, A.W.A., 1985. *Panorama of the Flying Reptiles Study in Brazil and South America*. Anais da Academia Brasileira Ciencias, 1985, 57 (4): 453-466; Rio de Janeiro.

Right: This is the skull of *Anhanguera blittersdorffi* from the Lower Cretaceous Santana Formation. It was a large pterodactyloid pterosaur with a considerable wing span. The bony crest on top of the snout is characteristic.

Anhanguera Skull Restoration (right)
The drawing shows a restoration of the skull of *Anhanguera blittersdorffi* which is pictured above. The lower jaw is not preserved. Skull length is 1·6ft (50cm).

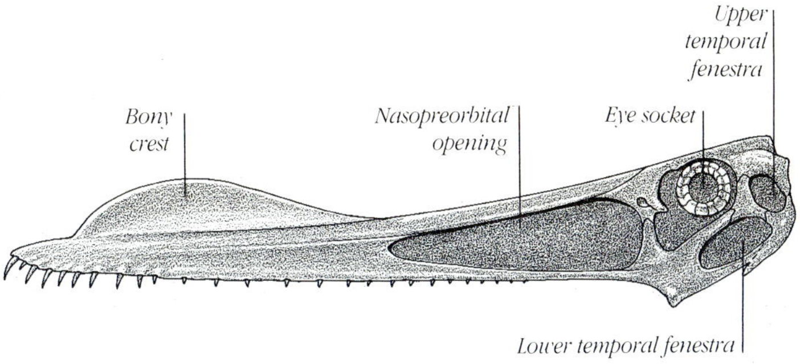

Above: This is *Anhanguera santanae*, a specimen in the collection of the American Museum of Natural History in New York. It was preserved in a large calcareous nodule from the Lower Cretaceous Santana Formation. The photograph shows the New York specimen after preparation, revealing skull, vertebral column, shoulder girdle, pelvis, and parts of the wings. When found, some of the bones were still in their natural articulation. This is the most complete skeleton of a pterosaur of this size range known. This individual had a wing span of 13·6ft (4·15m). Note how long the skull is in comparison to the body.

calcareous concretion and it was to a large extent possible to free them completely from the rocks. The proportions of this large pterosaur with a wing span of more than 13ft (4m) are unusual: in comparison with the actual body the skull is twice as long.[44] The great disproportion between wings and hind legs is expressed by the fact that the pectoral girdle is large and robust, but in comparison the pelvis is quite small.

Three-dimensional reconstruction of the pelvis of *Anhanguera* revealed that the hind legs could not be brought into a vertical position under the body, but were splayed slightly to the side. Thus bird-like, bipedal locomotion on the ground was scarcely possible. Orientation of the hip sockets obliquely upwards and the slight bend of the articular head of the thigh bone make quadrupedal locomotion on the ground more probable.[45]

So far only two species of *Anhanguera* are known. A separate family, the Anhangueridae, was suggested for them. They have a characteristic crest on the snout, and complete ossification of some elements of the skeleton (skull, pectoral girdle, notarium, carpus and pelvis) did not take place until very late in the growth of the individual, perhaps not until shortly before the adult stage. The different formation of the crest in the two species could also be interpreted as a sexual characteristic of male and female individuals.

Cearadactylus and Tropeognathus

A long pterosaur skull with lower jaw from the Santana Formation was also established as a new genus *Cearadactylus* (=Ceará-finger) in 1985.[46] The skull is not preserved intact on the

44 Wellnhofer, P., 1991. *Weitere Pterosaurierfunde aus der Santana-Formation (Apt) der Chapada do Araripe, Brasilien.* Palaeontographica, A, 215; Stuttgart.
45 Wellnhofer, P., 1988. *Terrestrial locomotion in pterosaurs.* Historical Biology, 1: 3-16.
46 Leonardi, G. and Borgomanero, G., 1985. *Cearadactylus atrox nov. gen., nov. sp.: Novo Pterosauria (Pterodactyloidea) da Chapada do Araripe, Ceará, Brasil.* D.N.P.M., Coletana de trabalhos Paleontologicos, Séria Geológica, 27: 75-80; Brasilia.

Above: The skull of the New York specimen of *Anhanguera santanae*. Unfortunately the bony cranial crest has broken off this particular specimen.

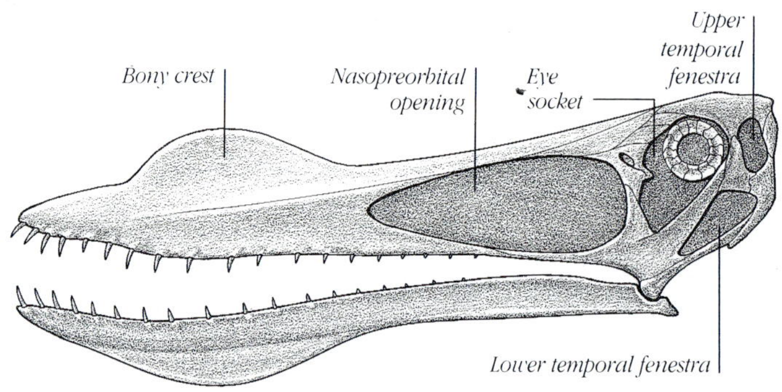

Anhanguera Skull (left)
This restoration of the skull of *Anhanguera santanae* is based on the New York specimen pictured above. Here the bony cranial crest has been restored in position.

Above: A series of neck vertebrae of *Anhanguera santanae* completely isolated from the rock. They are arranged in their natural positions. The lateral openings are weight-saving features.

Above: The shoulder girdle of *Anhanguera santanae* consisting of the shoulder blade and the coracoid bone which are not yet fused into a single element, but forming the articular socket for the upper arm.

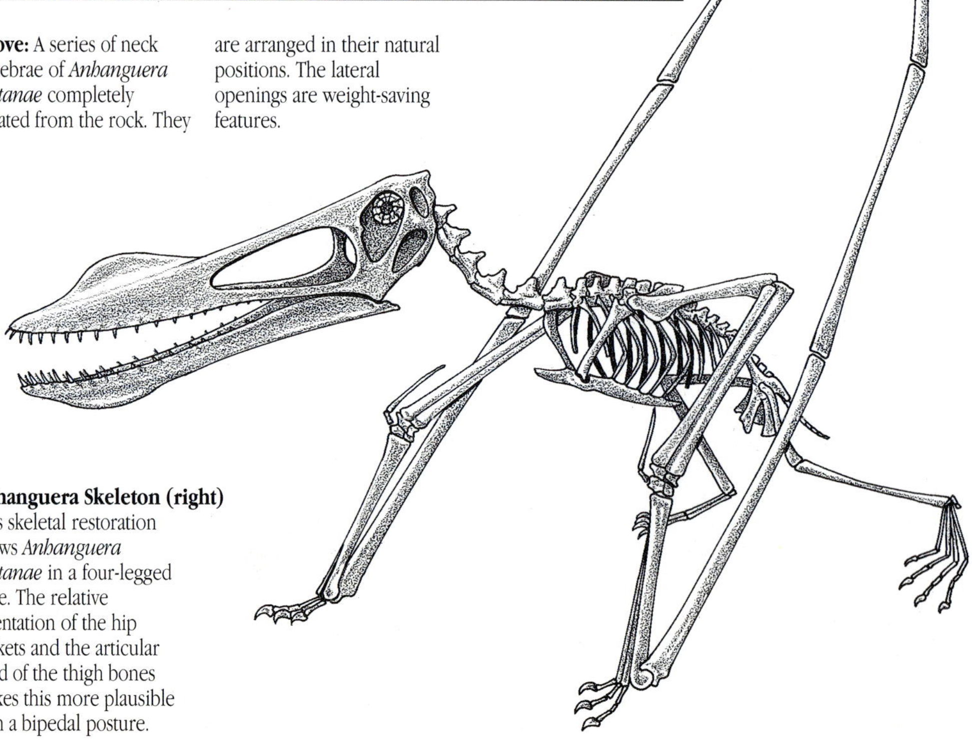

Anhanguera Skeleton (right)
This skeletal restoration shows *Anhanguera santanae* in a four-legged pose. The relative orientation of the hip sockets and the articular head of the thigh bones makes this more plausible than a bipedal posture.

Above: Left and right wrists of *Anhanguera santanae* with the bones arranged in their natural articulation. These bones can be seen with their associated pteroid bones in the skeletal restoration on the opposite page.

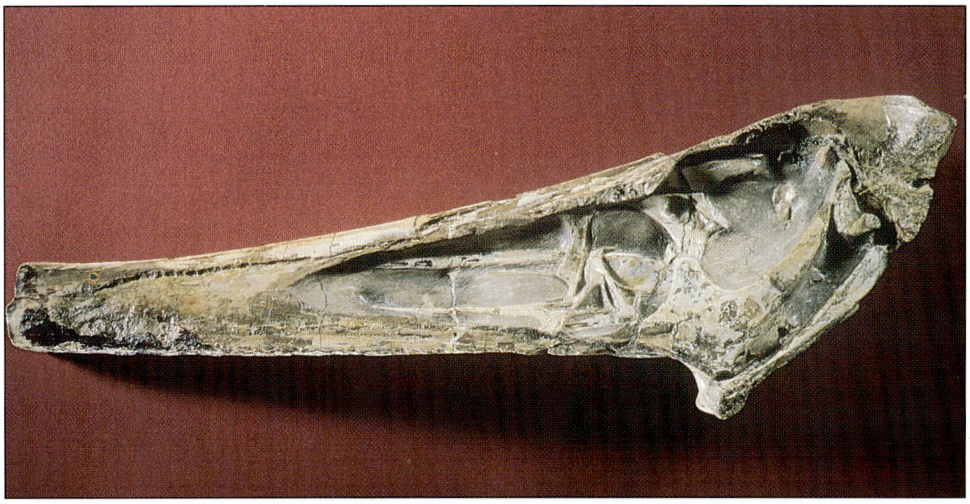

Right: The skull of the Munich specimen of *Anhanguera santanae* lacks the front end of the snout and the bony crest. This specimen was also collected from the Lower Cretaceous Sanatana Formation in north-eastern Brazil.

rear side. Its overall length must have been 1·9ft (57cm). The wing span can be estimated at 18ft (5·5m). A particular feature of *Cearadactylus* is its powerful dentition. The front teeth are much longer and much stronger than the back ones. When the snout is closed there was a gap in the front area. The long front teeth, set in jaws which broaden to a spoon shape at the front, suggest an excellent grip when catching slippery fish. Nothing is known of this genus except the skull. It is therefore perfectly possible that *Cearadactylus* is identical with other genera based only on post-cranial bones, perhaps with *Araripedactylus*, for example.

A complete skull with lower jaw and an isolated lower jaw documented another new pterosaur genus from the Santana Formation.[47] Its special characteristic is a tall, rounded medial crest at the front end of the snout and a similar crest on the lower side of the lower jaw, at the point where the two branches of the mandible have fused in the mid-line to form a symphysis. These crests are in the form of the keel of a ship, for which reason these ptero-

47 Wellnhofer, P., 1987. *New Crested Pterosaurs from the Lower Cretaceous of Brazil.* Mitteilungen der Bayerischen Staatssammlung für Paläontologie und historische Geologie, 27: 175-186; Munich.

Above: This incomplete skull from the Santana Formation is the only evidence for the genus *Cearadactylus.*

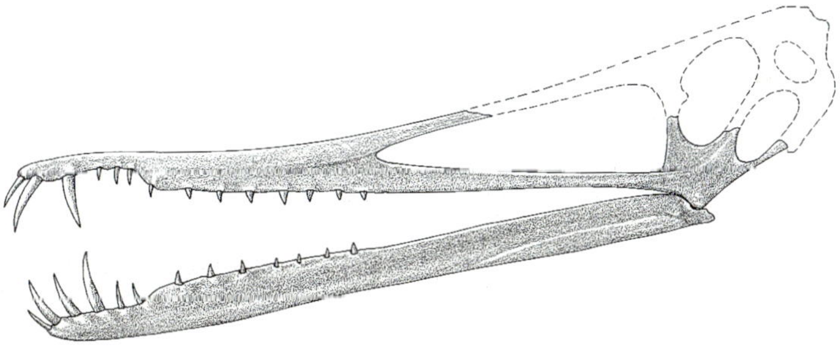

Cearadactylus Skull (left)
This is a tentative restoration of the skull of *Cearadactylus atrox*. Its length is about 1·9ft (57cm).

Left and below: These photographs show the skull and lower jaw of *Tropeognathus mesembrinus*, a large pterosaur from the Lower Cretaceous Santana Formation of Brazil. At the front end of the long jaws deep bony crests are developed. The photograph (left) shows these crests from the front. Their shape resembles the keel of a ship; perhaps they also served as stabilizing devices in the water when the pterosaur was skimming for fishes. The skull length is slightly over 2ft (63cm).

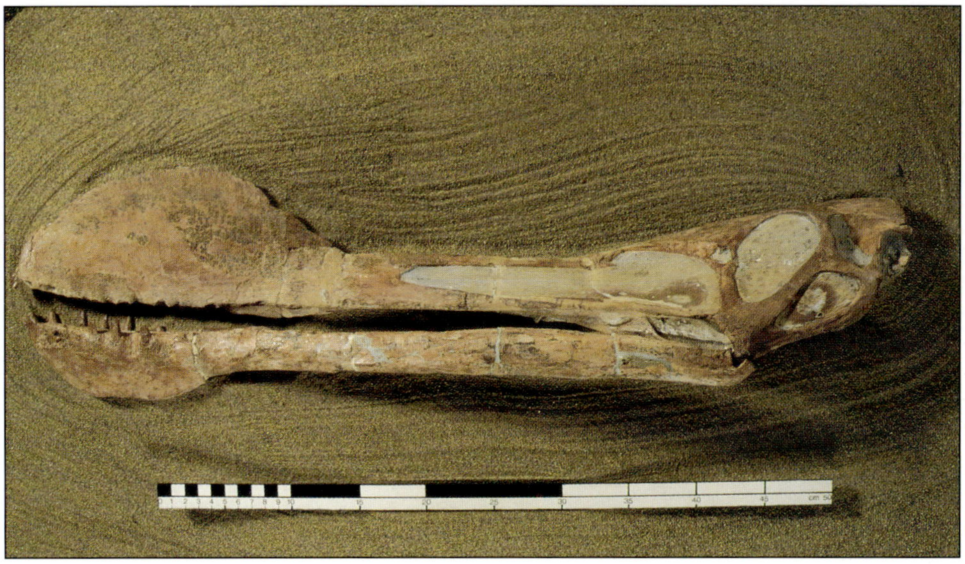

saurs were given the name *Tropeognathus* (=keel-jaw). Another shorter and blunter crest developed on the back of the skull. The dentition consists of a total of 26 teeth in the upper jaw and 22 in the lower.

The function of the crests on the tip of the snout was probably hydrodynamic. *Tropeognathus* also fed on fish. In flight the tip of the snout had to dip into the water and plough through the surface of the sea. The skull was so long that it had to be stabilized by the crest in this phase. This saved muscle mass on the neck, and thus weight.

Tropeognathus is so far known through only two species, represented by two specimens. The skull lengths are 2·0 and 2·2ft (63 and 67cm). *Tropeognathus robustus* had a wing span of 20ft (6·2m) and is thus the largest Santana pterosaur so far found.

The crest at the front end of *Tropeognathus*' snout is highly reminiscent of the high front end of the snout of *Criorhynchus* from the

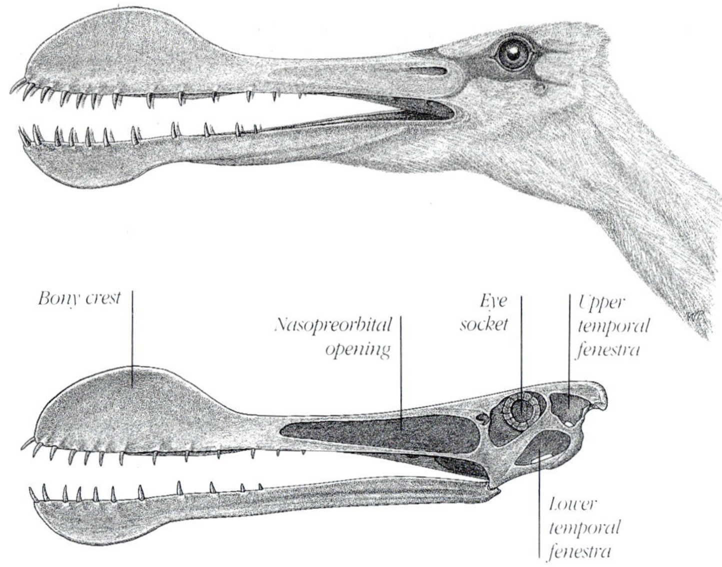

Tropeognathus mesembrinus Life Portrait and Skull (right)
Shown here is a life portrait of *Tropeognathus mesembrinus*, a large toothed and crested pterosaur from Brazil, and below that is a restoration of the skull based upon the specimen illustrated on the previous page. It was almost certainly a fish eater, and its dentition consists of 26 teeth in the upper jaw and 22 in the lower jaw. The bony crests at the front end of the upper and lower jaws would most likely have been covered with a horny sheath. Two species of *Tropeognathus* are known.

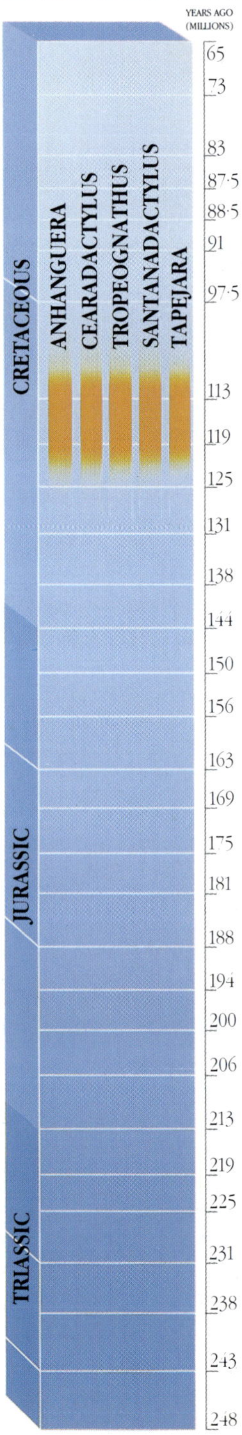

Time Chart (left)
During the early Cretaceous, in the Aptian age, a great variety of pterodactyloid pterosaurs lived in the region of modern north-eastern Brazil. Some of them may be related to English forms.

Cearadactylus (right)
Among the largest pterosaurs from the Santana Formation, *Cearadactylus* is viewed here sitting on a rock and holding its prey fish in a firm grip in its claws. Its long front teeth would have proved very effective tools for fishing in the shallows of near-shore lakes.

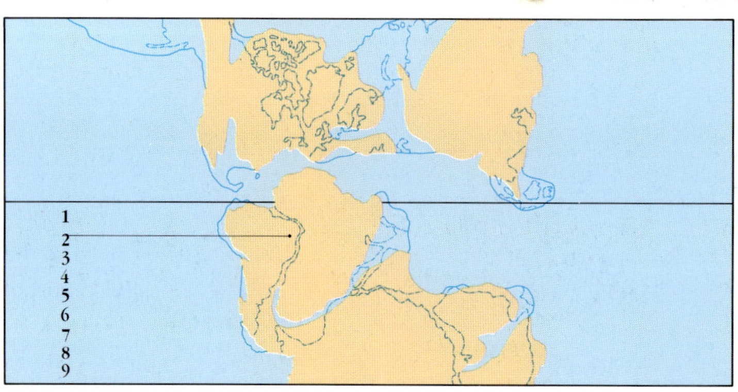

Map (left)
1 *Anhanguera*
2 *Cearadactylus*
3 *Tropeognathus*
4 *Santanadactylus*
5 *Araripedactylus*
6 *Araripesaurus*
7 *Brasileodactylus*
8 *Tupuxuara*
9 *Tapejara*

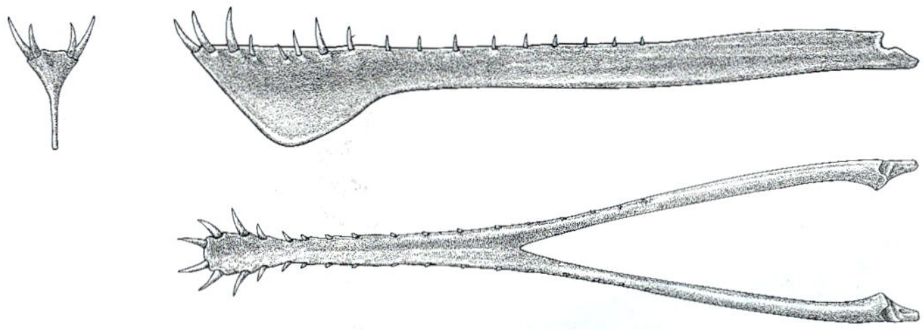

Tropeognathus robustus Lower Jaw (above)
Only the lower jaw of this second species of *Tropeognathus* is preserved. It is shown here from the front, from the side and from above.

Tropeognathus (left)
This is one of the most unusual pterosaurs known. It was named because of the resemblance of the crests on its jaws to a ship's keel (from the Greek word *tropis* meaning keel). These crests must have served to stabilize the head when the tips of the jaws were ploughing through the water. Two species are known, *Tr. mesembrinus* and *Tr. robustus*, which can be distinguished by their different crests and teeth.

Comparative Sizes (above)
1 *Anhanguera*: wing span 13ft (4m).
2 *Cearadactylus*: wing span 18ft (5·5m).
3 *Tropeognathus*: wing span 20ft (6.2m).

Anhanguera (left)
Anhanguera is one of the best known early Cretaceous pterosaurs. It was quite large with very long wings and peculiar low medial crests on top of the skull and below the lower jaw. Like *Tropeognathus*, these served to stabilize the head when fishing. In this scene *Anhanguera* has just caught a fish and the front of its head on its extremely flexible neck is still ploughing through the water. In the background can be seen turtles on the beach leading to the sea.

Cambridge Greensand in England. It was therefore suggested that *Tropeognathus* should be included in the family Criorhynchidae as well.

Toothless Forms

All species of pterosaur from the Santana Formation of Brazil so far mentioned were toothed forms, but in 1989 a toothless species was discovered for the first time. This was the front part of the skull with a medial crest and a few bones from the wing skeleton. The edges of the jaw are clearly toothless, thus indicating a different family from the ornithocheirids or the anhanguerids. A.W.A. Kellner and D.A. Campos named this pterosaur *Tupuxuara*, after a 'familiar spirit' from the culture of the Tupi, a tribe of Brazilian Indians.[48] The fossil remains are still too sparse to give more precise detail about this toothless pterosaur. They are however clearly different from other toothless genera of the Upper Cretaceous like *Pteranodon*, *Nyctosaurus* or *Quetzalcoatlus*.

A second toothless pterosaur from the Santana Formation in Brazil was not described until 1990, and was named *Tapejara* by A.W.A. Kellner.[48] The name means 'the old being' and comes from the mythology of the original inhabitants of Brazil, the Tupi Indians. So far only the skull is known, which like that of *Tupuxuara* has a tall crest in the mid-line. The eye sockets are relatively small, while the nasopreorbital opening on the other hand is very large. The ends of the jaws are directed downwards at the front, like a bird's beak. It was certainly one of the strangest Cretaceous pterosaurs that ever lived. Unfortunately little is known of the rest of the skeleton. Because of certain similarities *Tapejara* and *Tupuxuara* have been placed in a common family, the Tapejaridae. Their relationships to other toothless pterosaurs, like for example *Pteranodon* from the Upper Cretaceous, cannot be established, as yet.

The fossil sites of the Santana Formation of the Araripe Plateau in north-eastern Brazil seem to be inexhaustible. More completely new pterosaurs are being analysed by scholars at the time of writing, so that there will be additions to the nine genera and fourteen species so far known. Thus the Santana strata have similar significance for research into Cretaceous pterosaurs as the Solnhofen strata have hitherto had for pterosaurs of the Jurassic. From Solnhofen we so far know nine genera, with

48 Kellner, A.W.A. and Campos, D.A., 1989. *Sobre um Novo Pterossauro com Crista Sagittal da Bacia do Araripe, Cretáceo Inferior do Nordeste do Brasil.* Anais Academia Brasileira Ciencias (1988), 60 (4): 459-469; Rio de Janeiro.
Kellner, A.W.A., 1990. *A New Edentate Pterosaur of the Lower Cretaceous from the Araripe Basin, Northeast Brazil.* Anais Academia Brasileira Ciencias (1989), 61 (4): 1-7; Rio de Janeiro.

Left: The first toothless pterosaur discovered in the Santana Formation was called *Tupuxuara* after a mythical ghost in the culture of the Brazilian Tupi tribe. However, only a fragment of the upper jaw and a few fragments of the wing skeleton, metacarpals and wing phalanges (shown here) were preserved. The skull had a long bony crest above the nasal opening. The remaining parts of the skull are unknown so far.

Below: A second toothless pterosaur from the Sanatana Formation was described and named as *Tapejara* after another figure in Tupi mythology. Originally, only this fragment of a skull was discovered. It is characterized by a high bony crest on top of its front end. A second, more complete skull was subsequently discovered.

Left: This almost complete skull of *Tapejara* is the second specimen of this toothless pterosaur to be discovered in the Santana Formation of Brazil. It includes the lower jaw, and reveals the very unusual shape of the bony crest on top of the skull.

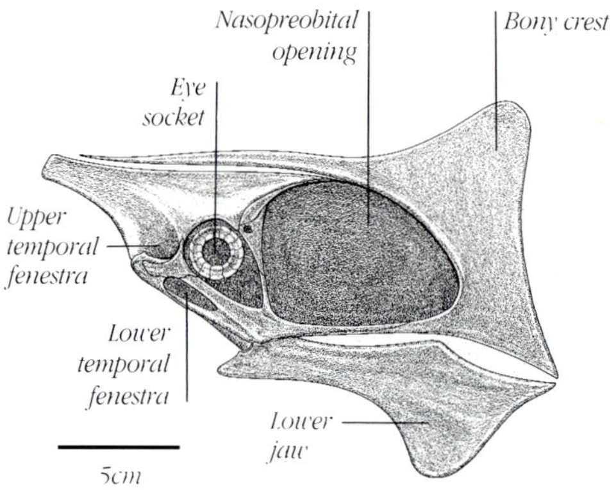

Tapejara Skull Restoration (above)
This is a restoration of the skull of *Tapejara*. The jaws are rather short and the eye socket is only small. The thin bony crest is separated from the back of the skull roof, rather like a cock's crest.

Above: *Pterodaustro* from early Cretaceous sediments of Argentina is the most extreme example of a filter feeder among pterosaurs discovered to date. This specimen is a disarticulated skeleton. The skull with its upwardly bent jaws can be seen at the bottom of the slab. The jaws are long and the lower jaw bears hundreds of long, seemingly flexible 'teeth' along its length.

eighteen different species. Many unusual and exciting new discoveries can be expected from the calcareous concretions of the Santana Formation in the future.

Argentina's 'South Wing'

In 1970 Dr José Bonaparte discovered one of the most unusual pterosaurs to have been found to date. At first only a skull fragment, vertebrae and some elements of the appendicular skeleton were found, then later a complete skull and a complete skeleton including the skull came to light. They came from Lower Cretaceous (Lagarcito Formation) strata in the province of San Luis in Argentina.[49]

49 Bonaparte, J.F., 1970. *Pterodaustro guinazui gen. et sp. nov. Pterosaurio de la formación Lagarcito, provincia de San Luis, Argentina.* Acta Geologica Lilloana, 10 (10): 207-226; Tucuman.
Bonaparte, J.F., 1971. *Descripción del Cranéo y Mandibulas de Pterodaustro guinazui (Pterodactyloidea – Pterodaustriidae nov.), de la formación Lagarcito, San Luis, Argentina.* Publicaciones des Museo Municipal de Ciencias Naturales de Mar del Plata, 1 (9): 263-272.
Sanchez, T.M., 1973. *Redescripción del Cranéo y Mandibulas de Pterodaustro guinazui Bonaparte (Pterodactyloidea, Pterodaustriidae).* Ameghiniana, 10 (4): 313-325; Buenos Aires.

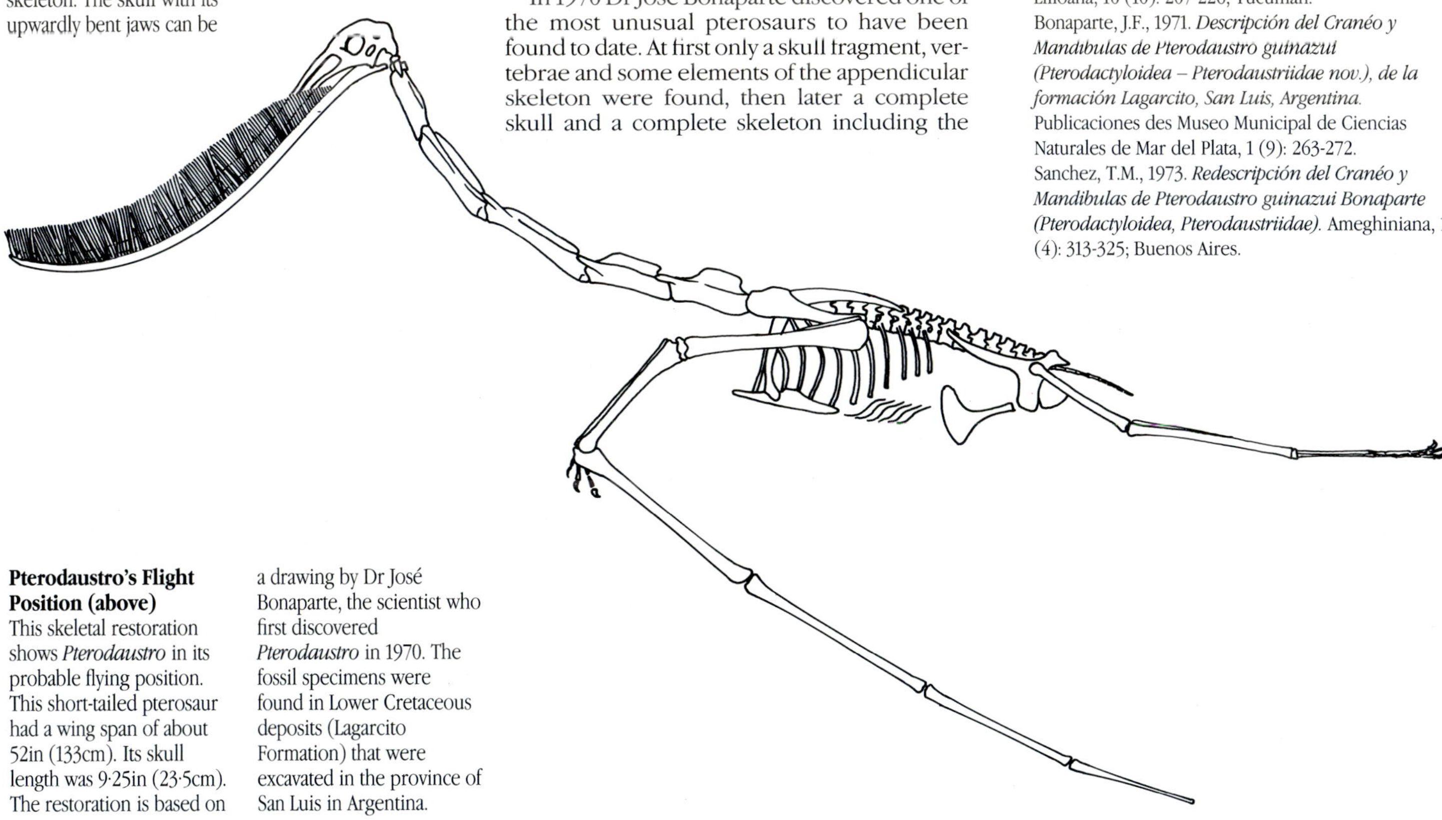

Pterodaustro's Flight Position (above)
This skeletal restoration shows *Pterodaustro* in its probable flying position. This short-tailed pterosaur had a wing span of about 52in (133cm). Its skull length was 9·25in (23·5cm). The restoration is based on a drawing by Dr José Bonaparte, the scientist who first discovered *Pterodaustro* in 1970. The fossil specimens were found in Lower Cretaceous deposits (Lagarcito Formation) that were excavated in the province of San Luis in Argentina.

Dr Bonaparte called this genus *Pterodaustro* (=south-wing); it is a short-tailed pterosaur with unique dentition. The skull is very markedly elongated and the front parts of the jaw are bent upwards. The lower jaw has a side groove in which are set a large number of long, tightly packed, apparently elastic 'teeth', which could more properly be called bristles. About 24 such lower jaw teeth occupy a centimetre of jaw length, which means that in a jaw 7·9in (20cm) long there are almost 500 teeth in each half of the jaw. *Pterodaustro*'s lower jaw was thus a highly effective sieving apparatus for filtering small organisms out of the water. The food content of this filter basket was chopped up into smaller bits by blunt, short teeth in the upper jaw.

Pterodaustro's skull was 9·25in (23·5cm) long, but its wing span only 4·36ft (1·33m). Its filter dentition made it far more specialized than *Ctenochasma* from Solnhofen, which had filter dentition in the upper and lower jaw, but

Left: Dr José Bonaparte of the Museo Argentino de Ciencias Naturales in Buenos Aires, the discoverer of *Pterodaustro*, one of the most unusual pterosaurs to have been found to date. Dr Bonaparte is here pictured in the main hall of the museum.

Pterodaustro (below) Dubbed the 'flamingo pterosaur', *Pterodaustro* was a remarkable filter feeder. It could not trap its food on the wing, but had to stand in the shallows and sieve small organisms out of the water with its filter basket.

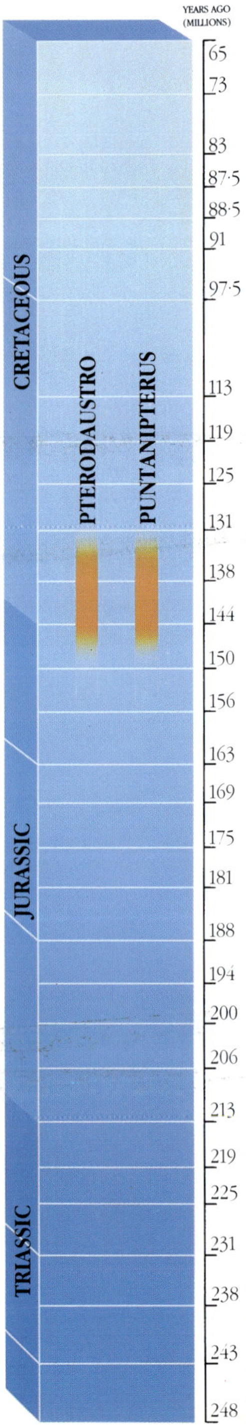

Time Chart (left) *Pterodaustro* is only known from the Lower Cretaceous of Argentina and possibly Chile. *Puntanipterus*, probably a dsungaripterid, was a contemporary of *Pterodaustro*.

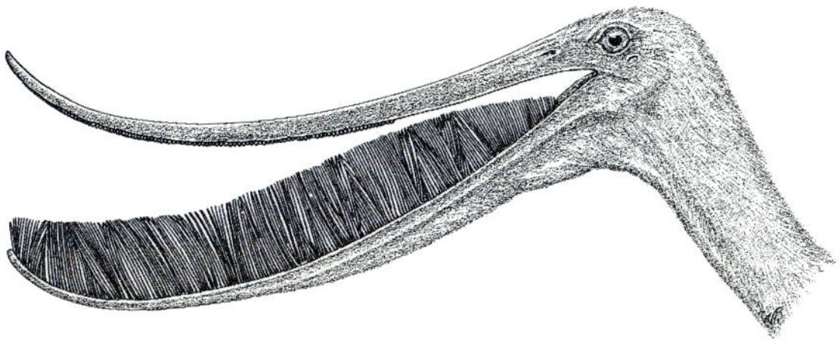

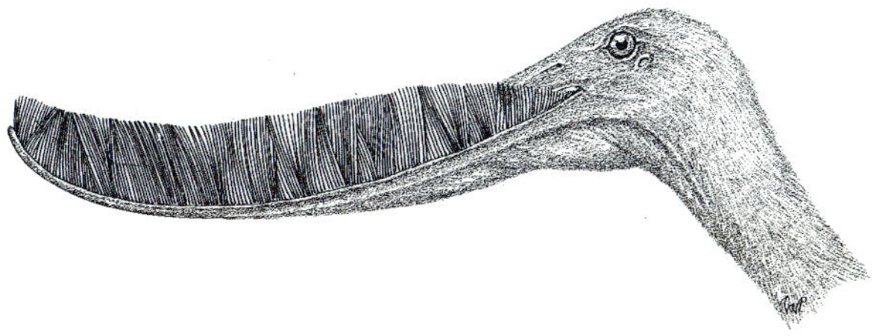

Pterodaustro Life Portraits (left and above) The head of *Pterodaustro* is here restored as in life with its beak open and closed. Its dentition consisted of a comb-like array of long, elastic bristles (not actually teeth) in the lower jaw for sieving creatures from the water. In its upper jaw was a series of short, blunt teeth for chopping up the food into smaller pieces to be swallowed.

Map (right)
1 *Pterodaustro*
2 *Puntanipterus*

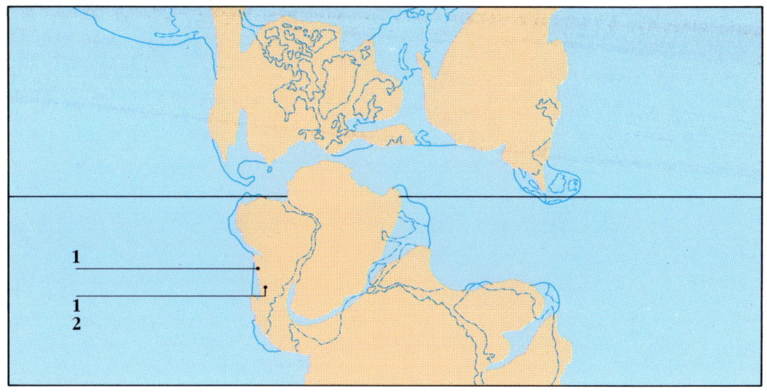

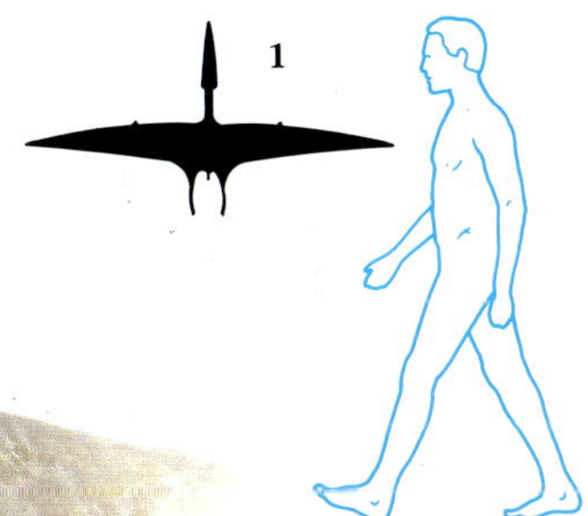

Comparative Sizes (above)
1 *Pterodaustro*: wing span 4·36ft (1·33m).

consisting of considerably fewer teeth. For this reason *Pterodaustro* was placed in a family of its own, the Pterodaustridae.

Skull fragments and a humerus of *Pterodaustro* have also been discovered in the Lower Cretaceous of Chile, in the province of Antofagasta.[50]

J.F. Bonaparte and T.M. Sanchez reported on a second pterosaur genus from the Lower Cretaceous (La Cruz Formation) of the province of San Luis in Argentina in 1975.[51] What was found was a 3·5in (9cm) long tibia, which broadened at the end to form a bird-like roller-joint. With this were only a dorsal vertebra, a wing phalanx and a foot digit phalanx. The particular features of the tibia, which is fused with the fibula, are different from those of *Pterodaustro*, and were seen as diagnostic of a new genus, *Puntanipterus* (named after natives of the province of San Luis, called *puntanos* in Spanish). In the opinion of Peter Galton of Bridgeport University, *Puntanipterus* belongs to the family Dsungaripteridae, which was distributed from China via East Africa to South America.

One of the most southerly occurrences of pterosaurs was found in Lower Cretaceous strata of the province of Santa Cruz in Patagonia. They are bone fragments of a small ulna, which were questionably classified as ornithocheirid.[52] These Patagonian pterosaurs must have lived in a paleolatitude of 51°S, though not as far south as the New Zealand Cretaceous pterosaurs mentioned earlier.

Part of a femur of a pterodactyloid pterosaur that cannot be any more precisely classified also came from Argentina. It was found in Lower Cretaceous deposits in Neuquén in northern Patagonia.[53]

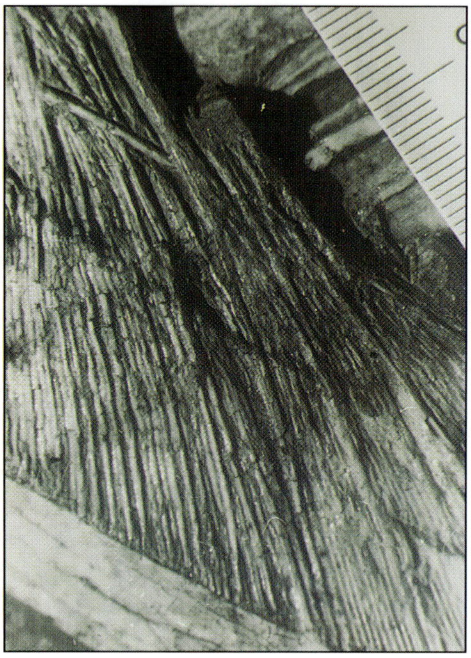

Above: This photograph shows in detail the bristle-like dentition in the lower jaw of *Pterodaustro*. A mere centimetre of jaw might hold 24 such 'teeth', which meant that *Pterodaustro*'s entire lower jaw could accommodate nearly 1,000 of these bristles. They acted like a fine sieve which trapped aquatic organisms.

In 1989 Christopher Bennett described a pterosaur humerus from the Chulec Formation (Lower Cretaceous, Albian) of Huanuco in Peru.[54] This well-preserved, fairly complete bone is 7·7in (19·6cm) long and suggests a wing span of about 13ft (4m). It is very similar to humeri from the Santana Formation of Brazil, like *Santanadactylus* for example, and thus proves that Brazilian Santana pterosaurs were distributed over large areas of South America in the Lower Cretaceous.

North American Finds

The first fossil remains of North American pterosaurs were described as *'Pterodactylus' oweni* by O.C. Marsh in 1871.[55] A year later Marsh changed the name to *'Pterodactylus' occidentalis*, as the specific name *oweni* had already been allotted to an *Ornithocheirus* from the Cambridge Greensand in England by Seeley in 1870.[56] During the 1870 expedition mounted by Yale College, New Haven, to the American West, a stop was also made on the way back in West Kansas, and a very successful search for fossils carried out in the Kansas Chalk of the Niobrara Formation (Santonian) on the Smoky Hill River. One of the participants in the 1870 Yale College Expedition reported on the events in this vein: '. . . and finally reached Fort Wallace in Kansas. The last geological expedition was to be made from this post, along the Smoky River, and, with a small escort of cavalry, we started on the 20th of November. The nights had now become bitterly cold, and to avoid the piercing wind our

50 Chong, D.G., 1976. *Los relaciones de los Sistemas Jurásico y Cretácico en la zona preandina del Norte de Chile.* Actas I Congresso Geologico Chileno, 1976, 1: A21-A42; Santiago.

51 Bonaparte, J.F. and Sanchez, T.M., 1975. *Restos de un pterosaurio Puntanipterus globosus de la formación La Cruz, provincia San Luis, Argentina.* Actas Primo Congresso Argentino de Paleontologia e Bioestratigraphia, 2: 105-113; Tucuman.

52 Urreta, M.B.A. and Ramos, V.A., 1981. *Estratigraphia y Paleontologia de la Alta Cuenca des Rio Roble Cordillera Patagonia – provincia de Santa Cruz.* VIII Congresso Geologico Argentino, San Luis, Actas III: 101-138.

53 Montanelli, S.B., 1986. *Sobre e primer resto des Pterosauria (Reptilia) de la Formación La Amarga (Cretácico inferior), Neuquén, Argentina.* Boletin informativo Associación Paleontologie Argentina, 15: 13.

54 Bennett, S.C., 1989. *A pteranodontid Pterosaur from the early Cretaceous of Peru, with comments on the Relationships of Cretaceous Pterosaurs.* Journal of Paleontology, 63 (5): 669-677.

55 Marsh, O.C., 1871. *Note on a new and gigantic species of Pterodactyle.* American Journal of Science, 1: 472.

56 Marsh, O.C., 1872. *Discovery of additional remains of Pterosauria etc.* American Journal of Science, 3: 241.

Right: The lower jaw of *Pteranodon*, incompletely preserved, from the Upper Cretaceous Niobrara Formation of Kansas, from above. Instead of teeth, the outer margins of the jaws are developed as sharp ridges. These were covered by horny beaks.

Left: Exposures of the Niobrara chalk along the Smoky Hill River in West Kansas. These marine strata are deposits of a 'midcontinental seaway' that ran through the North American continent in late Cretaceous times. Here O.C. Marsh and his field party from Yale College in Connecticut found the first fossil remains of the giant pterosaur *Pteranodon* in 1870. The Niobrara chalk has produced hundreds of pterosaur bones, fish and marine reptiles in the course of excavations since that time.

Right: The field party led by Othniel Charles Marsh (standing in the centre) to the Western Territories in 1872. This expedition was very successful in collecting additional and more complete fossil bones of the large pterosaur *Pteranodon* from the late Cretaceous Niobrara chalk of Western Kansas. In those days, this was still Indian territory, and rifle, revolver and bowie knife took their places alongside the geological hammer as part of the standard equipment for palaeontological field work. On the evidence of this picture, hats were also *de rigueur* in Marsh's company.

Below: The skull of *Pteranodon* collected from the Niobrara chalk by S.W. Williston, and described by O.C. Marsh as *Pteranodon longiceps* in 1876. This skull shows nicely the long toothless jaws. However, the cranial crest at the rear is missing.

Above: A partial skull of *Pteranodon*, again from the West Kansas Niobrara Formation, showing the posterior half with the long bony crest extending from the back of the skull (on the left).

camp was pitched under a high bank. About midnight a wolf, attracted by the scent of meat, jumped off this bank, into the midst of our mules, and frightened them to such a degree that about a dozen broke loose and stampeded. The night was dark and the greatest confusion followed; for until the sentinels told us the true cause of disturbance we instinctively thought of Indians. The mules, with broken halters and lariats flying, reached the fort early in the morning, and caused great consternation among the officers, who naturally concluded that the Cheyennes had attacked us, and sent a company of soldiers to our rescue. The troops appeared more disappointed at losing the expected fight than gratified at our safety.

'The search for fossils met with great success, and remains of Cretaceous reptiles and fishes were collected in great quantities. One trophy was the skeleton of a sea serpent, nearly complete, and so large that we spent four days in digging out and carrying it to camp . . .

'The Smoky River runs through the great Kansas hunting grounds. Every day herds of the buffalo were around us, and we enjoyed many an exciting 'run' across the prairie.

'The weather day by day grew colder, and at length we saw indications of an approaching storm. Knowing the danger of exposure to snow on these open plains, we reluctantly bade farewell to our geological diggings, and satiated even with buffalo hunting, turned back to Fort Wallace . . .

'On commencing the journey homeward, and entering the palace cars, our ruffianly appearance created consternation among sober railroad tourists. Months of hardship, labor and adventure had made many a rent in our well-worn clothes; and the buckskin breeches and army blouses of several members gave to the party a wild and warlike character, in keeping with the open display of revolver and bowie knife, and bronzed faces covered with the untrimmed stubble of a season. We reached New Haven on the 18th of December, after six eventful months, during which no serious illness or accident had happened to any of our party.'[57] Among the fossil remains of fish and marine reptiles that they had collected was a single bone fragment from a large pterosaur. It was half a metacarpal bone, from which Marsh deduced this must have come from a giant pterosaur with a wing span of 20ft (6m).

In the following summer, 1871, at the same site he found the missing second half of this bone, which was 15·7in (40cm) long, confirming Marsh's estimate of the size of this individual. Further fossil discoveries on the Smoky Hill River in Kansas produced a large quantity of pterosaur material, including skulls. It now turned out, however, that in contrast with the English ornithocheirids these giant pterosaurs were toothless, and that they had a long crest at the rear end of the skull, which must have made this flying reptile look very bizarre indeed.

57 'Harper's New Monthly Magazine', volume 43, pp.663-671.

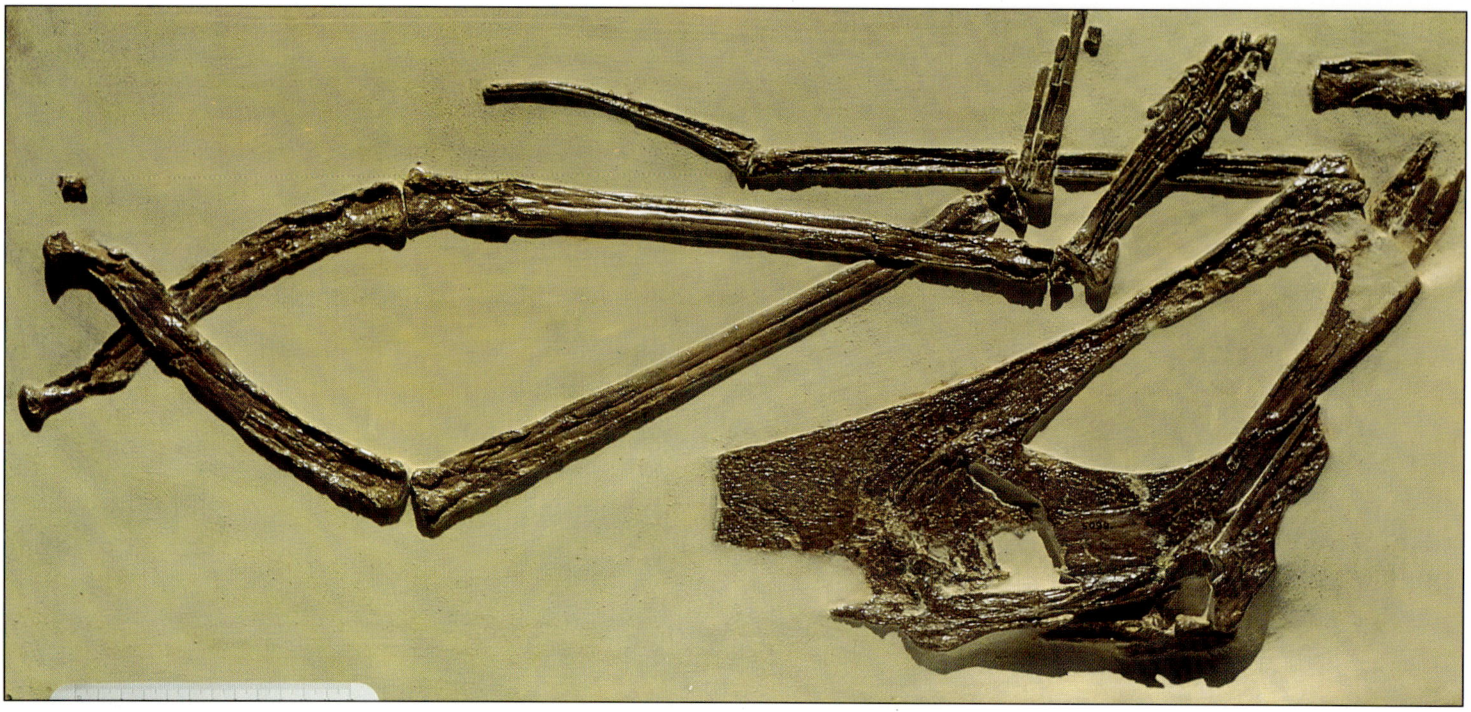

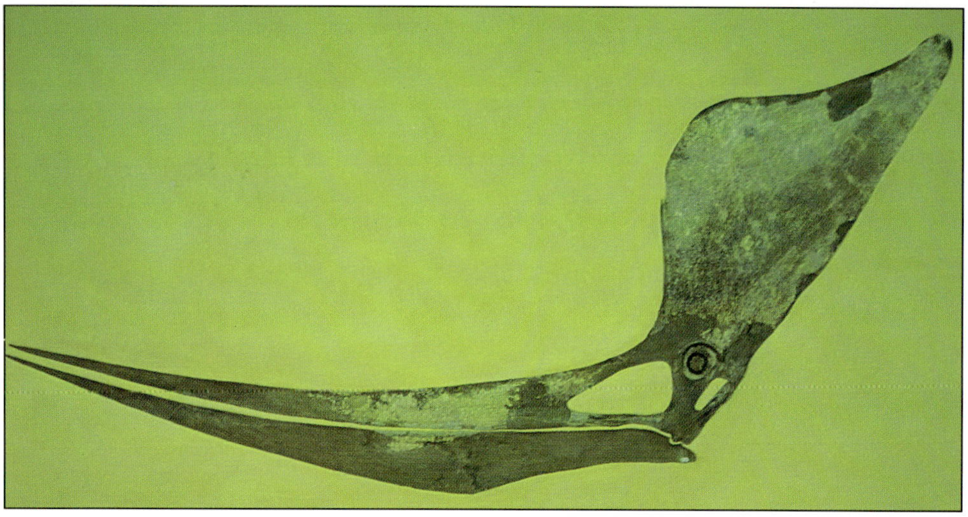

Above: These are incomplete skeletal remains of *Pteranodon* from the Upper Cretaceous of Kansas. This specimen comprises a fragment of the skull, the bones of the wing finger, which extend across the top of the picture from right to left, and both fairly complete hind legs. The skeleton of this large pterosaur is optimised for lightness. The bones are thin walled and hollow, and have openings which allow the penetration of air sacks which were connected to the lungs. Modern birds have a similar system.

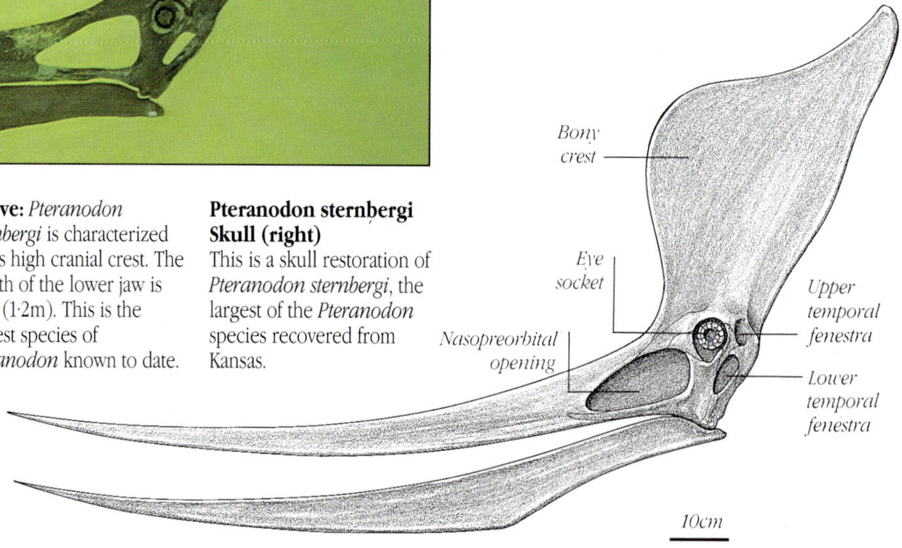

Above: *Pteranodon sternbergi* is characterized by its high cranial crest. The length of the lower jaw is 3·9ft (1·2m). This is the largest species of *Pteranodon* known to date.

Pteranodon sternbergi Skull (right)
This is a skull restoration of *Pteranodon sternbergi*, the largest of the *Pteranodon* species recovered from Kansas.

For this reason Marsh thought it necesssary in 1876 to distinguish the American Cretaceous pterosaurs from the English forms. He introduced the name *Pteranodon* (=toothless flier) for them,[58] and placed them in a family of their own, the Pteranodontidae, and even a suborder of their own, the Pteranodontia. He also included another genus here, *Nyctosaurus* (=naked reptile), which had been found with *Pteranodon* skeletal remains in the Kansas Chalk.

None of the skulls collected or described by Marsh was originally attached to the bones of the rest of the skeleton, with the exception of neck vertebrae. It is therefore difficult to assign skeletal bones to particular skulls. Despite this, Marsh distinguished various different species of *Pteranodon*, which differed from each other particularly in the morphology of their skulls, especially in the shape of the crest. For example, *Pteranodon ingens* had a skull 5·9ft (1·79m) long, of which almost half consisted of the crest, rising well back over the rump. *Pteranodon sternbergi* had a crest which rose steeply and was broader at the top. The lower jaw alone of this species is 3·9ft (1·2m) long, thus longer by a third than *Pteranodon ingens*, which had an estimated wing span of about 23ft (7m). *Pteranodon sternbergi* was thus one of the largest known pterosaurs, and must have had a wing span of over 30ft (9m). This was only exceeded by the azhdarchids, like *Quetzalcoatlus* or *Titanopteryx*.

In contrast with these pterosaurs with extremely long necks, *Pteranodon* had a relatively short neck with powerful but short cervical vertebrae. Its long, pointed jaws are toothless, the eye socket is relatively small and placed fairly high in the skull. There is always a crest on the back of the head, but not on the snout or lower jaw. All the bones are extremely thin walled and pneumatic, meaning that they are hollow and had small air vents, possibly to allow the penetration of air sacks, which were connected with the lung, as in modern birds. The vertebrae also have large lateral openings, and are very lightly constructed.

After O.C. Marsh, S.W. Williston in particular devoted himself to research into 'Kansas Pterodactyls',[59] and in 1910 G.F. Eaton, Curator of

58 Marsh, O.C., 1876. *Principal characters of American Pterodactyls.* American Journal of Science, 12: 479.

59 Williston, S.W., 1892-1893. *Kansas Pterodactyls. Part I and II.* Kansas University Quarterly, I: 1-13; II: 79-81.

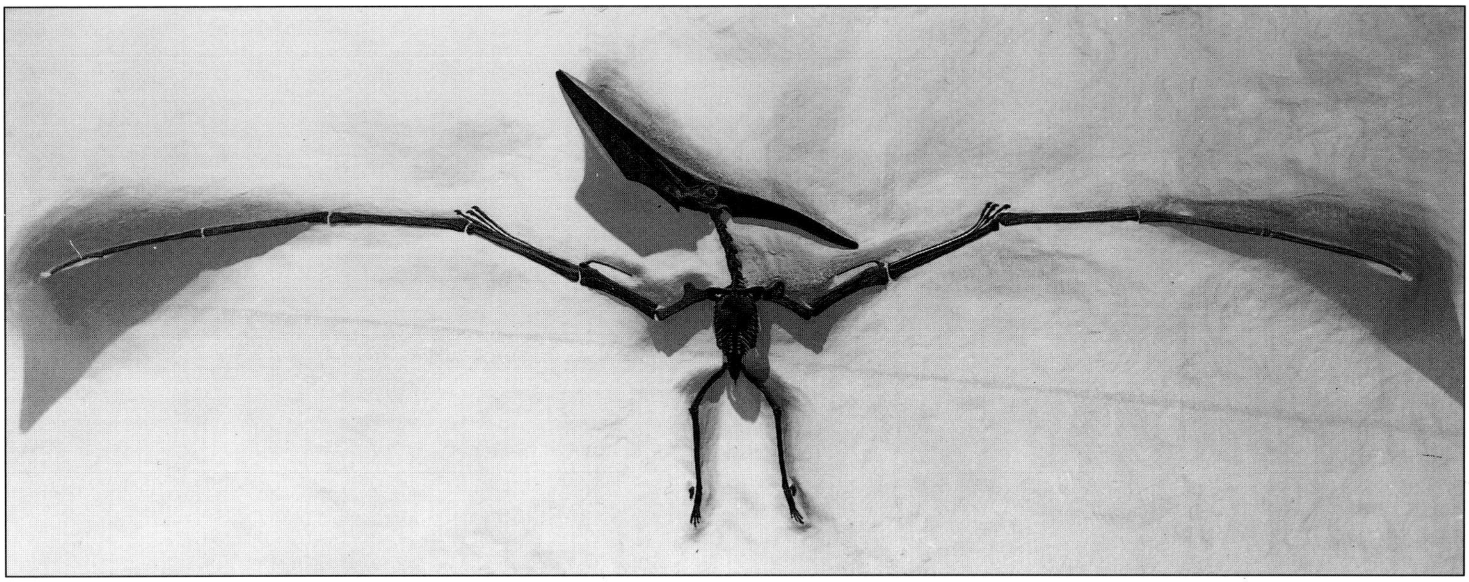

Above: A skeletal restoration of *Pteranodon* from the Upper Cretaceous of Kansas, mounted and on display in the Peabody Museum of Yale University in New Haven, Connecticut. The wing span of this particular specimen is 23ft (7m). The compactness of the body and hind legs when compared with the wings is remarkable in this view. The relative size of the skull with its long cranial crest is impressive.

Osteology and Vertebrate Palaeontology at the Peabody Museum of Yale College in New Haven could boast an inventory of *Pteranodon* skeletal remains featuring no fewer than 465 individuals.[60]

Pteranodon must have been able to sustain flying and soaring, ranging far out over the open sea to catch fish. The West Kansas strata in which finds were made are deposits of an extended sea that ran through the North

60 Eaton, G.F., 1910. *Osteology of Pteranodon.* Memoirs of the Connecticut Academy of Arts and Science, 2: 38 pp.; New Haven.

Pteranodon ingens (below right)
Shown here is a life portrait of *Pteranodon ingens* (above), and a comparable skeletal restoration (below). The skull of this particular species could measure as much as 5·9ft (1·79m) in length.

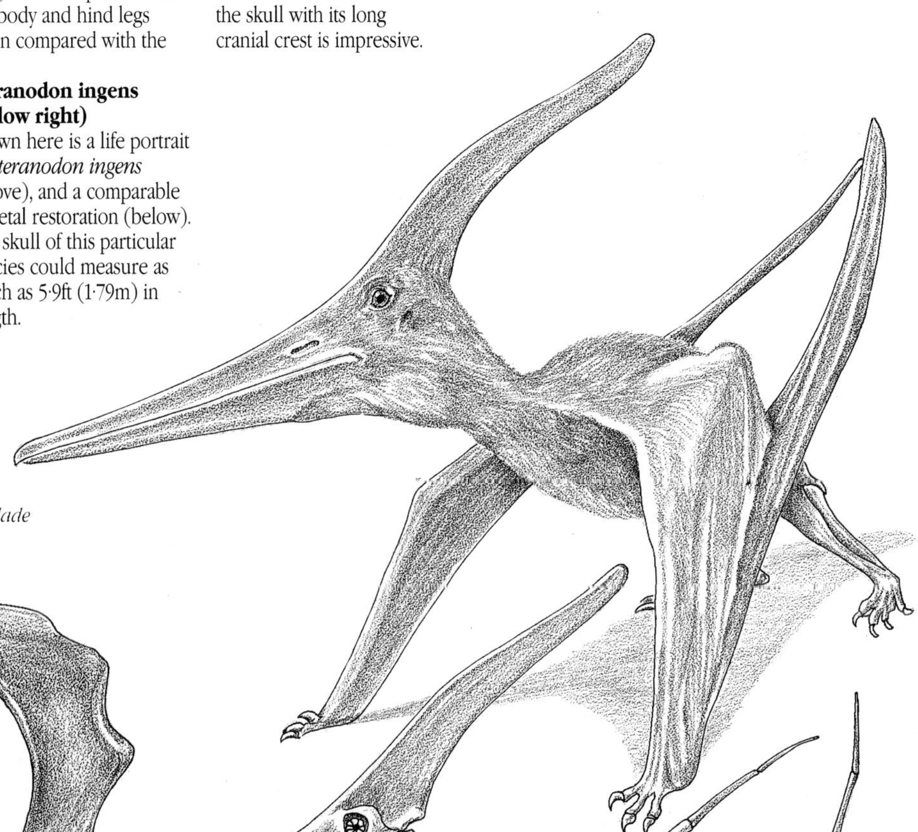

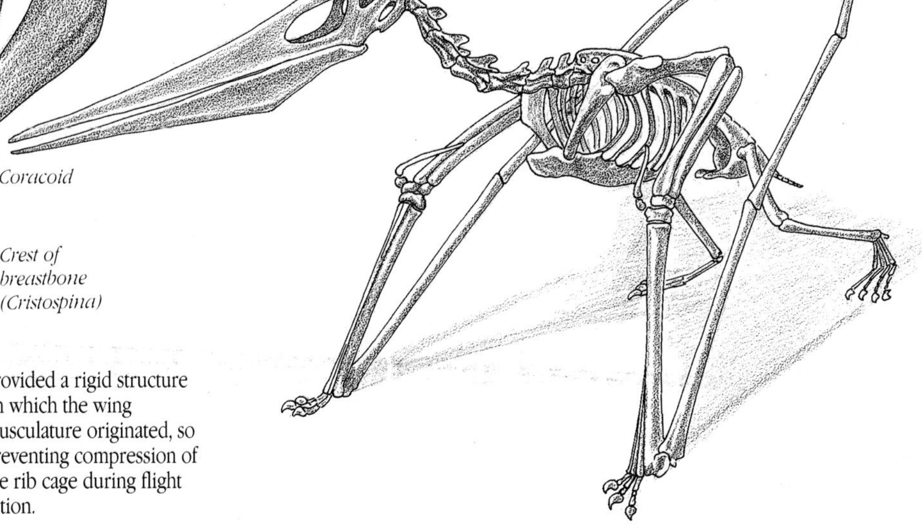

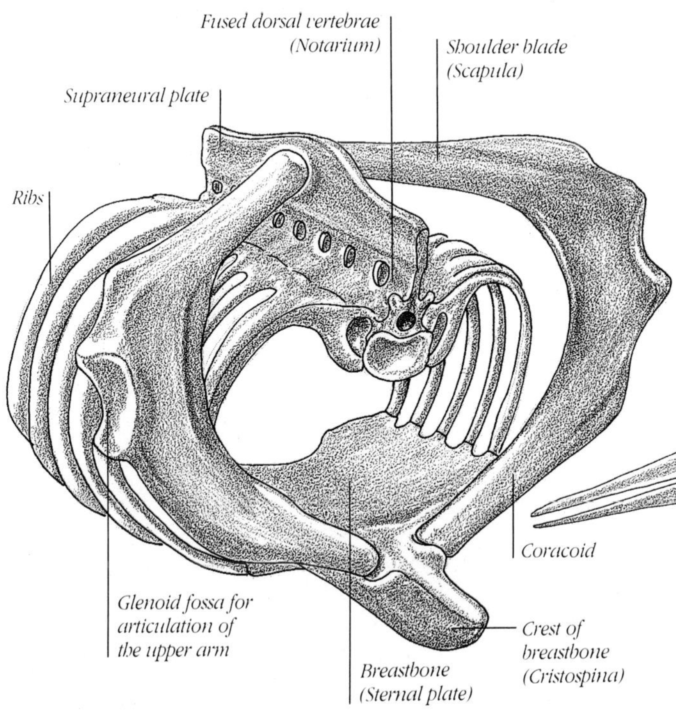

Pteranodon's Shoulder Girdle (above)
This drawing shows the shoulder girdle and the front section of the trunk of *Pteranodon*. The first eight dorsal vertebrae are fused and form a notarium providing strong support for the shoulder blades. In conjunction with the ribs and breastbone, this provided a rigid structure on which the wing musculature originated, so preventing compression of the rib cage during flight action.

American continent in the Cretaceous as a 'midcontinental seaway'. As the sites where *Pteranodon* remains were found are at least 100 miles (160km) from what was then the coast, it can be presumed that these pterosaurs perished in or over the water, far away from their nesting or resting places on the coast. Despite their large wings their actual bodies were rather small. Cherrie Bramwell and George Whitfield of Reading University calculated the body weight of *Pteranodon ingens* with a wing span of 22·8ft (6·95m) as a mere 36·6lb (16·6kg). With this the animal was able to achieve a maximum speed of 31mph (50km/h).

Nyctosaurus

When more pterosaur material from the Smoky Hill River in West Kansas was found, O.C. March recognized that it also included skeletal remains of a second, smaller form, which he called *Nyctosaurus* (=naked rep-

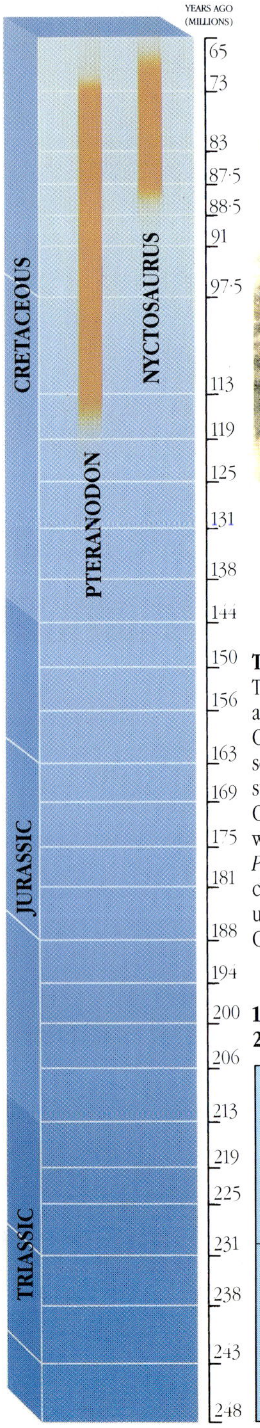

Time Chart (left)
The earliest pteranodontids are known from late Lower Cretaceous (Albian) sediments. *Pteranodon* survives into the late Cretaceous. *Nyctosaurus* was a contemporary of *Pteranodon*, but it is not clear whether it survived until the very end of the Cretaceous.

Map (below)
1 *Pteranodon*
2 *Nyctosaurus*

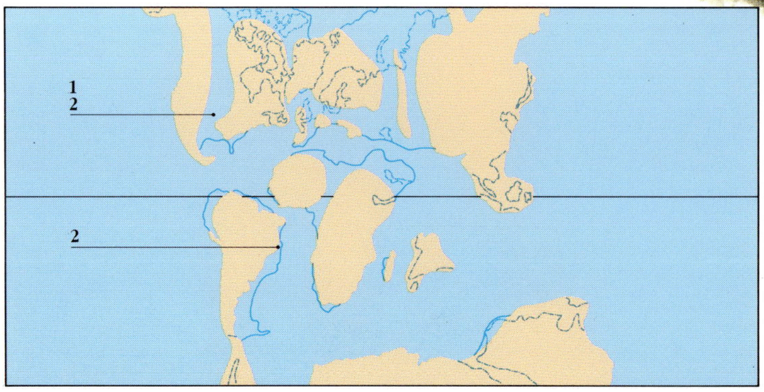

Pteranodon ingens (below left)
The scene here depicts the habitat of the late Cretaceous pterosaurs from Kansas with cliffs and nesting sites on the shore of the great mid-continental seaway which separated the western from the eastern part of North America at that time. *Pteranodon*, as the name suggests a 'toothless flier', was among the largest pterosaurs. The species seen here is *Pteranodon ingens* which, with a wing span of 23ft (7m), was smaller that *P. sternbergi*.

Nyctosaurus (below)
Nyctosaurus was also a toothless pterosaur from late Cretaceous deposits in Kansas. However, it was smaller than *Pteranodon*, and only had a short cranial crest at the rear of its head. In the background can be seen the diving bird *Hesperornis*, and the 'Cretaceous gull' *Ichthyornis*, both toothed birds.

Pteranodon ingens Life Portrait (right)
This is a life portrait of *Pteranodon ingens* from the Upper Cretaceous of Kansas which is also seen at top left of the colour artwork. The bony crest may have served as a counterbalance for the beak.

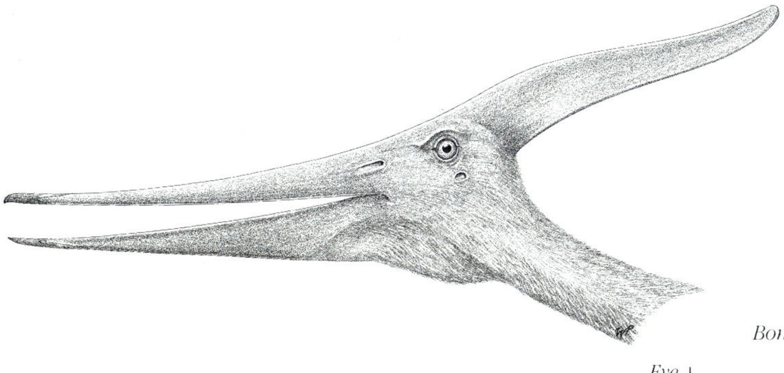

Pteranodon ingens Skull Restoration (below)
This restoration shows the skull of *Pteranodon ingens* which measured some 5·9ft (1·8m) in length. Despite its size, its overall body weight was probably remarkably small, perhaps only about 37lb (17kg).

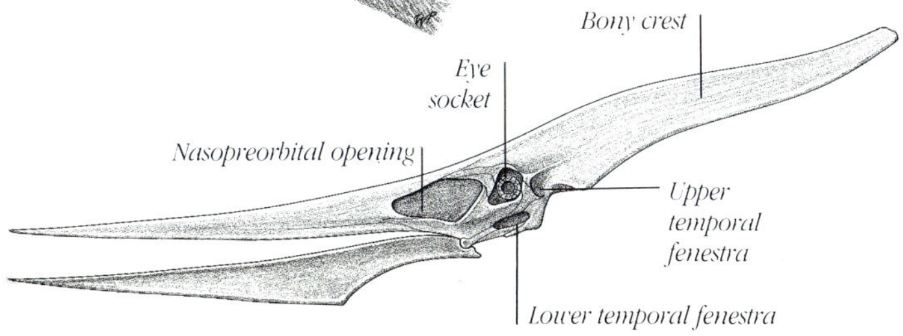

Pteranodon sternbergi (below)
Pteranodon sternbergi, the largest of the *Pteranodon* species, is characterized by the high upright crest on its skull. Its wing span could measure over 30ft (9m) in total. Almost certainly fish eaters, these pterosaurs seem to have led a life similar to today's soaring seabirds, such as albatrosses.

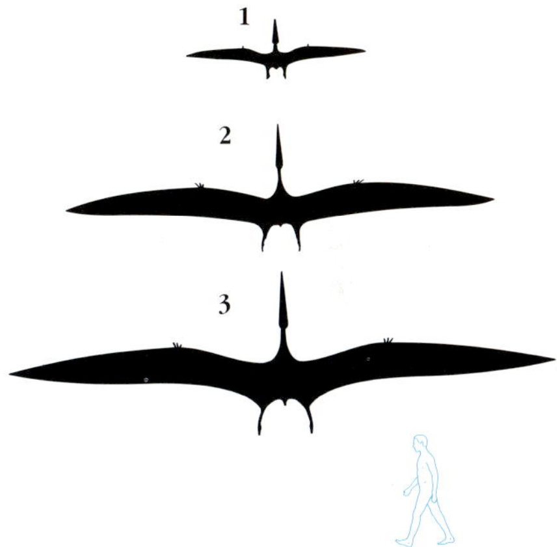

Comparative Sizes (above)
1 *Nyctosaurus*: wing span up to 9·5ft (2·9m).
2 *Pteranodon ingens*: wing span 23ft (7m).
3 *Pteranodon sternbergi*: wing span 30ft (9m).

tile).[58] Relatively complete skeletons were later described by S.W. Williston, who also provided a first reconstruction of the skeleton.[61] *Nyctosaurus* was considerably smaller than *Pteranodon.* It has a wing span of 7·9-9·5ft (2·4-2·9m). Its skull is low and elongated. It is also toothless, and has no crest. As it is also different with regard to other skeletal features, it has been grouped in a family of its own, the Nyctosauridae.[54] Another characteristic is that the flight digit consisted of only three, rather than the usual four, phalanges.[62]

A humerus and fused dorsal vertebra from the marine Lower Cretaceous (Albian) of Oregon is similar to *Nyctosaurus*, but clearly a larger species, described in 1928 by C.W. Gilmore as *'Pteranodon' oregonensis.*[63]

Pterosaur remains from Upper Cretaceous strata have also been found in other sites in North America, and it may be that they are pteranodontids: in Delaware (early Campanian),[64] Georgia (Santonian),[65] Montana (early Campanian),[66] and Alberta (early Campanian).[67]

Quetzalcoatlus

On the 'big bend' on the Rio Grande in western Texas on the border with Mexico is one of the lesser known but scenically most magnificent National Parks in the United States, the Big Bend National Park. Its landscape is made up to an equal extent of mountains of volcanic origin and also of terrestrial sediments from the late Cretaceous and early Tertiary, of which the soft, weathering forms are the typical 'badlands' of a dry, warm climate. There is little vegetation to prevent the erosion of these clayey sediments. For this reason fossil remains, dinosaurs, crocodiles and mammals are continually washed free by occasional violent rain storms, but they are also destroyed if they are not collected and saved.

The Cretaceous and Tertiary strata of the Big Bend National Park have long been known to contain fossils. They are particularly interesting because the transitional strata between the youngest Cretaceous and the oldest Tertiary, the Cretaceous/Tertiary boundary (K/T boundary for short), are revealed. This is one of the few places on Earth where the factors that led to the extinction of dinosaurs and other land vertebrates at the end of the Cretaceous, 65 million years ago, can be studied.

In 1975 *Science* published a short report entitled 'Pterosaur from the Latest Cretaceous of West Texas. Discovery of the Largest Flying Creature'.[68] Even palaeontologists, otherwise used to large prehistoric animals, could hardly

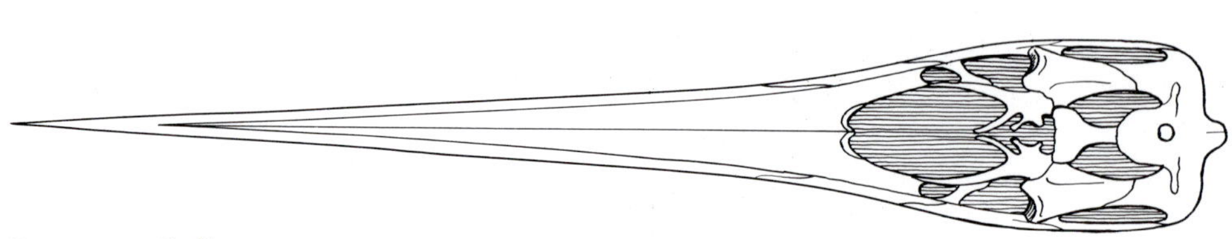

Nyctosaurus Skull (above)
We see here the skull of *Nyctosaurus gracilis* from below looking up into the palate. The length of this particular skull with its long, slender beak is about 1ft (30cm).

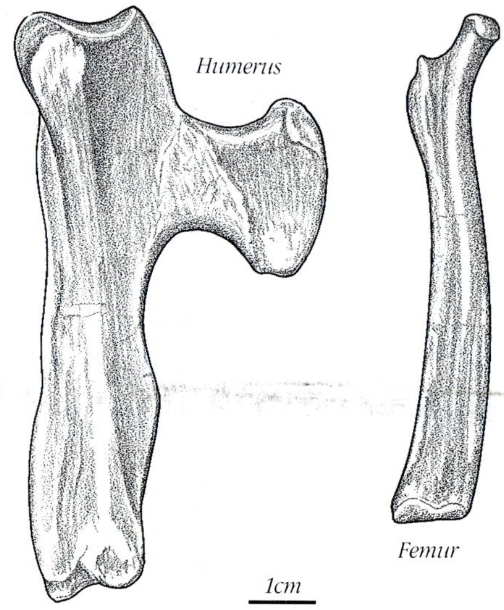

Nyctosaurus Arm and Leg Bones (right)
These drawings depict the upper arm (left) and upper leg of *Nyctosaurus gracilis* from behind. The humerus is characterized by its distinctive hatchet-like shape.

61 Williston, S.W., 1902. *On the skeleton of Nyctodactylus, with restoration.* American Journal of Anatomy, 1 (3): 297.
Williston, S.W., 1902. *On the skull of Nyctodactylus, an Upper Cretaceous Pterodactyl.* Journal of Geology, 10 (5): 520-531; Chicago.
Williston, S.W., 1903. *On the osteology of Nyctosaurus (Nyctodactylus), with notes on American pterosaurs.* Field Columbian Museum Publications, Geological Series, 2: 125-163.
62 Brown, G.W., 1986. *Reassessment of Nyctosaurus: new wings for an old pterosaur.* Proceedings, Nebraska Academy of Science, 1986: 47.
63 Gilmore, C.W., 1928. *A new pterosaurian reptile from the marine Cretaceous of Oregon.* Proceedings of the U.S. National Museum, 73 (24): 1-5.
64 Baird, D. and Galton, P.M., 1981. *Pterosaur Bones from the Upper Cretaceous of Delaware.* Journal of Vertebrate Palaeontology, 1 (1): 67-71.
65 Schwimmer, D.R., Padian, K. and Woodhead, A.B., 1985. *First Pterosaur Records from Georgia.* Journal of Palaeontology, 59, (3): 674-676.
66 Padian, K., 1984. *A large Pterodactyloid Pterosaur from the Two Medicine Formation (Campanian) of Montana.* Journal of Vertebrate Paleontology, 4 (4): 516-524.
67 Currie, P.J. and Padian, K., 1983. *A New Pterosaur Record from the Judith River (Oldman) Formation of Alberta.* Journal of Paleontology, 57 (3): 599-600.

68 Lawson, D.A., 1975. *Pterosaur from the Latest Cretaceous of West Texas. Discovery of the Largest Flying Creature.* Science, 187; 947-948.
Langston, W., Jr., 1978. *The Great Pterosaur.* Discovery 2 (3): 20-23; Austin.

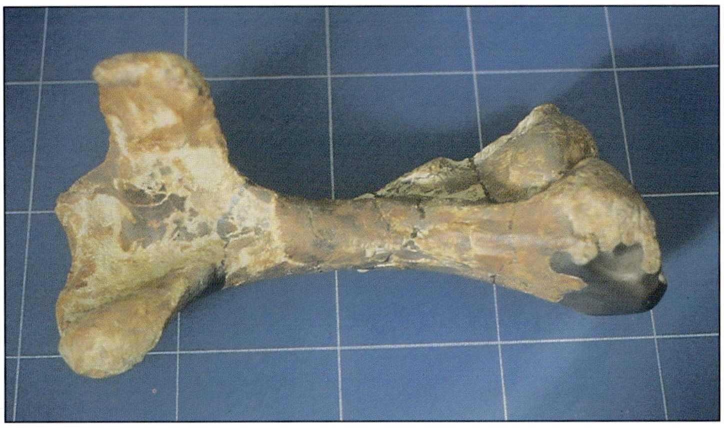

Left: The upper arm of *Quetzalcoatlus northropi* from the latest Cretaceous of West Texas. The bone is 1·7ft (52cm) long and very massively built. The wing span of this giant pterosaur has been calculated as up to 39ft (12m) in total.

Right: Dr Wann Langston Jr leading a field excursion in Big Bend National Park in 1989. Dr Langston has excavated many pterosaur bones from this important locality.

believe that this 'Texas pterosaur', as it came to be known, must have had a wing span of 51ft (15·5m), as the first calculation suggested. This was twice as big as the largest pterosaurs hitherto known, pteranodontids from the Kansas Chalk.

The history of its discovery began in 1971, when Douglas A. Lawson, a student at the University of Texas in Austin, was carrying out geological investigations for his master's thesis. He was especially interested in the sediments of the Javelina Formation, which were deposited at the end of the Cretaceous, about 65 million years ago: his aim was to find out about the environment and life conditions of the last dinosaurs, like the titanosaurids, for example. One afternoon, in a dry valley, Lawson found some bone fragments that must have been washed down from higher up. He followed the trail to a steep wall of rock, where he discovered the source of all these pieces of bone. He saw part of a bone 3ft (1m) long, still sticking in the rock. He brought part of this back to Austin with him and showed it to his professor, Dr Wann Langston Jr, as he had no idea what he had discovered. After thorough examination of the fragment and comparison with fossil remains in other museums, it rapidly became clear that this long, hollow, very thin-walled bone could only be a bone from a pterosaur wing.

There was no time to lose. Every day that passed would bring more weathering and destruction of the bones that had been revealed. Professor Langston and his student took the long road back from Austin to Big Bend, over 500 miles (800km), and investigated the site thoroughly. They dug out all the bones that were still to be found in the sediment. They even hoped they might rescue the entire skeleton of this giant of the air. But unfortunately it seemed that only a wing had survived. It must have been detached from the body after the creature died, and embedded further away.

But they still had the humerus and hundreds of small fragments of bones, which it was possible to assemble as lower arm and carpus in the laboratories of the Texas Memorial Museum. Other parts of the wing phalanges were added, one of which alone was 4ft (1·22m) long, but broken off at both ends. Lawson named this pterosaur *Quetzalcoatlus* after the Mexican deity Quetzalcoatl, who was worshipped by the Aztecs in the form of a feathered snake.

In subsequent years Dr Langston carried out regular investigations in the Javelina Formation of the Big Bend National Park, in the course of which he was able to collect numerous fossil bones and parts of the skeleton of smaller individuals in another part of the

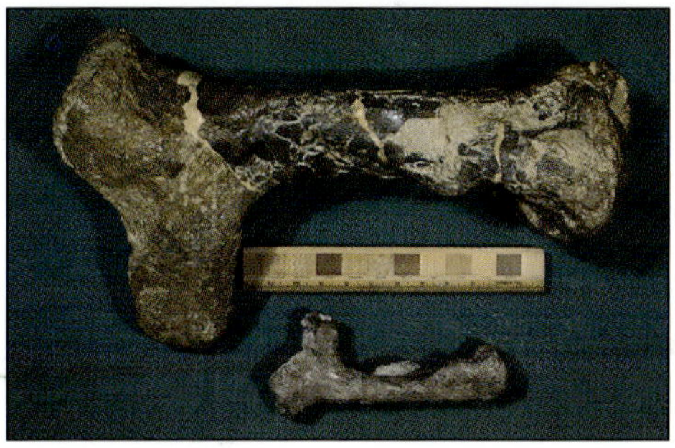

Above: A comparison of the upper arms of the giant *Quetzalcoatlus northropi* and of the smaller, almost half-sized species *Quetzalcoatlus* sp. (not yet named) which were both found in Big Bend NP.

Left: The fossil site where the giant Texas pterosaur *Quetzalcoatlus* was discovered in latest Cretaceous sediments of the Javelina Formation in Big Bend National Park, West Texas. This first discovery of a *Quetzalcoatlus* humerus was made in 1971 by Douglas Lawson, who was then a student working under the supervision of Dr Wann Langston of the University of Texas in Austin.

Right: Dr Wann Langston with a complete wing of the 'smaller' *Quetzalcoatlus* sp.. The bones are arranged in their natural articulation and suspended individually in order to allow them to be manipulated to aid study of their possible movements. These studies were undertaken to assist in the construction of a flying model of *Quetzalcoatlus* by Dr Paul MacCready.

park. Even so this 'little' *Quetzalcoatlus* had a wing span of at least 18ft (5·5m). These remains made it possible to establish a fairly good idea of the full skeleton of *Quetzalcoatlus*, and to calculate the probable wing span of the large individual at 36-39ft (11-12m).[69]

It is not yet clear whether these individuals, all found in a closely restricted area about 30 miles (50km) from the site at which the larger creature was found, were young individuals of the large species, or whether they represent a distinct, smaller species of *Quetzalcoatlus*. But because they are so complete they show us how these pterosaurs were constructed, and what they looked like. They were toothless, had sharp edges to their jaws, probably covered with horn, and long, narrow pointed beaks with a low, slender crest. The neck vertebrae were extremely long and show typical characteristics of the kind noted in *Titanopteryx* and *Azhdarcho*. The articulations of the cervical vertebrae allowed practically no lateral movement for the long neck. The long wing phalanges are constructed in a way that differs from other Pterodactyloidea. While the first of the four wing phalanges is hollow and oval in cross-section, the other three phalanges are made of solid bone tissue, with a T-shaped cross-section. Thus nature found a perfect solution for the technical problem of the accumulation of forces during the downstroke of the wing, and combined the highest possible strength with the lowest possible weight.

Aeronautical engineers quickly calculated that a pterosaur the size of the large *Quetzalcoatlus* must probably have weighed well over 220lb (100kg), and simply did not have enough muscular mass to raise this weight into the air and achieve continuous flapping flight. A glance at the massive bone crests on the humerus of *Quetzalcoatlus* is enough to show what powerful flight muscles must have been attached here.

The humerus is 1·7ft (52cm) long, and very robust in structure. One of the largest living birds, the wandering albatross, has a humerus 1·3ft (40cm) long and a wing span of 11ft (3·4m). On this basis *Quetzalcoatlus* would have had a wing span of only 14·4ft (4·4m). The fact that in reality its wings stretched almost three times as far is due to the completely different structure of the pterosaur skeleton. In *Quetzalcoatlus* the humerus is the shortest bone in the wing, whereas in the albatross it is the longest. Wing length in pterosaurs is determined above all by the metacarpus and the flight digit, whereas in birds the feathers protrude far beyond the skeleton of the wing. According to an estimate by Dr Langston, *Quetzalcoatlus* could have weighed 190lb (86kg). Bramwell and Whitfield's calculations gave a weight for *Pteranodon* of between 28·2 and 52·5lb (12·8 and

69 Langston, W., Jr., 1981. *Pterosaurs.* Scientific American, 244 (2): 122-136.

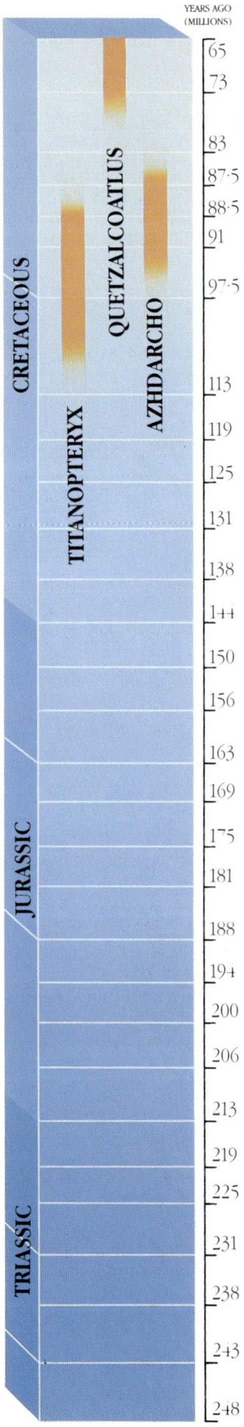

Quetzalcoatlus (below) This is the last known pterosaur which survived to the very end of the Cretaceous period. With a wing span between 36 and 39ft (11-12m), it was the largest flying creature of all time. Its neck was extremely long, its slender jaws were toothless, and its head was topped by a long, bony crest. Here individuals of *Quetzalcoatlus* are seen in a landscape that was probably typical of the environment in what is now West Texas about 65 million years ago. Stream channels flow into flood plains while a herd of titanosaurid dinosaurs wander slowly along. Opinions are divided as to whether *Quetzalcoatlus* was a carrion-feeder.

Time Chart (left) The azhdarchids were the last pterosaurs. *Quetzalcoatlus* survived to the very end of the Cretaceous. The two other azhdarchid genera lived in earlier Upper Cretaceous times. *Titanopteryx* may even have lived in the uppermost Lower Cretaceous.

Right: Technician Earl Yarmer of Texas Memorial Museum, Austin, is carefully chipping pterosaur wing bones out of the rock at Big Bend National Park. This is a partly articulated wing of a *Pteranodon*-sized *Quetzalcoatlus* in Upper Cretaceous floodplain sediments.

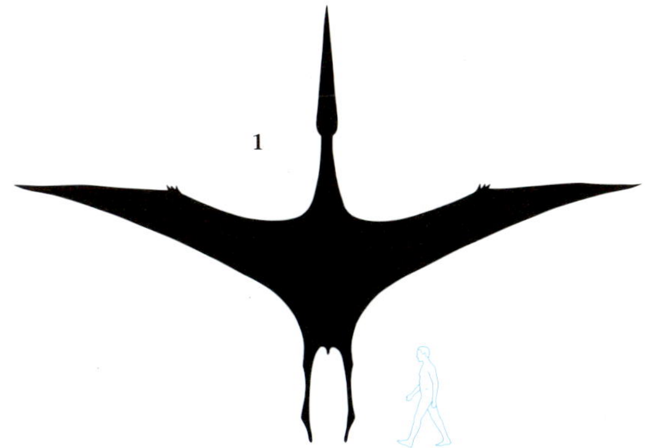

Comparative Sizes (above)
1 *Quetzalcoatlus northropi*: wing span 36-39ft (11-12m).

Map (below)
1 *Quetzalcoatlus*
2 *Titanopteryx*
3 *Azhdarcho*

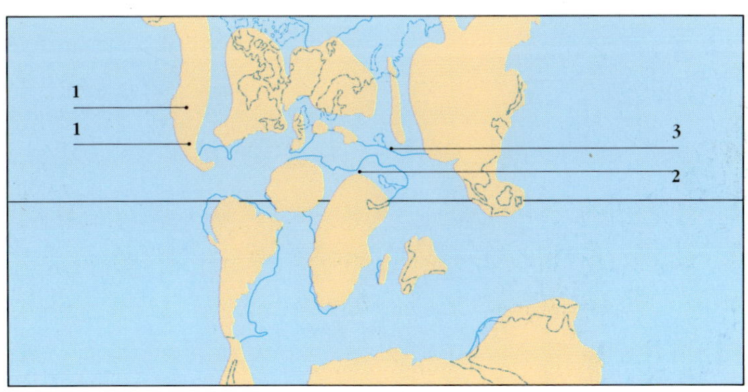

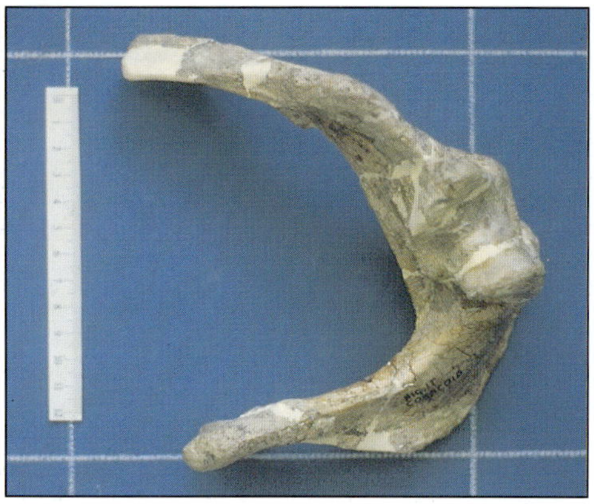

Above: This is the shoulder girdle of *Quetzalcoatlus* sp., a smaller species about half the size of the giant *Quetzalcoatlus northropi*. This specimen shows the glenoid fossa which was for articulation of the upper arm.

Feeding Habits (right)
Quetzalcoatlus lived inland from the sea. Perhaps, therefore, it was a carrion feeder, although its lack of teeth makes this seem improbable. Alternatively it may have probed for shellfish in small pools.

23·8kg), with a wing span of 23ft (7m).[70] Therefore *Quetzalcoatlus*, with a wing span of 36-39ft (11-12m), was possibly even lighter than the 190lb (86kg) that have been assumed, absolutely comparable in size and weight with a modern ultra-light aircraft.

Problems also arise when we try to imagine the habitat and way of life of these giant Texan pterosaurs. Unlike most other pterosaur fossils they were not found in marine strata, but in the sand and silt of the extensive flood plain of a former meandering river system which at that time, during the latest Cretaceous, was well inland, about 250 miles (400km) from the nearest sea coast. There is no geological evidence for large fresh-water lakes in the area. Lawson therefore took the view that *Quetzalcoatlus* might have lived rather like the modern vulture, that it was an eater of carrion who fed on the corpses of dinosaurs. Its long neck would have been well adapted for this. As a good soarer with considerable stamina it was certainly able to cover large distances in search of dead dinosaurs. Nevertheless many contradictions remain. Was the long, almost inflexible neck possibly not more of a hindrance for an eater of carrion? And could the pterosaur tear pieces of flesh from the corpse of an animal at all with its pointed, toothless jaws? On the

70 Bramwell, C.D. and Whitfield, G.R., 1974. *Biomechanics of Pteranodon*. Philosophical Transactions of the Royal Society London, (B), 267: 503-581.

Above: The front end of the jaws of the 'small' *Quetzalcoatlus* sp. from the latest Cretaceous of West Texas. The jaws are toothless. The length preserved is 31·5in (80cm).

Left: This long neck vertebra also belonged to *Quetzalcoatlus* sp. These long and slender vertebrae are diagnostic for inclusion in the family Azhdarchidae which has a worldwide distribution during the Cretaceous.

SUMMARY OF CRETACEOUS PTEROSAURS

Pterodactyloidea
Family Ornithocheiridae
Ornithocheirus compressirostris
Upper Cretaceous, Turonian, Chalk, Kent, England.[1,2]
Ornithocheirus cuvieri
Ornithocheirus fittoni
Ornithocheirus giganteus
Ornithocheirus microdon
Ornithocheirus sedgwicki and other species
Upper Cretaceous, Cenomanian, Cambridge Greensand, England.[3]
Ornithocheirus clifti
Ornithocheirus curtus
Lower Cretaceous, Wealden, Sussex, England.[6]
Ornithocheirus sagittirostris
Lower Cretaceous, Hastings Beds, Sussex, England.[6]
Ornithocheirus daviesi
Lower Cretaceous, Gault, Albian, Kent, England.[6]
Ornithocheirus diomedius
Upper Cretaceous, Turonian, Kent, England.[6]
Ornithocheirus bunzeli
Upper Cretaceous, Gosau Formation, Campanian, Austria.[13,14]
Ornithocheirus hlavatschi
Upper Cretaceous, Turonian, Bohemia, Czechoslovakia.[15]
Ornithocheirus hilsensis
Lower Cretaceous, Neocomian, Hannover, Germany.[16]
Ornithocheirus sp.
Lower Cretaceous, Gault, Albian, La Meuse, France.
Lower Cretaceous, Hauterivian, Haute-Marne, France.[17]
Lower Cretaceous, Albian, Aube, France.[18]
?Ornithocheirus
Lower Cretaceous, Toolebuc Formation, Albian, Queensland, Australia.[32]
Lower Cretaceous, Rio Belgrano Formation, Barremian, Santa Cruz, Argentina.[52]
Upper Cretaceous, Campanian-Maastrichtian, New Zealand.[34]
Upper Cretaceous, Cenomanian-Turonian, Zaïre, Africa.[30]
Santanadactylus brasilensis[39]
Santanadactylus araripensis[40]
Santanadactylus pricei[40]
Brasileodactylus araripensis[42]
Lower Cretaceous, Santana Formation, Aptian, Araripe Plateau, Brazil.
Family Anhangueridae
Anhanguera blittersdorffi[43]
Anhanguera santanae[40,44]
Lower Cretaceous, Santana Formation, Aptian, Araripe Plateau, Brazil.
Family Tapejaridae
Tupuxuara longicristatus
Tapejara wellnhoferi
Lower Cretaceous, Santana Formation, Aptian, Araripe Plateau, Brazil.[48]
Family Cearadactylidae
Cearadactylus atrox[46]
Lower Cretaceous, Santana Formation, Aptian, Araripe Plateau, Brazil.
Family Criorhynchidae
Criorhynchus simus and other species
Lower Cretaceous, Wealden, Sussex;
Upper Cretaceous, Cenomanian, Cambridge Greensand, England.[5]
Tropeognathus mesembrinus
Tropeognathus robustus
Lower Cretaceous, Santana Formation, Aptian, Araripe Plateau, Brazil.[8]
Family Ornithodesmidae
Ornithodesmus latidens
Lower Cretaceous, Wealden, Isle of Wight, England.[12]
Family Dsungaripteridae
Dsungaripterus weii
Lower Cretaceous, Tugulu Group, Xinjiang, China.[19,20]
?Dsungaripterus brancai
Upper Jurassic, Tendaguru, Tanzania, Africa.[22]
Noripterus complicidens
Lower Cretaceous, Tugulu Group, Xinjiang, China.[20]
Phobetor parvus
Lower Cretaceous, Zagan Zabsk Formation, West Mongolia.[21]
'Santanadactylus' spixi
Lower Cretaceous, Santana Formation, Aptian, Araripe Plateau, Brazil.[23]
Puntanipterus globosus
Lower Cretaceous, La Cruz Formation, San Luis, Argentina.[51]
Family Pterodaustridae
Pterodaustro guinazui
Lower Cretaceous, Lagarcito Formation, San Luis, Argentina.[49]
Pterodaustro sp.
Lower Cretaceous, Neocomian, Antofagasta, Chile.[50]
Family Pteranodontidae
Pteranodon longiceps[58]
Pteranodon ingens[56]
Pteranodon marshi[72]
Pteranodon occidentalis[56]
Pteranodon eatoni[72]
Pteranodon sternbergi[72]
Pteranodon walkeri[72]
Upper Cretaceous, Niobrara Formation, Santonian, West Kansas, USA.
Pteranodon sp.
Upper Cretaceous, Santonian-Campanian, Hokkaido, Japan.[28]
'Pteranodon' oregonensis
Lower Cretaceous, Hudspeth Formation, Albian, Oregon, USA.[63]
Ornithostoma seeleyi
Upper Cretaceous, Cenomanian, Cambridge Greensand, England.[73]
Ornithostoma orientalis
Upper Cretaceous, Senonian, Saratov, Petrovsk, USSR.[24]
Indeterminated Pteranodontidae
Lower Cretaceous, Toolebuc Formation, Albian, Queensland, Australia.[33]
Lower Cretaceous, Chulec Formation, Albian, Huanuco, Peru.[23]
Upper Cretaceous, Merchantville Formation, Campanian, Delaware, USA.[64]
Upper Cretaceous, Two Medicine Formation, Campanian, Montana, USA.[66]
Upper Cretaceous, Eutaw Formation, Santonian, Georgia, USA.[65]
Upper Cretaceous, Judith River Formation, Campanian, Alberta, Canada.[67]
Family Nyctosauridae
Nyctosaurus lamegoi
Upper Cretaceous, Gramame Formation, ?Maastrichtian or Campanian, Paraiba, Brazil.[35]
Nyctosaurus gracilis
Upper Cretaceous, Niobrara Formation, Santonian, West Kansas, USA.[58]
Family Azhdarchidae
Doratorhynchus validus
Lower Cretaceous, Purbeck Limestone, Berriasian, Dorset, England.[9,10]
Titanopteryx philadelphiae
Upper Cretaceous, Maastrichtian, Rosaifa, Jordan.[27]
Azhdarcho lancicollis
Upper Cretaceous, Turonian-Coniacian, Uzbekistan, USSR.[25]
Quetzalcoatlus northropi
Quetzalcoatlus sp.
Upper Cretaceous, Javelina Formation, Maastrichtian, Big Bend National Park, West Texas, USA.[68]
?Quetzalcoatlus sp.
Upper Cretaceous, Judith River Formation, Campanian, Alberta, Canada.[71]
Indeterminated Azhdarchidae
Upper Cretaceous, Campanian-Maastrichtian, Paki, Senegal.[31]
Indeterminated Pterodactyloidea
Araripesaurus castilhoi[37]
Araripedactylus dehmi[38]
Lower Cretaceous, Santana Formation, Aptian, Araripe Plateau, Brazil.

other hand, there is much evidence of burrowing animals in the rock in which the fossil remains were found. The occurrence of large quantities of fossil tree trunks in the area suggests periodic flooding at the time. All this allows the possibility that *Quetzalcoatlus* used its slender, pointed beak to search in the ground for the molluscs and crabs that lived in the shallow pools of water.

These and many other questions will perhaps be answered when scientific investigation of the rich skeletal material of *Quetzalcoatlus*, still under study at the time of writing, is finally concluded.

Finds in the Judith River Formation (Campanian) in the Dinosaur Provincial Park in Alberta, Canada, make it probable that *Quetzalcoatlus* did not just live in Texas.[71] However, fragments of a femur and a neck vertebra do not permit definite classification in the same genus as the Texas finds.

The Texan pterosaurs from Big Bend National Park are especially significant for another reason. They were not only the largest, but also the last of these fascinating flying reptiles to live on Earth. The strata in which the finds were made have been dated as latest Upper Cretaceous (Maastrichtian). They are only a few metres below the boundary layer with the Tertiary, the time marker between the Mesozoic and the Cenozoic Eras. Above this boundary neither dinosaurs nor pterosaurs are found. Thus both groups of reptiles became extinct at the same time. The causes that led to this 65 million years ago were probably the same for both of them.

71 Currie, P.J. and Russell, D.A., 1982. *A giant pterosaur (Reptilia: Archosauria) from the Judith River (Oldman) Formation of Alberta.* Canadian Journal of Earth Sciences, 19 (4): 894-897.

72 Miller, H.W., 1972. *The Taxonomy of the Pteranodon Species from Kansas.* Transactions of the Kansas Academy of Science, 74, 1: 1-19.

73 Lydekker, R., 1904. *Vertebrate Palaeontology of Cambridgeshire.* in: J.E. Marr and A.E. Shipley, *Handbook to the Natural History of Cambridgeshire,* pp.51-79; Cambridge.

In The Lost World, Arthur Conan Doyle tells how Professor Challenger and his companions met living pterosaurs, survivors from the Jurassic, on a remote highland plateau in South America: '. . . and (it) flapped its twenty-foot span of leathery wings as it soared up into the air . . . It was a wonderful sight to see at least a hundred creatures of such enormous size and hideous appearance all swooping like swallows with swift, shearing wing-strokes above us.'

Pterosaurs have always stimulated the imagination of naturalists working on their fossil remains. For example, the Rev. William Buckland, Professor of Geology at the University of Oxford, saw the creature as 'a monster resembling nothing that has ever been seen or heard-of upon earth, excepting the dragons of romance or heraldry.'[1] And in a contribution to the 'Bridgewater Treatises on the Power Wisdom and Goodness of God as manifested in the Creation' he wrote in 1836: 'In external form, these animals somewhat resemble our modern Bats and Vampires . . . Their eyes were of enormous size, apparently enabling them to fly by night . . . It is also possible that the Pterodactyles had the power of swimming.'[2]

Buckland was not the first person to see pterosaurs as bat-like. Thomas von Soemmerring advanced this view in 1812 and defended it against Cuvier, and it seems that pterosaurs have retained this bat-like image right down to the present day.

The last living pterosaurs disappeared from the Earth 65 million years ago, long before man existed. All that is left are fossilized remains of their bones, although in a few cases we also have impressions of their skin and body-covering in the rock, casts of cavities in the skull like the brain case and finally – extremely rarely – fossilized remains of food in the stomach. How then can palaeontologists reconstruct the life style of pterosaurs, particularly as there are no directly comparable creatures in the modern animal kingdom? Bats and birds are very different from pterosaurs in the way their skeletons are constructed, especially in the structure of their flight equipment. Furthermore, the surface of the wing is made up of feathers in birds, and in bats of thin membranes which were probably very different from those of pterosaurs.

The fossilization process causes considerable loss of information. Fossil remains of an individual are inevitably fragmentary. An additional factor is that the pterosaur skeleton was lightly built, and as a rule the bones were thin-walled and fragile. This means that fossil survival was only possible under favourable conditions in soft sediments. Up to the time of writing more than 50 genera of pterosaur comprising almost 100 different species are known. They lived from the late Triassic to the end of the Cretaceous, that is to say for a period of about 155 million years. There have been birds on Earth for about as long. We may assume that pterosaurs achieved as great a variety of forms in the most varied environments as birds. However, this also means that our knowledge of pterosaurs may be limited to a mere one per cent of the pterosaur fauna that in fact once existed.

Almost all pterosaur fossil finds come from marine deposits near the coast. The potential for fossilization in terrestrial upland areas was much too slight for delicate, fragile pterosaur skeletons to survive. They were crushed and destroyed by weathering, erosion and the action of flowing river water. Thus we are faced with a situation comparable with knowing only the shore birds of the 9,000 species of bird alive today.

We must take this limitation into consideration when attempting to reconstruct the life style of the pterosaurs so far known to us. In detail this means answering the following questions on the basis of the fossil material:

1) Locomotion: how did pterosaurs fly and how did they move on the ground?
2) Nutrition: how did they feed and what did they eat?
3) Reproduction: how did they reproduce themselves? Did they lay eggs or produce living young?
4) Physiology: were they warm-blooded or cold-blooded? Was their physiology bird-like or reptilian?

The fossils themselves provide no direct answers to many of these questions. They can often only be answered by indirect, analogous inferences. Naturally there are also alternative interpretations, which have led to different, even contrary views of the life style of pterosaurs. But often a single fossil find in good condition is enough to knock a conventional theory on the head, or to defend a traditional view against a new hypothesis. Thus our present ideas on the life style of pterosaurs will also be called into question or refined by future fossil finds.

Above: Georges Cuvier, the great Parisian comparative anatomist, who, in the early years of the nineteenth century, engaged in a lively debate with Thomas von Soemmerring about the nature of pterosaurs and their likely behaviour. Like William Buckland, Soemmerring thought that pterosaurs were bat-like, to be classified as mammals somewhere between bats and birds. For his part, Cuvier stressed the reptilian nature of these remarkable flying creatures.

Right: William Buckland reconstructed the habits and behaviour of pterosaurs for the 'Bridgewater Treatises on the Power Wisdom and Goodness of God' of 1836, from which this illustration is reproduced. He regarded them as animals of bat-like appearance living on cliffs by the Jurassic Sea. Shown here is the genus *Dimorphodon* which Buckland first described in 1829. It was not known at that time that *Dimorphodon* actually had a long tail.

Pterosaur Flight

Since the time of Georges Cuvier (1801) it has been known that pterosaurs were flying reptiles. Flying creatures need a propulsion system powerful enough for them to overcome Earth's gravity and at the same time drive themselves forwards. In tetrapod vertebrates evolution solved this problem by modifying the forelimbs to form wings. These wings are aerofoils that have to be moved by muscles to achieve active powered flight. In order to understand pterosaur flight we have to analyse first of all the structure of their wings, both the skeleton of the wing and its muscles, and also the aerofoil, the flight membrane.

The elements of the skeleton that are important for flight are powerfully developed in all pterosaurs. The shoulder girdle is fused into a hook-shaped bone made up of the scapula and the coracoid, the so-called scapulocoracoid. This, with its glenoid fossa, offers a strong abutment for the short, compact upper arm

1 Buckland, W., 1835. *On the discovery of a new species of pterodactyl in the Lias of Lyme Regis.* Geological Transactions, London, ser. 2, vol. 3: 217-222.

2 Buckland, W., 1836. *Geology and Mineralogy, Considered with reference to Natural Theology.* I, pp. 224-245, in volume 5 of the Bridgewater Treatises on the Power Wisdom and Goodness of God as manifested in the Creation, London.

bones of the wings. In the large Cretaceous pterosaurs this bone is still anchored over the scapula to the notarium, the fused dorsal vertebrae. Below, the shoulder girdle is articulated to the sternum via the coracoid, in a way similar to modern birds.

The sternum itself is a broad plate of bone which towards the front becomes a keel-shaped cristospine. As the most important flight muscles originated at the sternum, this braced structure of the shoulder girdle prevented compression of the rib cage during contraction of the flight muscles.

The shoulder joint is oriented laterally and in its normal position points slightly backwards. If the articular head of the upper arm bone is placed in the shoulder socket, it shows that in its optimum position and with maximum wing extension the humerus was directed 20° degrees above the horizontal. On the downstroke of the wing the humerus could be moved a maximum of 20° downwards from the horizontal plane, but at least 60° above it on the upstroke.

During the up- and downstroke of the wing the humerus could rotate around its longitudinal axis, thus altering the position of the wing. This is an essential prerequisite of powered flight, as it is the only way in which the flying body can move forward. Rotation of the wing in the nose-down sense was produced automatically by the muscles responsible for the downstroke of the wing, above all the pectoral muscles originating on the sternum and its protruding cristospine. They were inserted on the inside of the lateral delto-pectoral crest, thus causing a twisting movement forwards.

There are other muscle scars on the pterosaur humerus, marking the points at which the muscles for bending and stretching the lower arm and raising the upper arm on the upstroke were inserted. Thus the upstroke of the wing functioned on the same principle as that of a bird. In the latter it is brought about by two groups of muscles, one with its origin in the shoulder girdle that raises the humerus directly upwards (M. deltoidus), and a second, originating in the sternum, the coracoid and the wishbone (M. supracoracoideus). This supracoracoideus muscle is the second largest muscle in the front extremity of birds, after the great pectoral muscle (M. pectoralis).

Although this muscle lies below the humerus, it is responsible for raising it. This is done by means of a tendon that passes through a channel in the upper end of the coracoid and is anchored on the upper side of the humerus. Therefore a contraction of the supracoracoid muscles pulls the humerus upwards by means of this tendon, causing the wing upstroke.

In pterosaurs we find a similar channel on the coracoid as in birds, so that we can assume that in pterosaurs too the upstroke was brought about by the supracoracoid muscle and a tendon running over the coracoid as if over a pulley. Certainly pterosaurs did not have a wishbone like birds, which is created by the fusion of the clavicles. The pterosaurs' supracoracoid muscle thus originated principally at the sternum and probably partially at the coracoid as well, and was covered by the stronger pectoral muscle, which also originated at the sternum.[3] Deltoid muscles were certainly also responsible for the upstroke as well as the supracoracoid muscle; they originated at the shoulder blade and possibly at the notarium.

In comparison with the relatively short and compact humerus, the radius and ulna are longer and thinner. The ulna is always the thicker of the two bones, which are set close together and form a functional unit. No rotation was possible in the elbow joint, a hinge joint which could be opened up to 150° maximum extension. Flexion was only possible to an angle of 110°, to the front. These mechanical constraints are another reason for excluding the possibility that pterosaur wings could be as tightly folded and held close to the body, as is the case in birds.

3 Padian, K., 1983. *A functional analysis of flying and walking in pterosaurs.* Paleobiology, 9(3): 218-239.

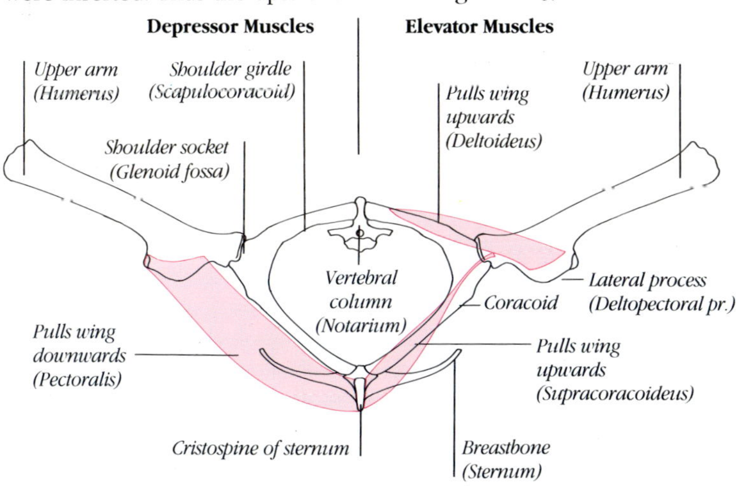

Flight Muscles (left)
The main flight muscles of a Cretaceous pterosaur are seen here from the front. Two muscle groups operate the wing, the depressor (left) and the elevator (right) muscles. The principal depressor muscle (M. pectoralis) originated from the sternum and was attached to the humerus. The main elevator muscles (deltoideus and supracoracoideus) originated from the shoulder blade, and the sternum and coracoid.

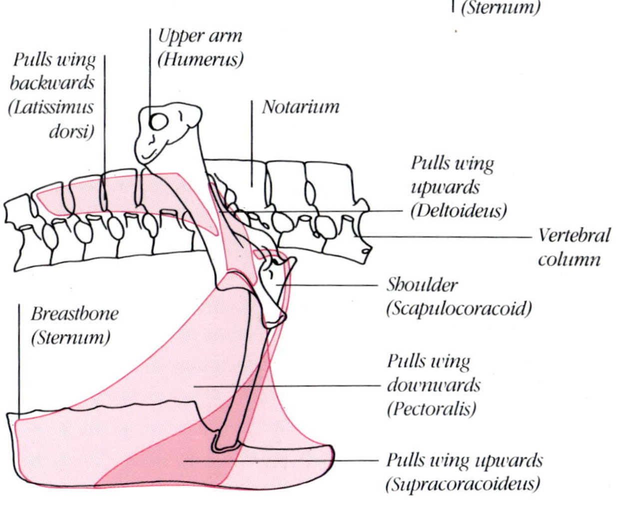

Flight Muscles (left)
The flight muscles are seen here from the side. The pectoralis muscle, the principal depressor, is the largest muscle of the flight apparatus. Underneath this the supracoracoideus pulled the humerus upwards by means of a tendon which ran around a pulley near the shoulder socket. The deltoideus muscle helped this movement. The latissimus dorsi muscle pulled the humerus, and so the wing, backwards.

The Pterosaur Hand

The range of movement in the wrist encompassed sliding forwards and backwards, and rotation around the longitudinal axis. Thus the flight hand could be angled both backwards and downwards, although not very much, by about 30° in both directions. In advanced forms the wrist is composed of three carpal bones, a proximal carpal, articulated with the lower arm, a distal carpal, set between the proximal carpal and the metacarpus, and a lateral (actually medial) carpal articulated with the distal carpal at the front. All these carpal bones have very complex articular surfaces, to allow fine adjustment of the wing.

The Pteroid Bone

A peculiarity of the wing skeleton of the pterosaur is a small bone articulated with the lateral carpal. This long, thin bone is always directed towards the body in undisturbed skeleton finds. This suggests that the pteroid was connected by a tendon to a muscle in the area of the shoulder, probably at the coracoid. Its function was apparently firstly to strengthen and tighten the leading edge of a small, triangular flight membrane between the base of the neck and the upper and lower arm, the so-called propatagium, and secondly to alter the position of the propatagium and achieve an aerodynamic effect in certain flight manoeuvres by means of various angles of attack.

The pteroid originates from the wrist but is regarded as an accessory bone, found only in pterosaurs. There are various hypotheses about its function, like that of E. Frey and J. Riess, who work on the assumption that in the flight position the pteroid was originally directed forwards and downwards. This leads them to a new reconstruction of the pterosaur wing.[4] According to this hypothesis the pteroid would have spread a large pre-wing membrane, extending between shoulder, upper and lower arm down to the small fingers and along the front side of the flight digit. The pteroid would have made this pre-wing adjustable, and assisted the camber. The leading edge may have been reinforced with a tendon, which could have extended along the whole of the leading edge of this pre-wing.

Another view of the function of the pteroid was put forward by C. Pennycuick.[5] According to him the pteroid could have been directed in turn both towards the body and also forwards and downwards. It would have been in the first, more streamlined position in the leading edge of the propatagium during rapid flight, and the second position, extended forwards and

4 Frey, E. and Riess, J., 1981. *A new Reconstruction of the Pterosaur Wing.* Neues Jahrbuch für Geologie und Paläontologie, Abhandlungen, 161(1): 1-27; Stuttgart.
5 Pennycuick, C.J., 1988. *On the reconstruction of pterosaurs and their manner of flight, with notes on vortex wakes.* Biological Review, 63: 299-331.

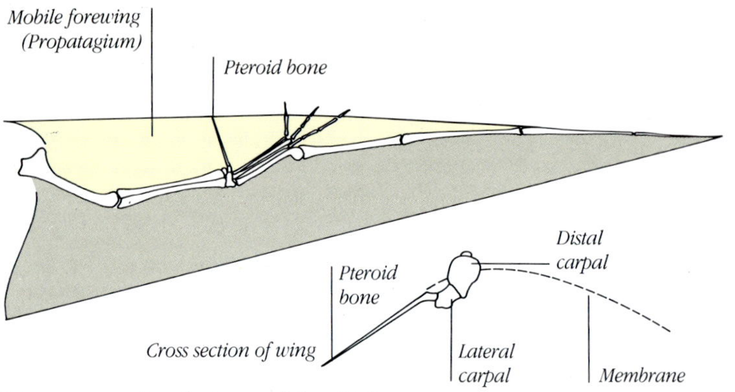

Wing reconstruction after Frey and Riess (1981)

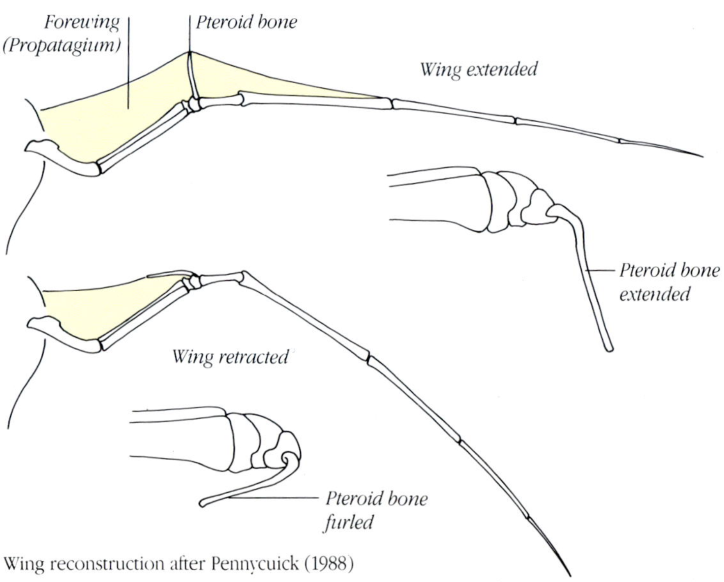

Wing reconstruction after Pennycuick (1988)

Wing Reconstructions (left)
Shown here are alternative wing reconstructions as proposed by E. Frey and J. Riess in 1981 (above) and C.J. Pennycuick in 1988 (below). Frey and Riess suppose that an enlarged forewing could be stretched by the pteroid bone which would have pointed forwards and downwards. The small digits of the hand would have been included in this forewing. Pennycuick's reconstruction assumes an automatic snap action of the pteroid bone whereby protraction of the wing finger deployed a drooped leading edge (upper drawing), while retraction would have furled the forewing. In this way, pterosaurs might have had the equivalent of variable geometry aircraft wings which could be modified in shape according to flight speed requirements.

Below: The first phalanx of the fourth finger, the wing finger, articulates in a pulley-like joint with the metacarpal bone (here partly covered). This was the main wing folding joint in pterosaurs.

downwards, during slow flight, that is to say when landing and taking off. Pennycuick assumes that in this case a tendon was stretched from the shoulder area over the tip of the pteroid to the joint between the first and second wing phalanges. If the flight digit was swept back in rapid flight, then under this hypothesis the pteroid would have snapped automatically from a position in which it was directed forwards, to a furled position.

However, in pterosaur finds, particularly from the Solnhofen limestones, in which the imprints of the wing membranes are well preserved, there are never signs of a front wing extending over the propatagium and outwards as well over the pteroid down to the fingers. Besides, the pteriod is usually such a delicate, long and slender bone (up to 80 per cent of the length of the lower arm) that it could not have withstood the stress of a leading edge tendon of this kind.

The pterosaur metacarpus is an extended element of the wing skeleton, still relatively short in the long-tailed Rhamphorhynchoidea, but relatively long in the short-tailed Pterodactyloidea; in the extreme forms of the Upper Cretaceous it was even longer than the lower arm. The dominant bone is the fourth metacarpal, the flight digit metacarpal. The other three metacarpals supported the three small claw-bearing digits and are thinner, sometimes even reduced to splint-shaped rods of bone.

The powerful flight digit metacarpal had a shallow, convex articular facet with a pivot-shaped process that fitted into a corresponding socket in the distal carpal. This connection allowed limited rotation of the metacarpal around its longitudinal axis and thus adjustment of the flight digit, but not flexion. The distal joint of the flight digit metacarpal, however, is in the form of a pulley and made it possible for the flight digit to be folded a long way backwards. This is the principal folding joint in the pterosaur wing. To make this possible the proximal joint of the first wing phalanx is built asymmetrically. The slightly oblique arrangement of this joint made it possible for the first phalanx to slide in part laterally above the metacarpal when the flight digit was folded back to its maximum extent.

The Digits

A strong process on the front side of the first wing phalanx joint served for attachment of a strong extensor tendon, running along the front side of the metacarpal via a bone channel to an extensor muscle on the lower arm (M. extensoris digiti). At the same time this extensor process on the digit joint prevented forward hyperextension of the flight digit. If the flight digit was at its maximum extension it was oriented at an angle of 165° to the metacarpus.

Between the four enormously elongated and reinforced phalanges of the flight digit are relatively shallow, oval, concave-convex articular facets at which little flexion was possible. The joint connections were relatively tight, and probably appropriate ligaments ensured a large degree of rigidity in the four-phalanged flight digit.

The first three small fingers could be moved freely and rose forwards out of the wing. They had scarcely any role to play in flight. Their sharp, bent claws had well-developed flexor processes to which strong flexor tendons were attached. Thus the digital claws were ideally suited for gripping and climbing steep surfaces like rocks, cliffs and tree trunks.

Wing Extension (below)
The extensor digiti was a special muscle for the extension of the wing finger. The pteroid bone was operated by the pteroid muscle at the coracoid. Its tendon formed the leading edge of the forewing. Over the pteroid muscle the angle of attack of the wing, and thus its camber, could be altered. The wing profiles show the camber at different cross sections.

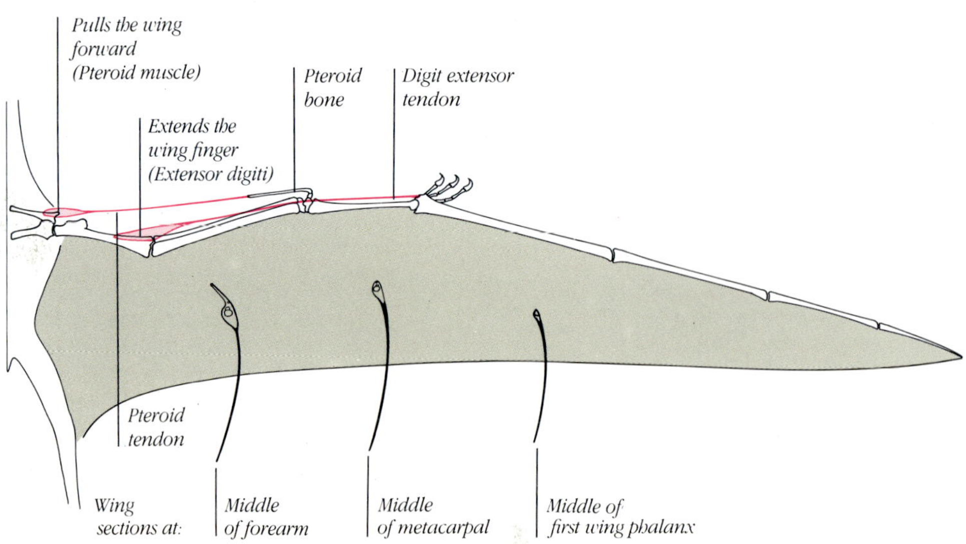

Bone Structure

The bones of the pterosaur wing skeleton are hollow. Only in the area of the joints, at the ends of the long bones, is there an internal zone of spongy bone tissue, appropriate to the increased load on the articular facets. The wall thickness of the tubular bones of the wing is very slight, often only that of a postcard. In the wing phalanges the bone wall on the longitudinal edges is somewhat thicker, especially at the front, on the leading edge of the wing. This structure afforded optimal bending strength to these bones, which were under particular stress in flight. In the case of *Quetzalcoatlus*, the giant Texan pterosaur, the three outer wing phalanges were no longer hollow and thin-walled, but T-shaped in cross-section. They were constructed like T-bars and thus resisted the ever-increasing bending momentum on the outer sections of the wing.

Other bones in the pterosaur wing were not completely hollow either. In the large Cretaceous species in particular the bones of the upper and lower arm were reinforced internally by a system of thin bony struts. This can be seen particularly clearly in a specimen of *Santanadactylus* from a calcareous nodule in the Brazilian Santana Formation.[6] The humerus is 6·7in (17cm) long and has an average wall thickness of only 0·55mm. Numerous bony support struts run transversely through the shaft cavity, in part interconnected with each other. They run in the direction of the major mechanical forces that were exerted on the bones during the strokes of the wing.

This wing bone structure, similar to that of modern birds, resulted in maximum weight reduction combined with optimum rigidity and strength. This was the only way in which the long bones, particularly of large pterosaurs, could be so thin-walled. They were protected against breaks by internal supports. These were only as thick as was absolutely necessary, and were actually no more than materialized lines of force. An extremely lightly built skeleton was the secret of the great evolutionary success of the pterosaurs of the Mesozoic Era.

Flight Membranes

In 1817, S.T. von Soemmerring published a reconstruction of a small *Pterodactylus* from the Solnhofen limestones, in which he drew the outline of the flight membranes. As he thought the pterodactyles were bats, it was only logical that he should give the Solnhofen *Pterodactylus* bat's wings. Flight membranes extended back to the feet, between the hind legs and the short tail, and in front of the arms to the neck. This was long before pterosaur flight membranes were discovered as detailed imprints on the surface of Solnhofen limestones.

The so-called 'Zittel-wing', a *Rhamphorhynchus* wing with an extremely well-preserved flight membrane, described by Munich professor Karl A. Zittel in 1882, was a famous specimen.[7] In the same year O.C. Marsh, who was friendly with Zittel, brought out a publication on a *Rhamphorhynchus* specimen that had survived complete, in which not only the flight membranes of the wings but also a membrane on the end of the long tail, the so-called tail vane, had been preserved.[8]

Pterosaur finds with wing membrane imprints are not as rare as is often thought. They are known from specimens from the Lower Jurassic of Holzmaden, the Upper Jurassic of Solnhofen in southern Germany, from the Upper Jurassic of Kazakhstan in the USSR, and from the Lower Cretaceous of Brazil, in the genera *Dorygnathus, Rhamphorhynchus* and *Sordes*, and in *Pterodactylus* and a genus from the Santana Formation in Brazil, as yet undetermined. The Solnhofen specimens in particular have made a great contribution to our knowledge of the extent and constitution of pterosaur flight membranes.[9]

Fossil finds show that pterosaur flight membranes were not bat-like in appearance. This is a consequence of the different construction of the hand. In bats the flight membrane is spread between the elongated second, third, fourth and fifth digits and the feet, but in the case of pterosaurs the flight membrane is spread by a single digit, the fourth, or flight digit. This flight digit is much thicker and longer than the wing digits of the bat. The extreme length of the flight digit alone means that the flight membrane must have been fairly narrow. Whether this was only attached to the sides of the body, so that the hind legs remained completely free,[10] or whether the flight membrane was also attached to the upper leg, or even to part of the lower leg,[9] or reached to the ankle and was spread by the fifth foot digit,[5] is a subject of scientific controversy.

The best-preserved flight membrane so far, in a Solnhofen *Pterodactylus* in the Naturhistorisches Museum in Vienna, the 'Vienna specimen', suggests the following conclusions: the flight membrane of the wing, the brachiopatagium, was attached to the upper leg and ex-

6 Wellnhofer, P., 1985. *Neue Pterosaurier aus der Santana-Formation (Apt) der Chapada do Araripe, Brasilien.* Palaeontographica (A), 187, p.178; Stuttgart.

7 Zittel, K.A. von, 1882. *Über Flugsaurier aus dem lithographischen Schiefer Bayerns.* Palaeontographica, 29: 47-80; Stuttgart.

8 Marsh, O.C., 1882. *The wings of Pterodactyls.* American Journal of Science, 23: 251-256.

9 Wellnhofer, P., 1987. *Die Flughaut von Pterodactylus (Reptilia, Pterosauria) am Beispiel des Wiener Exemplares von Pterodactylus kochi (Wagner).* Annalen des Naturhistorischen Museums Wien, 88, A: 149-162; Vienna.

10 Wellnhofer, P., 1975. *Die Rhamphorhynchoidea (Pterosauria) der Oberjura-Plattenkalke Süddeutschlands. III. Palökologie und Stammesgeschichte.* Palaeontographica (A), 149: 1-30; Stuttgart.

Padian, K., 1979. *The Wings of Pterosaurs: A New Look.* Discovery, 14(1): 20-29; New Haven.

Right: This humerus of *Santanadactylus* from Brazil has been split open to reveal the system of bony struts that strengthened the hollow, very thin-walled shaft of the bone. The outer wall of this bone is only 0·02in (0·5mm) thick. The internal struts can be regarded as materialized lines of force, and provided maximum lightness combined with optimum strength.

Left: This is a *Rhamphorhynchus* specimen from Solnhofen with the impressions of the wing membranes still attached to the wing fingers clearly visible. When this pterosaur died, its membrane must have been embedded in the Solnhofen sediment in a relaxed state. It is still possible to see folds parallel to the wing fingers. Finds such as this which preserve the imprints of the wing membrane are rather more common than might first be imagined. Several are known from Germany, while others come from the Soviet Union and Brazil.

Left: The Vienna specimen of *Pterodactylus kochi* from the Solnhofen limestone of Bavaria has the impressions of the wing membranes superbly preserved. The animal, which has a wing span of 15in (38cm), was embedded lying on its stomach. The specimen is now housed in the Naturhistorisches Museum in Vienna.

Pterodactylus Flight Membranes (right)
This restoration of the flight membranes of *Pterodactylus*, based on the Vienna specimen illustrated at left, shows the extent of the main wing membrane, the brachiopatagium, and of the forewing, the propatagium, with the arrangement of the internal strengthening fibres, the aktinofibrils. There is no evidence of a uropatagium, a membrane between the hind legs and tail. The wing membrane did not reach the ankle, and generally was quite narrow in shape.

Pteranodon's Tail (right)
The tail of *Pteranodon* consisted of a series of short vertebrae with a double condyle articulation terminating in a pair of greatly elongated rod-like bones. Chris Bennett of the University of Kansas at Lawrence has suggested that these served for the attachment of the wing membrane, thus leaving the hind legs completely free of the wing.

The Vienna Pterodactylus (left)
This specimen of *Pterodactylus kochi* indicates that the wing membrane of this pterosaur genus extended to the upper leg rather than down to the ankle. The wings are preserved in a folded and relaxed position.

Right: A rhomboid vane was developed at the distal end of the long vertebral tail of rhamphorhynchoid pterosaurs. In life it was oriented vertically, and stretched by internal, transverse stiffening zones, the function of which was presumably to keep the tail membrane permanently spread. It probably served as a drag-rudder in flight in order to stabilize the position of the animal in the air.

tended to the side of the upper part of the lower leg. Despite this the wing was very narrow and pointed. There is no indication of a broader membrane between the hind legs or between legs and tail, of a uropatagium in other words. A small front wing membrane, a propatagium, was spread on the inside of the arm between wrist and shoulder. This reconstruction corresponds fairly precisely with the version developed by the Viennese palaeobiologist Othenio Abel as early as 1919.[11]

The wing geometry of *Pterodactylus* was not necessarily valid for all Pterodactyloidea. Thus, as a result of new observations on *Pteranodon*, Chris Bennett came to the conclusion that the flight membrane of this large Cretaceous pterosaur was attached at the tail.[12] The last section

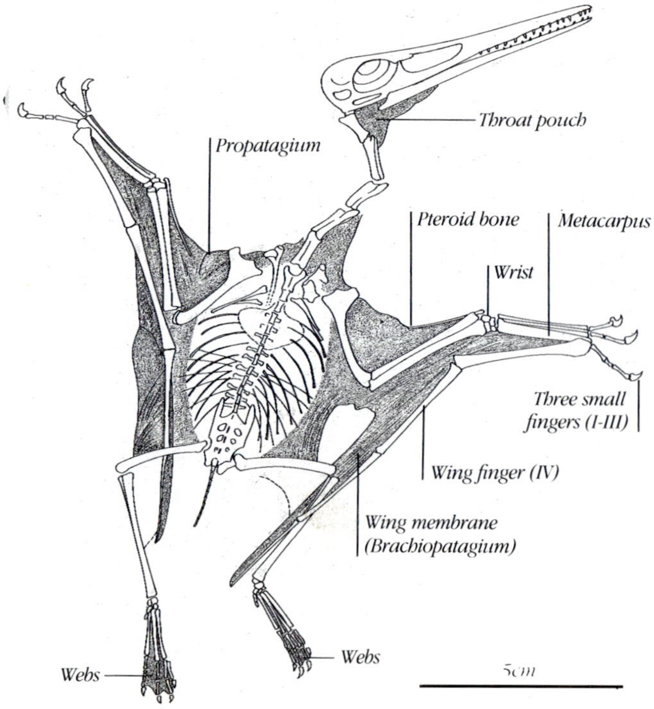

11 Abel, O., 1919. *Neue Rekonstruktion der Flugsauriergattungen Pterodactylus und Rhamphorhynchus.* Die Naturwissenschaften, 7, Heft 37: 661-665; Berlin.

12 Bennett, C., 1987. *New Evidence on the tail of Pterosaur Pteranodon (Archosauria: Pterosauria).* Short Papers of the Fourth Symposium on Mesozoic Terrestrial Ecosystems (ed. P.M. Currie and E.H. Koster): 18-23; Drumheller.

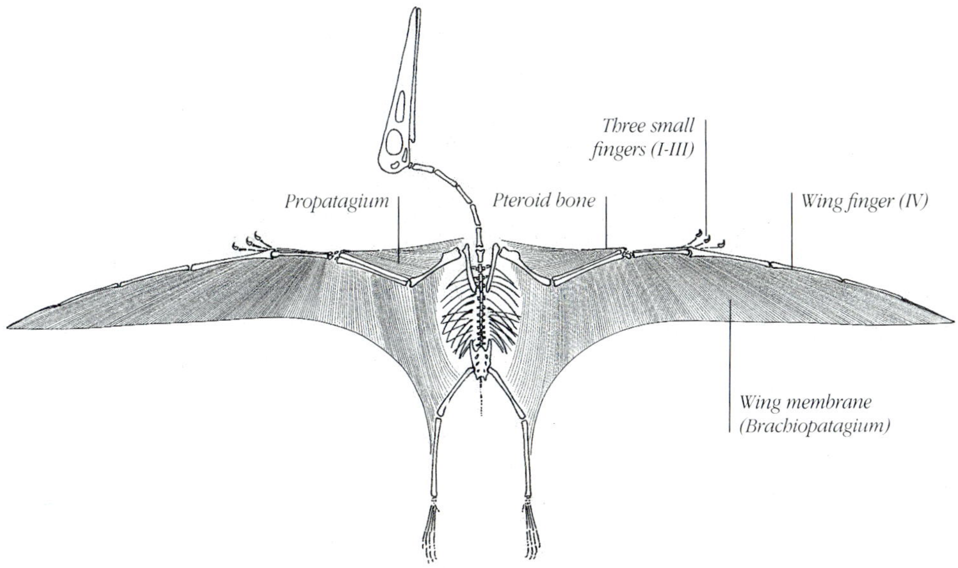

Tail of Pteranodon from above and from the side

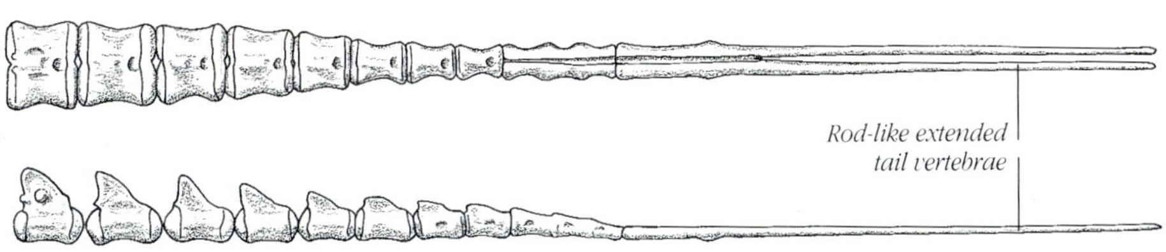

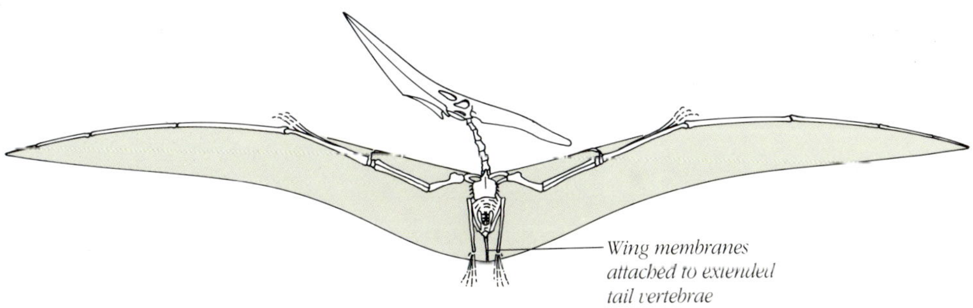

of the caudal vertebral column in *Pteranodon* in fact consisted of a pair of long rods, possibly embedded in the horizontal extension of the flight membranes. They would have been long and thick enough to control the flight membrane. When the tail was moved up and down the membrane in the tail area must have functioned as a pitch control device. In this model the hind legs would have been completely free and independent of the flight membranes. In flight they would have been drawn right in towards the body in the direction of the centre of gravity.

In the long-tailed Rhamphorhynchoidea the long vertebral tail had a small additional membrane at the end, the terminal tail membrane. In the case of *Rhamphorhynchus* it was rhomboid or triangular and somewhat asymmetrical. It was apparently a different shape in different species. A typical feature are bands running transversely, suggesting a system of reinforcement by means of which the tail membrane was kept permanently spread. In life and in flight the tail membrane must have been oriented vertically. It was very firmly anchored on the last 15 to 17 caudal vertebrae.

The tail of the Rhamphorhynchoidea was stiffened and not very elastic. The caudal vertebrae could only move in relation to each other in the front section. Vertebral processes on the sides indicate particularly great lateral mobility. The vertical terminal tail membrane could thus function like a rudder in flight, and possibly when swimming in the water as well.

Internal Wing Structure

As early as 1882 Zittel observed fine parallel striations in the flight membrane of *Rhamphorhynchus*, which he took to be the result of a particular internal structure. What can be seen are the sharp imprints of very fine 'fibres', running through the flight membrane in tight sequence, in a longitudinal direction near the flight digit and increasingly obliquely towards the trailing edge of the flight membrane. The 'fibres' are 0·002in (0·05mm) thick, and a uniform 0·008in (0·2mm) apart. Thus within a millimetre width of skin there are four to five of these fibres, called 'aktinofibrils'. It is not known of what material they consist, but it may have been keratine, the horny material hairs, scales and claws are made of. Towards the edge of the flight membrane there are frequently additional aktinofibrils intercalated, so that when the wing was unfolded and the flight membrane spread these fibres were spread out like a fan, but an even arrangement was guaranteed overall.

Aktinofibrils have been observed in the

Tail Membranes (below) The tail membrane is preserved in several specimens of *Rhamphorhynchus* from the Solnhofen limestone. They show that its size and shape varied between different species, although all had transverse stiffening. From left to right: *Rhamphorhynchus longicaudus*, *Rh. intermedius*, *Rh. gemmingi*, *Rh. muensteri*, and a large, as yet unnamed species.

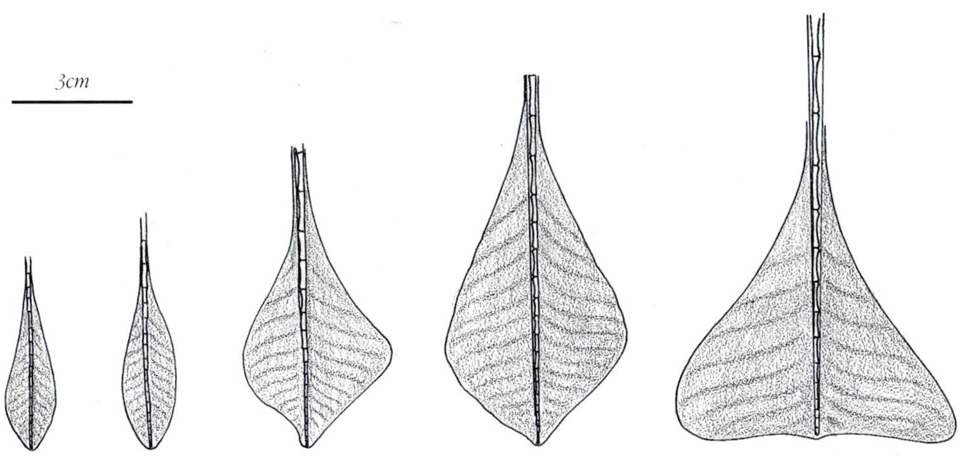

flight membrane of *Rhamphorhynchus* and *Pterodactylus*,[9] but also in Pterodactyloidea from the Santana Formation of Brazil.[13] In no case did folding take place at right angles to these supporting fibres. The flight membrane could only be folded parallel to the course of the aktinofibrils. This suggests an aerodynamic function for these structures. They were not just a system for reinforcing the flight membrane, but also prevented wobbling in flight, especially on the trailing edge of the wing, which the fibres met at right angles.

For this reason the pterosaur flight membrane was also called an 'aktinopatagium'. The internal fibres were covered with an upper and a lower layer of skin. It was a sandwich structure. The covering skin layers may have had fine muscles running through them to exert active tractive power to bend the aktinofibrils. This gave the cambering necessary to provide lift in flight.[14]

The stripes visible on the fossil flight membrane imprints have also been interpreted differently, as wrinkles in the relaxed flight membrane. According to this interpretation these wrinkles were formed by inner elastic fibres running diagonally.[5] But these fine structures are so sharply imprinted on the fossil itself that they could hardly have been produced by soft parts like skin. Surviving flight membrane in a Santana pterosaur does not show this ray system in relief, but in the form of dark coloured stripes. This would hardly be possible if we were dealing merely with wrinkles in the skin.

Finally a third model has been developed for the internal structure of the pterosaur wing membrane.[15] It is based on a small section of wing membrane near the lower arm of a pterodactyloid from the Lower Cretaceous of Brazil. Here the flight membrane in soft part preservation shows various layers of skin under the electron microscope: on the outside is a thin horny epidermis, below this a layer obviously with numerous capillary blood vessels, under this a layer with a network of organic

13 Campos, D.A., Ligabue, G. and Taquet, P., 1984. *Wing membrane and wing supporting fibres of a flying reptile from the Lower Cretaceous of the Chapada do Araripe (Aptian, Ceará State, Brazil).* Short Papers of the Third Symposium on Mesozoic Terrestrial Ecosystems: 37-39; Tübingen.

14 Schaller, D., 1985. *Wing Evolution.* The Beginnings of Birds. Proceedings of the International *Archaeopteryx* Conference Eichstätt 1984: 333-348; Eichstätt.

15 Martill, D.M. and Unwin, D.M. *Exceptionally well preserved pterosaur wing membrane from the Cretaceous of Brazil.* Nature, 340, No.6229: 138-140.

Below: A close-up view of the 'Zittel' wing reveals a system of fine, parallel fibres, called aktinofibrils, which were sandwiched between layers of tough skin. The fibres or rays can be clearly distinguished from the larger wrinkles which were caused by relaxation of the wing after death.

Above: The best preserved pterosaur wing membrane is still the *Rhamphorhynchus* wing described by K.A. Zittel in 1882, here seen in a section detail. It reveals particularly clearly the internal structure of the wing which shows up as sharp impressions on the surface of the limestone slab.

Internal Strengthening (below)
This is a model of the internal structure of the wing membrane in pterosaurs. The strengthening fibres, the aktinofibrils, were only 0·002in (0·05mm) thick. Folding of the wing membrane could only occur parallel to the course in which the aktinofibrils lay.

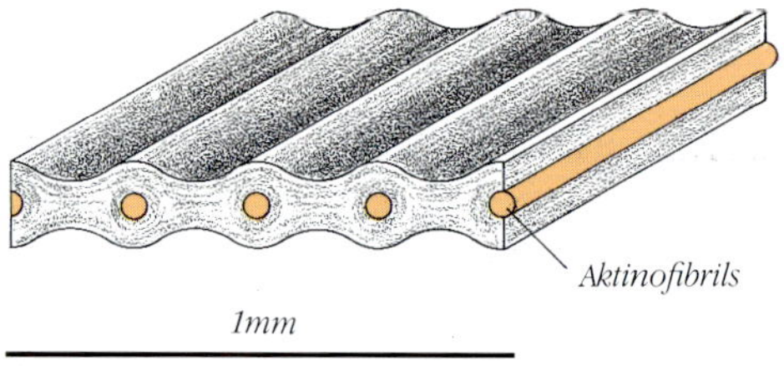

Left: A section of the wing membrane of a large pterodactyloid pterosaur from the early Cretaceous of Brazil shows the mannner in which the aktinofibrils were embedded in the membrane.

Right: A section of the trailing edge of the wing membrane of a Solnhofen *Rhamphorhynchus*. This shows that the aktinofibrils met the hind margin of the wing membrane at right angles to prevent it from flapping due to turbulence.

elastic fibres and finally the bottom layer with striated muscle fibres. The whole layer of skin must have been approximately 1mm thick.

Of course these observations only permit statements about a very small section of wing membrane, not from the actual aktinopatagium. Beyond this the flight membrane of pterosaurs seems to have been particularly well supplied with blood near the body, and it was assumed that here dilatation of the blood vessels developed a physiological mechanism to disperse excess metabolic heat.

Problems of the fine structure and histology of the pterosaur wing membrane have not been finally cleared up, but it is possible to establish that the skin structure of the membrane was stout, tough, relatively thick but elastic and fairly complex, of a kind that is not found in bats or other known modern animals.

Pterosaurs' Ability to Fly

In the animal kingdom there are three different principles of aerial locomotion: gliding, soaring and powered flight. Gliding is a means of passive locomotion with the aid of enlarged body surfaces which transform vertical falling into a tranverse mode of transport, but always accompanied by loss of height. Examples are the flying lizard *Draco* or the flying squirrel *Petaurista*. But many birds glide between the active flapping phases.

Soaring is achieved by means of air currents, wind or rising columns of air in which height can be gained. Many modern birds use this flying technique.

Powered flight is always flapping flight, in which the energy comes from muscles that move the wings up and down, as in birds and bats. The wing surfaces have to fulfil the aerodynamic principles of an aerofoil, i.e. they have an upper side with convex camber and a flat or concave lower side. If an aerofoil with a wing profile of this kind is moved through the air, the airstream produces reduced pressure on the upper side and increased pressure on the lower side. According to this principle, discovered by physicist Daniel Bernoulli as many as 200 years ago, the aerofoil is subject to lift. The greater the camber on the aerofoil, the greater the lift.

It is assumed that pterosaurs were capable of powered flight because of the large areas for the origin and attachment of muscles on the breast bone and upper arm. Some, particularly the large Cretaceous pterosaurs, certainly also used soaring, although the areas for the origin of muscles on the breastbone are not as large as in modern flying birds.

To have functioned in accordance with the Bernoulli effect, pterosaurs' wings must have had cambered surfaces. Indeed they must have been in a position actively to control and alter this camber. Bats manipulate the camber of the flight membrane with the long digits of their hands. Pterosaurs, with only a single flight digit, had a system of aktinofibrils, fibre structures embedded in the flight membrane, that could be cambered by muscle power.

However, the greatest lift came from the section of the wing membrane nearest the body, in other words at places at which the front membrane, the propatagium, could alter its angle by means of the pteroid muscle at the front, and where the wing grew into the upper part of the hind leg, at the back. The outer section of the wing had, as in birds, a high speed profile with a more shallow camber,[16] at least in the more advanced pterosaurs with long wings. Thus the digital wing had a more shallow profile than the arm wing, and was the more important for active flight.

It is worth noting that aerononautical engineers were involved in early investigations of pterosaur flight mechanics. First E.H. Hankin and palaeontologist D.M.S. Watson published a paper entitled 'On the flight of pterodactyls' in the Aeronautical Journal in 1914,[17] and G.H. Short wrote in the same magazine on 'Wing adjustments of pterodactyls'.[18]

Hankin and Watson investigated in particular the mechanical structure of *Pteranodon*'s skeleton, especially possibilities of mobility in the individual joints of the wing skeleton. They believed that this large Cretaceous pterosaur was primarily a soarer, as they thought its flight muscles too weak to have been adequate for persistent flapping flight. This first biomechan-

16 Herzog, K., 1968. *Anatomie und Flugbiologie der Vögel.* 180 pp.; G. Fischer Verlag, Stuttgart.

17 Hankin, E.H. and Watson, D.M.S., 1914. *On the flight of Pterodactyls.* The Aeronautical Journal, 72: 1-12.

p.11: 'The weakness of the flapping muscles makes it highly probable that their habitual mode of flight was by soaring rather than by flapping . . . The implied suggestion that their flight was like that of an albatross agrees well with the little we are able to infer about their habits.'

18 Short, G.H., *Wing adjustments of pterodactyls.* The Aeronautical Journal, 72: 13-20.

Below: The flight of a soaring albatross with a wing span of about 10ft (3m) can best be compared to the flight of a large pterosaur like *Pteranodon.* Albatrosses are able to stay in the air for a long time, soaring very long distances over the sea without needing to land. *Pteranodon*'s wings would also have given it excellent soaring ability.

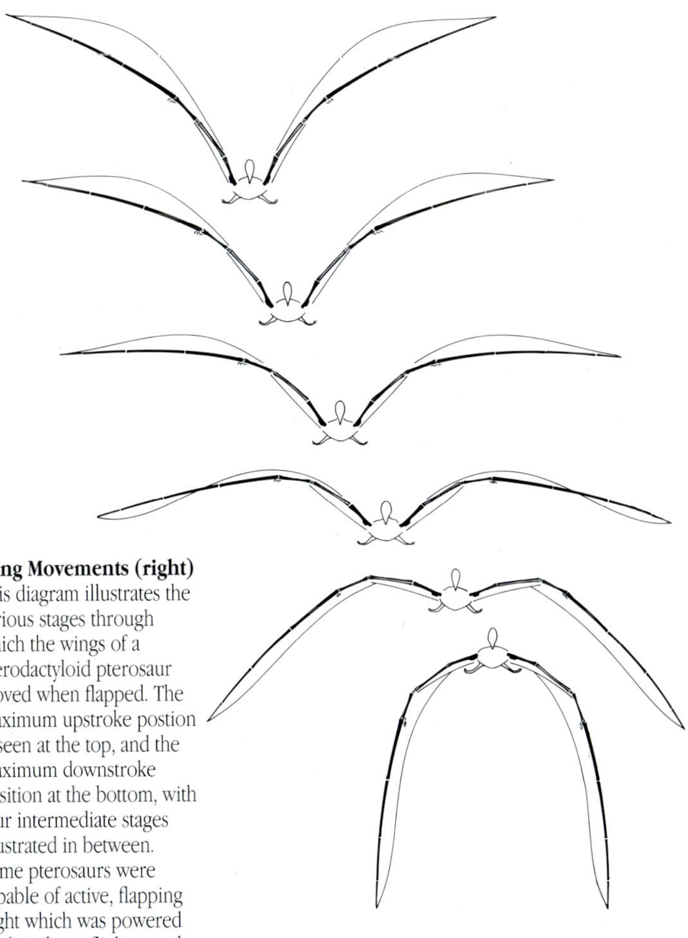

Wing Movements (right)
This diagram illustrates the various stages through which the wings of a pterodactyloid pterosaur moved when flapped. The maximum upstroke postion is seen at the top, and the maximum downstroke position at the bottom, with four intermediate stages illustrated in between. Some pterosaurs were capable of active, flapping flight which was powered by their large flight muscles.

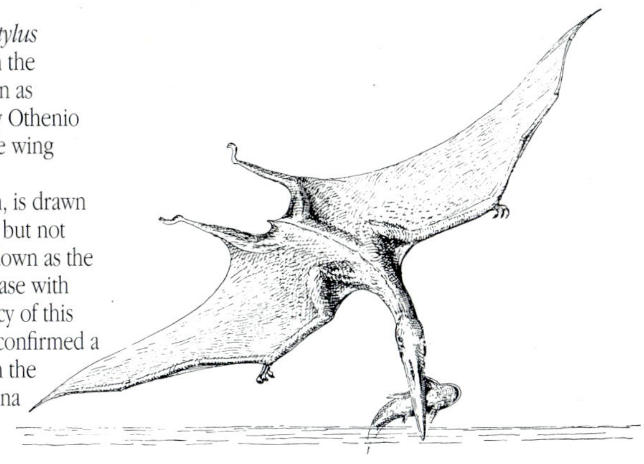

Right: *Pterodactylus* catching a fish in the Solnhofen lagoon as reconstructed by Othenio Abel in 1925. The wing membrane, the brachiopatagium, is drawn to the upper leg but not reaching as far down as the ankle, as is the case with bats. The accuracy of this restoration was confirmed a few years ago on the basis of the Vienna specimen of *Pterodactylus.*

Soaring Mastery (left)
Pteranodon had a wing span of about 23ft (7m). The long, narrow shape of these wings enabled *Pteranodon* to soar effortlessly in light winds. It may have used rising air in thermals on the upwind side of cliffs for lift, as we see in this restoration. Cherrie Bramwell and George Whitfield of the University of Reading made a study of the likely flight mechanics of *Pteranodon* and they concluded that it also could have soared over the sea in search of food. Because of its low sinking speed and its ability to stay airborne at low speeds, a temperature difference of just 1°C between the air and the water would have been sufficient for the formation of thermals in which *Pteranodon* could have soared. It may even have been able to alter the wing camber and so affect lift.

ical study of pterosaur flight is considered a classic, and was for a long time the only one on the subject.

Othenio Abel, the Viennese palaeobiologist, thought *Pterodactylus* was a flapping flyer which flew in a similar way to bats. He saw *Rhamphorhynchus*' long tail with its terminal vane as a pitch control.[19] However, anatomical findings show that it cannot have been oriented horizontally, but vertically.[20] Thus it was a rudder, probably a drag rudder, used to stabilize the flight position. The short-tailed pterosaurs, the Pterodactyloidea, show that pterosaurs could fly without a tail vane.

Abel also assumed that *Rhamphorhynchus*' normal resting position was on the ground. Taking off would have been assisted by striking the tail against the ground. However, it is now thought that the tail muscles would not have been strong enough for this.

The first mathematical calculations of flight technique data were made by ornithologist D. von Kripp in 1943.[21] He designed a restoration of *Pteranodon* and compared it with the aerodynamics of tailless aircraft. With a wing span of 23ft (7m) its wing area was 37·7 sq ft (3·5m^2). Its body weight was 66lb (30kg). This gives a wing loading of 1·75lb/sq ft (8·5kg/m^2). The high aspect ratio, i.e. long, narrow wings, endowed *Pteranodon* with outstanding soaring qualities: it could soar skilfully, quickly and over long distances, indeed even better than the albatross. An updraft of only 2 to 3ft/sec (0·6 to 1m/sec) would have been enough for it to maintain level flight. Over the open sea updrafts of this strength could easily occur above the crests of the waves.

The sites where fossils were found in the Niobrara Chalk of Kansas were at least 100 miles (160 km) from the coast at the time *Pteranodon* lived, so it must have flown long distances. The animals discovered presumably flew out over the open sea and died there. Incidentally, Kripp thought it impossible that *Pteranodon* could land on the water, swim and take off again. He therefore believed that it must have caught mainly flying fish above the surface of the sea.

It was not until the seventies that *Pteranodon*'s flying abilities were investigated again. First, in 1971, W.B. Heptonstall produced computer calculations giving a weight of 50lb (22·7kg) for a wing span of 22·3ft (6·8m).[22] According to this estimate the pectoral muscles would not have been strong enough for the creature to take off from the ground. The muscles alone would have had to weigh 220lb (100kg). The weight of a bird increases with the square of the wing area, for which reason large birds like the albatross with a weight of 19lb (8·5kg) and a wing span of 11ft (3·4m) cannot sustain flapping flight but can only soar.

Pteranodon, or *Quetzalcoatlus*, which was almost twice as big, and other large pterosaurs could probably only take off with the assistance of a headwind or from an elevated site.

At about the same time Cherrie Bramwell and George Whitfield of the University of Reading turned their attention to *Pteranodon*'s flight mechanics.[23] They arrived at a body weight of only 36·6lb (16·6kg) with a wing span of 22·8ft (6·95m). They calculated a sinking speed of only 1·4ft/sec (0·42m/sec) with an optimum soaring flight speed of 26·25ft/sec (8m/sec) (18mph; 29km/h). According to this *Pteranodon* could fly extremely slowly and soar in thermals and rising air currents by cliffs. Its top speed was 46ft/sec (14m/sec) (31mph; 50km/h), its stalling speed was calculated at 22ft/sec (6·7m/sec) (15mph; 24km/h). This made a safe and gentle landing possible. A headwind of somewhat more than 22ft/sec (6·7m/sec) was sufficient for take-off: *Pteranodon* only needed to spread its wings to raise itself into the air.

Probably the wing area that forms the basis of these calculations, 46·07 sq ft (4·28m^2) is somewhat too high. It assumes that the flight membrane extended to the feet, but more recent research suggests that this was not the case. It is possible that the wing membrane was only attached to the upper part of the hind legs, to just below the knee at most, or, as Chris Bennett suggested for *Pteranodon*, to the tail.[12]

Finally R.S. Stein even carried out wind tunnel tests for a biomechanical analysis of *Pteranodon*'s flight dynamics.[24] He concluded from this that *Pteranodon* was suitable for soaring and gliding flight, but primarily adapted to slow flapping flight. Stein calculated its weight at only 33lb (15kg), its stalling speed at 15ft/sec (4·5m/sec) (10mph; 16·2km/h) and its top speed at 49·2ft/sec (15m/sec) (33·55mph; 54km/h). Its muscular performance was 0·1hp. One stroke of the wings per second was enough to take off from the ground.

Stein also worked on the basis that the camber of the wings, and thus lift, could be controlled by flexion or extension of the flight digit. Stein also assumed that the wings extended as far as the feet, and relied on the reconstruction given by Eaton in 1910.

J.C. Brower, Professor of Geology at Syracuse University, New York, based his aerodynamic computer-supported calculations on a very narrow flight membrane, which left the legs completely free; this goes back to Kevin Padian's idea.[25] Despite the smaller wing area assumed by this hypothesis, Brower also came up with very low stalling speeds for *Pteranodon*. It was only capable of short periods of flapping flight, but was primarily a soarer. The smaller *Nyctosaurus* did have enough muscle power for continuous level flapping. However, its flight was probably alternating flapping and gliding flight, as in modern seagulls.

How can the giant Texas pterosaur *Quetzalcoatlus* with its wing span of 36 to 39ft (11 to 12m) and a weight of 165 to 190lb (75 to 86kg) have flown? Its muscle power can scarcely have been sufficient for continuous powered flight. It too was a highly specialized soarer.[26]

---

19 Abel, O., 1927. *Lebensbilder aus der Tierwelt der Vorzeit.* 2. Auflage, 714 pp.; G. Fischer Verlag, Jena.
20 Holst, E. von, 1957. *Der Saurierflug.* Paläontologische Zeitschrift, 31: 15-22; Stuttgart.
21 Kripp, D. von, 1943. *Ein Lebensbild von Pteranodon ingens auf flugtechnischer Grundlage.* Nova Acta Leopoldina, N.F. 12, Nr. 82: 217-246; Halle.
22 Heptonstall, W.B., 1971. *An analysis of the flight of the Cretaceous pterodactyl Pteranodon ingens (Marsh).* Scottish Journal of Geology, 7(1): 61-78.

23 Bramwell, C.D., 1970. *The first hot-blooded flappers.* Spektrum, 69: 12-14; Oxford.
Bramwell, C.D., 1970. *Those flappers again!* Spektrum, 72: 7; Oxford.
Bramwell, C.D. and Whitfield, G.R., 1970. *Flying Speed of the Largest Aerial Vertebrate.* Nature, 225, No. 5233: 660-661.
Bramwell, C.D. and Whitfield, G.R., 1974. *Biomechanics of Pteranodon.* Philosophical Transactions of the Royal Society of London, B, 267: 503-592.
24 Stein, R.S., 1975. *Dynamic analysis of Pteranodon ingens: a reptilian adaptation to flight.* Journal of Paleontology, 49(3): 534-548.

25 Brower, J.C., 1980. *Pterosaurs: How they flew.* Episodes, 1980 (4): 21-24.
Brower, J.C., 1982. *The Aerodynamics of an Ancient Flying Reptile.* Syracuse Scholar, 45-57; Syracuse, N.Y.
Brower, J.C., 1983. *The Aerodynamics of Pteranodon and Nyctosaurus, two Large Pterosaurs from the Upper Cretaceous of Kansas.* Journal of Vertebrate Palaeontology, 3(2): 84-124.
Brower, J.C. and Veinus, J., 1981. *Allometry in Pterosaurs.* University of Kansas Palaeontological Contributions, 105: 1-32; Lawrence.
See also:
Cox, B.C., 1980. *Trimming the pterosaur's wings.* Nature, 284: 400-402.
26 Langston, Jr., W., 1981. *Pterosaurs.* Scientific American, 244 (2): 122-136.

Overall pterosaur flying ability must have been very varied. The early Rhamphorhynchoidea still had relatively short, broad wings, the later ones already had narrow, long wings. Their long tail with a vertical terminal vane served as a rudder, probably as a drag rudder, to stabilize the animal in flight. They always had a relatively broad sternum with cristospina for the origin of powerful flight muscles. They were capable of continuous flapping flight. Later, advanced long-tailed pterosaurs, like *Rhamphorhynchus* of the Jurassic, had long, narrow wings, and thus good soaring ability, combined with low weight, calculated at 1·07lb (484g) for a *Rhamphorhynchus* with a wing span of 2·9ft (89cm).[27] A herring gull of similar span weighs more than double.

Pterosaurs' very low weight is due mainly to the fact that their bodies were relatively small in comparision with their wing span. Thus the trunk length of *Rhamphorhynchus* mentioned above is only about 4in (10cm) with a wing span of 2·9ft (89cm). Proportional differences are even more extreme in the large pterosaurs of the Cretaceous. An *Anhanguera* from the Lower Cretaceous had a trunk only 9·5in (24cm) long, but a wing span of 13·6ft (4·15m), a difference of more than seventeen times.

The small Pterodactyloidea of the Jurassic were short-tailed, powered flapping fliers. They were less stable in flight, but more able to manoeuvre, and more agile.

The large Cretaceous pterosaurs were all short-tailed. They were perfectly adapted to continuous gliding and soaring flight; they weighed very little, flew very slowly and thus used little energy. They were ideally built for gliding and soaring under mild and calm climatic conditions.

Locomotion on the Ground

The problem of how pterosaurs were able to move on the ground is almost as old as the discovery of the first fossil remains of these flying reptiles.[28] Connected with this of course is also the question of how they took off from the ground, and how they could land on the ground.

Often the problem of pterosaur locomotion was polarized into two contrary views: bat-like versus bird-like, in other words quadruped versus biped. Hankin and Watson saw pterosaurs as completely helpless once they were on the ground: 'Perhaps the most feasible method of progression for them on land is that, having alighted on their feet, they fell over on their stomachs and pushed themselves along, after the manner of penguins, by means of the hind legs, perhaps with the occasional slight lift from the wings for surmounting an obstacle.'[17]

Othenio Abel, at the time the leading authority in the field of palaeobiology and the reconstruction of fossil vertebrates, explained: 'When a *Pterodactylus* moved on the ground, which can in any case only have happened very rarely, we will have to assume exactly the same position of the body for this means of locomotion as in a crawling bat, with the belly resting on the floor and only raised when the hind legs are pushed under the body and the rear part of the body thus lifted a little.'[29]

The opposing view was put by Carl Stieler, who thought the long-tailed Liassic pterosaur *Dorygnathus* capable of reaching the necessary speed for take-off by running with short steps (but with its legs wide apart) on its toes.[30] And Kevin Padian discussed the function of the pelvis and the hind legs of the Liassic pterosaurs *Dimorphodon* and *Campylognathoides*, and came to the conclusion that all pterosaurs were bipedal and did not crawl on all fours like bats, but could walk on two legs like birds.[3]

What characteristics of fossil pterosaur skeletons are significant for the one or the other hypothesis? To answer this we must first examine the structure of the pelvis and the hind legs.

Pterosaurs all have a very reptilian pelvis with an elongated ilium, and beneath this a broad plate of bone made up of the ischium and pubis fused, the so-called ischio-pubic plate. This is more like the construction of an archosaur pelvis as in *Euparkeria* than the pelvis of a bird. The primitive *Euparkeria* from the Triassic as a rule had sprawling or semi-erect stance and gait. The pelvic muscles were used both to draw the hind legs up to the body and to swing the lower leg backwards or forwards.

In birds the pelvis is markedly elongated, and ischia and pubes are not fused at the bottom, but wide open. The pelvic muscles work in such a way that the bird is suspended over its hind legs as in a seesaw, and the legs swing forwards and backwards. Thus the femur has to be articulated in the hip socket in such a way that it can be moved in this vertical plane parallel with the longitudinal axis of the body. In birds this is possible because the articular head of the femur is set almost at a right angle inwards from the bone shaft, and the sideways oriented hip socket is covered by a bony protuberance. Thus the weight taken by the hind legs is absorbed at the top.

Rhamphorhynchus the Quadruped (left) Here we see *Rhamphorhynchus* as reconstructed in a quadrupedal stance and gait, after the idea of Manfred Reichel. The ungainly pose suggests somewhat clumsy terrestrial movements.

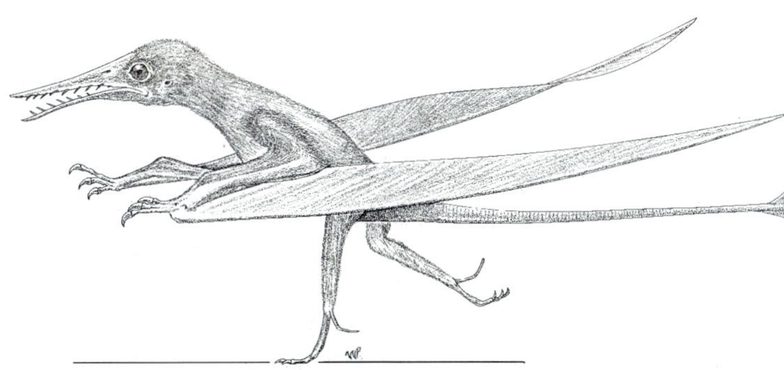

Rhamphorhynchus the Biped (left) Here, by contrast, is *Rhamphorhynchus* reconstructed as a biped. The hind legs are positioned under the body in bird-like manner enabling the pterosaur to run on two legs. The wings, however, could not have been folded as closely to the body as in birds.

Right: As this picture of a roadrunner (*Geococcyx californianus*) clearly shows, birds are digitigrade; that is, they walk on their middle toes rather than on the soles of their feet. This makes them capable of quite rapid motion on two legs. The situation with pterosaurs is different. The structure of their hip joints and feet is not well adapted to motion on two legs. It is more likely that they moved on four legs on the ground.

27 Wellnhofer, P., 1982. *Zur Biologie der Flugsaurier.* Natur und Museum, 112 (9): 278-291; Frankfurt.

28 Seeley, H.G., 1870. *The Ornithosauria: An Elementary Study of the Bones of Pterodactyles.* 130 pp.; Deighton, Bell and Co., Cambridge. In this study Seeley gave a short review of the various points of view in the controversy. He noted that Soemmerring (1812) regarded *Pterodactlyus* from the Solnhofen limestone as an unknown kind of bat with comparable locomotion on the ground, and also quoted Goldfuss (1831) who argued for quadrupedal, bat-like locomotion, as follows: 'This animal was enabled by means of the pelvic bones and the long hind-legs to sit like the squirrels. We should regard this position as natural but for the long wing-finger hanging far down the sides. If it were to creep along it would have the same difficulties as a bat, and the length and weight of the head, as well as the proportional weakness of the hind limb, make it improbable that they progressed by leaping. These animals made use of their claws only to hang on to rocks and trees and to climb up steep cliffs . . .'
An early exponent of the opposing theory of bipedal, bird-like walking, a view strongly supported by Seeley himself, was Quenstedt (1855), who thought 'that the animal was able to walk upright, being probably still more upright than birds, since the great disproportion between the neck on the one hand, and the thigh on the other, could not have allowed a more appropriate position.'

29 Abel, O., 1925. *Geschichte und Methode der Rekonstruktion vorzeitlicher Wirbeltiere.* 327 pp.; G. Fischer, Jena.

30 Stieler, C., 1922. *Neuer Rekonstruktionsversuch eines liassischen Flugsauriers.* Naturwissenschaftliche Wochenschrift, N.F. 21, Nr. 20: 273-280; Jena.

Furthermore birds walk on their three middle toes, thus they are digitigrade, rather then plantigrade, which means walking on the soles of the feet. Thus many birds are capable of running rapidly and effectively, for example when escaping from danger or to achieve the speed necessary for take-off.

If we consider the construction of pelvis and hind legs in pterosaurs, we find that things are quite different. The hip sockets in the pelvis are not only oriented sideways, but also somewhat upward and backwards. There is no rim above the hip socket to support the femur. The articular head of the femur is never at right angles, as in birds, but in the best case bent at 120° to the bone shaft, usually at 130° to 160°. If the femur of a pterosaur is placed in the hip socket of the pelvis, the bone is splayed out and cannot be moved into a vertical position.[31]

However, the pterosaur femur was relatively mobile, and could rotate backwards into the horizontal plane, i.e. into the plane of the wings. This is the position that was adopted in flight, in order to spread the flight membrane, to give it camber and by alternate raising and lowering to steer the animal around its roll axis. The front edge of the hip socket is thickened to support the articular head. But the femur could also rotate to the side and the front for quadrupedal locomotion on the ground or for landing and climbing on trees and cliffs.

The knee joint also seems to have been relatively flexible and allowed the lower leg to turn inwards and the foot to move forwards and backwards parallel to the middle vertical plane when striding. The whole foot met the ground, thus the gait was plantigrade. The digits were of unequal length, the last penultimate phalanx being long, and they had sharp, pointed claws; they are very different from the three-toed foot of bipedal dinosaurs and birds.[32] These creatures also walk only on their toes, and are thus digitigrade. The middle digit is always the longest in such cases, and the first digit is reversed, i.e. directed backwards. The structure of the skeleton of the foot is thus fundamentally different in birds and pterosaurs.

Pterosaurs' long foot digits always have sharp claws with needle-fine points. They seem to have been better suited to gripping and climbing than walking on the ground.

A more recent reconstruction of the pelvis and hind legs of a pterodactyloid from the Brazilian Santana Formation suggests that the animal was biped, with an upright body, standing almost like a penguin.[33] This assumption refers only to short-tailed pterosaurs, but does run into great difficulties as far as the position of the wings is concerned.

Pterosaurs could not draw their wings close to the body and fold them, as birds can. The

31 Wellnhofer, P. and Vahldiek, B.-W., 1986. *Ein Flugsaurier-Rest aus dem Posidonienschiefer (Unter-Toarcian) von Schandelah bei Braunschweig.* Paläontologische Zeitschrift, 60: 329-340; Stuttgart. Wellnhofer, P., 1988. *Terrestrial Locomotion in Pterosaurs.* Historical Biology, 1: 3-16.

32 Unwin, D.M., 1987. *Pterosaur Locomotion. Joggers or Waddlers.* Nature, 327: 13-14. Unwin, D.M., 1988. *New remains of the pterosaur Dimorphodon (Pterosauria: Rhamphorhynchoidea) and the terrestrial ability of early pterosaurs.* Modern Geology, 13: 57-68.

33 Bennett, S.C., 1990. *A Pterodactyloid Pterosaur from the Santana Formation of Brazil: Implications for Terrestrial Locomotion.* Journal of Vertebrate Paleontology, 10(1): 80-85.

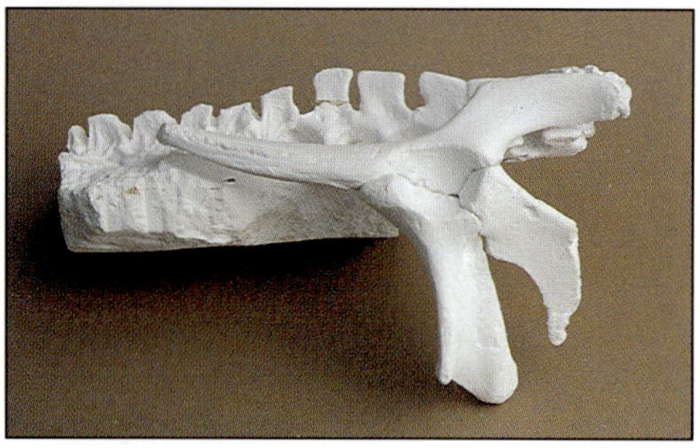

Above: The pelvis of *Anhanguera* was able to be reconstructed three-dimensionally in plaster. The hip socket points sideways and upwards, a position that would have made vertical orientation of the upper leg impossible.

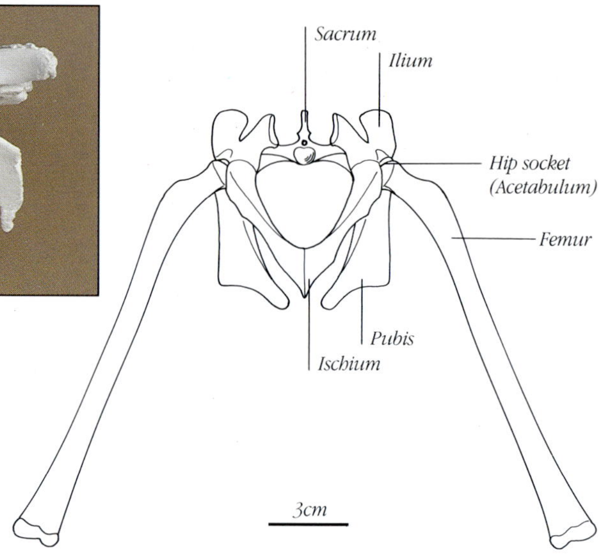

The Pelvis of Anhanguera (above)
The reconstructed pelvis of *Anhanguera* is seen here from behind with the upper leg bones articulating in the hip sockets. This is the maximum extent to which the upper legs can be moved into the vertical plane. Evidently *Anhanguera*'s legs would have been splayed out sideways if it had tried to stand bipedally.

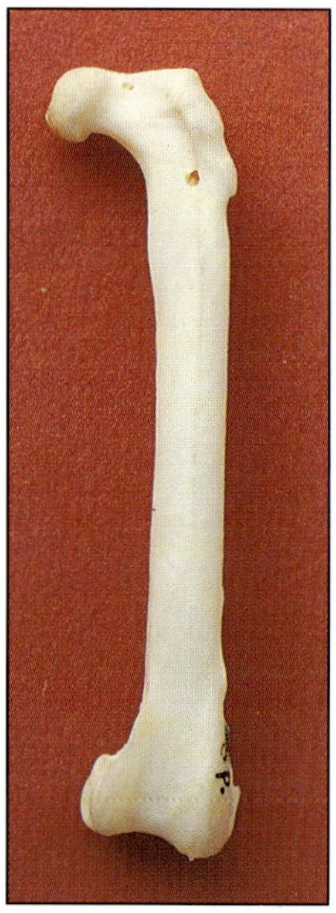

Above and above right: These photographs allow us to compare the femur of a pterosaur (*Rhamphorhynchus*, above) with that of a contemporary bird, a toucan (above right). Careful comparison reveals the different orientation of the femur head which articulates in the hip socket. In the case of the pterosaur the head is set at only a slight angle to the shaft, whereas in the bird it is set at a right angle. This makes it possible for the bird's legs to be swung vertically beneath its body, enabling it to walk and run bipedally.

Anhanguera the Quadruped (above)
This drawing shows *Anhanguera*, a pterodactyloid pterosaur from Brazil with a wing span of about 13ft (4m), restored in a quadrupedal stance. The belly is lifted off the ground and the hind legs are splayed out to the sides.

Anhanguera the Biped (right)
This is *Anhanguera* from the early Cretaceous of Brazil restored according to the ideas of Chris Bennett. By adopting a rather steep position of the body, the pterosaur might have been able to walk bipedally in an upright posture. However, the fact that its wings could not have been folded as closely to its body as those of birds today, and the difficulty of balancing its large head in this position casts doubt on the likelihood of this kind of two-legged locomotion on the ground.

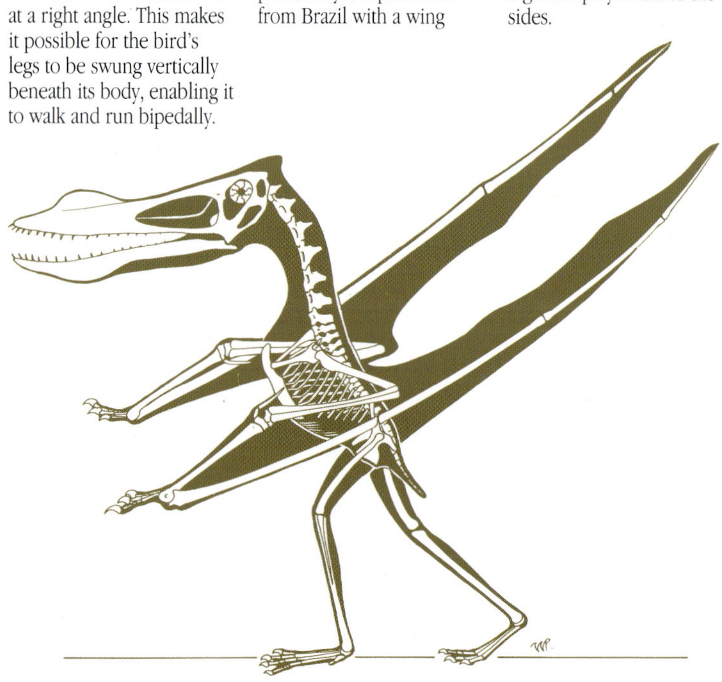

only point at which the wing could be folded back was at the pulley joint between metacarpal and flight digit. Even if the humerus was pulled as far back and the elbow and hand joint flexed as far as possible, bipedal pterosaurs must to an extent have had to walk with arms outspread.

It is obvious that this required both an open space as habitat and also that the creatures must have had considerable balancing problems on the ground. A small gust of wind would have been enough to blow them over.

For all these reasons we must assume that pterosaurs, if they ever landed on level ground, must have moved using four feet, in other words in a quadrupedal fashion. Of course the forelimbs were not well adapted for this, on the one hand because the humerus could not be moved into a vertical position, and on the other because the hand with its three small, clawed fingers was better adapted to climbing than to walking. But the humerus could rotate forward around its longitudinal axis, thus bringing the lower arm and above all the metacarpus into a vertical position. Perhaps pterosaurs did not support themselves on their small fingers, but on the knuckles, with the flight digit folded back.

Thus we have a picture of a quadrupedal pterosaur bracing its body upwards when walking, pushing the body forwards with laterally directed hindlimbs and moving forwards slowly, somewhat clumsily, with its broad and bulky forelimbs. This was certainly the exception and not the rule, but was needed for finding hiding places, when breeding, when building a nest and when looking after young.

Thus pterosaurs were not bat-like on the ground, but pterosaur-like. Between the early rhamphorhynchoids with their relatively long hind legs, and the later forms with hind legs that had developed rather less strongly, there were differences in proportion suggesting that locomotion on the ground was quadrupedal in all cases, but better executed in some than in others. Pterosaurs' hind legs were primarily an aid to flight, because the upper leg at least was integrated into the flight membrane, and fulfilled an important aerodynamic function.

Pterosaurs were adapted to flight to an extreme extent. They spent their active life in the air and were presumably pretty helpless if they had to land on flat ground. Probably they simply avoided landing on a flat surface. Of course this did not condemn them to a 'Flying Dutchman existence'. They must have rested and bred on cliffs and rocks, where they could hang with their sharp, hook-shaped claws on hands and feet. To take off they only needed to swing themselves into the air. They did not need to achieve the necessary take-off speed by taking a run on two legs.

Pterosaurs could probably take off from the ground by standing on their hind legs and facing into the wind with outspread wings. A simultaneous jump and stroke of the wings raised them into the air. This was more easily possible for early pterosaurs like *Dimorphodon* and *Pterodactylus*, as they were smaller and lighter and had relatively long hind legs.

As a rule however, they too, like the great pterosaurs of the Cretaceous, probably took off when hanging from a raised point by their strong digital claws. It is not certain whether they could also hang head down on branches and rocky protrusions, like bats. The long-tailed Rhamphorhynchoidea would have found their long, stiff tails a great hindrance in doing this. Possibly the small pterosaurs of the Jurassic, as reconstructed very convincingly by Othenio Abel, could have rested in this position.[11]

Direct proof of the nature of pterosaurs' terrestrial locomotion would be provided by their tracks preserved in fossil form. Palaeoichnology is the branch of science concerned with the study of such track impressions, which are known in the cases of many reptiles, including dinosaurs. Fossil trackways provide much information about the maker of the footprints, e.g. whether it walked on two or four legs.

On All-Fours (below)
This is a skeletal reconstruction of *Anhanguera* in a quadrupedal stance. This pterosaur had a wing span of 13ft (4m), and the drawing shows the marked disproportion between fore and hind legs. Such an animal would have moved clumsily on all-fours, but locomotion on the ground would have been possible.

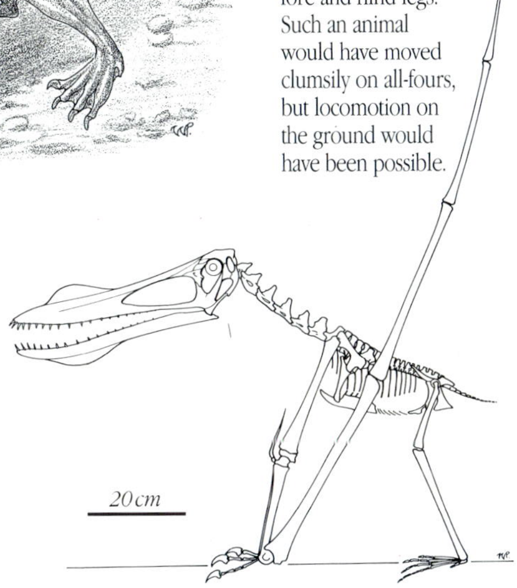

Below: On the ground a bat is almost helpless, and can only creep along on all-fours pushing its body forward on its stomach. Pterosaurs could certainly move more confidently on the ground than bats. They could lift their bodies off the ground, at least, and walk in a semi-erect manner on four legs.

Above: That *Pterodactylus* did not move like a bat on the ground is shown by this life restoration suggested by the ideas of Manfred Reichel. Walking on the ground must have been possible on all-fours, although the forelimbs, especially the hands, were designed for climbing rather than walking.

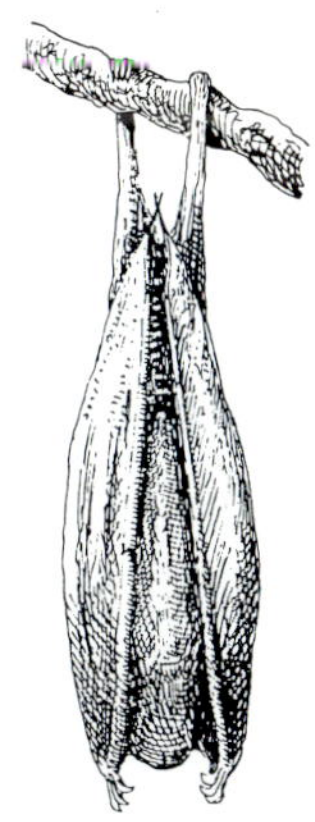

Just Hanging Around (left)
Othenio Abel also published this life restoration showing a sleeping *Pterodactylus* enveloped in its wing membranes and hanging upside down from a branch by its feet in a bat-like manner. This seems a convincing posture for a short-tailed pterosaur like *Pterodactylus*. A long-tailed rhamphorhynchoid would have had greater problems, however.

An Upside-Down Existence (below)
The manner in which the flying fox (*Pteropus celaeno*, a species of fruit bat) can climb upside down (below right) led the Viennese palaeobiologist Othenio Abel to the suggestion that *Pterodactylus* (below) could have climbed in the same way. It is shown in this drawing from 1925 hanging upside down from the branches of a tree, holding on securely with the claws of its feet and hands.

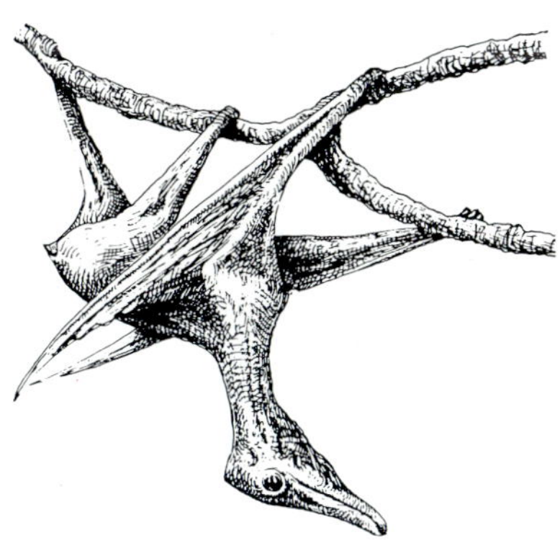

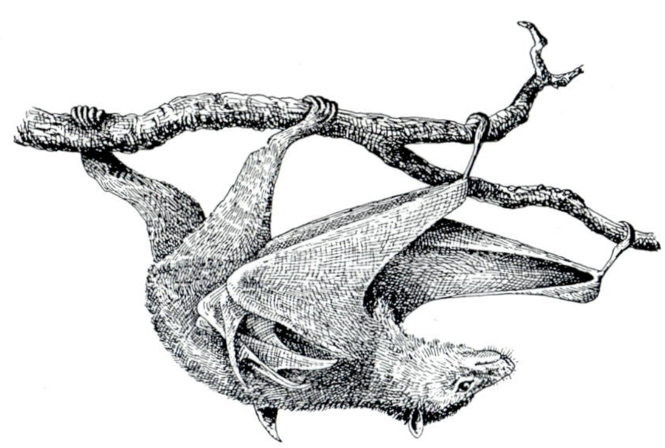

Rhamphorhynchus on All-Fours (left)
One of the earliest restorations of *Rhamphorhynchus* walking on all-fours was this one by Riou that was reproduced in L. Figuier's *La Terre avant le Déluge*, a book published in Paris in 1863. This interpretation was inspired by fossil trackways found in the Solnhofen limestone in Bavaria which seemed to show foot and tail prints of pterosaurs.

Left: *Pterodactylus* about to take off from its resting position. This painting by Neave Parker assumed that this pterosaur could hang upside down; to take off, it only needed to drop from the branch, develop forward speed and become airborne.

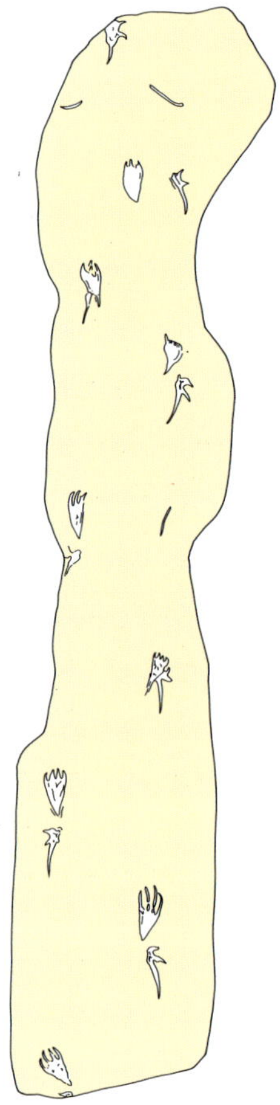

A Fossil Track (right)
This is the supposed pterodactyle trackway of *Pteraichnus* (wing track) identified by W.L. Stokes from the Morrison formation of Arizona.-Later the tracks were interpreted as those of a crocodilian rather than those of a pterosaur.

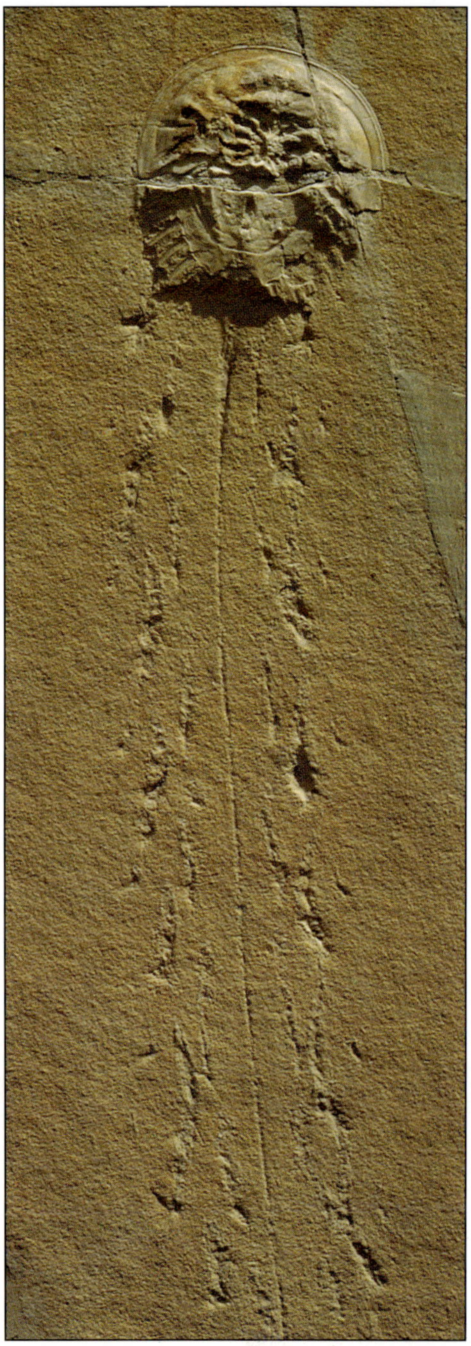

Left: This fossil track helped the American K.E. Caster clear up further confusion. The trackmaker was the horseshoe crab, which is here fossilized at the end of the track. The middle line was made by its spiny tail, and not by that of *Rhamphorhynchus* as previously believed.

Even in the nineteenth century tracks on the surface of the Solnhofen limestone were interpreted as foot- and tailprints of pterosaurs.[34] It later emerged that they were made by the horseshoe crab: the creature itself is sometimes found at the end of its trail.

Various fossil footprints from the North American Jurassic have been interpreted as pterosaur tracks. In 1957 W.L. Stokes described a track from the Upper Jurassic Morrison strata of Arizona as *Pteraichnus saltwashensis*, and ascribed it to a quadruped pterodactyloid.[35] Later however it was convincingly argued that these tracks were possibly not made by a pterosaur, but by a member of the crocodile family.[36] It has also been proved that other ostensible pterosaur tracks could not have been made by pterosaurs.[37]

Could Pterosaurs Swim?

H.G. Seeley was the first to propose that pterosaurs could swim. As a proof of this, webs can be seen between their long toes: they survived as prints in some of the Solnhofen specimens.[38]

It is to be expected that flying animals that feed on fish must inevitably have to come into contact with water, and are sometimes compelled to settle upon it. It can hardly be assumed that pterosaurs are an exception to this rule and fed only on flying fish, which they caught in flight just above the surface of the

Below: This is a close-up of one of the *Pteraichnus* tracks (see drawing above), which was interpreted as the imprint of a four-toed foot and three fingers of the hand of a pterodactyle. However, one would expect pterosaur hands to be more widely spaced than this.

34 Figuier, L., 1863. *La Terre avant le déluge.* Paris.
35 Stokes W.L., 1957. *Pterodactyl tracks from the Morrison Formation.* Journal of Paleontology, 31: 952-954.
Stokes, W.L. and Madsen Jr., J.H., 1979. *Environmental significance of pterosaur tracks in the Navajo Sandstone (Jurassic) Grand County, Utah.* Brigham Young University Studies, 26 (2): 21-26.
Logue, T.J., 1977. *Preliminary investigation of pterodactyl tracks at Alcora, Wyoming.* The Wyoming Geological Association Earth Science Bulletin, 10 (2): 29-30.
36 Padian, K. and Olsen, P.E., 1984. *The fossil trackway Pteraichnus: not pterosaurian, but crocodilian.* Journal of Paleontology, 58 (1): 178-184.
37 Unwin, D.M., 1989. *A Predictive Method for the Identification of Vertebrate Ichnites and its Application to Pterosaur Tracks.* In Gilette, D.D. and Lockley, M.G. (eds.) *Dinosaur Tracks and Traces,* Cambridge University Press: 259-274.
38 Broili, F., 1927. *Ein Exemplar von Rhamphorhynchus mit Resten von Schwimmhaut.* Sitzungsberichte der Bayerischen Akademie der Wissenschaften, math.-naturwiss. Abt., 1927: 29-48; Munich.
Döderlein, L., 1929. *Über Rhamphorhynchus und sein Schwanzsegel.* Sitzungsberichte der Bayerischen Akademie der Wissenschaften, math.-naturwiss. Klasse, 1929: 1-46; Munich.

water. It is even suggested that some long-tailed pterosaurs dived into the water like gannets and hunted fish beneath the surface.[39] But the argument against this is that pterosaurs could not hold their wings as close to their bodies as divers today. Thus the flight membrane would have been a great hindrance. Who would put on a flapping raincoat then dive in for a swim?

Fishing in flight just above the water was probably achieved by a rapid forward thrust of the long head and seizing the prey with the pointed jaws. In the course of this the tip of the beak dipped in the water at times and was drawn through a stretch of water. Many of the large pterosaurs, like *Tropeognathus* for example, even developed keel-like bony crests on their upper and lower jaw, in order to be subject to the lowest possible water resistance during this phase.

Another kind of fishing was achieved by skimming the surface of the water with the lower jaw, the long, narrow horn sheath of which dipped into the water in flight, rather like the modern black skimmer.

If pterosaurs landed on the water in the course of fishing, they could raise themselves into the air again from the crest of a wave with frog-like, simultaneous swimming thrusts of their feet. The foot digits were opened wide during this procedure to spread the webs. In the rhamphorhynchoids the very long, sideways pointing fifth digit spread an additional web which allowed the feet to function as an effective paddle. It is possible that the long tail was also used as a paddle in the water, but less probable that it was of any use when taking off from the water. The vertically oriented tail membrane was unsuitable for this.

39. Wild, R., 1978. *Die Flugsaurier (Reptilia, Pterosauria) aus der oberen Trias von Cene bei Bergamo, Italien.* Bolletino della Società Paleontologica Italiana, 17 (2): 237.

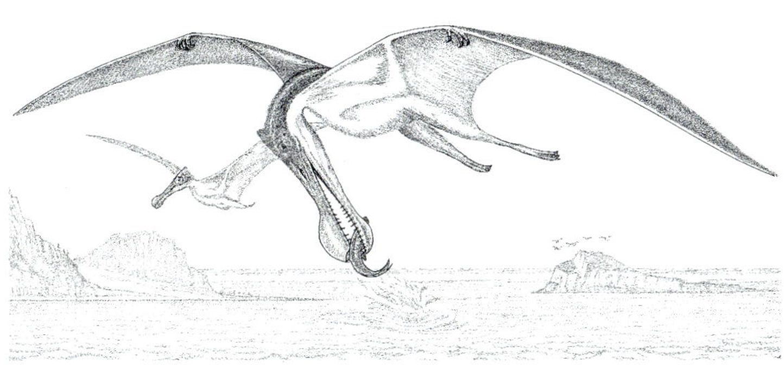

Fishing on the Wing (right)
This scene shows *Tropeognathus* fishing near the coast of the early Cretaceous sea in South America. It is thought that the bony crests that were developed on its upper and lower jaws served to minimize hydrodynamic resistance as its beak dragged through the water when it plucked a fish from beneath the surface.

Feeding Habits

Most pterosaurs so far known lived near water, on the sea coast and on islands and lakes. They fed on aquatic organisms. Probably they populated the coasts of the Mesozoic seas.

Judging by their dentition, some of them must have been specialized feeders, like for example the Jurassic pterosaurs *Ctenochasma, Gnathosaurus* and *Huanhepterus*. They had numerous long, slender teeth in their long jaws, and used them to filter small aquatic organisms like the larvae of crustaceans out of the water. An extreme example of this was the short-tailed pterosaur *Pterodaustro* from the Lower Cretaceous of Argentina. Its lower jaw had a set of long, bristle-like 'teeth', which must have functioned like the baleen of a whale. The jaw is bent upwards at the front and formed what amounted to a filter basket. *Pterodaustro*'s upper jaw contained only short, blunt teeth, used to break down the content of this filter basket into even smaller pieces.

It is not likely that these filter feeders collected their food in flight, pulling the lower jaw through the water like a fishing net. The forces exerted would inevitably have forced the lower jaw backwards, causing stalling. It is more probable that filter feeders simply stood in the water and dabbled for small animals and algae, as flamingos do.

Fossilized stomach contents of some pterosaurs such as *Eudimorphodon, Rhamphorhynchus, Pterodactylus* and *Pteranodon*, have been found, these being the remains of the last meal before they died. In all cases they are remains of fish. In the case of the Triassic pterosaur *Eudimorphodon* the contents of the stomach were the hard, shiny scales of small ganoid fish.[39] In many cases the teeth of this long-tailed pterosaur are extraordinarily worn by chewing. This is easily explained by the fact that the hard armour of scales had to be bitten through when the fish were caught. The indigestible hard parts, bones and scales, were spat out again as pellets.

Right: A foot skeleton of *Pterodactylus* from Solnhofen with impressions of webs between the long toes preserved. The toes could be widely spread and formed a sturdy paddle which might propel the pterosaur on water.

Webbed Feet (below)
The foot skeleton of *Rhamphorhynchus* (left) and *Pterodactylus* (right). The shaded area shows where the imprints of webs were preserved.

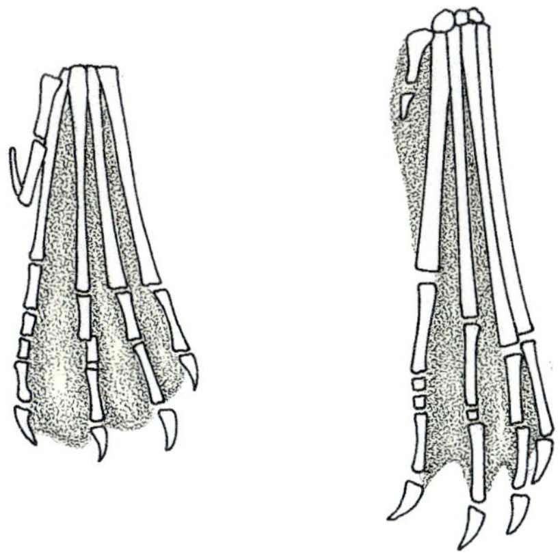

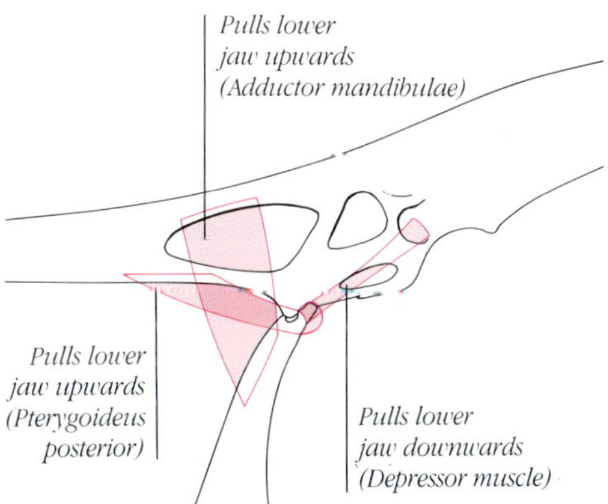

Jaw Muscles (above)
The jaw musculature of pterosaurs consisted of especially strong adductor muscles for pulling the lower jaw upwards. This was vital to close the beak when fishing on the wing. The joints of the jaw permitted the beak to be opened to a considerable width.

Below: Some layers of Solnhofen limestone are covered by small fishes of the genus *Leptolepides*. These must have been an easy prey for pterosaurs fishing on the wing. The indigestible hard parts of those bony fishes were probably vomited up as pellets by pterosaurs after the nutritious parts had been digested. The hard scales covering such fish were evidently quite hard to bite through, as the teeth of some fish-eating pterosaurs show signs of much wear.

Left: It is very rare to find stomach contents fossilized with the skeleton of a pterosaur. However, in the case of this small *Rhamphorhynchus* from Solnhofen, its last supper is preserved with the body. The remains of a small fish, probably the common bony fish *Leptolepides* which is pictured above, can be identified. A drawing of this specimen which identifies the various elements is included on the following page.

The dsungaripterids from the Lower Cretaceous of Asia, *Dsungaripterus* and *Phobetor*, with their tweezer-like pointed beak and strong crushing teeth either fed on hard-scaled fish as well, or like shore birds looked for molluscs, snails and crabs, whose hard shells they broke open with their strong teeth.

Half-digested remains of a small bony fish that had apparently been swallowed whole, head first, were found in the stomach of a *Rhamphorhynchus* from the Solnhofen limestone. There were also other remains of food in the stomach of the pterosaur, in the form of small, elongated objects that have so far not been identified. They suggest that *Rhamphorhynchus* did not eat only fish.

Remains of fish were also found in the great *Pteranodon* from the Cretaceous of Kansas. Barnum Brown, at the time Curator Emeritus of the American Museum of Natural History in New York, reported on this in 1943: 'A lower jaw recently prepared in the American Museum contained the remains of a last supper — backbones of two species of fishes and the joint of a crustacean, lying in the position of the throat pouch when death overtook the animal.'[40]

The jaws of fish-eating pterosaurs are always long. The *Rhamphorhynchus* species have long teeth directed forwards, making rapid seizure of slippery fish possible. It is striking that the front ends of the long jaws of many pterosaurs are very narrow and laterally compressed, in both the toothed and toothless forms. The tips of the jaws must have had horn beaks above them that have not survived in fossil form.

Examination of the mechanics of the articulation of the jaw have shown that pterosaurs could open their beaks very wide, and that when this happened the branches of the jaw splayed out, probably to increase the size of the throat pouch opening, as in the pelican.[41] Perhaps food was predigested in this, or stored, and brought back to the nest to feed the young.

Many pterosaurs seem to have fished with the lower jaw hanging, like the black skimmer, and many others by rapidly thrusting their pointed jaw into the water. *Pterodactylus* with its very long slender jaws and short, cone-

40 Brown, B., 1943. *Flying Reptiles*. Natural History, 52 (3): 104-111; American Museum, New York.
41 Wellnhofer, P., 1980. *Flugsaurierreste aus der Gosau-Kreide von Muthmannsdorf (Niederösterreich) – ein Beitrag zur Kiefermechanik der Pterosaurier*. Mitteilungen der Bayerischen Staatssammlung für Paläontologie und historische Geologie, 20: 95-112; Munich.

Fishing Techniques (below)
Some pterosaurs, such as *Rhamphorhynchus*, may have used fishing techniques similar to those practised by the black skimmer, which ploughs its lower jaw through the surface of the water during flight.

Above: This skull of *Rhamphorhynchus* from the Solnhofen limestone shows its long, curved, sharply pointed, intermeshing teeth which must have formed an extremely effective gripping tool for snatching slippery prey, like fish, out of the water. The jaws of fish-eating pterosaurs are always long like this.

Left: The lower jaw of *Rhamphorhynchus muensteri* is equipped with long teeth which point forwards, and has a pointed front end which was probably covered by a horny beak. This shape of jaw suits a creature that must stab it in the water after fish.

Above: The lower jaw of *Rhamphorhynchus longicaudus* has a flattened, laterally compressed front end, which suggests that it served the same purpose as that of the black skimmer which dips its lower jaw in the water as it skims over the surface.

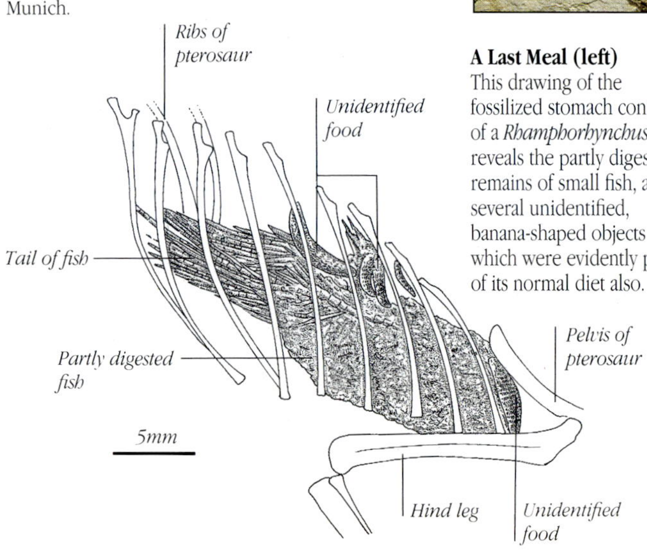

A Last Meal (left)
This drawing of the fossilized stomach contents of a *Rhamphorhynchus* reveals the partly digested remains of small fish, and several unidentified, banana-shaped objects which were evidently part of its normal diet also.

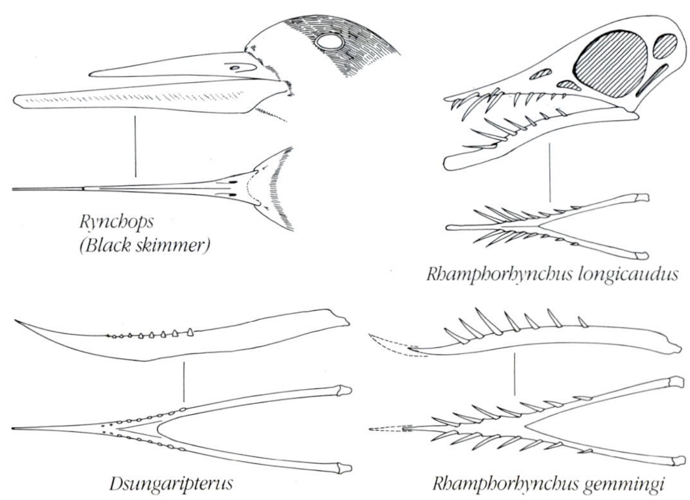

Above: Fishing pterosaurs had a throat pouch similar to pelicans for storing and predigesting fish, and possibly carrying it long distances to feed their young. Impressions of skin below the lower jaw of this *Pterodactylus* indicate the presence of such a throat sack. When its jaws were opened, the branches of the lower jaw splayed out causing the pouch to open.

Above: Pterosaurs with broad mouths and short heads, such as *Anurognathus* and *Batrachognathus*, were probably insect feeders. Fossil insects such as this dragonfly (*Cymatophlebia*) are abundant in the same fossil deposits.

Specialized Beaks (left)
The lower beak of the black skimmer is laterally compressed and flattened, and extends beyond the upper beak. So, the knife-like front end of the lower beak encounters little resistance when cleaving through the water for small fish and crustaceans. The front end of the lower jaw of *Rhamphorhynchus longicaudus* is similar in shape. Other pterosaurs, like *Rh. gemmingi* or *Dsungaripterus*, also exhibit lower jaws whose front ends are elongated and pointed.

shaped teeth could also have fished, but probably lived mainly on small marine creatures. R.T. Bakker even thought it possible that it was an 'airborne worm tweezer'. 'It may well have probed the sand flat like a Jurassic sandpiper, poking its long snout into the burrows of polychaete worms, shrimp-like crustaceans, and sand fleas.'[42]

Whether the giant Texan pterosaur *Quetzalcoatlus* was a carrion eater who fed on dinosaur carcasses has never been proved, and is in fact rather improbable. Its long neck with extremely elongated cervical vertebrae was certainly not as flexible as would have been desirable for vulture-like gutting of dinosaur cadavers. Its long, tweezer-pointed and toothless jaws are much more suggestive of a diet of fish, though against this assumption must be set the great distance of the Texan fossil sites from the coast at the time, which was 250 miles (400km) away. Wann Langston therefore thought that *Quetzalcoatlus* used its slender beak to probe for molluscs and arthropods in shallow flood basins. There are numerous traces of burrowing animals in the strata in which fossils were found.[43]

A third group of nutrition specialists among the pterosaurs were the insectivores. Modern vertebrates that catch flying insects all have a broad mouth, like microchiroptera, or small bats, and various insectivorous birds like swallows, swifts and nightjars. Two Upper Jurassic pterosaur genera, *Anurognathus* and *Batrachognathus*, have a tall, short skull and a broad mouth slit with short, peglike teeth. It is to be assumed that both genera fed on flying insects which they caught in their mouths in flight. They must have been agile and skilful fliers to do this. There was certainly no lack of insects. Numerous insects, including mayflies (Ephemoptera), dragonflies, cicadas, beetles, wood wasps, caddice flies and flies, have been found in the same strata in the Solnhofen limestone and also in the Karatau lake deposits in Kazakhstan.

So far there are no known fruit-eating pterosaurs. It has been assumed that, especially in the Cretaceous when the first higher flowering plants appeared, it was pterosaurs in particular who distributed the seeds, as bird fauna were still relatively sparse at the time. Such fruit-eating pterosaurs probably lived further inland in higher regions subject to weathering and erosion, and thus with less chance of fossilization. If there were fruit-eaters — which is highly probable — we are unlikely ever to find their fossil remains.

42 Bakker, R.T., 1986. *The Dinosaur Heresies*. 481 pp.; W. Morrow and Co., Inc., New York.

Probing for Food (below)
Pterodactylus probing for a worm, after the idea of R.T. Bakker. Its narrow beak may well have been used in this way, although the walking abilities required were probably not sufficiently developed.

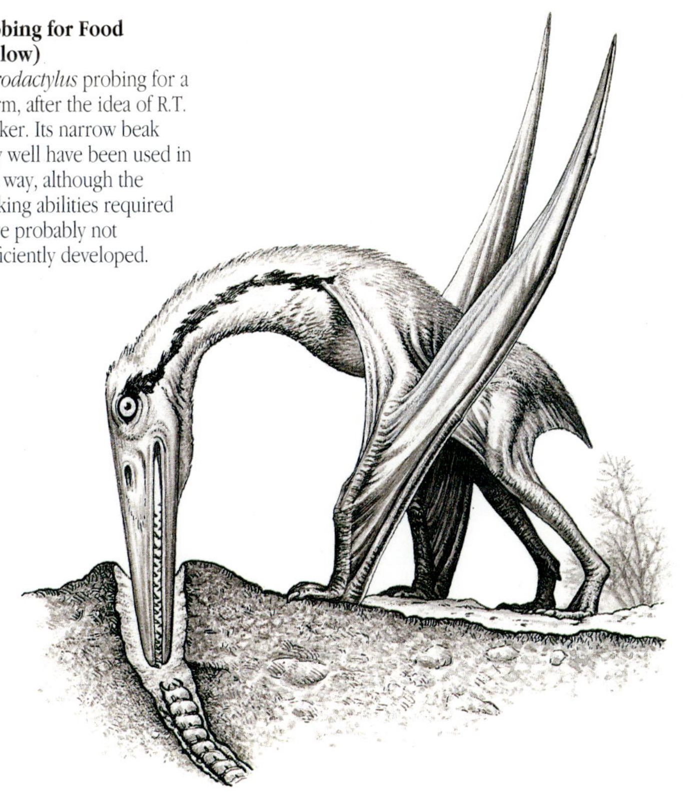

Reproduction

Did pterosaurs lay eggs, or give birth to living young? The likely answer is that they laid eggs, a method of reproduction typical of both cold-blooded reptiles and warm-blooded birds. Even in the nineteenth century some fossil eggs were interpreted as pterosaur eggs.

Thus as early as 1860 Professor J. Buckman reported the discovery of fossilized eggs in a quarry near Cirencester in Gloucestershire, England, in Great Oolite (Middle Jurassic) strata.[44] Buckman described the find 'as a cluster of the remains of at least eight eggs, of a uniformly ovate, not ovoid shape.' A single egg was 1·75in (44mm) long, and 1·1in (28mm) broad. The shell was only 0·01in (0·3mm) thick, and they seem to have been pliable. They were filled with calcite crystals. Buckman himself did not associate these eggs with pterosaurs, but he recognized they were reptile eggs, and named them *Oolithes bathonicae*.

In 1871 fossil eggs were found in the Stonesfield slate (Middle Jurassic) of Oxfordshire, England, and interpreted as tortoise eggs.[45] But as pterosaur bones (*Rhamphocephalus*) had also been found in the Stonesfield slate, H.G. Seeley assumed that they could also be the eggs of these pterosaurs. They are globular in form and about 0·75in (19mm) in diameter. They were named *Oolithes sphaericus*.

43 Langston Jr., W., 1981. *Pterosaurs*, Scientific American, 244 (2): 122-136.
44 Buckman, J., 1860. *On some fossil reptilian eggs from the Great Oolite of Cirencester*. Quarterly Journal of the Geological Society of London, 16: 107-110.
45 Carruthers, W., 1871. *On some supposed vegetable fossils*. Quarterly Journal of the Geological Society of London, 27: 443-449.
The fossil eggs from the Stonesfield Slate were originally taken to be fruit: '. . . and they so closely resembled the aspect of the ripe seed of a chestnut that it is not to be wondered that they are always placed among vegetable fossils in museums.'

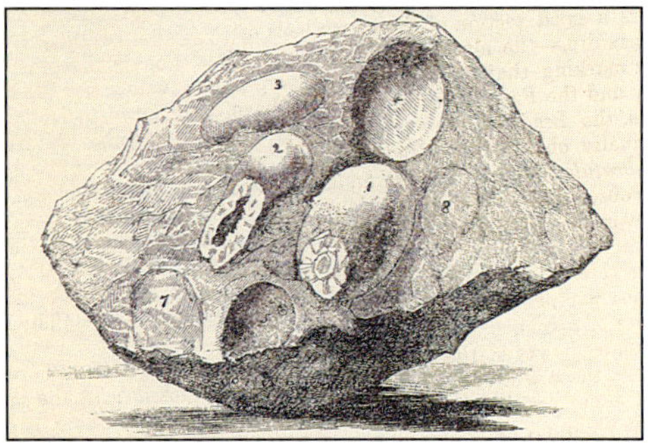

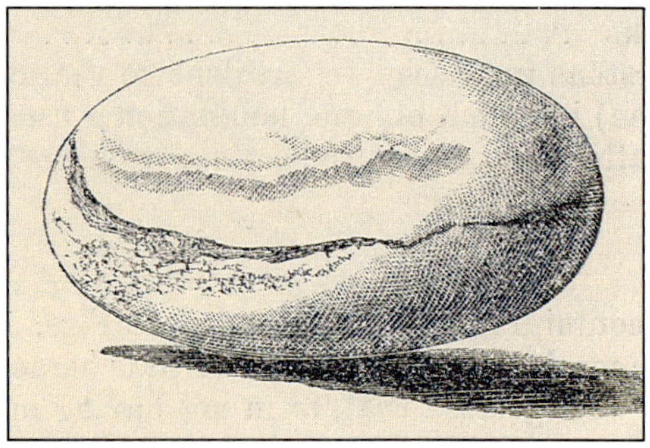

Fossil Eggs (above)
A block with eight fossil eggs from the Great Oolithe (middle Jurassic) of Gloucestershire was described by Professor J. Buckman in 1860. They were initially thought to be the fossil remains of pterosaur eggs.

A Single Egg (above right)
One of the eggs from the Great Oolithe was separated from the rock and measured accurately. There is no convincing evidence, however, that these eggs were actually laid by female pterosaurs.

Feeding the Young (right)
This reconstruction shows a conjectural posture (after a drawing by Peter Schouten, 1983) illustrating how *Ornithocheirus* may have fed its young. The parent pterosaur is seen having just returned from fishing with the prey stored in its throat pouch. The young individual thrusts its beak deep into its parent's jaws to pick the food out of the pouch. It may also be fed by food that is regurgitated by the adult. Such parental care suggests a well developed intelligence.

Examination of the shell structure of these eggs in 1928 revealed that they were certainly reptile eggs, but that it is not possible to assign them to pterosaurs.[46]

Fragments of eggshells which may come from giant pterosaurs came to light for the first time in 1989 at the site at which *Quetzalcoatlus* sp. was found, in the Upper Cretaceous sediments of the Big Bend National Park in Texas, but at the time of writing there is no positive proof that they were associated with them.

So far there have been no fossil finds of pterosaurs with eggs or embryos in their bodies, as is known in ichthyosaurs of the Jurassic. We do not even have a direct indication of the sex of the individuals, of sexual dimorphism, to enable us to decide whether we are dealing with male or female animals.

However, it is very strongly supposed that in the case of certain species of *Rhamphorhynchus* from the Solnhofen strata, of which we have enough skeleton specimens for statistical investigation, that there are two forms for the two sexes. One type has a relatively long skull and wings, the other has a relatively short skull and wings.[47] As both types occur with approximately equal frequency, one can conclude that here we can distinguish male animals, perhaps those with a longer skull and wings, and female animals, perhaps those with a shorter skull and wings.

Sexual dimorphism has also been assumed in the case of *Pteranodon* from the Upper Cretaceous of Kansas. On the basis of pelvic morphology it was possible to identify a smaller form with a smaller cranial crest and large pelvic canal as female and a large form with a large cranial crest and small pelvic canal as male.[48] The small, presumably female individuals occur three times as frequently as the large, presumably male ones.

It can also be assumed that in other Cretaceous pterosaurs as well, like *Anhanguera* from the Lower Cretaceous of Brazil, for example, the cranial crest was of a different size in male and female, and thus possibly a display structure for courtship purposes.

When one considers the small body and relatively small pelvic canal of pterosaurs, the eggs must have been relatively small as well. Probably the females could produce few, or maybe only a single, egg. It would be illogical for a flying animal, organized for extreme lightness of body, to be weighed down by a large quantity of eggs.[49]

As they were warm-blooded and hairy, it is entirely possible that they hatched their eggs themselves. This would mean that pterosaurs would have paired for a season, or maybe for their whole lives, as birds do, and that the male would have had to feed the female from his throat pouch. This highly developed social behaviour required higher intelligence than is displayed by modern reptiles, who hardly practise care for the brood at all. This assumption is well supported by the bird-like structure of the brain.

But pterosaurs could equally well have laid their eggs in a nest built in a high niche in the rock facing the sun. The warm Mesozoic climate would then have hatched the eggs. In either case, a pterosaur hatchling will have been very small, and certainly not capable of flight. Thus they could not have gone in search of food themselves. This means in its turn that the parents had to deal with bringing up their young, fetch food for them and feed them. A.J. Desmond assumes that pterosaurs must have had hatching colonies, in which the vulnerable chicks were protected by one or two guard pterosaurs.[50]

The Growing Pterosaur

It has been possible to reconstruct the growth pattern of pterosaurs on the basis of statistical investigations of *Pterodactylus* species from the Solnhofen limestone. In the case of *Pterodactylus kochi* for example, the smallest, and therefore youngest, individual had a wing span of only 7·5in (19cm), the largest, and thus oldest, a wing span of 2·2ft (67·5cm). A continuous growth series can be assembled in between. It turned out that not all elements of the skeleton grew at the same speed, but that for example the skull grew more quickly in the young animal and is thus relatively larger than in the older and adult individuals. Also the eye socket is much larger in proportion to the skull in the young animals than in the older ones. A growth pattern of this kind corresponds entirely to the customary scheme in vertebrates, and is similar in crocodiles, birds and man.

As well as these changing proportions in individual parts of the skeleton the degree of ossification (bone formation) also increases as the individual grows. For example in young pterosaurs the tiny phalanges in the third and fourth digits of the foot were still cartilagenous, for which reason they are not preserved as fossils, and so corresponding gaps are left in these digits. It was not until the animals were older that they ossified and were then preserved as fossils.[51]

Other skeletal bones as well, like carpals and tarsals, scapulocoracoid, pelvis, notarium and the bones of the skull were still separate in youth and are fused firmly with each other only with increasing age. Thus juvenile and adult animals can be distinguished by the degree of

46 Straelen, V. van, 1928. *Les oeufs de reptiles fossiles.* Palaeobiologica, 1: 295-305.

47 Wellnhofer, P., 1975. *Die Rhamphorhynchoidea (Pterosauria) der Oberjura-Plattenkalke Süddeutschlands. Teil III: Palökologie und Stammesgeschichte.* Palaeontographica (A), 149: 1-30; Stuttgart.

48 Bennett, S.C., 1987. *Sexual dimorphism in the pterosaur Pteranodon.* Journal of Vertebrate Paleontology, 7, Supplement to Number 3: 9.

49 Wiman, C., 1924. *Aus dem Leben der Flugsaurier.* Bulletin of the Geological Institute Uppsala, 19: 115-127.

50 Desmond, A.J., 1975. *The Hot-Blooded Dinosaurs.* p.171; Blond and Briggs, London.

51 Wellnhofer, P., 1970. *Die Pterodactyloidea (Pterosauria) der Oberjura-Plattenkalke Süddeutschlands.* Abhandlungen der Bayerischen Akademie der Wissenschaften, math.-naturwiss. Klasse, Neue Folge, 141: 113 pp.; Munich.

Right: The extremely good fossil preservation in the Solnhofen limestone has made it possible not only to distinguish the relatively tough wing membranes of pterosaurs, but also the imprints of body skin and hairs. This is a section of a *Rhamphorhynchus* skin with clearly identifiable imprints of short hairs and hair clusters. Fine, needlepoint impressions indicate hair papillae.

Above and right: These photographs allow us to compare the skulls of a half-grown baby (above) and a fully grown adult *Pterodactylus antiquus* (right). The skulls exhibit markedly different proportions. The juvenile individual has a shorter beak and a relatively larger eye socket than the adult. The skull lengths are 1·73in (44mm) and 4·25in (108mm) respectively. Apparently the skulls of young *Pterodactylus* grew at a faster rate than other parts of the skeleton.

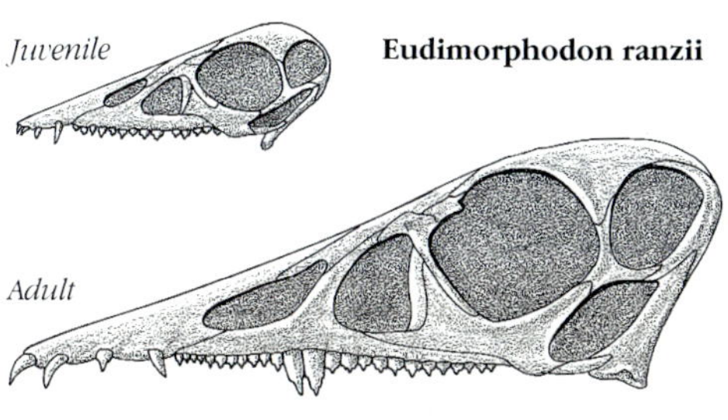

Differing Skulls (above) The skulls of young and adult individuals of the Triassic *Eudimorphodon* from Northern Italy were also found to exhibit different proportions. In the young animal the skull is smaller, the beak shorter, the eye socket relatively larger, and the number of teeth less than in the adult.

ossification, and species with small adults distinguished from those with large adults.

A striking juvenile feature is also dentition. Because of the shortness of the jaw young pterosaurs have fewer and smaller teeth than adults. In the Italian Upper Triassic *Eudimorphodon* the teeth of the young animals are hardly worn in contrast with those of the adults. This leads to the assumption that these pterosaurs, unlike their parents, did not eat fish with hard scales, but may have been insectivorous.[52]

At the end of a number of long bones of non-adult pterosaurs are so-called epiphyses, which as in birds are centres of ossification at which the longitudinal growth of long bones takes place. This leads one to assume that pterosaurs grew like birds, and reached their specific final size relatively early. Reptile-like growth throughout the whole life span would sooner or later have led to a point at which factors like body weight, muscular performance and wing area, significant parameters in flying animals, would have exceeded a critical point.

Thus pterosaurs can definitely be compared with birds in their reproductive behaviour and growth. All the signs point to their having been egg-laying, warm-blooded, intelligent flying animals with a growth pattern and social behaviour similar to modern birds.

Physiology

It was conjectured even in the nineteenth century that pterosaurs were warm-blooded. Many naturalists took them to be water-birds, bats or marsupials. August Goldfuss, the Bonn professor of zoology and mineralogy, thought he found prints of hair in a Solnhofen pterosaur in 1831. He wrote: 'According to this *Pterodactylus* was not like the reptiles, with scales and plates, but had a pelt of soft hairs almost an inch long, perhaps even covered with feathers in some places.'[53]

In 1870 H.G. Seeley also suggested that pterosaurs had a covering of hair, like bats, and he concluded that they were warm-blooded because of their bird-like behaviour.[54]

In 1908, Karl Wanderer of the University of Munich examined the pterosaurs in the Dresdener Museum. He was familiar with the specimen on which Goldfuss thought that he had seen hair, but reasoned that these impressions were nothing more than unevenness on the surface of the stone. On the other hand he did find a Solnhofen *Rhamphorhynchus* with numerous pits like needle pricks between which fine, short stripes ran, but did not at first associate them with hair.[55] It was not until Ferdinand Broili examined this specimen again in 1927 that these hollows and stripes on the surface of the stone were interpreted as prints of hair follicles and tufts of hair.[56] Individual hairs, apparently stuck together, could be followed up to a length of about a third of an inch (7-8mm). He established that there was body skin with hair at several points on the body. These were to be found above the skull in the form of a hairy crest of skin about 1·6in (40mm) long and 0·7in (18mm) high, on both sides of the neck, on the wing between the humerus and the lower part of the arm, and on the rear side of the lower arm down to the first wing phalanx. There were no hair follicles on the flight membrane itself, which seems to have been largely naked. Later Broili was also able to confirm a covering of hair in *Dorygnathus* from the Liassic and *Pterodactylus* from Solnhofen. Several more specimens of *Rhamphorhynchus* from Solnhofen have come to light in which remains of skin with hair have been found. Here the hairs are very short, only 0·1in (2-3mm) long and between 0·004in and 0·002in (0·1mm – 0·05mm) thick. They seem to have had a thin central channel running through them.[47]

In 1970 the zoologist A.G. Sharov of the Moscow Academy of Science discovered the skeleton of a pterosaur during fossil investigations in Upper Jurassic lake deposits in the Karatau mountains in the Soviet republic of Kazakhstan. The body skin and flight membranes of this skeleton had survived in extraordinarily good condition. The Karatau fossil strata are fine-grained sediments similar to the flaggy limestones of Solnhofen and Eichstätt in Bavaria, and like these have provided a large quantity of fossils in outstanding condition.

Close examination of the Russian pterosaur revealed that the creature's body must have been covered with thick fur, consisting of fairly thick hairs up to a quarter of an inch (6mm) long. They grew somewhat more sparsely on the flight membrane, the digits of the hand and on the webs between the foot digits, while the

52 Wild, R., 1984. *Flugsaurier aus der Obertrias von Italien.* Naturwissenschaften, 71: 1-11; Springer Verlag.

53 Goldfuss A., 1831. *Beiträge zur Kenntnis verschiedener Reptilien der Vorzeit.* Nova Acta Academiae Leopoldinae Carolinae, 15: 61-128; Breslau and Bonn.

54 Seeley, H.G., 1870. *The Ornithosauria: An elementary study of the bones of Pterodactyles.* 130 pp.; Cambridge.

55 Wanderer, K., 1908. *Rhamphorhynchus Gemmingi H. v. Meyer. Ein Exemplar mit teilweise erhaltener Flughaut aus dem kgl. Mineralog.-Geol. Museum zu Dresden.* Palaeontographica, 55: 195-216; Stuttgart.

56 Broili, F., 1927. *Ein Rhamphorhynchus mit Spuren von Haarbedeckung.* Sitzungsberichte der Bayerischen Akademie der Wissenschaften, math.-naturwiss. Abt., 1927: 49-67; Munich.

Broili, F., 1938. *Beobachtungen an Pterodactylus.* Sitzungsberichte der Bayerischen Akademie der Wissenschaften, math.-naturwiss. Klasse, 1938: 139-154; Munich.

Broili, F., 1939. *Ein Dorygnathus mit Hautresten.* Sitzungsberichte der Bayerischen Akademie der Wissenschaften, math.-naturwiss. Klasse, 1939: 129-132; Munich.

long tail was apparently completely naked. Sharov named the animal *Sordes pilosus*, which means something like 'hairy devil'.[57]

Direct proof of a hair-like body covering seems to have confirmed the warm-bloodedness of pterosaurs once and for all, as only mammals, i.e. warm-blooded creatures, have hair today. What has not been explained is the nature of pterosaur hairs, and how they came into being. Probably they developed from reptile scales, like the feathers of birds, and are not homologous with mammal hairs. Their function must have been the same, however: protection against heat loss.

The majority of pterosaurs were actively flying animals. Flight requires high energy levels. This energy can only be generated by efficient circulation of the blood, effective respiration and thus a high metabolic rate. It is possible that the heart already had four completely separate chambers, in order to provide the muscles with sufficient oxygen-rich blood. In flying vertebrates, birds and bats, this kind of physiology goes hand in hand with a high body temperature, which has to be kept stable for optimum biological functioning.

Typical cold-blooded reptiles, like lizards, for example, have a body temperature corresponding to the ambient external temperature. They are stiff and lethargic when it is cool, and are only fully agile when they are warmed up by the sun. Warm-blooded creatures need an insulating body covering, to avoid heat loss if the ambient air temperature is lower. This was especially important for pterosaurs with their large wing membranes and small bodies, as they had an unfavourable surface/volume ratio from a thermophysiological point of view. In addition to this, their flight activities have to be considered: as Carl Wiman put it, this was so expensive that it could only be financed by warm-bloodedness.

Something else that speaks for the warm-bloodedness of pterosaurs could be the so-called pneumaticity of their bones. In birds air flows out of the lungs into a system of air sacks which surround all the internal organs of the body. Extensions of the air sacks often continue into the vertebrae and the limb bones, by means of small openings in the walls of the bones, the so-called pneumatic foramina. In pterosaurs, and in large ones in particular, air pores of this kind are also to be found in vertebrae and long bones. Presumably through them air sacks entered the bones, as in birds.

The significance of these air sacks was probably that inhaled air was warmed up, thus increasing lift force and reducing the specific weight of the bones. As well as this they cushioned the internal organs and cooled the interior of the body, i.e. freed it from excess heat and prevented overheating as a result of muscular activity while flying.

It is further striking that most pterosaurs have been found with nothing in their stomachs. Because of their high energy requirements when flying, birds require large quantities of nourishing food. Bats too eat incredible quantities of insects relative to their body weight. Pterosaurs also had to eat a great deal because of their flying, which required high energy levels. This means that digestion must have been rapid, more rapid than in cold-blooded reptiles, which explains why the stomachs of pterosaur fossils rarely contain the remains of food.

If we add to observations such as the existence of hair and air sacks, and high food requirements, the fact that pterosaurs grew rapidly, as seen from the presence of epiphyses, then we are establishing characteristics typical of modern warm-blooded creatures, birds and mammals, but not of reptiles. These are cold-blooded, have scales, no air sacks, need less food and grow slowly.

Thus there are strong reasons for thinking that pterosaurs were warm-blooded. If that was the case, they were able to settle in higher geographical latitudes, as is shown by recent fossil finds from Australia and New Zealand, and live in a cooler climate. On the other hand they would have had to undertake parental care sooner, warming their young with their bodies and keeping them under their protecting and warming flight membranes, and perhaps even hatching their eggs themselves.

The Pterosaur Brain

Pterosaurs differ from typical reptiles not only in their warm-bloodedness, but also in the form of their brain. The size and shape of the brain of fossil vertebrates can be reconstructed from casts of the skull cavity. Under favourable conditions the cast of the brain itself occurs in fossil form, as an endocast, when the brain case has filled with sediment which then petrifies. In pterosaurs the shape of such an interior cast is so like a brain that one can assume that as in birds, the brain filled the whole brain case. In modern reptiles it only fills half its cavity.

Even in the nineteenth century H.G. Seeley used fossil skull fragments from the Cretaceous Cambridge Greensand to show that the

57 Sharov, A.G., 1971. *New flying reptiles from the Mesozoic deposits of Kazakhstan and Kirghizia.* Trudy of the Paleontological Institute of the Academy of Sciences U.S.S.R., 130: 104-113; Moscow. (In Russian)

Left: A section of the fossil slab with a specimen of *Sordes pilosus* showing part of the wing finger with a narrow strip of the wing membrane, and a dark area representing body skin covered with hairs. These reptilian hairs must have derived from scales, and so are unlike the hairs of mammals. Nevertheless, they must have had the same function: to protect the animal from heat loss.

Left: Several specimens of *Sordes* have been found associated with parts of the wing membranes, the body skin and the body covering consisting of a dense fur of short hairs. The first of these spectacular discoveries was described by the Moscow zoologist A.G. Sharov in 1971.

Right: The presence of pneumatic foramina in the long bones of larger pterosaurs, such as in this *Santanadactylus* wing metacarpal, suggests an air bag system similar to that of birds. Extensions of the air sacks which fill the body cavities enter the hollow bones.

pterosaur brain was bird-like. Since then there have been numerous other finds pointing in the same direction, these finds including those of the genera *Parapsicephalus*, *Rhamphorhynchus*, *Pterodactylus* and *Pteranodon*. Tilly Edinger in particular was able to demonstrate that many features of the pterosaur brain are very like those of birds.[58]

As in modern birds the pterosaur brain rises obliquely forwards in the skull. The foremost protrusion of the fore brain, the olfactory lobes, important for the sense of smell, is small. This suggests very limited development of the sense of smell. In crocodiles for example these olfactory lobes are very long, but short in birds as well. The fore brain consists of two large, upwardly bulging hemispheres. It grew broader as pterosaurs evolved. In early pterosaurs its proportions were more reptilian, the fore brain was longer than it was broad, but in more advanced pterosaurs like *Rhamphorhynchus* and *Pteranodon* it is broader than it is long, thus reaching similar proportions to modern birds. In them it is the seat of an association centre, in which complex instinctive actions are controlled. *Pterodactylus*' fore brain had two furrows, which are a definite avian characteristic. It is possible that they had formed in older pterosaurs as well.

Another definitely bird-like quality is that the optic lobes of the mid-brain are no longer in contact on their mid-line, as in reptiles, but have been forced apart by the cerebellum, which has extended between them, and therefore they have been shifted to both sides and downwards. Their relatively large volume indicates that pterosaurs must have had outstanding optical capabilities. This is also indirectly confirmed by the generally large size of the eye sockets. For flying animals who catch living prey in flight this is a requirement of life, and a decisive selective advantage.

The cerebellum of pterosaurs bulges up between the optic lobes and is like the cerebellum of birds in this respect. This is quite clearly one of the characteristics which distinguishes pterosaurs from all other reptiles. However, the pterosaur cerebellum is smaller than that of birds in volume. It also does not have the transverse furrows typical of birds. Side appendages to the cerebellum, the so-called flocculi, are developed as in birds. They are used to regulate balance in flight.

The fact that the form of the pterosaur brain is more like the brain of birds than that of reptiles is evidence to support the bird-like lifestyle of the pterosaurs. In pterosaurs the size of the cerebellum is at the upper limit of variabilty for reptiles, and at the lower limit for birds. It is extraordinarily important for the regulation and co-ordination of motor activity. Apparently very similar brain structures necessary for a flying lifestyle originated independently in birds and pterosaurs.

However, in terms of the size of their brain relative to body volume, pterosaurs were still in the realms of modern reptiles. There is probably no connection between relative brain size and body temperature. All the same, the brains of modern warm-blooded creatures, mammals and birds, are on average ten times bigger than those of modern living cold-blooded vertebrates, in other words reptiles, amphibians and fish. Ptersosaurs come in between. Chicago palaeontologist J. Hopson conjectured that relative brain size 'reflects the plasticity of behaviour and also the intelligence of an animal', and that this is 'perhaps directly related to the overall activity characteristic of the animal' and thus to its energy balance.[59]

58 Edinger, T., 1927. *Das Gehirn der Pterosaurier.* Zeitschrift für Anatomie und Entwicklungsgeschichte, 83 (1/3): 105-112; Munich and Berlin.
Edinger, T., 1941. *The Brain of Pterodactylus.* American Journal of Science, 239: 665-682; Washington.

59 Hopson, J.A., 1977. *Relative brain size and behavior in archosaurian reptiles.* Annual Review of Ecology and Systematics, 8: 429-448.

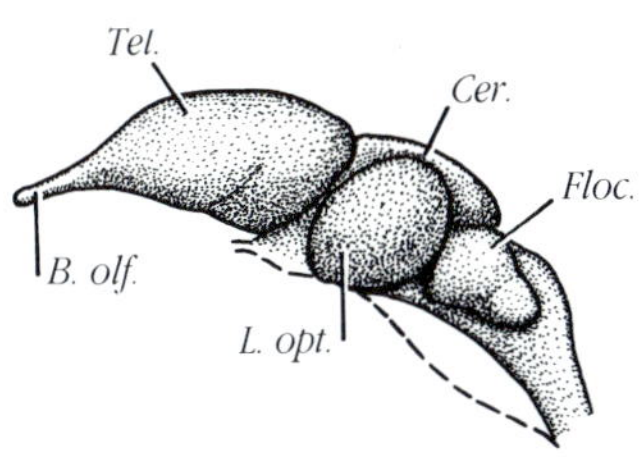

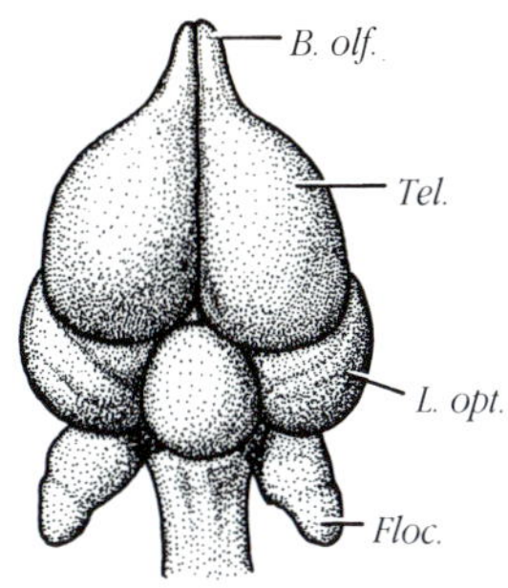

A Pterosaur Brain (above)
On the basis of the endocranial cast, the brain of *Parapsicephalus* can be restored. It is seen here from its left side (left) and from above (right). The abbreviations annotating the drawing mean: B. olf.: olfactory bulbs, responsible for the sense of smell; Cer.: cerebellum, the hindbrain, responsible for the co-ordination and regulation of movement; Floc.: flocculi, important for the maintenance of equilibrium, direction and orientation during flight; L. opt.: optic lobes, midbrain, responsible for the sense of vision; Tel.: telencephalon, forebrain, centre of association and control of complex instinctive actions.

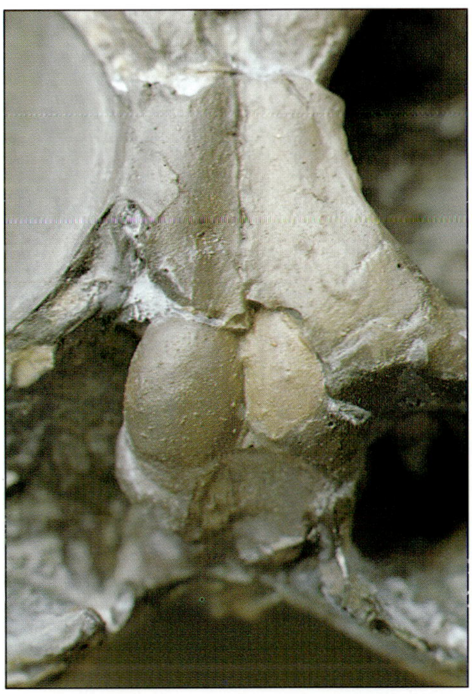

Above: Endocranial casts of pterosaurs sometimes reveal a duplicate copy of the brain, because they are the petrified sedimentary infillings of the brain case. One of the best examples is the skull of *Parapsicephalus* found in the 1880's in the Liassic Alum shales near Whitby on the Yorkshire coast. The natural cast of its brain is seen here from above, and shows both hemispheres of the forebrain. The forebrain and hindbrain of pterosaurs are larger than those of reptiles such as crocodiles, but less developed than those of birds, as the diagram (right) reveals.

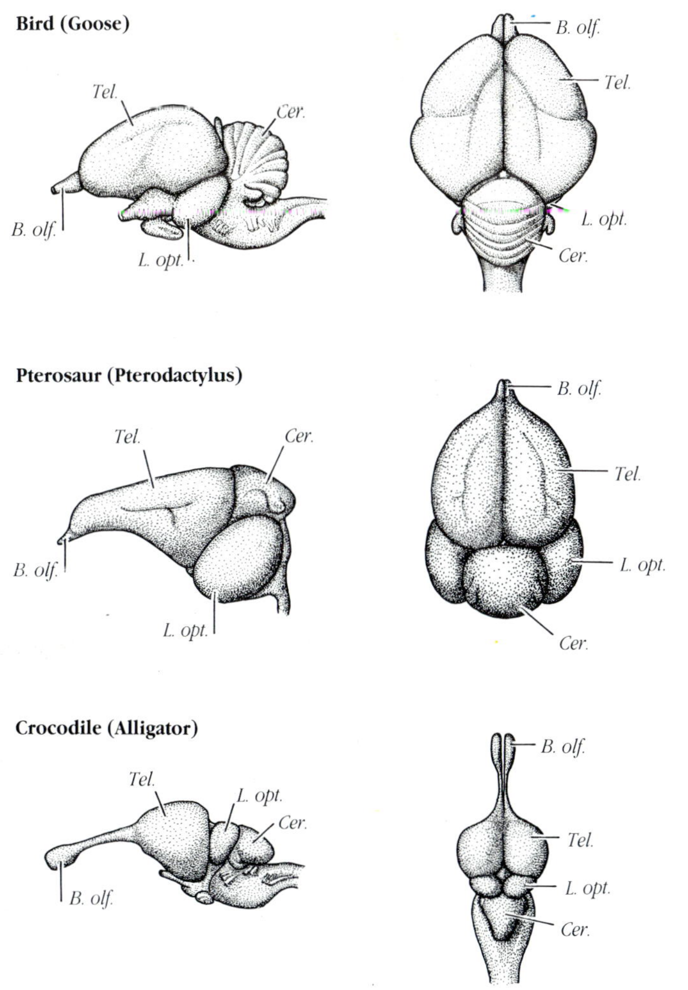

Brain Comparisons (right)
These drawings compare the brains of a goose, a pterosaur (*Pterodactylus*), and an alligator in left lateral and dorsal views. The pterosaurian brain is intermediate in relative size and shape between reptiles and birds.

EXTINCTION OF THE PTEROSAURS

Extinction of animals, plants and evolutionary lines is a quite normal and natural process in the history of life on Earth. It is a natural balance to the emergence and evolution of new species and life forms. Palaeontologists can show, on the basis of the fossil record, that extinction occurred so frequently and regularly in the course of the history of the Earth that it must be seen as a necessary factor in the evolution of life.

However, as well as this more or less constant process of dying out over millions of years there have also been times in which the extinction rate has been higher; evolutionary crises, as it were, in which many groups of organisms disappeared from the Earth within a relatively short space of time. This often seems to have happened so quickly and so thoroughly that one is tempted to think of global catastrophes that might have caused 'mass extinction' of this kind. There have been many such periods of crisis in the geological past. Certainly one of the most spectacular was at the turn from the Mesozoic to the Cenozoic Era. Pterosaurs were victims of this striking extinctión of many animal forms on the Cretaceous-Tertiary boundary.

It was not just the pterosaurs, however, but also dinosaurs, marine reptile groups like plesiosaurs, ichthyosaurs and mosasaurs, and many invertebrate marine animals from various molluscs to protozoa, who died out 65 million years ago. The reasons for this 'mass extinction' at the end of the Cretaceous have not yet been completely established. Do we have to assume a single, possibly global catastrophe like the impact of an asteroid from space for all these animal groups?[1] Or did pterosaurs die out for different reasons from dinosaurs?

In order to answer this we must look at the evolution of pterosaurs, their rise, the period in which they flourished, and their decline. Here the main problem is the paucity of the fossil record, especially for pterosaurs; their bones were very fragile, and generally only became fossilized in aquatic sedimentation basins; they are simply not documented in terrestrial habitats.

In evaluating the approximately 40 genera and 100 species of pterosaur known from the Triassic, Jurassic and Cretaceous, one thing is clear: after modest beginnings in the late Triassic (220 million years ago), they first flourished, in very diverse forms, in the late Jurassic (150 million years ago). Long-tailed pterosaurs, the Rhamphorhynchoidea, had in various evolutionary lines reached the pinnacle and at the same time the end of their evolution. At the same time short-tailed pterosaurs, the Pterodactyloidea, appeared and occupied various 'ecological niches' in several adaptive types. These 'more modern', more successful, because better adapted Pterodactyloidea of the Jurassic were the parent group of the later Cretaceous pterosaurs. The Rhamphorhynchoidea were driven out by them, and finally became extinct at the turn from the Jurassic to the Cretaceous, after a few lines had already died out earlier, like Eudimorphodontidae and Dimorphodontidae, for example.

Pterosaurs became more diverse in the Lower Cretaceous, and flourished again in the middle Cretaceous, with a large number of toothed forms. Toothless pterosaurs, which were to dominate in the late Cretaceous, only appear as isolated specimens. Before this all the toothed forms had died out completely. A typical evolutionary trend of the toothless Pterodactyloidea of the Upper Cretaceous is a general increase in size. Three families can be shown to have existed with certainty in the latest Cretaceous period: Pteranodontidae, Nyctosauridae and Azhdarchidae. The azhdarchids were the last to survive: the genus *Quetzalcoatlus* was extant at the end of the Cretaceous. As we have seen, the Azhdarchidae were large to very large, toothless and highly specialized pterosaurs.

Thus the history of pterosaurs included two periods in which they flourished, in the late Jurassic and the middle Cretaceous, and three extinction events, at the Jurassic/Cretaceous boundary, in the early Cretaceous and finally at the Cretaceous/Tertiary boundary.[2]

Thus a clearly differentiated picture of pterosaur extinction patterns emerges, with an

1 Alvarez, L., Alvarez, W., Asaro, F. and Michel, H.V., 1980. *Extraterrestrial cause for the Cretaceous-Tertiary extinction.* Science, 208: 1096-1108.
Taking as their departure point the unusually high concentrations of the rare metal iridium in Cretaceous/Tertiary boundary strata, Alvarez and his team concluded that an extra-terrestrial asteroid six miles (10km) in diameter crashed into the Earth 65 million years ago. They suggest that dust swept up into the atmosphere by this collision would have blocked out the sun for months or even years, leading to consequent cooling of the Earth and damage to vegetation as a result of interrupted photosynthesis in plants. This catastrophic event would have had so deleterious an effect on the Earth's biosphere in so short a time that it could have brought about the extinction of dinosaurs and other animal groups at this time.

2 Unwin, D.M., 1987. *Pterosaur extinction: nature and causes.* Mémoires de la Société géologique de France, N.S. 150: 105-111.

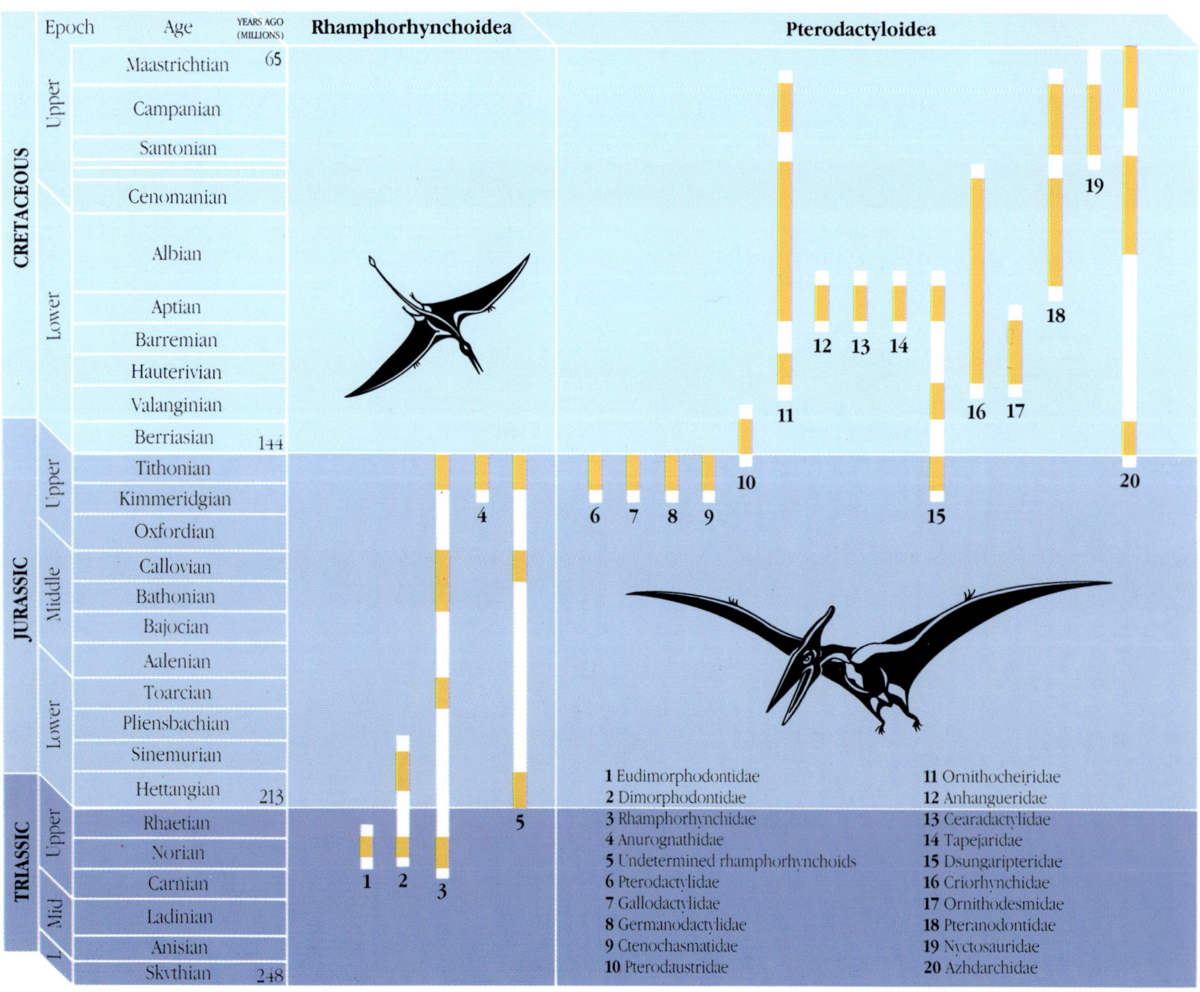

Extinction Patterns (left)
The fossil record of pterosaurs demonstrates a gradual pattern of origins and extinctions throughout their evolutionary history. In this diagram each bar represents a pterosaur family, the black sections showing where it is documented by fossils. The first pterosaurs, the rhamphorhynchoids, appear by late Triassic times with three different families. Two of these survive into the Jurassic, one until the late Jurassic. At the same time the pterodactyloids appear. They must be derived from unknown Jurassic rhamphorhynchoids. At the end of the Jurassic, a first extinction event can be recognized. During the Lower Cretaceous at least ten families of pterodactyloids can be recorded, indicating a high degree of diversity in Middle Cretaceous times. However, all these familes bar one, the Azhdarchidae, had become extinct prior to Maastrichtian times. It seems that the extinction pattern was a gradual decline rather than one single catastrophe.

evident but gradual decline towards the end. There is nothing to suggest a sudden 'mass extinction' brought about by a particular catastrophe. These findings correspond completely with the course of evolution palaeontologists have observed in other animal groups, like dinosaurs, for example.

Thus we must formulate the question about the extinction of pterosaurs rather more precisely: 'Why did the *last* pterosaurs become extinct on the Cretaceous/Tertiary boundary'? This event, however natural it may have been, was not an inevitable fate that can be taken for granted. Other reptile contemporaries of the pterosaurs, like crocodiles and turtles, did not become extinct at that time, but have survived to the present day. Thus there must have been factors peculiar to the pterosaurs.

We are possibly nearer to the truth if we ask: 'Why did crocodiles survive, but not pterosaurs?' Modern crocodiles are not essentially different from crocodiles of the Cretaceous, or even the Jurassic, i.e. their structure has hardly changed over all those millions of years. The evolution of crocodiles did not lead to any extreme special adaptations, they can be considered a 'conservative' reptile group. The lack of extreme specialization was their survival strategy, allowing them to accommodate themselves more easily to changing conditions.

Things were very different for the pterosaurs. They were not only generally adapted to a flying lifestyle to an extreme extent, but they also developed special adaptations to quite particular environmental conditions as well. Towards the end of the Cretaceous came the additional factor of their 'megalomania', which led to an increase in size that touched the boundaries of the possible for a flying animal. Evolution that led to such gigantic, but successful flying animals, was only possible under optimum life and environmental conditions.

Even relatively minor changes to these external conditions could disturb this equilibrium and endanger the continued existence of populations and entire species. The great long-necked, toothless azhdarchids with wing spans from 13 to 40ft (4 to 12m), had long bones that were extremely thin-walled and hollow. This structure was intended solely for weight reduction. The bones were hardly suitable to support high mechanical loads. Apparently they were adapted to climatic conditions appropriate to this skeletal structure.

As Bramwell and Whitfield discovered, *Pteranodon*, with flight speeds of 23ft/s to 46ft/s (7m/s to 14m/s), was outstandingly adapted to soaring in a light wind.[3] It is to be assumed that this was similarly true of *Quetzalcoatlus*, *Titanopteryx* and *Azhdarcho*, and that an increase of wind speed of only 16·4ft/s (5m/s) would have been enough to make flying absolutely impossible for giant pterosaurs.

A change in the average wind speed could have been caused by global cooling, with clear temperature differences between the poles and the equator, and the consequent emergence of seasonality. Longer periods of the year with higher wind speeds, during which giant pterosaurs were unable to fly, would have sufficed to weaken populations to the extent that they would have sunk beneath the critical minimum number of individuals, and thus that they were condemned to extinction.

3 Bramwell, C.D. and Whitfield, G.R., 1974. *Biomechanics of Pteranodon.* Philosophical Transactions of the Royal Society London, (B), 267: 503-581.

Above: The sedimentary sequence in the Big Bend National Park in West Texas reveals a continuous transition from the uppermost Cretaceous Javelina Formation to the lowermost Tertiary Black Peaks Formation, indicated by the purple-coloured bands. Sometime during the deposition of the Javelina Formation, the last pterosaurs went extinct.

And indeed there are geological and palaeontological signs of deterioration of the climate in the late Cretaceous.[4] It is assumed that in the period between the late Cretaceous and the early Tertiary the average annual temperature on Earth dropped by 50°F (10°C). This cooling need not necessarily have had extraterrestrial causes. The geological history of the Earth teaches us that it was just at the Cretaceous/Tertiary boundary that large areas of continental shelf, shallow seas in other words, dried up, or, as geologists put it, marine regression took place. This was caused by plate tectonics, movements of parts of the Earth's crust, which brought new oceans into being. Thus for example South America and Africa drifted apart (which they are incidentally still doing), and in the same way North America and Eurasia split apart. Thus the North and South Atlantic were created, and with this a system of marine currents which had a considerable influence on climate and weather conditions, as they still do today. It is in these climatic changes that we should look for the cause of the extinction of the last pterosaurs.

The evolutionary process that led to the giant pterosaurs of the late Cretaceous was a one-way street that finally became a cul-de-sac, or indeed a trap, from which there was no way back. In order to survive, pterosaurs would have had to renounce their specialization. In particular they would have had to become smaller, develop bones with thicker walls, and acquire flapping flight again. This was not possible. It would have contradicted what is recognized as a basic rule of biology, that of the non-reversibility of evolution.

After the disappearance of the last pterosaurs 65 million years ago, birds began to assert themselves more strongly in the Tertiary. Of course there was already a whole series of birds in the late Cretaceous, which were clearly little affected by the climatic crisis at the Cretaceous/Tertiary boundary. It was often assumed that birds had driven out the last pterosaurs in competition and thus brought about their extinction. Certainly the number of species of bird and their diversity had increased towards the end of the Cretaceous, while that of pterosaurs had steadily declined.[5] But probably Cretaceous birds, like the diver *Hesperornis*, had a different way of life from pterosaurs, and were thus not really in biological competition with them. Birds like gulls or petrels did not appear until the early Tertiary.

Thus the extinction of pterosaurs is marked by a gradual decline in their diversity during the Upper Cretaceous. Late Cretaceous pterosaurs were highly specialized giant forms, and their end was brought about by deterioration of the climate. This is adequately explained by geological factors which resulted from the internal dynamics of the Earth. Even if a meteor or an asteroid had crashed on Earth 65 million years ago, it could probably scarcely have influenced the long-term, entirely normal process of pterosaur extinction.

4 Van Valen, L. and Sloan, R.E., 1977. *Ecology and the extinction of dinosaurs.* Evolutionary Theory, 2: 37-64.

5 Unwin, D.M., 1987. *Extinction and survival in birds.* In *Extinction and Survival in the Fossil Record* (ed. G.P. Larwood), Systematics Association Special Volume No. 34; Clarendon Press, Oxford.
Unwin suggested that birds probably began to compete with pterosaurs at about the time of the early Cretaceous. 'It would appear that, owing to their more flexible bauplan, birds were more successful, and by the late Cretaceous they appear to have filled virtually all niches, except for those where their avian bauplan was of no particular advantage. These were the niches occupied by large ocean-going forms such as *Pteranodon* and possibly *Quetzalcoatlus*.'

PTEROSAUR RECONSTRUCTIONS

Right: Life in the Liassic Sea as reconstructed by de la Bèche, and, in 1846, reproduced in F.J. Pictet's *Traité élémentaire de Paléontologie*. The sea is populated by numerous types of marine animal, while pterosaurs fly in the air. They are regarded as bat-like, with their wing membranes extending down to their feet and tails.

Left: In this restoration by Riou we see a Jurassic landscape. Individual *Pterodactylus* are here shown flying to catch a dragonfly on the wing, and sitting upright in a resting position in the background. This scene was repeatedly published in several editions of L. Figuier's *La Terre avant le Déluge* from 1863 on, and in O. Fraas' *Vor der Sündflut* in 1866. Both were popular books of their time.

Attempts to reconstruct pterosaurs as they appeared in life were first undertaken in the very early nineteenth century. We have reconstruction drawings by August Goldfuss dating from the year 1831,[1] by William Buckland in his contribution to the *Bridgewater Treatises* of 1836,[2] and by Thomas Hawkins, who represented pterosaurs as well as great marine reptiles in fatal conflict in his *Book of the Great Sea-Dragons* dating from 1840.[3] Curiously they have bat-like wings and are apparently behaving as carrion feeders. Othenio Abel, Professor of Palaeobiology at the University of Vienna and an expert on the reconstruction of prehistoric vertebrates, was reminded by this picture of 'a ghostly figure of one of Breughel's hells'.

Right: The frontispiece of Thomas Hawkins' *Book of the Great Sea-Dragons* of 1840 is an engraving by John Martin which reveals his vision of the primeval antediluvian world. Marine saurians are locked in mortal combat, and toothed pterosaurs with bat-like wings are shown as carrion feeders scavenging on a carcass. Such early reconstructions look more like grotesque caricatures rather than life restorations based on careful scientific research.

1 Goldfuss, A., 1831. *Beiträge zur Kenntnis verschiedener Reptilien der Vorwelt.* Nova Acta Academiae Leopoldinae, 15: 61-128. Goldfuss, who had studied the *Scaphognathus* from the Solnhofen strata of Bavaria, was of the opinion that pterosaurs used their claws to climb rock walls and cliffs, or trees as well. He also believed that they flew close to the water, to catch insects or aquatic creatures.

2 Buckland, W., 1836. *Geology and Mineralogy, Considered with reference to Natural Theology*, I: 224-225, in volume 5 of *The Bridgewater Treatises on the Power Wisdom and Goodness of God as manifested in the Creation.* London.

3 Hawkins, T., 1840. *The Book of the Great Sea-Dragons, Ichthyosauri and Plesiosauri, Gedolim Taninim of Moses, Extinct Monsters of the Ancient Earth.* London.
The frontispiece to Hawkins' book was engraved by John Martin. Hawkins was a tireless collector, who carried away specimens by the quarryload. We owe some of the best marine reptile fossils from the Dorset coast of southern England to him; they found their way in to the Natural History Museum in London, where they can still be seen today. Hawkins was an eccentric and considered these fossil creatures to be an early creation by Jehovah. 'They perpetuate a Designe no longer in use', he wrote. In his opinion marine reptiles did not even belong to the animal kingdom. Thus he created a new kingdom for them, the 'Gedolim Taninim'.

Another, likewise very imaginative representation of living pterosaurs was given by de la Bèche. This was a so-called menagerie picture of life in the Liassic ocean, reproduced by F.J. Pictet in 1846.[4] The water is absolutely teeming with ichthyosaurs, plesiosaurs, crocodiles and fish, and there are a few pterosaurs in the air, with flight membranes that were again bat-like, extending right to their feet. They are also shown with a flight membrane between their hind legs and tails This reconstruction can be seen at the top left of the facing page.

Two very peculiar portraits of pterosaurs had appeared at an even earlier stage, intended to express that pterosaurs were not reptiles. One was a drawing by Munich zoologist Johann Wagler dating from 1830, showing *Pterodactylus* as a swimming creature, using its wings like long penguin flippers.[5] Wagler believed that pterosaurs and other marine reptiles belonged to a separate class of vertebrate, which he named Gryphi.

The other portrait was by the English zoologist Edward Newman, who in 1843 classified pterosaurs as carnivorous flying marsupials.[6] After A. Goldfuss thought he had found hair on his Solnhofen *Scaphognathus* in 1831, Newman was convinced that pterosaurs could not be reptiles, as reptiles have scales by definition. He was additionally aware of the controversy over part of the jaw of an opossum from the Jurassic Stonesfield Slate that Buckland had described 30 years earlier (in fact the first mammal found in Mesozoic strata), in the same strata as the pterosaurs appeared. He therefore believed that he should be permitted to reconstruct pterosaurs as hairy, flying marsupials. Thus in his view they were mammals, a kind of hairy marsupial bat, warm-blooded, with pretty little ears, and certainly not cold-blooded, lethargic reptiles.

Later life restorations of pterosaurs on a more scientific basis appeared. Representations worthy of mention are those by O.C. Marsh dating from 1882,[7] by K.A. Zittel in the same year,[8] and in particular by H.G. Seeley, who in his book *Dragons of the Air* dating from

4 Pictet, F.J., 1846. *Traité élémentaire de Paléontologie.* Geneva.

5 Wagler, J.G., 1830. *Das natürliche System der Amphibien.* München, Stuttgart, Tübingen.

6 Newman, E., 1843. *Note on the pterodactyle tribe considered as marsupial bats.* The Zoologist, I: 129-131.

7 Marsh, O.C., 1882. *The wings of Pterodactyles.* American Journal of Science, 23: 251-256.

Left: The life restorations of pterosaurs by Zdeněk Burian are still among the best reconstructions of their habits and patterns of behaviour as we imagine them to be. Here Jurassic *Pterodactylus* is shown fishing.

Above: A life restoration of pterosaurs from N. Hutchinson's book *Extinct Monsters* of 1910. The pterosaurs shown, (l to r) *Pterodactylus*, *Rhamphorhynchus* and *Dimorphodon* are not all actually from the same geological age.

Below: A reconstruction of *Rhamphorhynchus* by the late Professor Manfred Reichel of Basel University. Reichel's restorations were based on thorough scientific study informed by a life-long passion for bird-watching.

1901 produced numerous portraits of various pterosaurs of the Jurassic and the Cretaceous known at the time.[9]

There were, however, great divergences of view about matters like the extent of the pterosaur flight membrane, and scientists are still not agreed about this question. An example of a large flight membrane is the image given by N. Hutchinson in his book *Extinct Monsters*.[10] This is also a 'menagerie picture', in which various species from different geological ages were thrown together: *Dimorphodon*, from the earliest Jurassic here becomes a contemporary of *Rhamphorhynchus* and *Pterodactylus* from the late Jurassic, while in reality they lived at least 50 million years apart. In all of them a flight membrane extending to the feet and including the tail was assumed (see the illustration at top right).

A highlight in the reconstruction of prehistoric vertebrates are the standard works of Othenio Abel, who from 1907 concerned himself with the structure, appearance and life style of pterosaurs.[11] More recently the Czech artist Z. Burian, working with Professor Augusta in Prague, and also M. Reichel, professor at the University of Basel, have produced outstanding portraits of pterosaurs, which also appear correct from a biological point of view, and thus life-like.[12]

Left: Othenio Abel (1875-1946) established palaeobiology as an independent field of research. He also studied the pterosaur fossils in detail and strove to understand their life styles and behavioural habits as living animals.

8 Zittel, K.A., 1882. *Über Flugsaurier aus dem lithographischen Schiefer Bayerns.* Palaeontographica, 29: 47-80; Stuttgart.
9 Seeley, H.G., 1901. *Dragons of the Air.* Reprinted 1967, Dover paperback; New York.
10 Hutchinson, N., 1910. *Extinct Monsters.* London.
11 Abel, O., 1907. *Bau und Lebensweise der Flugsaurier.* Verhandlungen der Kaiserlichen und Königlichen zoologisch-botanischen Gesellschaft Wien, pp. 253-254; Vienna.
Abel, O., 1919. *Neue Rekonstruktion der Flugsauriergattungen Pterodactylus und Rhamphorhynchus.* Naturwissenschaften, 7 (37); 661-665; Berlin.
Abel, O., 1925. *Geschichte und Methode der Rekonstruktion vorzeitlicher Wirbeltiere.* 327 pp.; G. Fischer, Jena.
Abel, O., 1927. *Lebensbilder aus der Tierwelt der Vorzeit.* 2nd edition, 714 pp.; G. Fischer, Jena.
12 Augusta, J. and Burian, Z., 1961. *Flugsaurier und Urvögel.* Artia, Prague.
Reichel, M., 1985. *1896-1984, dessins.* 60 pp., edited by L. Hottinger, Geologisches Institut der Universität Basel.

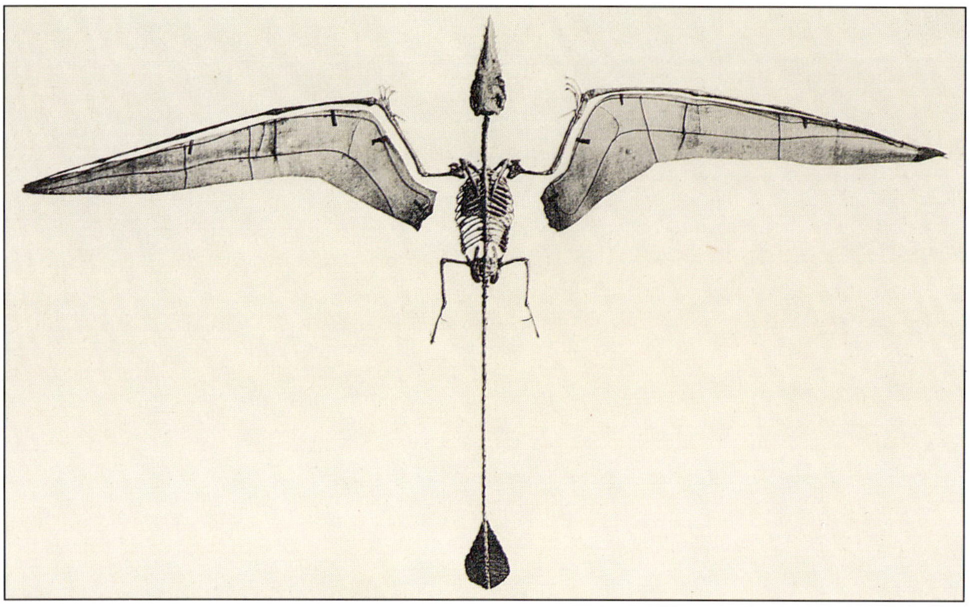

Above: The first attempt to build a model of a pterosaur was made by Munich Professor Ernst Stromer in 1913. His model was based on fossils of *Rhamphorhynchus* from the Solnhofen limestone.

All these reconstructions were in the form of drawings, but an important step was taken when pterosaurs were reconstructed three-dimensionally as well, both their skeletons and their bodies, and indeed even as flying models, as passively gliding or finally also actively flying 'artificial' pterosaurs.

Skeletal Models

In 1913 the Munich palaeontologist E. Stromer was one of the first to attempt to reconstruct the skeleton of the Solnhofen long-tailed pterosaur *Rhamphorhynchus*.[13] His model was made of wood, rubber and modelling wax, and mounted on a wire frame.

C. Bramwell and G. Whitfield made a skeletal model of *Pteranodon* in 1974, to test the possibilities of movement in the wing skeleton on a small scale. It was a purely functional model, made of balsa wood.[14]

M. Reichel of the University of Basel in Switzerland was one of the first to make a life-size model of a *Pteranodon* skeleton, made of wood and now exhibited in the Naturhistorisches Museum in Basle.

A true-to-life skeleton was also built of the Brazilian Lower Cretaceous genus *Santanadactylus* with a 20ft (6m) wing span, for the Geological Museum of the Univeristy of Amsterdam. It was constructed by Dr P.H. de Buisonjé.[15]

In 1972 the author made a life-size skeletal model of a Solnhofen *Rhamphorhynchus*. This was based on a fossil from the Carnegie Museum of Natural History in Pittsburgh, USA, bought in 1903 with a large fossil collection belonging to the Belgian Baron de Bayet.[16] The skeletal model consists of 50 bones and skeletal sections. They were shaped first of all from wax, on wire with cotton wrapped round it. Plaster, wood and lead were also used to make skull, teeth and pelvis. After this the shaped

13 Stromer, E., 1910. *Bemerkungen zur Rekonstruktion eines Flugsaurier-Skelettes.* Monatsberichte der deutschen Geologischen Gesellschaft, 62 (1): 85-91.
Stromer, E., 1913. *Rekonstruktion des Flugsauriers Rhamphorhynchus Gemmingi H.v.M.* Neues Jahrbuch für Mineralogie, Geologie und Paläontologie, 2: 49-68.

14 Bramwell, C.D. and Whitfield, G.R., 1974. *Biomechanics of Pteranodon.* Philosophical Transactions of the Royal Society London, B, 267: 503-581.

15 Buisonjé, P.H. de, 1981. *Santanadactylus brasilensis: Skelet-reconstructie van een vliegend reptiel met zes meter vlucht.* Gea, 14 (2): 37-49.

16 Wellnhofer, P., 1973. *Flying Reptiles in the Bayet Collection.* Carnegie Magazine, January 1973: 11-15; Pittsburgh.

Above: Professor Dr Manfred Reichel (1896-1984) of the University of Basel in Switzerland. Reichel was both a scientist and a trained artist. This combination of talents meant that his restorations of pterosaurs were not only biologically accurate, but also lively and convincing. He was also one of the first scientists to attempt to build a life-size model of a pterosaur skeleton, that of *Pteranodon* which is pictured left.

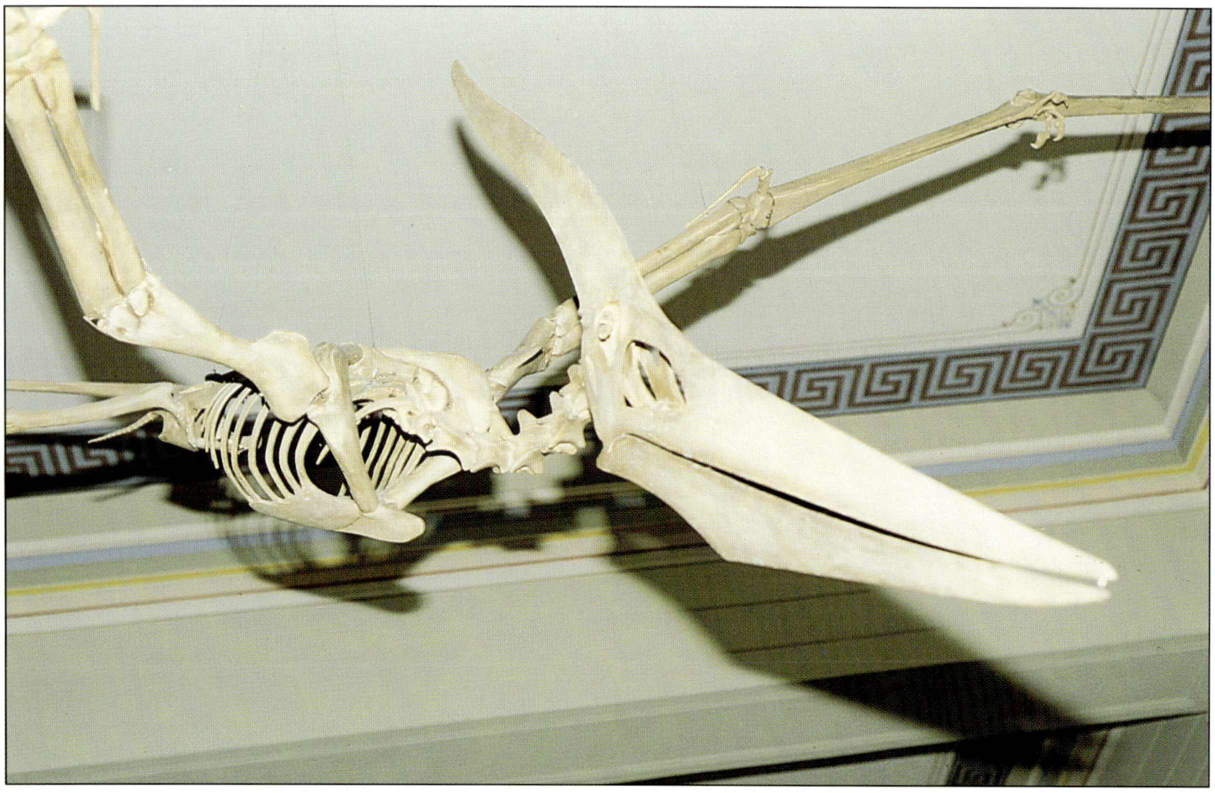

Left: This is the skeletal restoration of a life-sized *Pteranodon* that was built by Professor Manfred Reichel. The skeleton is carved out of wood and mounted in the flight position. It is now on display in the Naturhistorisches Museum in Basel, Switzerland, which is where this photograph was taken.

Right: A skeletal model of *Rhamphorhynchus* based on a Solnhofen specimen in the Carnegie Museum of Natural History, Pittsburgh. It was built by the author in 1972. It consists of fifty bones and skeletal sections, which were modelled out of beeswax mounted on a cotton and wire, wood and lead framework. The model was finally cast in epoxy resin.

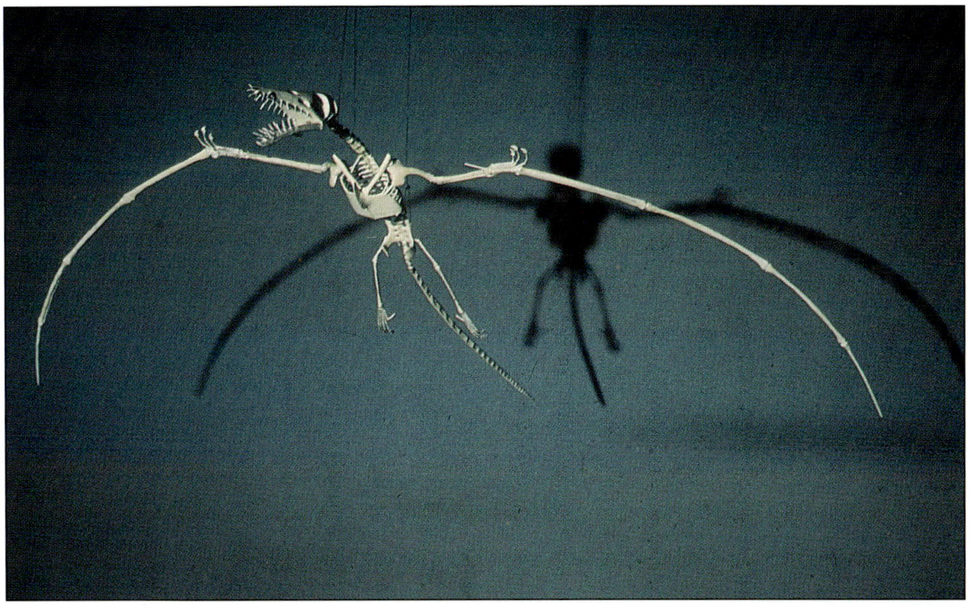

bones were cast in artificial resin, and assembled to make a skeleton in flying position. To put together a dinosaur skeleton you need heavy lifting gear, crane and welding torch, but the *Rhamphorhynchus* skeleton was assembled using tweezers, magnifying glass and glue. The pterosaur had a wing span of about 3ft (98cm) and was used as the basis for a later flesh restoration, showing the animal in flight position, at the moment when it is plunging down to catch a fish.

This three-dimensional model also made it possible to calculate the weight of a living *Rhamphorhynchus*. First the volume was determined, and multiplied by the specific weight of living tissue, 0·9. The result, 17oz (484g), was surprisingly low. A herring gull with the same wing span weighs twice as much!

In 1981 a model of a *Pteranodon* skeleton with a wing span of 23ft (7m) was made for exhibition purposes in the Paläontologisches Museum in Munich, and it still hangs in the great hall of this museum, in flight position. As a test it was previously lifted into the air by a builders' crane, so that for a short time the skeleton could be seen flying over the roofs of Munich. It is not until you see it high above you in the air that you realize how small the actual body of *Pteranodon* was, and how extremely long the wings were in comparison.

Life Restorations

The Crystal Palace, the largest iron-framed glass building of its time, was built in London's Hyde Park to house the Great Exhibition of 1851. At the end of this international trade fair for technical products the Crystal Palace was dismantled and rebuilt in Sydenham in South London. It was set in a landscaped park, and

Above: Mounting of a life-sized skeletal reconstruction of *Pteranodon* in the workshop of the Munich Palaeontological Museum. Preparator Leonhard Bimmer (left) did much of the modelling of the bones and skull, and Ernst Schmieja (right) constructed the internal metal supporting framework. The skeleton is seen shortly before a 'test flight' in the spring of 1981 during which it was lifted into the air by a construction crane in the vicinity of the Museum.

Left: The *Pteranodon* skeleton replica in its final destination: the Great Hall of the Palaeontological Museum in Munich. It is on display hanging high above visitors' heads, suspended by a thin steel wire from the metal frame of the Museum's glass roof.

intended to serve as a permanent exhibition for arts and science.

It is said to have been Prince Albert, Queen Victoria's consort, who suggested that the park in Sydenham should be decorated with reconstructed prehistoric animals. Thus the animal painter and sculptor Benjamin Waterhouse Hawkins was commissioned to take on this task. After studying Richard Owen's monographs he decided to bring the giant reptiles of the Mesozoic back to life. Hawkins and Owen worked closely together, with Hawkins making the reconstructions to Owen's instructions. He made enormous clay moulds, from which he produced the casts. Thus the dinosaurs known at the time gradually came into being, *Iguanodon*, *Hylaeosaurus* and *Megalosaurus*, marine reptiles like ichthyosaurs, plesiosaurs and mosasaurs, and also crocodiles. Hawkins also constructed two large pterosaurs, which still sit majestically on a rock today, one with outspread wings, as if about to fly away, the other with folded wings. Queen Victoria and Prince Albert reopened the Crystal Palace in Sydenham on 10 June 1854, and they and 40,000 visitors greatly admired the animals reconstructed by Hawkins and Owen.[17]

The two Sydenham pterosaurs dating from 1854 were the first attempt at three-dimensional life restoration. Since then many museums and dinosaur parks have made life restorations of pterosaurs for exhibition purposes, like for example models of *Pteranodon* in the Haus der Natur in Salzburg, Austria, in the Museo Civico di Storia Naturale in Milan, Italy,[18] and in the Museum of Victoria in Melbourne, Australia. There is also a large *Pteranodon* hanging in the entrance hall of the Bäumlihof grammar school in Basel, Switzerland, with a wing span of about 26·3ft (8m). This excellently reconstructed model was built in aluminium and various plastics by D. Oppliger and C. Schärler, working to ideas by Professor Manfred Reichel. It only weighs 44lb (20kg), and must thus be fairly close to the weight of a live *Pteranodon*.

In the Zigong Dinosaur Museum in Dashanpu, Sichuan Province, China, is a life model

17 Desmond, A.J., 1975. *The Hot-Blooded Dinosaurs. A Revolution in Palaeontology.* Blond and Briggs, London.

18 Pinna, G., 1973. *La riconstruzione di uno Pteranodonte (Pteranodon ingens Marsh) esposta nel Museo Civico di Storia Naturale di Milano.* Natura, 64 (1): 35-39; Milan.

Above: Waterhouse Hawkins' studio as it appeared in 1853. Here Hawkins and Owen must have built the pterodactyls, the first three-dimensional life restorations of pterosaurs, which were put on display in the Crystal Palace Park at Sydenham.

Above: These are life restorations of *Pteranodon* which are to be seen hanging in the McCoy Hall of the Museum of Victoria in Melbourne, Australia. It is interesting to compare the width of the wing with that of another *Pteranodon* model, seen below.

Above: This model of *Pteranodon* was built by D. Oppliger and C. Schärler under the direction of Professor Manfred Reichel. Its weight is only 44lb (20kg) which is reckoned to be close to the actual weight of the living animal. It hangs in the hall of Bäumlihof Gymnasium in Basel, Switzerland.

Left: The pterodactyls built by Owen and Hawkins as illustrated in the *Guide to the Crystal Palace and Park* by S. Phillips in 1856. These models were part of a display opened by Queen Victoria and Prince Albert at Sydenham in 1854.

of an *Angustinaripterus* from the middle Jurassic of Zigong, on display.

Life-size models of *Rhamphorhynchus* can be seen moving as a mobile in the hall of the Paläontologisches Museum in Munich.

There is a model of an 'automatic', that is to say moving *Dimorphodon* from the Lower Lias of Dorset in 'The Changing Earth' gallery of the City of Bristol Museum and Art Gallery in England. Nicknamed Didi, it was the star of a 1985 BBC TV 'Wildlife on One' series.

There is a model of a *Dorygnathus* from the Upper Lias of Banz in Upper Franconia in the 'Petrefaktensammlung' of the former monastery of Banz. It is based on pterosaur bones found by Carl Theodori in 1830 in the neighbourhood of the monastery and described as *'Pterodactylus' banthensis.*

There is a life restoration of the filter eater *Pterodaustro* in the Museo Argentino de Ciencias Naturales in Buenos Aires, Argentina. This model was built by the Argentinian artist José

Left: These models of *Quetzalcoatlus* were built by Matt B. Smith, and are now on display in the Museum of the Rockies.

Above: 'Didi', the mechanically movable life model of *Dimorphodon* that appeared in the BBC's 'Wildlife on One' series.

Above: This life-size model of *Quetzalcoatlus* was built a few years ago in Washington DC, and now hangs in the US Museum of Natural History.

Left: *Pterodaustro*, the 'flamingo pterosaur' from the Lower Cretaceous of Argentina. This life model of the filter feeder is on display in the Museo Argentino de Ciencias Naturales in Buenos Aires.

Below: The author modelling a life restoration of *Rhamphorhynchus*. It is constructed with fully extended wings on the basis of the skeletal restoration shown on page 171.

Luis Gómez on the basis of the original fossil material of this 'flamingo pterosaur' from the Lower Cretaceous strata of San Luis.

Finally, life restorations of the giant Texan pterosaur *Quetzalcoatlus* have been constructed in recent years, in the National Museum of Natural History, Smithsonian Institution, in Washington DC, USA, in the New Mexico Museum of Natural History in Albuquerque, New Mexico, USA, and in the Museum of the Rockies in Bozeman, Montana.

Flying Models

The first attempt to make a model of a pterosaur that actually flew was made by the zoologist and behavioural researcher Erich von Holst in 1956.[19] On the basis of the 'artificial' birds he built a flying model of a Solnhofen *Rhamphorhynchus* of balsa wood, wire and Japanese paper. The wings were moved up and down by means of an ingenious but simple rubber band mechanism, so that it could use flapping flight. The wing span was 4ft (1·2m). E. von Holst wrote as follows about building his *Rhamphorhynchus* model: 'For those interested in technical matters, here in conclusion is a short sketch of the motor that made the *Rhamphorhynchus* model capable of flight. It was placed in the shell of the body, which is made of Japanese paper stiffened with glue and consists essentially of a balsa wood tube, through which runs a strong rubber band attached at one end to the axis of the pulley, around which a nylon thread is wrapped. The thread runs over a wheel which is held in a

19 Holst, E. von, 1957. *Wie flog Rhamphorhynchus.* Natur und Volk, 87: 81-87; Frankfurt.

flexible position to a plate, the ends of the axis of which run out in two cranks which are connected by a rod to the arm section of the wing. When it is wound up the crank is turned, the thread taken from the pulley on to the plate and the rubber band subjected to torsion, causing the wing to flap up and down and at the same time swinging the wing arm around its horizontal axis, thus creating the twisting necessary for propulsion. The form of the plate regulates the distribution of force in the flap of the wing, the conical form of the pulley balances the tension in the rubber, so that all the 30 to 40 beats of the wing are executed with equal force. The model flies horizontally or at a climbing angle of up to 30° with two to three beats of the wing per second. This is followed by gliding flight. Turns are achieved by various positional angles of the wings.'

At the annual meeting of the Paläontologische Gesellschaft in Wilhelmshaven in autumn 1956 the model flew successfully. This experiment was not intended to produce a spectacular effect, but to clear up a question that was troubling scientists at the time, whether the terminal tail vane of *Rhamphorhynchus* was oriented horizontally or vertically in flight. The only possible solution to this question seemed to be by experiment. The result was that the model only flew when the tail membrane was horizontal, but crashed when it was vertical. Holst concluded from this that *Rhamphorhynchus*' tail membrane was not a rudder, but an elevator, regulating height.

In 1978 Stephen Winkworth, an Englishman who made model planes as a hobby, built a model of a *Pteranodon* which was capable of gliding like a model glider. Winkworth succeeded in building and flying first a small version with a wing span of 5ft (1·5m), and then a model that was twice as big. Then in 1985 he built a radio-controlled *Pteranodon* model with a wing span of 15ft (4·6m). His pterosaur models were static gliders, not capable of flapping flight.[20] The model was made of balsa-plywood, the feet were oriented vertically with spread digits, and served as a diagonal rudder. The radio-controlled *Pteranodon* flew excellently, and was the star of the BBC's scientific programme *Pterodactylus lives*, broadcast in January 1985. Winkworth's *Pteranodon* mastered its film role perfectly, and in good weather with a perfect wind for hang gliding, it flew by the Dorset cliffs where pterosaurs like *Dimorphodon* had circled 200 million years earlier.

But by far the most spectacular attempt to reconstruct a flying pterosaur was without a doubt the *Quetzalcoatlus* project, launched in America in 1984. The aim was to build a half-life-size flying model of the giant Texas pterosaur, thus with a wing span of 18ft (5·5m). It was intended not only to soar, but actively to move forwards by flapping its wings. This extraordinarily difficult project was taken on by the American aeronautical engineer Dr Paul MacCready and his team from AeroVironment Inc. in Monrovia, California.[21]

The actively flying *Quetzalcoatlus* model was intended for a film called 'On the Wing' for the Smithsonian Air and Space Museum in Washington D.C. The project was 'a large-format IMAX film exploring the dynamic relationship between natural and mechanical flight, contrasting the biological evolution of winged creatures with the technical innovation of man.'

The life-like, flying model of the giant pterosaur was to be the climax of the film, intended to fly realistically, move forwards by flapping

A Flying Rhamphorhynchus (right)
In 1956 Professor Erich von Holst built a flying model of *Rhamphorhynchus* out of balsa wood, wire and rice paper. This drawing shows the power unit that flapped the wings. The mechanism was driven by a rubber band that could deliver up to 30 or 40 wings beats.

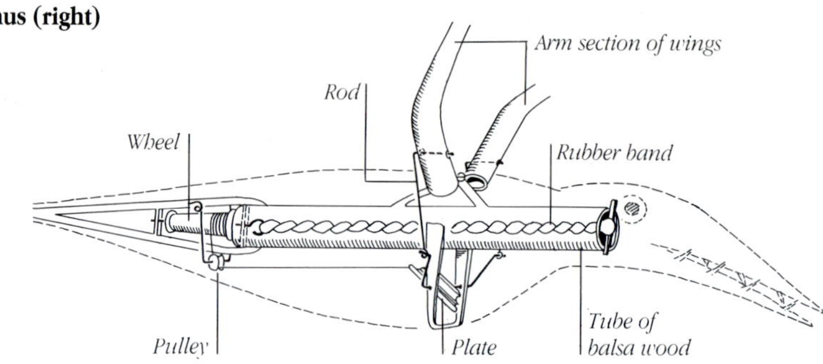

Flapping Action (right)
These drawings show how von Holst's model of *Rhamphorhynchus* flapped its wings in flight. They beat at a frequency of two or three flaps a second. For aerodynamic reasons the tail membrane had to be oriented horizontally to keep the model airborne.

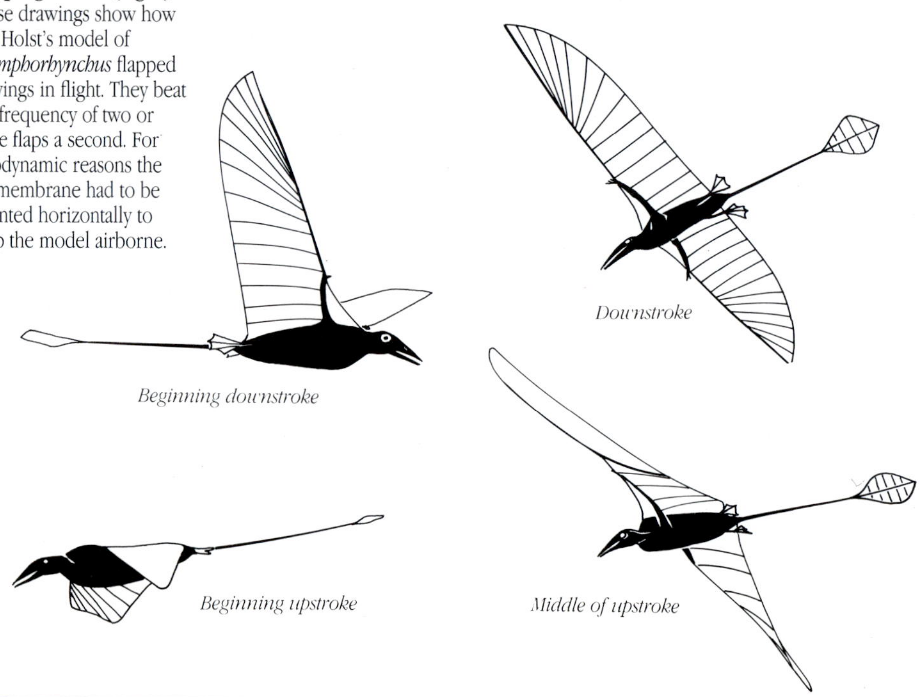

Left and below: These photographs show Stephen Winkworth and his model of *Pteranodon* which was constructed as a glider with radio-controlled steering. The model was made of balsa wood and had a wing span of 15ft (4·6m). It flew successfully, and featured in a BBC TV programme in 1985 when it was seen soaring above the cliffs of the Dorset coast in southern England. Whether *Pteranodon* actually used its feet in life for steering itself, as the model did, remains open to question, however.

20 Winkworth, S., 1985. *Pteranodon*. Flug und Modelltechnik, 359, 990-993, Verlag für Technik und Handwerk, Baden-Baden.
Winkworth, S., 1985. *Pteranodon flies again*. New Scientist, 3 January 1985: 32-33.
21 MacCready, P., 1985. *The Great Pterodactyl Project*. Engineering and Science, November 1985: 18-24; CIT, Pasadena.
Dr Paul MacCready became internationally known in 1977 as the 'father of human-powered flight' when his 'Gossamer Condor' made the first controlled flight achieved by human muscle-power. Two years later he created the 'Gossamer Albatross', a 55lb (25kg) flying machine with a 96ft (30m) wing span, which crossed the English Channel using muscle power.
Parrish, M., 1986. *Flying as they did 65 million years ago*. Smithsonian, 16 (12): 72-81; Washington.

Above, right and below: In 1985 Dr Paul MacCready and his team from AeroVironment Inc. built a half-life-size replica of *Quetzalcoatlus* which flew by flapping its wings. The model had a wing span of 18ft (5·5m), weighed 44lb (20kg), and was capable of flying at 35mph (56km/h). It was flown successfully many times at Death Valley in California, but crashed on its first public flight in 1986.

its wings, be completely under control in a normal flying position, and fly for a few minutes under electric power.

In order to fulfil these conditions the AeroVironment engineers had to develop a special, computer-controlled auto-pilot system. To maintain a stable flight position the pterosaur had to move its head sideways, stretch its digits on the leading edge of the wing, and not only move its wings up and down, but also twist them and move them forwards and backwards.

After a series of small test models the 18ft (5·5m) flapping version was finally built. The *Quetzalcoatlus* model made its first successful flight on 1 December 1985, and early in 1986 made over 20 trouble-free flights in Death Valley in California. Each flight lasted about four minutes. These impressive, remote-controlled flights were captured for the film 'On the Wing', and can still be seen in the Air and Space Museum in Washington D.C. Unfortunately, on the occasion of its first public demonstration flight at Andrews Air Force Base near Washington on 17 May 1986, *Quetzalcoatlus* crashed shortly after take-off and was smashed to pieces.[22] Dr Wann Langston's eye witness report tells us that *Quetzalcoatlus*' last flight at Andrews Air Force Base went like this: the pterosaur took off without any problem in front of thousands of spectators. The model was winched into the air, assisted by a tail boom in the form of an aircraft tailplane for this climbing phase. This auxiliary element was normally discarded by remote control once the desired altitude had been reached and the reptile flew independently on, flapping its wings. For some unknown reason the tailplane came off a few seconds too early, at a height of about 400ft (120m). At this point, however, the auto-pilot in the head of the model was not switched on. Without automatic pilot the pterosaur got out of control and looped the loop. When it came out of this circle the automatic pilot finally started to work. But at the same moment a violent gust of wind blew the head so hard to one side that the neck broke, and *Quetzalcoatlus* crashed. The emergency parachute in the rear of the model did open, but only 60ft (18m) above the ground, so that a hard crash-landing was unavoidable. The model has since been restored, and is on show in the National Air and Space Museum, Washington D.C. There are no plans to make it capable of flight again. It is hoped that one day it will be possible to build an actual-size model of *Quetzalcoatlus*.

The model weighed 44lb (20kg), and flew at 35 mph (56km/h). It carried a radio receiver, autopilot system, sensors, 56 batteries and two electric motors to move the wings.[23]

The greatest difficulty in controlling the giant pterosaur was caused by the fact that these Cretaceous forms were tailless. Thus aerodynamically effective steering aids had to be developed. For taking off, which was achieved by means of a winch, the model needed an undercarriage and a tail-boom that was dropped when the correct height was reached, and floated down on a parachute.

Despite the final failure, Paul MacCready's *Quetzalcoatlus* was a wonderful enterprise, and it generated fascinating pictures of a flying giant pterosaur from the dim and distant past. You felt as though you have travelled 65 million years into the past, and like Conan Doyle's Professor Challenger were seeing a pterosaur flying with slowly flapping wings. Dr Langston summed it up: 'I think that in the air the *Quetzalcoatlus* model was probably very close to the living creature.'

22 Wellnhofer, P., 1986. *Die Saurier fliegen wieder.* Fossilien, 2/86: 73-79.
Wellnhofer, P., 1986. *Der Saurier flog wieder.* Fossilien, 6/86: 258-264.
Paul, G.S., 1987. *Pterodactyl habits – real and radio-controlled.* Nature, 328: 481.

23 Technical details of the *Quetzalcoatlus* model from a press release of the Smithsonian Institution's National Air and Space Museum:
Control system: On-board, eight channel radio receiver with an autopilot custom-built for QN (*Quetzalcoatlus northropi*). Three axis autopilot system. The autopilot receives information from the sensors during flight. The sensors consist of: a yaw vein below the neck; a pitch-rate gyro, and a yaw-rate gyro inside the body. The system uses a specially modified transmitter.
Power system: 56 sub-C NiCad batteries (6lb (2·7kg) total battery weight). Wing flapping by two 1-HP samarium-cobalt DC motors.
Flapping rate: 1·2 flapping cycles per second.
Pilot commands by radio control: Pitch attitude, turn rate, flapping amplitude, autopilot on/off, tailboom elevator, tailboom drop, head up/down, emergency parachute deploy.
Construction: The main structural elements are made from thin wall carbon fiber tubing. The wing spars are constructed with carbon fiber C-channel. The airfoil streamline shape of the wings is expanded polystyrene foam. Body, neck and head shells are modeled from Kevlar. Skin covering for the wings is rubber sheeting, ·003in thickness. Parts of the body use synthetic fur covering with the backing removed.

OTHER FLYING VERTEBRATES

Coelurosauravus (left)
Coelurosauravus was a lizard-like, insect-eating reptile that could use its wing membranes supported by 21 elongated ribs for gliding from trees. Skeletal remains of this gliding reptile, the oldest in the fossil record, were discovered in Upper Permian marine deposits.

Icarosaurus (right)
Like *Coelurosauravus*, *Icarosaurus* was able to glide by means of a membrane of skin that was supported by a series of elongated ribs. Dr E.H. Colbert described the fossil, naming the genus after Icarus, the ill-fated son of Daedalus in Greek mythology.

In the course of their evolutionary history vertebrates often succeeded in conquering gravity and – at least for a time – raising themselves from the ground and using the air to move about in. Fish evolved species of flying fish and amphibians flying frogs, which could glide through the air for short distances, but it was above all reptiles, birds and mammals which produced effective flying forms in the course of the history of the Earth.

Gliding Reptiles

Small gliding reptiles are known as early as the Palaeozoic (late Permian), named *Daedalosaurus* by Dr Robert Carroll of McGill University in Montreal, after Daedalus in Greek legend, who managed to escape from captivity in Crete by flying with wings he had built himself. The fossilized skeletal remains of *Daedalosaurus* were found on Madagascar. They were not quite complete, but showed very characteristically elongated ribs, 21 on each side of the body, used to support a gliding membrane.[1] This lizard-like animal had a long tail and was about 16in (40cm) long. The span of the open 'wings' was about 13in (33cm). In its body structure *Daedalosaurus* is reminiscent of the modern flying dragon *Draco* from the agamid family, which today lives in South-East Asia. *Draco* has large wing-like flaps of skin on its flanks, supported on each side by five to seven elongated ribs. These skin wings are usually folded sideways flat against the body, so that they do not hinder the animal when it is climbing trees. They are spread by pulling the moveable ribs forwards, and then the flying dragon can glide from tree to tree. Flights of up to 200ft (60m) have been observed.

1 Carroll, R.L., 1978. *A Gliding Reptile from the Upper Permian of Madagascar.* Palaeontographica Africana, 21: 143-159.
Evans, S.E., 1982. *The gliding reptiles of the Upper Permian.* Zoological Journal of the Linnean Society, 76: 97-123; London.

Fossil remains have been found in Germany, England and Madagascar of another gliding reptile from the late Permian that is also related to *Daedalosaurus*.[2] It was named *Coelurosauravus* (hollow-tail-reptile). The rare skeletal remains from the so-called Kupferschiefer (late Permian) of Germany were formerly known as *Weigeltisaurus*, after Prof. J. Weigelt of Halle.

Similar too were two other fossil gliding lizards discovered in late Triassic rocks, *Kuehneosaurus*, named after Prof. W. Kühne, from the Bristol Channel area in Great Britain,[3] and *Icarosaurus*, named after Daedalus' son Icarus, from New Jersey, USA.[4] Their wings were also supported by long ribs, although there were only 10 to 11 pairs, meaning that the rib wings were narrower in shape. Perhaps they could also fold their wings back when climbing up tree trunks to catch insects. In any case this great area of skin, which certainly contained blood vessels, must have had a significant effect on heat regulation. The living *Draco* uses its wings, magnificently coloured in orange and red and blue with black patches on the underside, in courtship display.

In the same article in which Moscow zoologist A.G. Sharov described the hairy pterosaur *Sordes pilosus* from the Upper Jurassic of Kazakhstan in the Soviet Union, he also reported on two fossil reptile finds from late Triassic sediments in Kirghizia.[5] These two were gliding reptiles, but they had developed completely different 'wings' from the rib

2 Schaumberg, G., 1976. *Zwei Reptilneufunde (Weigeltisaurus Kuhn (?), Lepidosauria (?), Reptilia) aus dem Kupferschiefer von Richelsdorf (Perm, Hessen).* Philippia, 3 (1): 3-8; Kassel.
Evans, S.E. and Haubold, H., 1987. *A review of the Upper Permian genera Coelurosauravus, Weigeltisaurus and Gracilisaurus (Reptilia, Diapsida).* Zoological Journal of the Linnean Society, 90: 275-303; London.
3 Robinson, P.L., 1962. *Gliding lizards from the Upper Keuper of Great Britain.* Proceedings of the Geological Society London, 1601: 137-146.
4 Colbert, E.H., 1970. *The Triassic gliding reptile Icarosaurus.* Bulletin of the American Museum of Natural History, 143 (2): 85-142; New York.
5 Sharov, A.G., 1971. *New flying reptiles from the Mesozoic of Kazakhstan and Kirghizia.* Trudy of the Palaeontological Institute, Akademia Nauk, USSR, 130: 104-113; Moscow (in Russian).

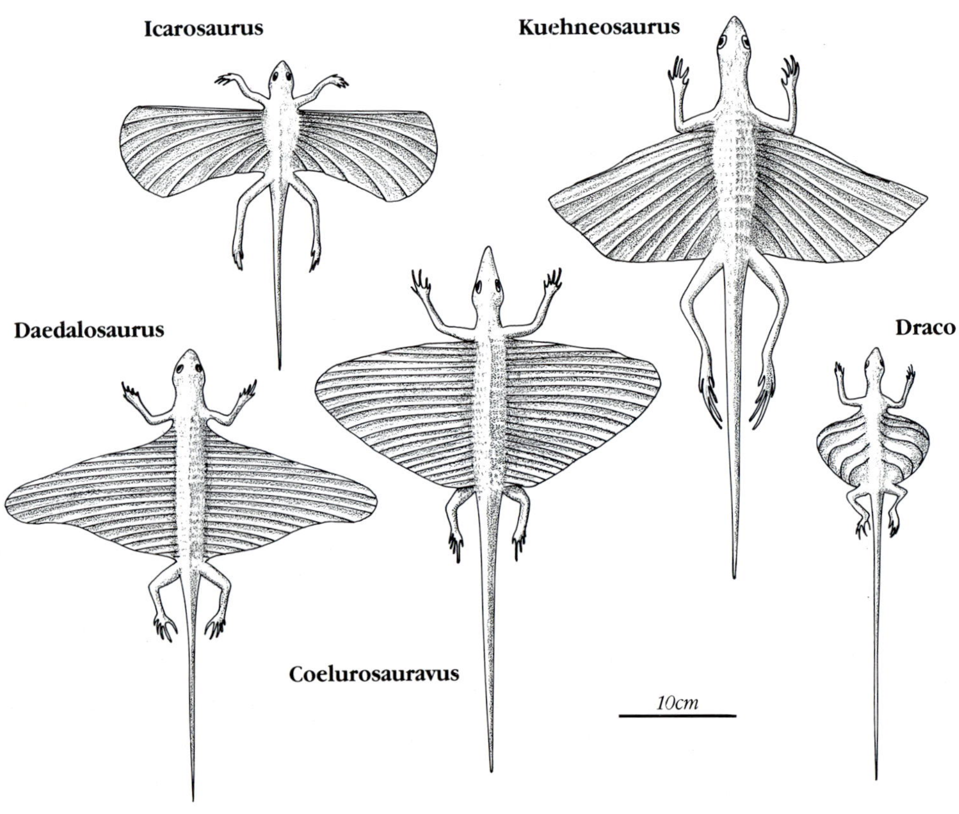

Ancient Gliding Reptiles (left)
These drawings compare gliding reptiles of the late Permian, *Daedalosaurus* and *Coelurosauravus*, and of the late Triassic, *Icarosaurus* and *Kuehneosaurus*, with the living *Draco*, the Flying Dragon, an agamid lizard from South-East Asia. The thoracic ribs of all these diapsid reptiles are greatly elongated in order to support a gliding membrane which is stretched across them, so enabling these arboreal animals to glide from one tree to another, or from a branch down to the ground.

Above: Three schoolboys exploring an old quarry in New Jersey, across the Hudson River from Manhatten, New York. It was studied by E.H. Colbert. found this specimen of *Icarosaurus siefkeri* in a piece of Upper Triassic shale while they were

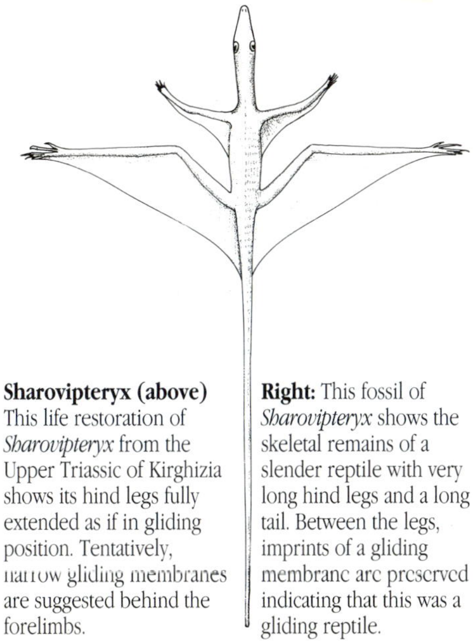

Sharovipteryx (above) This life restoration of *Sharovipteryx* from the Upper Triassic of Kirghizia shows its hind legs fully extended as if in gliding position. Tentatively, narrow gliding membranes are suggested behind the forelimbs.

Right: This fossil of *Sharovipteryx* shows the skeletal remains of a slender reptile with very long hind legs and a long tail. Between the legs, imprints of a gliding membrane are preserved indicating that this was a gliding reptile.

Longisquama (below) *Longisquama* had a unique gliding adaptation. A double series of long scale-like appendages were developed along its back. They could be folded and unfolded like the wings of a butterfly to form a continuous wing area.

gliders *Daedalosaurus*, *Coelurosauravus*, *Kuehneosaurus* and *Icarosaurus* described above.

Sharov named the first reptile *Podopteryx* (leg wing), because it had flight membranes between its long hind legs and tail. The flight membrane survived as imprints on the rock surface of the fossil slab. Later however it turned out that the name *Podopteryx* had been assigned to a fish a hundred years before. Thus the reptile from Kirghizia had to be given a new name. In honour of Dr Sharov it was very suitably named *Sharovipteryx* (Sharov wing).[6]

Sharovipteryx was a slim reptile about 10in (25cm) long, with a relatively long neck and enormously long hind legs, but very small front legs. A triangular flight membrane could be stretched between the long hind legs and the front section of the tail, real leg wings in fact. Certainly *Sharovipteryx* was not in a position to flap these leg wings, and fly actively, but it could probably glide for a certain distance. In flight the upper legs were spread slightly forwards and the lower legs and feet spread at a right angle from the body to the side. It is possible that flaps of skin had developed on the flanks or the back edge of the forelimbs, to stabilize gliding flight.[7]

The second reptile described by Sharov from the late Triassic of Kirghizia he named *Longisquama* (long scale), because of a row of enormously elongated scale-like appendages along the back. Its osteological characteristics suggested that the reptile was a small pseudosuchian. The animal is relatively small, about 4 to 5in (10 to 12·5cm) long. The appendages on the back are longer than the body and arranged in a double row one behind the other. They could apparently be folded upward like a butterfly wing and folded down to the sides. In this position they formed horizontal gliding surfaces, a kind of wing.[8] Each scale-shaped appendage consists of a very long, very thin shaft, broadening towards the distal end and bent backwards. Probably each pair of these appendages corresponds to a dorsal vertebra, thus 10 pairs, in close sequence and overlapping at their edges, which could create a continuous wing surface.

Longisquama was presumably also an arboreal reptile that certainly did not use its wings for flapping flight, but probably for gliding. It thus documents a unique solution to the problem of gliding flight, and together with the other gliders *Daedalosaurus, Coelurosauravus, Icarosaurus, Kuehneosaurus* and *Sharovipteryx*, shows that even in the later Permian and Triassic there was a very wide range of aerial adaptations among the various reptile

6 Cowen, R., 1981. *Homonyms of Podopteryx.* Journal of Palaeontology, 55: 483.

7 Gans, C., Darevski, I. and Tatarinov, L.P., 1987. *Sharovipteryx, a reptilian glider?* Paleobiology, 13 (4): 415-426.

8 Haubold, H. and Buffetaut, E., 1987. *Une nouvelle interprétation de Longisquama insignis, reptile énigmatique du Trias supérieur d'Asie central.* Comptes Rendus Académie des Sciences, Paris, 305 (II): 65-70.

groups, long before the radiation of the actively flying vertebrates, i.e pterosaurs, birds and bats.

Birds

Early adaptations of vertebrates to flying in the air produced only gliding, i.e. passive exploitation of lift and drag forces, created by static aerodynamic surfaces on the body or the limbs. But in the course of vertebrate evolution the ability to fly actively emerged as well, in other words powered flapping flight achieved by muscle power.

Active vertebrate fliers are always distinguished from passive gliders by the fact that their pectoral girdle and forelimb are turned into flying apparatus, or wings. Gliders retain their four legs, and the wings are an additional feature. Active fliers always acquired wings at the cost of their forelegs.

As is shown in this book, pterosaurs were the first vertebrates to develop active flight. The second group were the birds, who did not appear until the pterosaurs had already achieved a high degree of diversity and worldwide distribution. The oldest fossil bird known today was *Archaeopteryx* (ancient wing). It lived in the late Jurassic, about 150 million years ago, and was a contemporary of pterosaurs and dinosaurs. To date six remains of fossil skeletons of this primeval bird have been found. They all came from the Solnhofen limestone of Bavaria and are considered the most important and famous fossil finds of all for evolutionary research.

As early as 1860 workers in the Solnhofen quarry found the imprint of a small bird's feather. At the time this was the first sign that birds must have existed in the Jurassic. The oldest fossil remains of birds were known only from much more recent Tertiary strata. The Solnhofen bird's feather was about 2·5in (6cm) long, with the same structure as a modern feather, down to the finest details. Hermann von Meyer, then the leading expert on fossil vertebrates in Germany, named the bird from which this feather had fallen *Archaeopteryx lithographica*, the 'ancient wing from the lithographic limestone'.[9] At the time Solnhofen limestone slabs were used for lithography more than anything else.

In the very next year, 1861, a complete skeleton of this primeval bird was found near Solnhofen. Feathers were again imprinted on the surface of the rock in the region of the wings and the long vertebral tail. The rare and valuable fossil immediately came into possession of Carl Haeberlein, a country doctor in nearby Pappenheim, who shortly afterwards sold it to the British Museum (Natural History) in London for £700. There it was studied by Richard Owen, who published a precise description of it in 1863.[10]

Further specimens of *Archaeopteryx* came to light in 1876, 1951, 1956 and 1987. They are now in museums in Berlin, Eichstätt and Solnhofen.[11] One find is still in a private collection. There are also skeletal remains from Solnhofen limestone interpreted as a pterosaur by Hermann von Meyer in 1857, and given the name *Pterodactylus crassipes*.[12] This is the 'Haarlem specimen', first recognized as *Archaeopteryx* by John Ostrom in 1970.[13]

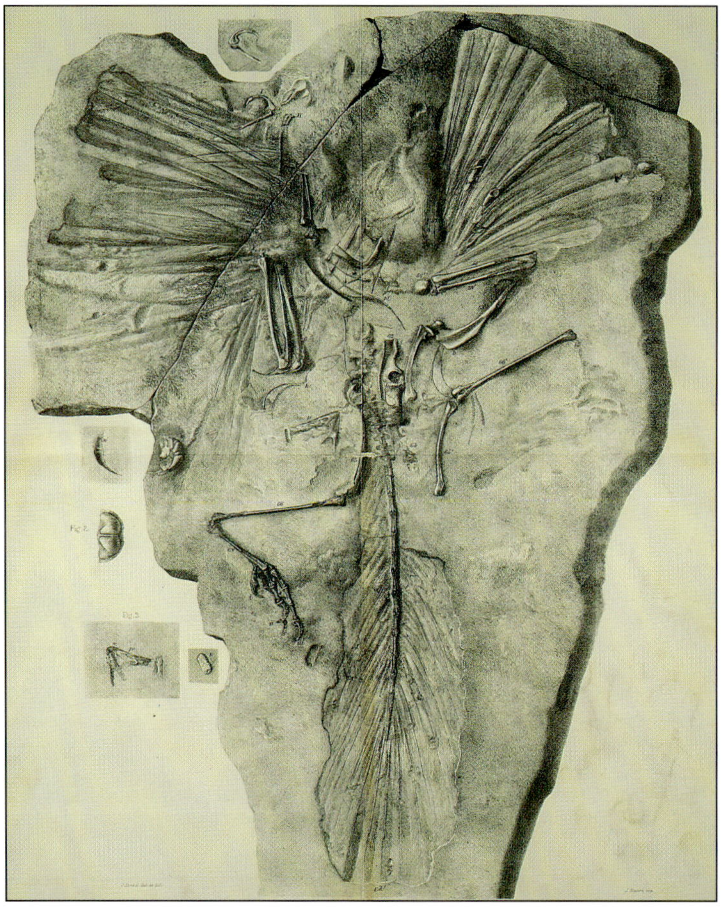

Left: This is Sir Richard Owen's lithograph of the London specimen of *Archaeopteryx* which appeared in his 1863 description of that specimen. It was found in 1861 near Solnhofen, purchased by Dr Carl Haeberlein in Pappenheim, and eventually sold to the British Museum (Natural History) in London. The skeletal remains were associated with impressions of feathers and so Owen classified it as a bird, though a primitive one, rather than as a transitional form halfway between reptiles and birds as the Darwinists saw it.

Below: This fossil feather of a primeval bird, later named *Archaeopteryx*, was discovered in the Solnhofen limestone in 1860. It was the first evidence of the existence of birds in periods older than the Tertiary. The details of the feather are preserved as black dendritic material rather than by impression in the sedimentary surface.

What are the particular features of this primeval bird *Archaeopteryx*? As it is the oldest bird known to man, one naturally expects it to provide indications of its origin and the origin of avian flight. In fact this bird has many characteristics that could be evidence of its evolutionary history, and its possible ancestors. Even early researchers noticed that the creature had bird's feathers, so must have been a bird, but it also had small teeth in its jaws, claws which it could move freely on its wings and a long tail made up of many vertebrae, all reptilian characteristics not known in modern birds. This mosaic of primitive and advanced characters led Sir Gavin de Beer to suggest a mosaic mode of evolution.[14] This made *Archaeopteryx* a transitional form, or a 'missing link' between two different animal classes, reptiles and birds. It is easy to under-

9 Meyer, H. von, 1861. *Archaeopteryx lithographica (Vogelfeder) und Pterodactylus von Solnhofen.* Neues Jahrbuch für Mineralogie, Geologie und Paläontologie, 1861: 678-679; Stuttgart.

10 Owen, R., 1863. *On the Archaeopteryx of von Meyer, with a description of the fossil remains of a long-tailed species from the lithographic stone of Solnhofen.* Philosophical Transactions, 153: 33-47; London. This is the first description of the 'London specimen' housed in the Natural History Museum.

11 Dames, W., 1884. *Ueber Archaeopteryx.* Palaeontologische Abhandlungen, 2 (3): 119-198; Berlin. Description of the 'Berlin specimen' in the Naturkundemuseum of Humboldt University, Berlin.
Heller, F., 1959. *Ein dritter Archaeopteryx-Fund aus den Solnhofener Plattenkalken von Langenaltheim/Mfr.* Erlanger Geologische Abhandlungen, 31: 1-25; Erlangen. Description of the 'Maxberg specimen', still in a private collection.
Mayr, F.X., 1973. *Ein neuer Archaeopteryx-Fund.* Paläontologische Zeitschrift, 47: 17-24; Stuttgart.
Wellnhofer, P., 1974. *Das fünfte Skelettexemplar von Archaeopteryx.* Palaeontographica, A, 147: 169-216; Stuttgart. Description of the 'Eichstätt specimen' in the Jura-Museum in Eichstätt, Bavaria.
Wellnhofer, P., 1988. *Ein neues Exemplar von Archaeopteryx.* Archaeopteryx, 6: 1-30; Eichstätt. Description of the 'Solnhofen specimen' in the Bürgermeister Müller Museum in Solnhofen.

12 Meyer, H. von, 1857. *Beiträge zur näheren Kenntnis fossiler Reptilien.* Neues Jahrbuch für Mineralogie, Geologie und Paläontologie, 1857: 437; Stuttgart.
Meyer, H. von, 1860. *Zur Fauna der Vorwelt. Reptilien aus dem lithographischen Schiefer in Deutschland und Frankreich.* Frankfurt. Description of *Pterodactylus crassipes*, later to be recognized as *Archaeopteryx*, the 'Haarlem specimen' by J.H. Ostrom.

13 Ostrom, J.H., 1970. *Archaeopteryx: Notice of a 'new' specimen.* Science, 170: 537-538; Washington.
Ostrom, J.H., 1972. *Description of the Archaeopteryx specimen in the Teyler Museum, Haarlem.* Proceedings of the Koninklijke Nederlandse Akademie van Wetenschappen, B, 75: 289-305; Amsterdam.

14 De Beer, G., 1954. *Archaeopteryx lithographica: a study based on the British Museum specimen.* 68 pp., London, British Museum (Natural History).

Left and below left: The 'Berlin specimen' of *Archaeopteryx* (left), discovered in 1876, reveals the sharp imprints of the feathered wings and tail in fine detail. The 'Solnhofen specimen' (below left) was discovered in the private collection of the former mayor of Solnhofen, F. Müller, in 1987. It had been mistaken for the small dinosaur *Compsognathus*.

stand that the discovery of such important proof of Charles Darwin's theory of evolution, only two years after the appearance of his famous book *On the Origin of Species* in 1859, caused a considerable stir.[15] *Archaeopteryx* immediately unleashed heated debate between anti-Darwinists like Richard Owen or Andreas Wagner in Munich, and Thomas Henry Huxley, Darwin's 'bulldog'.[16]

Precise analysis of the skeletal remains of *Archaeopteryx* so far known shows a high degree of correspondence in many osteological details with certain predatory dinosaurs, the Theropoda ('beast feet'). John Ostrom of Yale University in New Haven was from 1973 a particular proponent of the ideas of theropod ancestry,[17] and he produced many sound reasons for this. The structure of pelvis and hind legs, and osteological features of arm, foot and skull in particular suggest a clear relationship with dinosaurs like *Deinonychus, Ornitholestes* or *Compsognathus*. *Archaeopteryx*' dinosaur characteristics are so convincing that some palaeontologists would like to classify the birds as Dinosauria.[18] Despite this some specialists are of the opinion that *Archaeopteryx* and thus birds as well are descended from another group of reptiles, namely the pseudosuchians of the Triassic. As the pseudosuchians are also considered ancestors of the later archosaurs, and thus also of dinosaurs, this dispute is reduced to the question of whether birds were descended directly from the pseudosuchians, or indirectly, via the dinosaurs. These and other opposing viewpoints were discussed at the first International *Archaeopteryx* Conference in Eichstätt, not far from Solnhofen in 1984.[19]

Origin of Feathers

The only genuine and exclusive bird characteristic of *Archaeopteryx* is its feathers. The wing feathers are already very modern, and asymmetrical in form, which only makes sense if they had an aerodynamic function, as in modern birds capable of flight.[20] Feathers played a key role in the development of flight. It is generally accepted that birds' feathers developed from reptile scales. *Archaeopteryx*' still unknown ancestor, the hypothetical *Proavis*, must also have had a feather-like body-covering, and that even before flight was achieved. The question therefore arises, what was the primary function of these feathers? Were they a protection against cooling for creatures that were already warm-blooded, or the reverse, a shield against excessive irradiation by the sun for cold-blooded animals, a protection against overheating? Were the long, feathered forearms intended for display and fighting, or as fly-swatters for catching insects?

Above: Charles Darwin had hypothesized that birds must have developed from reptiles. The discovery of *Archaeopteryx* in 1861 provided evidence of just such an evolution: the intermediate reptile-bird was the 'missing link' in the fossil record.

15 Darwin, C., 1859. *On the Origin of Species by Means of Natural Selection, or the Preservation of Favoured Races in the Struggle for Life.* London.
16 The debate on the significance of *Archaeopteryx* is presented in a most interesting and readable fashion in the following two books:
Desmond, A.J., 1975. *The hot-blooded dinosaurs.* p.134 ff., Blond and Briggs, London.
Wilford, J.N., 1985. *The Riddle of the Dinosaurs.* p.71 ff., Knopf Inc., New York.

17 Ostrom, J.H., 1973. *The ancestry of birds.* Nature, 242: 136; London.
Ostrom, J.H., 1976. *Archaeopteryx and the origin of birds.* Biological Journal of the Linnean Society, 8: 91-182; London.
18 Bakker, R.T. and Galton, P.M., 1974. *Dinosaur monophyly and a new class of vertebrates.* Nature, 248: 168-172; London.
19 Hecht, M.K., Ostrom, J.H., Viohl, G. and Wellnhofer, P. (editors), 1985. *The Beginnings of Birds. Proceedings of the International Archaeopteryx Conference Eichstätt, 1984.* 38 contributions, 382 pp., Freunde des Jura-Museums, Eichstätt.

20 Feduccia, A., and Tordoff. H.B., 1979. *Feathers of Archaeopteryx: asymmetric vane indicates aerodynamic function.* Science, 203: 1021-1022; Washington.

Archaeopteryx

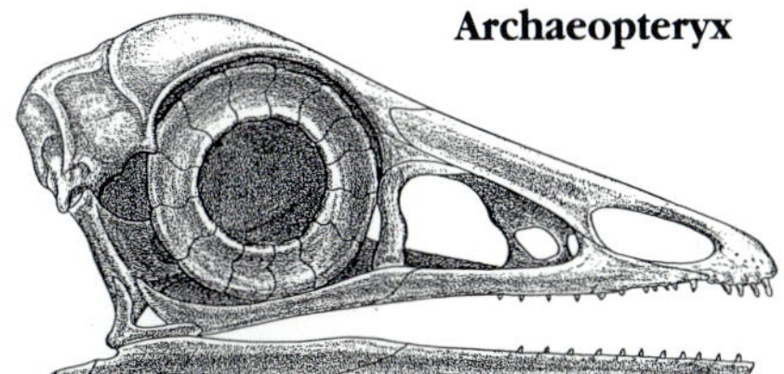

Compsognathus

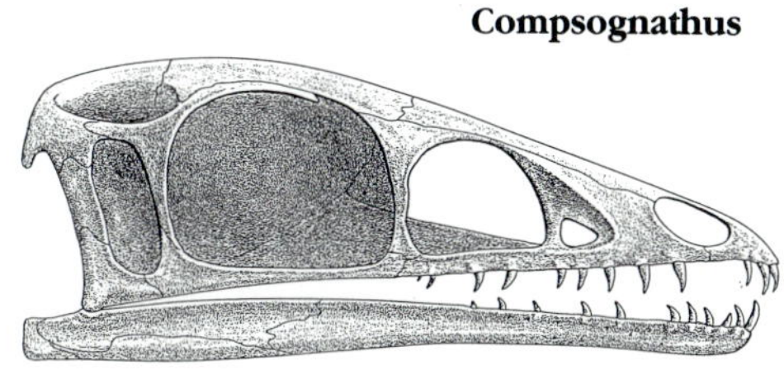

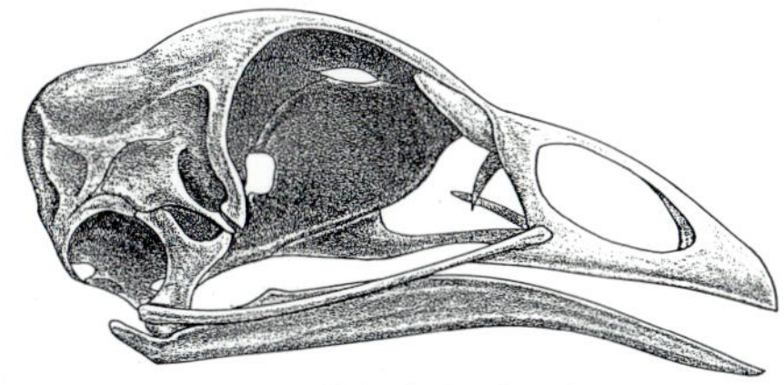

Gallus (Chicken)

Skull Comparisons (above right)
These drawings allow us to compare the reconstruction of the skull of *Archaeopteryx* (above), with the skull of a small theropod dinosaur, *Compsognathus* (above right), and an extant bird, a chicken (lower right). The teeth and the openings in front of the eye socket that we see in the skull of *Archaeopteryx* are reptilian features i.e. archosaurian characteristics. They are also present in *Compsognathus*, while modern birds by contrast lack teeth and have horny bills. However, the large eye socket and roomy braincase appear more bird-like than reptilian in nature, although small theropod dinosaurs also had large eye sockets. *Archaeopteryx* thus seems to occupy an intermediate evolutionary position between dinosaurs and birds, a view which is now widely accepted.

Were the outspread wings a kind of canopy to shade the surface of the water and so eliminate reflections when looking for food below the surface, or were the feathers primarily intended to have a water-repellent effect? There is still no satisfactory answer to these questions.

Evolution of Flight in Birds

There are different opinions about the evolution of bird flight, just as there are about the ancestry of birds. Essentially today there are two opposing theories, the arboreal theory and the cursorial theory.

The cursorial theory postulates that avian flight developed from the ground up. Its principal proponents like S.W. Williston (1879), Baron F. Nopcsa (1907) and J.H. Ostrom today argue that osteological peculiarities of pelvis and hind legs in particular of *Archaeopteryx* show that it was a typical biped. In John Ostrom's words (1985): 'In summary, it is my conviction that *Archaeopteryx* was still learning to fly – from the ground up – and that avian flight began in a running, leaping, ground-dwelling biped.'[21] The primeval bird began to flap its wings as it leapt up to catch insects in flight. Enlargement of the feathered area led to bigger and longer jumps, bringing with it a selection advantage that gradually led to active flapping flight.

This cursorial theory is completely opposed to the conventional arboreal theory, according to which flight must have developed from the trees down. This theory was postulated by the Danish scientist G. Heilmann in particular,[22] and is recognized by many biologists today. Its supporters' principal objection to the cursorial theory is that the energy requirements for fast running and simultaneous wing flapping are very high, and additionally that flight from the ground up is against the force of gravity. Gliding from the trees down needs much less energy, as it exploits the Earth's pull. Transport costs in terms of energy are further reduced if the animal extends its gliding flights by flapping its wings, and can thus enlarge its radius of action, for example when looking for food.

A further point is that *Archaeopteryx*' digital claws are highly curved, and had extremely sharp points. The toe claws were less curved but also had pointed horn claws, ideally adapted to climbing up tree trunks. *Archaeopteryx* could support itself with its long rigid tail when doing this. Birds that live exclusively on the ground and run on two legs have very worn claws.

Walter Bock, professor at Columbia University, New York, has postulated several stages that could have led to active flight in birds.[23] He assumes that small reptiles, 'protobirds', started to live in trees to hide, to sleep and possibly to nest. In the cool environment of the tree-tops they developed warm-bloodedness and feathers as insulation. Elongation of the feathers on the arms at first only served to soften falls to the ground by an increase in surface area. This was the start of the evolution of wings, which finally led via a parachute and gliding stage to active flapping flight, a stage

21 Ostrom, J.H., 1985. *The Meaning of Archaeopteryx.* in ref. 19: 161-176.

22 Heilmann, G., 1926. *The Origin of Birds.* 210 pp., Appleton, New York.

23 Bock, W.J., 1985. *The Arboreal Theory for the Origin of Birds.* in ref. 19: 199-207.

Archaeopteryx in Action (above)
This life restoration shows *Archaeopteryx* flying down from a tree, running on the ground, and climbing up another tree assisted by the sharp claws on its fingers and toes. In this way of life both arboreal and cursorial adaptations could be exploited to advantage, and were in fact mutually complementary.

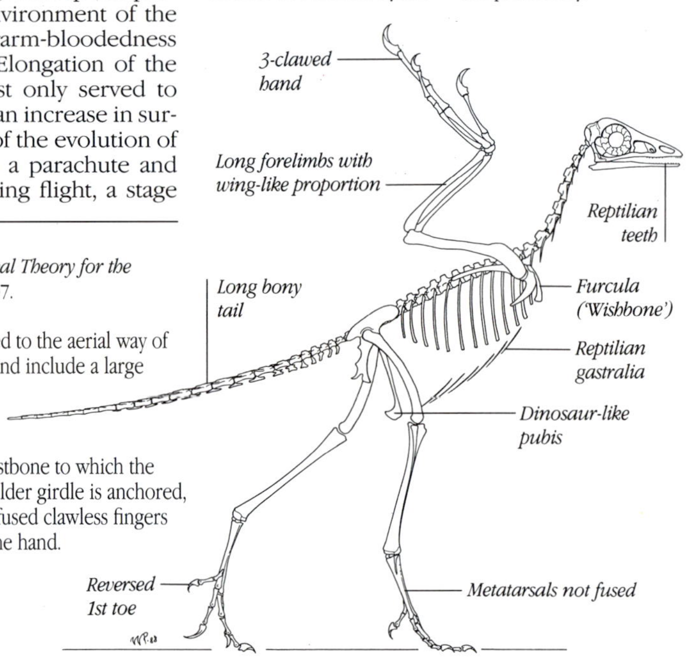

Archaeopteryx

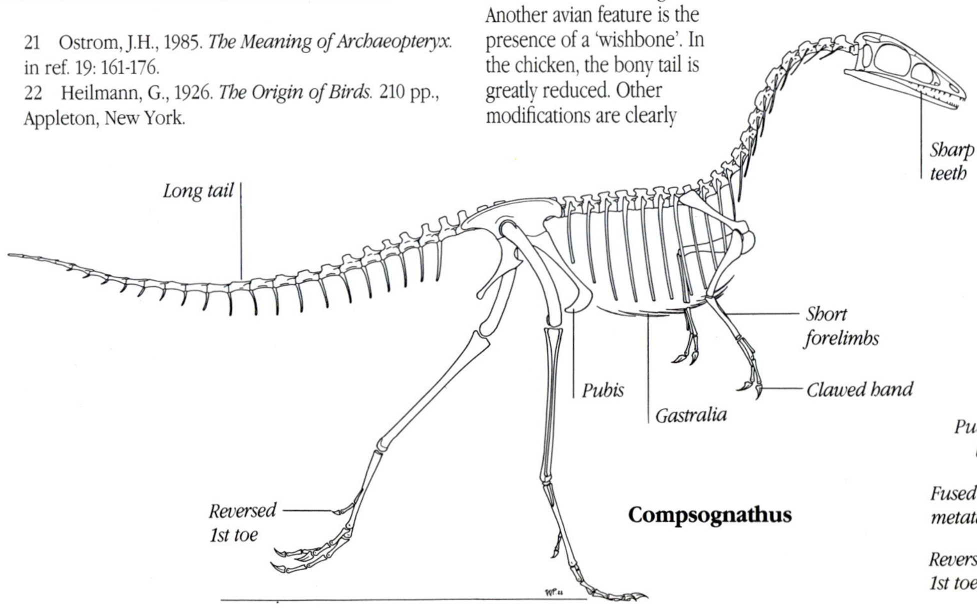

Compsognathus

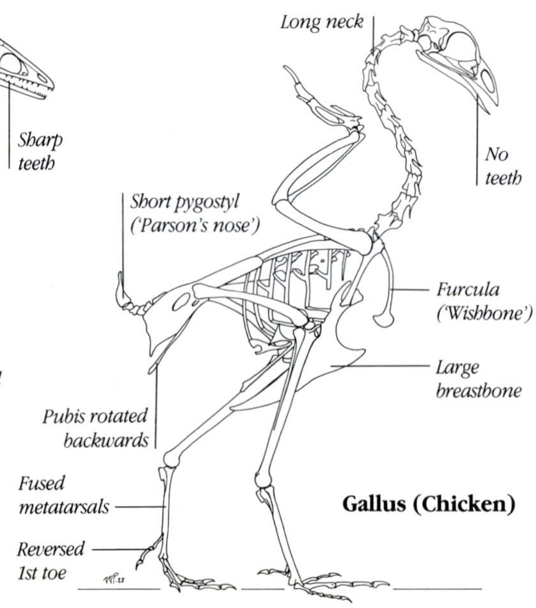

Gallus (Chicken)

Skeletal Comparisons (right and below)
Here *Archaeopteryx* is compared to *Compsognathus* and a chicken. *Archaeopteryx* more closely resembles *Compsognathus* in build, especially with regard to the skull, vertebral column, pelvis and hind legs. The main difference lies in the longer forelimbs which have developed to bear feathers and form wings. Another avian feature is the presence of a 'wishbone'. In the chicken, the bony tail is greatly reduced. Other modifications are clearly linked to the aerial way of life and include a large breastbone to which the shoulder girdle is anchored, and fused clawless fingers on the hand.

that had certainly already been reached in the case of *Archaeopteryx*.

It is nevertheless to be assumed that the Solnhofen primeval birds were still poor fliers. Their flight muscles cannot have been as well developed as those of modern birds, as they still did not have a breastbone for attachment of the pectoral muscles responsible for the downstroke of the wings.[24]

Other Fossil Birds

How did the evolution of birds proceed after *Archaeopteryx*? Unfortunately in the early Cretaceous there are great gaps in the fossil record. Recently remains of fossil skeletons of birds, some of them with feathers, were discovered in Spain.[25] They are about 125 million years old, thus 25 million years younger than *Archaeopteryx*. It was a small bird, already with an ossified sternum, and also a shortened tail, in which the last vertebrae had fused to form a Parson's nose (a so-called pygostyl), although the pelvis and other characteristics were still 'primitively' reptilian. This can be regarded as an intermediate form between *Archaeopteryx* and modern birds.

Another primitive bird, the genus *Ambiortus*, was found in Lower Cretacous sediments in Mongolia. It too shows a mosaic pattern of archaic and advanced characteristics. *Ambiortus* must already have been able to fly well, as it too had an ossified sternum.[26]

Analysis of these early fossil skeletons makes it quite clear that the evolution of birds was directed primarily at improving flight qualities. It enabled birds to spread world-wide even in the Cretaceous: fossil remains of Cretaceous birds have been found in America, Europe, Africa, Asia and Australia.

Skeletal remains that were in good condition and fairly complete were removed from the Niobrara Chalk of West Kansas, USA. These were the late Cretaceous strata that produced the great pterosaurs *Pteranodon* and *Nyctosaurus*. O.C. Marsh found the first bird bones during his first expedition to the Western USA in 1870.[27] More complete skeletons came to light in subsequent years, of two different types, *Ichthyornis* (fish bird), the Cretaceous 'gull', which was capable of flight, and the diver *Hesperornis* (bird of the West), which was incapable of flight. Both had teeth in their jaws, for which reason O.C. Marsh classified them as toothed birds (Odontornithes) in a major monograph written in 1880. Another genus, *Baptornis* was recognized from there, besides *Hesperornis*.[28]

Even in the Upper Cretaceous ancestors of modern bird orders are found. But it is not until the early Tertiary, about 50 million years ago, that the first great radiation, or splitting into many different orders and families, took

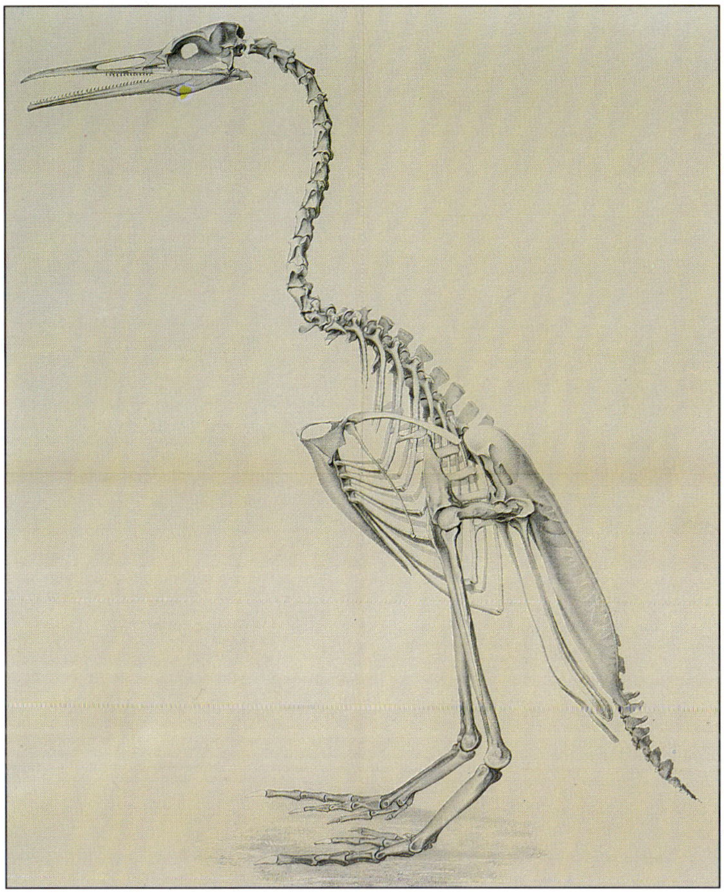

Left: The toothed diving bird *Hesperornis* occurs in the same fossil deposits in the Niobrara Chalk of West Kansas as *Ichthyornis* (right). It was a large, flightless bird, well adapted for diving. This restoration is from O.C. Marsh's 1880 monograph *On the Extinct Toothed Birds of North America*.

Right: Another skeletal restoration from Marsh's 1880 monograph, *Ichthyornis* was a primitive flying bird of the size and life style of a modern gull. These birds must have lived in large flocks on the coast of the Cretaceous mid-continental seaway.

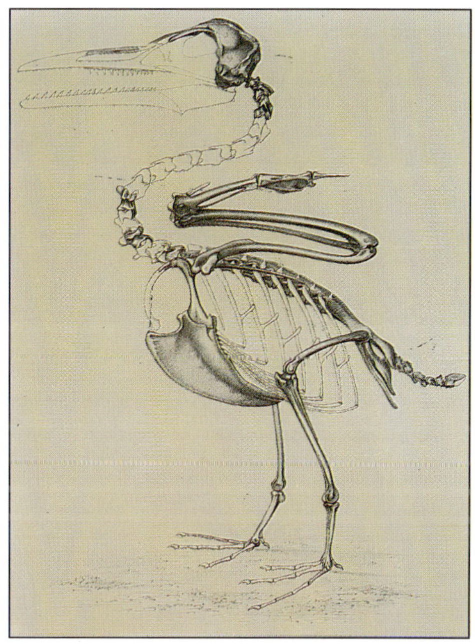

Baptornis (below)
The toothed diving bird *Baptornis* from the Kansas Niobrara Chalk was closely related to *Hesperornis* which was recognized by O.C. Marsh in 1877. It had a very long neck, reduced wings, and large webbed feet that it used to propel itself through the water.

24 Feduccia, A., 1980. *The Age of Birds.* Harvard University Press, Cambridge, Mass. German translation entitled *Es begann am Jura-Meer. Die faszinìernede Stammesgeschichte der Vögel.* Gerstenberg-Verlag, Hildesheim, 1984.
Wellnhofer, P., 1989. *Archaeopteryx.* Spektrum der Wissenschaft, 9, 1989.
Wellnhofer, P., 1990. *Archaeopteryx.* Scientific American, 262 (5): 70-77; Washington.

25 Sanz, J.L., Bonaparte, J.F. and Lacasa, A., 1988. *Unusual Early Cretaceous birds from Spain.* Nature, 331: 433-435; London.

26 Kurochkin, E.N., 1985. *A True Carinate Bird from Lower Cretaceous Deposits in Mongolia and Other Evidence of Early Cretaceous Birds in Asia.* Cretaceous Research, 6: 271-278; London.

27 Marsh, O.C., 1880. *Odontornithes: a monograph on the extinct toothed birds of North America.* Report of the US Geological Exploration of the Fortieth Parallel, no. 7; Washington.

28 Martin, L.D. and Tate, J., Jr., 1976. *The skeleton of Baptornis advenus (Aves: Hesperornithiformes).* Smithsonian Contributions to Paleobiology, 27: 35-66; Washington.

Above: *Icaronycteris* from early Tertiary lake deposits in Wyoming is regarded as the oldest known bat. The fossil shows the complete skeleton preserved with the long wing bones folded, and an unusual long tail. This fine specimen resides in the Yale Peabody Museum.

Right: Several genera of bats have been found in Lower Tertiary lake deposits in Germany. This is *Archaeonycteris* from the famous Messel pit near Darmstadt. In this oil shale soft parts, including even stomach contents, are preserved. The Messel bats were insect eaters.

place in birds. Very different habitats were occupied in the course of this, including those vacated by pterosaurs at the end of the Cretaceous. The last to emerge were the Passeriformes, in the late Tertiary.

From modest beginnings in the Jurassic, 150 million years ago, documented by the unique *Archaeopteryx* specimens from Solnhofen, 8,700 species of bird have now evolved, from the tiny humming-bird with a wing span of 3in (7·5cm) to the albatross with a wing span of 11ft (3·4m). During the last Ice Age there were far larger birds, however. The largest so far known is the extinct giant vulture *Argentavis* (Argentina bird). From head to foot it was 5ft (1·5m) long and had a wing span of 24ft (7·3m).[29] Thus it was just as big as *Pteranodon* from the Niobrara Chalk of Kansas, but certainly heavier than these pterosaurs.

Bats

There are various mammal groups that developed powers of gliding independently of one another, like for example flying lemurs and flying squirrels, or marsupials like honey gliders and greater gliding phalangers. However, bats were the only mammals to have developed active flapping flight. It is assumed that they are descended from gliding, arboreal insectivores, which in the course of evolution transformed their arms into wings. Bats' direct ancestors are not yet known. 'Bats have no *Archaeopteryx* yet,' as L. van Valen put it.[30]

In the fossil record bats do not appear until the early Tertiary. The oldest bat is *Icaronycteris* from the lake deposits of the Green River Formation in Wyoming, USA, which are about 50 million years old.[31] Some bat genera from Eocene deposits in Germany, like the lignite mines of the Geiseltal near Halle, and from lake sediments from the Messel pit near Darmstadt, the genus *Archaeonycteris*, for example, are almost as old. Even in these early forms one finds structures in the auditory region that prove that they already had echo location or sonar, like modern bats. This made them able to hunt for insects at twilight or even at night.

Bats are able to emit pulses of very high-pitched sound at short intervals, usually 10 to 200 times a second. When these sound signals meet an obstacle or a flying insect, they are reflected, and received by the bat's sensitive ears. This echo serves to localize prey, and is an aid to navigation in a three-dimensional airspace. This extreme specialization meant that bats could conquer their own ecological niches, and assert themselves *vis-à-vis* birds.[32]

Even the oldest bats have the typical skeletal structure of their order. The four outer digits (II-V) of their hand are enormously elongated and form an inner support for the wing surface. This is made up of a thin, almost naked membrane, extending down to the feet. Because the long fingers are an integral part of the wing, and can spread and fold it, bats have much greater wing manoeuvrability than pterosaurs, in which the flight membrane is spread by a single digit. Bats' flight membrane contains blood vessels and has muscle fibres running through it. They do not have the internal reinforcement fibres typical of pterosaurs. There is also a membrane between the hind legs and the tail, supported by a special bone, the calcar, originating at the ankle of the foot.

The first hand digit, the thumb, is short in all bats, and has a claw. The normal rest position is hanging by the feet, with the head downwards. They sleep in this position during the day, and hibernate in moderate zones. Then the wings are pulled tight round the body, which is completely wrapped up in the flight membranes.

On the ground bats become genuine quadrupeds. They fold their flight membranes and

29 Campbell, K.E., Jr., 1980. *The world's largest flying bird.* Terra, 19 (2): 20-23; Los Angeles.
Campbell. K.E., Jr. and Tonni, E.P., 1980. *A new genus of Teratorn from the Huayquerian of Argentina (Aves: Teratornithidae).* Contributions in Science, 330: 59-68; Los Angeles.

30 Van Valen, L., 1979. *The evolution of bats.* Evolutionary Theory, 4: 103-121.

31 Jepsen, G.L., 1966. *Early Eocene bat from Wyoming.* Science, 154: 1333-1339; Washington.

32 Novacek, M.J., 1985. *Evidence for echolocation in the oldest known bats.* Nature, 315: 140-141; London.

Further reading on bats:

Norberg, U.M. and Rayner, J.M.V., 1987. *Ecological morphology in bats (Mammalia; Chiroptera): wing adaptations, flight performance, foraging strategy and echolocation.* Philosophical Transactions of the Royal Society of London, B, 316: 335-427.

Fenton, M.B., Racey, P.A. and Rayner, J.M.V. (editors), 1987. *Recent Advances in the Study of Bats,* Cambridge University Press.

Nachtigall, W. (editor), 1986. *Biona Report 5, Bat flight – Fledermausflug,* Gustav Fischer Verlag, Stuttgart.

rest their fore feet on the ground on an upholstered pad on the wrist and support themselves on the soles of their hind feet. The hind legs stand off laterally from the body. Bats are by no means as awkward when moving on the ground as is often assumed. Many species are even nimble runners. But above all bats are excellent climbers. The clawed, freely-moving thumb plays an important part in this. On vertical walls the animal pulls itself upwards by its thumb claws and supports itself with the clawed toes.

Bats (Chiroptera) are a distinct order of mammal. Two suborders are distinguished, Microchiroptera, predominantly insectivorous, and Megachiroptera, mainly fruit eaters, like fruit bats and flying foxes. The latter do not appear in the fossil record until the middle to late Tertiary. Both suborders probably go back to a common ancestral form, as yet unknown, in the earliest Tertiary or latest Cretaceous.

About 800 species of bat are known today. They live all over the world except in the polar regions, but preferably in the tropics. As well as fruit and insect eaters there are nectar drinkers, carnivores which eat small animals or even other bats, and fish eaters, who catch their prey from the water with their feet in flight. Genuine vampire bats are bloodsuckers.

Bats vary considerably in size. Pipistrelles have a wing span of only 7in (18cm), while the Java flying fox reaches a wing span of 4ft (1·2m) and more. Bats are mammals, and thus give birth to living young, usually only one, on rare occasions two. At first they are entirely dependent on the mother, and are suckled. Later they make themselves independent of the mother and learn to fly and hunt for food for themselves.

Today there are many endangered species of both birds and bats. This is due to the activities of man who, within a very short time, introduced a new factor into the evolutionary process of the biosphere, also affecting many natural environments of birds and bats. It is to be hoped that these fascinating flying vertebrates with their long evolutionary history will be saved by an increasing awareness of our human responsibility for life on Earth.

Below: This portrait of a Bechstein's bat shows very clearly the wing membrane, supported by the elongated fingers, joining the body and hind legs down to the ankle, the so-called chiropatagium. A small, triangular membrane, the propatagium, extends between the upper and lower arm, while a further membrane, the uropatagium, joins the legs and tail.

Icaronycteris (right)
The skeletal restoration of *Icaronycteris*, the oldest known bat in the fossil record, shows that its bauplan had quite a modern appearance. Although about 50 million years old, it has no 'primitive' or intermediate features which would point to a particular ancestor for bats. The wing skeleton typically shows the greatly elongated bones of the upper arm, the lower arm and the four outer fingers which in life were enclosed within a thin membrane of skin that extended to the body and hind legs. The long tail was probably totally enclosed in a membrane between the hind legs. The thumb bearing a claw remained free, and could be used for climbing and hooking on to rocks or trees. The dentition points to a diet of insects. Recent studies indicate that even these oldest bats had an echolocation system like modern bats for orientation in the dark and the location of prey insects.

Icaronycteris

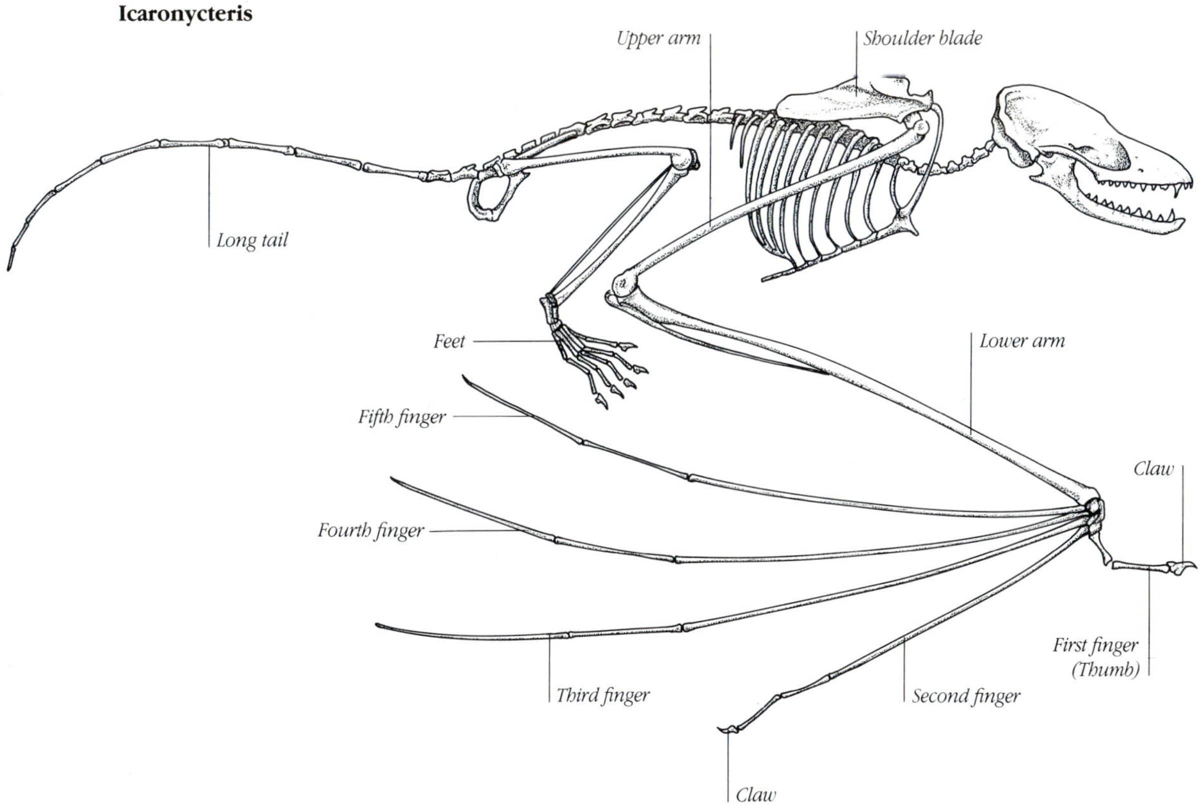

Systematics is the arrangement of organisms into a certain system, thus a classification. The aim of this is to present the subdivisional units of the system in a hierarchic sequence, and to use this rank order to make family connections, the natural links, as clear as possible. In palaeontology it is inevitable that new fossil finds will make the systematics more precise, or even change them, with the result that the system for a particular group of organisms can only represent the present state of knowledge.

This is certainly true of pterosaurs. We have a fossil record of their evolution for a period of almost 160 million years, but their very diverse evolutionary lines can hardly be reliably connected together. Gaps in the fossil record are far too large, and important connecting links are missing. An additional problem is that in the case of pterosaurs there are no living representatives upon which we can draw for comparison, and we usually have to rely on very fragmentary fossil material. Thus attempts to reconstruct pterosaur family connections are always hypothetical.

Systematic arrangement of organisms is done by taxonomy. By naming the rank order within the system, taxonomy allows us to establish family connections. To this end taxonomy has certain concepts at its disposal, so-called taxa, which are allotted in hierarchical order, like for example 'class', 'order', 'family', 'genus' and 'species'. Thus taxonomy is used to express, for example, that species that are similar to each other, and which thus seem to be related, are put together in a genus, and in the same way genera into families, families into orders and orders into classes. To an extent these categories are not adequate, and individual taxa have been further subdivided, for example into 'subclass', 'suborder', 'subfamily' and 'subgenus'. As in biology, the standard unit in palaeontology is the 'species'.

It has already been shown in the earlier chapter on 'What Are Pterosaurs?' that there are and always have been various views about the systematics and taxonomy of pterosaurs. Thus pterosaurs were classified as an independent 'class' alongside the reptile class, as a 'subclass' alongside the archosaur subclass and as an 'order' alongside other orders of the subclass Archosauria. In order not to confuse the reader, I intend to list here all hitherto named taxa of pterosaurs, down to the genus, using the conventional systematics with the order Pterosauria within the subclass Archosauria, within the class Reptilia. In each case the name of the author is given after the name of the taxon, and the year in which he established it.

Order PTEROSAURIA Kaup 1834
- Suborder Rhamphorhynchoidea Plieninger 1901
  - Family Dimorphodontidae Seeley 1870
    - Genus *Dimorphodon* Owen 1859
    - Genus *Peteinosaurus* Wild 1978
  - Family Eudimorphodontidae Wellnhofer 1978
    - Genus *Eudimorphodon* Zambelli 1973
  - Family Rhamphorhynchidae Seeley 1870
    - Genus *Preondactylus* Wild 1984
    - Genus *Dorygnathus* Wagner 1860
    - Genus *Campylognathoides* Strand 1928
    - Genus *Rhamphorhynchus* H. von Meyer 1847

The purpose of this listing is to show the major museums and collections in the world where pterosaurs are housed and on display, either fossil specimens or models of skeletons and life restorations. This list cannot be complete, however, and includes naturally only institutions which are open to the public, and not the many private collections which may also contain important specimens. Wherever possible, the pterosaur genera which that particular collection holds are indicated, although these may not always be on open display.

THE AMERICAS

Argentina

Fundación-Instituto Miguel Lillo,
Universidad Nacional
Tucuman
Pterodaustro
Puntanipterus

Museo Argentino de Ciencias Naturales
'Bernardino Rivadavia'
Buenos Aires
Pterodaustro
Herbstosaurus

Museo Municipal de Ciencias Naturales
Mar del Plata
Pterodaustro

Brazil

Departmento Nacional Produción Mines (DNPM)
Rio de Janeiro
Anhanguera
Araripesaurus
Brasileodactylus
and other pterosaurs from the Santana formation of Brazil

Canada

National Museum of Natural Sciences
Ottawa,
Ontario

Royal Ontario Museum
Toronto,
Ontario

Tyrrell Museum of Paleontology
Drumheller,
Alberta

United States of America

Academy of Natural Sciences
Philadelphia,
Pennsylvania
Pteranodon

American Museum of Natural History
New York, N.Y.
Anhanguera
Nesodactylus
Pteranodon
Pterodactylus
Rhamphorhynchus
Santanadactylus
Tapejara

Carnegie Museum of Natural History
Pittsburgh,
Pennsylvania
Campylognathoides
Nyctosaurus
Pteranodon
Pterodactylus
Rhamphorhynchus

Field Museum of Natural History
Chicago,
Illinois
Nyctosaurus
Pteranodon

Fort Hays Kansas State University, Sternberg Museum
Hays,
Kansas
Nyctosaurus
Pteranodon

Museum of Comparative Zoology
Harvard University
Cambridge,
Massachusetts
Pterodactylus
Rhamphorhynchus

Museum of Natural History
University of Kansas,
Lawrence,
Kansas
Nyctosaurus
Pteranodon

Museum of Paleontology
University of California,
Berkeley,
California
Pteranodon

Museum of the Rockies
Montana State University,
Bozeman,
Montana
Quetzalcoatlus, model

New Mexico Museum of Natural History
Albuquerque,
New Mexico
Rhamphorhynchus, models
Pteranodon, model
Quetzalcoatlus, model

Peabody Museum
Yale University,
New Haven,
Connecticut
Comodactylus
Dermodactylus
Dimorphodon
Nyctosaurus
Pteranodon
Rhamphorhynchus

Texas Memorial Museum
University of Texas,
Austin,
Texas
Quetzalcoatlus

U.S National Museum of Natural History
Smithsonian Institution,
Washington, D.C.
Pteranodon
Rhamphorhynchus

Utah Museum of Natural History
Salt Lake City, Utah
Pteraichnus

ASIA AND AUSTRALIA

Australia

Museum of Victoria
Melbourne
Pteranodon, model

Queensland Museum
Fortitude Valley,
Brisbane
Pterodactyloidea from the Lower Cretaceous of Queensland

China

Natural History Museum
Institute of Vertebrate Palaeontology and Palaeoanthropology,
Beijing
Dsungaripterus
Huanhepterus
Noripterus

Palaeontological Museum Zigong
Beipei Museum,
Dashanpu,
Sichuan Province
Angustinaripterus

EUROPE

Austria

Haus der Natur
Salzburg
Pteranodon, model

Naturhistorisches Museum Wien
Vienna
Dorygnathus
Pterodactylus

Denmark

Geologisk Museum
University of Kopenhagen,
Copenhagen
Rhamphorhynchus

France

Musée d'Histoire Naturelle
Lyon
Pterodactylus

Musée National d'Histoire Naturelle
Paris
Campylognathoides
Gallodactylus

Germany

Bayerische Staatssammlung für Paläontologie und historische Geologie
Paläontologisches Museum München,
Munich
Anurognathus
Araripedactylus
Campylognathoides
Ctenochasma
Dorygnathus
Germanodactylus
Gnathosaurus
Pteranodon, model
Pterodactylus
Rhamphorhynchus
Santanadactylus
Tropeognathus

Bürgermeister-Müller-Museum
Solnhofen
Pterodactylus
Rhamphorhynchus

Geologisch-Paläontologisches Institut der Universität Bonn
Bonn
Scaphognathus

Geologisch-Paläontologisches Institut der Universität Heidelberg
Heidelberg
Rhamphorhynchus

Hessisches Landesmuseum
Darmstadt
Rhamphorhynchus

Institut und Museum für Geologie und Paläontologie der Universität Tübingen
Tübingen
Campylognathoides
Dorygnathus
Pterodactylus
Rhamphorhynchus

Jura-Museum
Eichstätt
Ctenochasma
Gnathosaurus
Pterodactylus
Rhamphorhynchus

Maxberg-Museum beim Solnhofener Aktienverein
Solnhofen
Pterodactylus
Rhamphorhynchus
Scaphognathus

Museum Bergèr
Harthof near Eichstätt
Pterodactylus
Rhamphorhynchus

Museum Hauff
Holzmaden
Dorygnathus

Museum Mensch und Natur
Munich
Rhamphorhynchus, model
Tropeognathus

Museum für Naturkunde
Humboldt Universität,
Paläontologisches Museum
Berlin
Dorygnathus
Pterodactylus
Rhamphorhynchus

Naturmuseum Senckenberg
Frankfurt
Pterodactylus
Rhamphorhynchus

Staatliches Museum für Naturkunde
Museum am Löwentor,
Stuttgart
Campylognathoides
Dorygnathus

Genus *Odontorhynchus* Stolley 1936
Genus *Rhamphocephalus* Seeley 1880
Genus *Parapsicephalus* Arthaber 1919
Genus *Scaphognathus* Wagner 1861
Genus *Sordes* Sharov 1971
Genus *Angustinaripterus* He, Yan & Su 1983
Family Anurognathidae Kuhn 1967
Genus *Anurognathus* Döderlein 1923
Genus *Batrachognathus* Rjabinin 1948
Undetermined family of Rhamphorhynchoidea
Genus *Rhamphinion* Padian 1984
Genus *Herbstosaurus* Casamiquela 1975
Genus *Nesodactylus* Colbert 1969
Genus *Comodactylus* Galton 1981

Suborder Pterodactyloidea Plieninger 1901
Family Pterodactylidae Bonaparte 1838
Genus *Pterodactylus* Cuvier 1809
Family Gallodactylidae Fabre 1974
Genus *Gallodactylus* Fabre 1974
Family Germanodactylidae Young 1964
Genus *Germanodactylus* Young 1964
Family Ctenochasmatidae Nopcsa 1928
Genus *Ctenochasma* H. von Meyer 1852
Genus *Gnathosaurus* H. von Meyer 1834
Genus *Huanhepterus* Dong 1982
Family Pterodaustridae Bonaparte 1971
Genus *Pterodaustro* Bonaparte 1970
Family Dsungaripteridae Young 1964
Genus *Dsungaripterus* Young 1964
Genus *Noripterus* Young 1964
Genus *Phobetor* Bakhurina 1986
Genus *Puntanipterus* Bonaparte & Sanchez 1975
Family Ornithocheiridae Seeley 1870
Genus *Ornithocheirus* Seeley 1869
Genus *Araripesaurus* Price 1971
Genus *Santanadactylus* Buisonjé 1980
Genus *Brasileodactylus* Kellner 1984
Family Anhangueridae Campos & Kellner 1985
Genus *Anhanguera* Campos & Kellner 1985
Family Tapejaridae Kellner 1990
Genus *Tapejara* Kellner 1990
Genus *Tupuxuara* Kellner & Campos 1989
Family Cearadactylidae Wellnhofer 1991
Genus *Cearadactylus* Leonardi & Borgomanero 1985
Family Criorhynchidae Hooley 1914
Genus *Criorhynchus* Owen 1874
Genus *Tropeognathus* Wellnhofer 1987
Family Ornithodesmidae Hooley 1913
Genus *Ornithodesmus* Seeley 1887
Family Pteranodontidae Marsh 1876
Genus *Pteranodon* Marsh 1876
Genus *Ornithostoma* Seeley 1871
Family Nyctosauridae Williston 1903
Genus *Nyctosaurus* Marsh 1876
Family Azhdarchidae Nessov 1984
Genus *Azhdarcho* Nessov 1984
Genus *Titanopteryx* Arambourg 1959
Genus *Quetzalcoatlus* Lawson 1975
Genus *Doratorhynchus* Seeley 1875
Undetermined family of Pterodactyloidea
Genus *Dermodactylus* Marsh 1878
Genus *Mesadactylus* Jensen & Padian 1989
Genus *Araripedactylus* Wellnhofer 1977

Pterodactylus
Rhamphorhynchus

Staatliches Museum für Mineralogie und Geologie
Dresden
Rhamphorhynchus

Petrefaktensammlung Kloster Banz
Banz near Lichtenfels
Dorygnathus

Italy

Museo Civico di Scienze Naturali
Bergamo
Eudimorphodon
Peteinosaurus

Museo Civico di Storia Naturale di Milano
Milan
Eudimorphodon
Pteranodon, model

Museo Civico di Storia Naturale di Venezia
Venice
Pterodactyloidea from the Santana formation of Brazil

Netherlands

Geological Institute of the University
Amsterdam
Santanadactylus, model

Rijksmuseum van Geologie en Mineralgie
Leiden
Pterodactylus
Rhamphorhynchus

Teyler's Museum
Haarlem
Pterodactylus
Rhamphorhynchus

Sweden

Palaeontological Museum, Uppsala University
Uppsala
Dorygnathus
Pterodactylus

Switzerland

Naturhistorisches Museum
Basel
Pteranodon, model

Paläontologisches Institut und Museum der Universität Zürich
Zurich
Pterodactylus
Rhamphorhynchus
Santanadactylus

United Kingdom

City Museum and Art Gallery
Bristol
Dimorphodon, model

Crystal Palace Park
Sydenham, London
Hawkin's pterodactyl models

Lyme Regis (Philpot) Museum
Lyme Regis, Dorset
Dimorphodon

Natural History Museum
London
Criorhynchus
Dimorphodon
Doratorhynchus
Ornithocheirus
Ornithodesmus
Parapsicephalus
Pterodactylus
Rhamphocephalus
Rhamphorhynchus

Sedgwick Museum
Cambridge University, Cambridge
Criorhynchus
Ornithocheirus

USSR

Palaeontological Museum, Akademia Nauk
Moscow
Batrachognathus
Phobetor
Sordes

Zoological Museum, Akademia Nauk
Leningrad
Azhdarcho

Above: This is the central hall of the Bavarian State Collection of Palaeontology and historical Geology in the city of Munich. This institute houses one of the finest collections of pterosaur fossils in the world, many of the specimens coming from Solnhofen limestone which is quarried in Bavaria.

A

Adaptation
Fitness in structure or function for a particular kind of environment; the process of becoming so adjusted.
Adaptive radiation
Diversification of organisms along various evolutionary lines adjusted to different environments.
Aerodynamics
The study of the physical conditions of air currents and their effect on moving bodies.
Age
Geological time unit as a subdivision of the periods, for example Maastrichtian Age of the Cretaceous Period.
Agnathans
Primitive, jawless, fish-like, aquatic animals, the earliest vertebrates.
Aktinofibrils
Strengthening fibres embedded in the flight membrane of pterosaurs.
Alveoli
Tooth sockets in the jaw bone.
Ammonites
Extinct tentacled molluscs related to squids, but with a coiled and chambered shell similar to the *Nautilus* today. Abundant in Mesozoic seas and used as index fossils.
Amnion
A membraneous sac filled with watery fluid that encloses the developing embryo in reptiles, birds and mammals.
Amniotes
Tetrapods whose embryos develop within an amniotic membrane, i.e. reptiles, birds and mammals.
Amphibians
Tetrapods whose embryos do not develop within an amniotic membrane and which have to pass through a larval stage in the water breathing through gills, e.g. toads, frogs and salamanders.
Anapsids
A reptile group characterised by having no skull openings behind the eye socket, including the living turtles and tortoises, but also the earliest reptiles.
Anterior
Pertaining to the front of the body as opposed to posterior.
Arboreal
Living in trees.
Archosaurs
A grouping of reptiles based on certain shared skeletal features comprising dinosaurs, pterosaurs, thecodonts and crocodiles.
Arthropods
Invertebrate animals with jointed legs, e.g. insects, spiders, crabs, shrimps and some extinct groups.
Articular facet
A joint surface.
Articulation
A joint connecting bones.
Atlas
The first vertebra of the neck which connects directly with the skull (from the Greek mythology, where the giant Atlas supports the world on his shoulders).
Axial
Trunk muscles in general are called axial muscles. In fishes much of the bulk of the body is formed by axial musculature.
Axis
The second neck vertebra about which the atlas rotates to turn the skull.

B

Baleen
The Baleen Whale *(Mystacoceti)* is a very large whale, up to 108ft (33m) in length, and has peculiar filtering lamellae on the palate in its mouth.
Bauplan
Construction plan or pattern.
Belemnites
Extinct molluscs, related to squids with a bullet-shaped, calcified guard or rostrum.
Biosphere
Zones of the earth in which living organisms exist, i.e. from the depth of the oceans up to the highest mountains.
Biostratigraphy
The study of the rocks by dating their origin with containing fossils, i.e. index fossils.
Biozone
Geological time unit defined by the life time of the species of an index fossil.
Biped
An animal that stands and walks on its hind legs.
Bitumen
Natural inflammable substance composed of a mixture of hydrocarbons, for example petroleum, asphalts and natural mineral waxes.
Bivalves
Mussels, two-shelled molluscs, e.g. clams and oysters.
Botany
The science dealing with the study of plants.
Brachiopatagium
The main flight membrane of the pterosaur wing.
Brachiopods
Shelled sea creatures that look a little like mussels but are not closely related.

C

Carboniferous
A geological time period in the Palaeozoic Era; it lasted from 360-286 million years ago. In this period the main coal deposits were formed.
Carina
A keel, here on the breastbone of pterosaurs.
Carpals
Small bones forming the wrist.
Carpus
The wrist.
Cartilage
Organic tissue connected with skeletal functions, for example in joints. In embryonic and juvenile stages of vertebrates bones are preformed as cartilage and develop into bone substance during growth.
Caput
The articular upper end (head) of the upper arm (humerus) or the upper leg (femur).
Caudal
Pertaining to the tail, or posterior part of the body.
Cenozoic
'Recent life': the geological era after the Mesozoic, comprising the Tertiary and Quaternary Periods, i.e. the last 65 million years.
Cerebellum
The hindbrain, which is principally concerned with co-ordination of movements.
Cervical
Pertaining to the neck.
Chevrons
V- or Y-shaped bones attached to the lower side of the tail vertebrae. Anatomists call them haemapophyses.
Cladogram
A diagram representing the family tree of groups of organisms.
Class
A major category of organisms below the Phylum and above the Order, as for example the classes Amphibia, Reptilia, Mammalia and Aves (birds).
Clavicle
A bone in the shoulder girdle, the collar bone.
Concretion
A hard, compact, rounded mass of rock, also called a nodule, formed by orderly and localized precipitation from aqueous solution, often about a centre, such as a bone.
Coprolite
Fossilised dung.
Coracoid
A prominent bone in the shoulder girdle of reptiles and many other vertebrates. In pterosaurs and birds it articulates with the breastbone.
Cornea
The horny skin of the eye.
Cretaceous
The last period in the Mesozoic Era. It lasted from 144-65 million years ago.
Cristospina
A crest at the lower side of the breastbone extending to the front into a spine, typical for pterosaurs.
Crurotarsal ankle
Ankle structure in which the hinge between leg and foot has a sharp twist to allow the foot to swivel.
Cursorial
With slender limbs adapted to running.

D

Dendrochronology
The study of annual growth rings of trees for dating of the recent past.
Dentition
Teeth.
Description
Here the scientific description of fossil specimens, the results of their detailed study with conclusions concerning their anatomy determination, relationships, origin and functional morphology.
Diapsids
A reptile group characterised by having two skull openings behind each eye socket. Diapsids include the pterosaurs, crocodiles, dinosaurs, thecodonts as well as snakes and lizards and their ancestors.
Digit
Finger (on the hand) and toe (on the foot).
Digitigrade
Walking on the digits, as opposed to plantigrade.
Dimorphism
The characteristic of having two forms, usually sexual. If males and females of the same species look different, then the species exhibits sexual dimorphism.
Dinosaurs
A group of archosaurian land-living reptiles with an erect gait. They lived in the Mesozoic Era and became extinct at the end of the Cretaceous Period 65 million years ago. Dinosaurs can be classified in two orders: Saurischia and Ornithischia. The name means 'terrible lizard'.
Distal
Furthest from the point of attachment to the body (opposite of proximal).
Dogger
The middle epoch of the Jurassic Peroid (=Middle Jurassic), following the Lias and followed by the Malm. It lasted from 188-163 million years ago.

E

Endocast
A sediment in-fill of a cavity, for example the brain cavity, then also called endocranial cast.
Eosuchia
A group of early diapsid reptiles which gave rise to archosaurs, lizards and snakes.
Epaxial
Dorsal trunk muscles running the length of the body are called epaxial muscles.
Epidermis
The outer layer of the skin.
Epiphysis
An accessory centre of ossification at the ends of long bones forming the articular regions. Between epiphysis and shaft there is a long persistent band of cartilage which is continuously growing and being replaced by bone until epiphysis and shaft are united by bone and growth is over. This growth pattern is found in mammals, birds and to a limited extent also in reptiles.
Era
Geological time unit, for example Palaeozoic, Mesozoic or Cenozoic Eras.
Ethnology
Science dealing with the history and culture of peoples.
Euryapsids
A reptile group, exclusively aquatic and now extinct, e.g. plesiosaurs and ichthyosaurs, characterised by having a single opening (temporal fenestra) high up on the side of the skull behind the eye socket.
Evolution
The development of plants and animals through geological time, and the way that this development has come about. Organisms evolve, or develop, as a result of changes in their living conditions.
Exapophyses
Bony processes low on the front and back of the neck vertebrae of pterosaurs which form additional joints for the strengthening of the long neck.
Extensor tendon
The tendon of a muscle which straightens a joint, as opposed to a flexor tendon.

F

Family
A grouping of similar closely related genera.
Fauna
Animals, as opposed to flora (plants).
Femur
The long bone in the upper leg or thigh-bone.
Fenestra
Window-like opening in the skull.
Fibula
The smaller of the two bones of the lower leg or shin.
Flexion
Bending of a joint, as opposed to extension.
Flexor tendon
The tendon of a muscle which bends a joint, as opposed to an extensor tendon which straightens it.
Flocculi
Small, lateral appendages of the hindbrain (cerebellum), especially concerned with equilibrium and closely connected with the inner ear.
Flora
Plants, as opposed to fauna (animals).
Foramen magnum
The opening in the brain case of the skull through which the spinal cord penetrates into the spinal canal.
Fossil
The preserved remains of plants or animals that lived in the geologic past turned to stone. Fossils can be millions of years old.
Fossiliferous
Fossil bearing.

G

Ganoid fish
A group of fishes with enamel-like scales, very common in the Mesozoic Era.
Gastralia
Belly ribs.
Gastric pellet
Stomach pellet, containing indigestible parts of the diet, for example bones, which is vomited up.
Genus (pl. genera)
A group of closely related species of plants or animals.
Geology
The science which treats of the history of the Earth, the composition and structure of the Earth's crust.
Gingko
The maidenhair tree of East Asia; the sole survivor of a once abundant group of gymnosperm trees.
Glenoid fossa
The shoulder joint on the shoulder girdle.

H

Humerus
Upper arm bone.

I

Ichthyosaurs
Marine reptiles of the Mesozoic Era with streamlined, dolphin-shaped bodies.
Ilium
One of the bones of the pelvis; it is connected to the vertebral column via the sacral ribs.
Index fossil
Fossils of species with a relatively short life span and a wide geographical distribution, thus being useful as geological time-markers (see biostratigraphy).
Interclavicle
A bone in the anterior median part of the chest region between the proximal ends of the two clavicles ('collar bones').
Ischium
One of the bones of the pelvis; it points downward and backward from the hip socket.
Isotope
One of a set of chemically identical types of atom which differ in their weight and stability. Unstable isotopes are radioactive and 'decay' to form more stable isotopes. Isotope analysis is used in dating some types of rocks.

J

Jurassic
The middle period of the Mesozoic Era, following the Triassic and followed by the Cretaceous Periods. The Jurassic lasted from 213-144 million years ago.

K

Keratine
A horny substance which is the basic material of horns, claws and hairs in animals.

L

Lepidosaurs
Snakes and lizards.
Lias
The first epoch of the Jurassic Period (=Lower Jurassic); it lasted from 213-188 million years ago.
Ligaments
Tough sheets or threads of collagen (protein) which support joints between bones.
Limestone
Rock largely composed of calcium carbonate ($CaCO_3$).
Loess
Windblown dust of the Ice Age, a highly calcareous, fine-grained blanket deposit of mostly yellowish marl or loam, covering areas extending from North-central Europe to eastern China as well as in the Mississippi Valley and Pacific Northwest of the US.

M

Malm
The third epoch of the Jurassic Period (=Upper Jurassic); it lasted from 163-144 million years ago.
Mammal
A warm-blooded animal with hair that gives birth to young and produces milk to feed them. Examples are mice, rabbits, elephants, and humans.
Marl
Limestone with a high content of clay.
Marsupials
Mammals that bear live young that develop in a pouch on the mother's body. Examples include kangaroos, wombats, and koalas in Australia, and opossums in South America.
Median
The middle plane of a body.
Mesotarsal ankle
Ankle structure in which the hinge line between leg and foot runs between the proximal and distal series of the tarsal bones allowing a fully-erect leg posture.
Mesozoic
'Middle life': the middle Era between the Palaeozoic and the Cenozoic Eras. It lasted from 248-65 million years ago and is referred to colloquially as the 'Age of Reptiles'. The Mesozoic Era incorporates the Triassic, Jurassic and Cretaceous Periods.

Metabolic level
The intensity of the chemical processes within the body, i.e. the breakdown of food to release energy. A high metabolic level entails the ability to release more energy and with greater speed, involving an increased uptake of oxygen to 'burn' the food. Chemical reactions are slower if the temperature drops. Most reptiles are capable of short bursts of fast metabolism, but they are unable to sustain it because their body temperature is dependent on the surrounding air temperature. They have a low metabolic level. On the other hand, the warm-blooded mammals and birds are able to maintain a constant high body temperature, and can therefore sustain an increased energy output; they have a higher metabolic level.
Metacarpals
Long bones in the upper part of the hand, between the wrist and the digits (fingers). In pterosaurs the wing metacarpal supports the wing finger, is strongly developed and, in pterodactyloids, greatly elongated.
Metatarsals
Long bones in the upper part of the foot, between the ankle and the digits (toes).
Monophyletic
A single origin for a group of animals, originating from a common ancestor.
Morphology
The study of the form and shape.
Mosasaurs
Marine reptiles which lived in the Cretaceous Period. Large predators of the sea, these forms appear to be relatives of modern monitor lizards. They were first excavated near the River Maas in the Netherlands, hence their name.

N

Neural spine
A bone spine rising above a vertebra providing surface areas for the attachment of muscles and ligaments.
Notarium
A fusion of several anterior dorsal vertebrae to form a solid vertebral block. A notarium is commonly developed in large Cretaceous pterosaurs in order to provide a firm base for the strongly developed shoulder girdle and for the origin of wing muscles.

O

Occipital condyle
A ball-like process of a bone constituting a part of the back of the skull. The occipital condyle fits in a socket in the front of the anterior neck vertebrae thus forming the joint between the neck and the skull.
Olfactory lobes
The region of the vertebrate forebrain concerned with the sense of smell.
Orbital
Pertaining to the eye socket (orbit).
Order
A grouping of animals that includes a variety of similar, related families.
Ornithosuchids
Large thecodontian archosaurs of the Late Triassic.
Ossified
Composed of bone.
Osteology
The study of bones and skeletons.
Ovate
Equally oval in cross- and length-sections.
Ovoid
Egg-shaped.

P

Palaeogeography
The study of the geographical distribution of continents and oceans in each geological period of the history of the Earth.
Palaeoichnology
The study of footprints of fossil animals as preserved in rocks.
Palaeolatitude
The changing position of the continents during the history of the Earth means that a particular site had also a different latitude on the globe in former geological periods, called palaeolatitude.
Palaeoneurology
The study of fossil brains.
Palaeontologist
A scientist who studies fossils as documents of the history of life on Earth.
Palaeontology
Science dealing with the life of past geologic periods as based on the study of fossils.
Palaeozoic
'Ancient life': the era before the Mesozoic, between 590-248 million years ago, comprising the Cambrian, Ordovician, Silurian, Devonian, Carboniferous and Permian Periods.
Parietal
A bone (or pair of bones) forming the roof of the skull, between the frontal and the occipital.
Passeriformes
Order of birds, the sparrows.
Pectoral girdle
The shoulder girdle.
Pelvis
The hip bones collectively, the basin.
Pelycosaurs
Mammal-like reptiles of the Carboniferous and Permian Periods; some have distinctive 'sails' on their backs.
Permian
The last period of the Palaeozoic Era, following the Carboniferous and followed by the Triassic Periods. It lasted from 286-248 million years ago.
Petrification
'Turning to stone': the replacement by minerals, of the original tissues of a fossilised organism, so that it becomes stone-like in nature.
Phalanx (pl. phalanges)
The (usually) small bones in the fingers and toes of vertebrates. In pterosaurs the phalanges of the wing finger are greatly elongated and strongly developed in order to support the wing membrane.
Phylogeny
Evolutionary development of organisms during the history of their family tree.
Piscivorous
Feeding upon fish.
Placodonts
Marine reptiles of the Triassic Period that feed on shellfish, which they crushed with their flattened, plate-like teeth.
Plantigrade
Walking on the soles of the feet; in contrast to digitigrade (=walking on the toes).
Plesiosaurs
Aquatic reptiles of the Mesozoic Era.which swam by using their flipper-like feet.
Pneumatic foramen
A small opening in some bones of pterosaurs and birds through which extensions of the lung sacs penetrated.
Pneumaticity
The condition of hollow bones with pneumatic foramina.
Posterior
Pertaining to the back of the body, as opposed to anterior.
Preorbital
Pertaining to the front of the orbit or eye socket. The preorbital opening lies in front of the orbit and is a diagnostic character of archosaurs.
Prepubis
A pelvic bone attached to the lower end of the pubis in pterosaurs.
Primeval
Ancient, original, ancestral.
Process
A bony projection (in an anatomical sense).
Procoelous
The condition in a vertebra when the surface of the joint at the front is concave and convex at the back (opposite to opisthocoelous).
Profile
A sequence of rock strata (in a geological sense).
Prolacertilians
Primitive lizard-like, diapsid reptiles of the Early Triassic which are considered by some to be ancestral to the lizards.
Propatagium
A small wing membrane in pterosaurs and bats between neck, upper and lower arm.
Protozoans
Single-celled, microscopically small animals; the earliest and most primitive animals known on Earth.
Proximal
Nearest to the point of attachment to the body (opposite of distal).
Pterodactylid
Pertaining to the family Pterodactylidae.
Pterodactyloid
Pertaining to the suborder Pterodactyloidea.
Pterodactyloidea
A major grouping (suborder) of pterosaur families, the short-tailed pterosaurs, which lived from the Late Jurassic to the end of the Cretaceous.
Pteroid bone
A small slender bone arising from the wrist in pterosaurs. It supported the leading edge of the propatagium.
Pterosaurs
The flying reptiles of the Mesozoic Era considered to be an order of the archosaurs and thus distant cousins of the dinosaurs and crocodiles. Pterosaurs include the two suborders Rhamphorhynchoidea, the long-tailed pterosaurs, and the Pterodactyloidea, the short-tailed pterosaurs.
Pubis
One of the bones of the pelvis, usually pointing forward and downward from the hip socket. In pterosaurs this element is mostly fused with the ischium forming the ischiopubis.
Pygostyl
Fused tail vertebrae of birds, also called 'Parson's nose'.

Q

Quadrate
A bone in the skull of vertebrates with which the lower jaw articulates.
Quadruped
An animal that stands and walks on all fours.
Quaternary
The recent prehistoric past, the period following the Tertiary Period; the last 2 million years.

R

Radiation
Branching out from a common centre; diversification along various evolutionary lines from a common ancestral stock.
Reptile
Usually cold-blooded, egg-laying vertebrates covered with scales, for example the living crocodiles, turtles, lizards and snakes. In the Mesozoic Period reptiles were much more diverse and included very different forms, such as ichthyosaurs, plesiosaurs, placodonts, dinosaurs and pterosaurs. Some of them may even have been warm-blooded, as the pterosaurs.
Rhamphorhynchid
Pertaining to the family Rhamphorhynchidae.
Rhamphorhynchoid
Pertaining to the suborder Rhamphorhynchoidea.
Rhamphochynchoidea
A major grouping (surborder) of pterosaur families, the long-tailed pterosaurs, which lived from the Late Triassic to the Late Jurassic.
Rhynchocephalia
A group of lizard-like reptiles widely distributed in the Mesozoic Era. The only living representative is the tuatara of New Zealand.
Rhynchosaurs
Squat, pig-like, plant-eating reptiles with a hooked beak that lived in late Triassic times.

S

Sacral
Pertaining to the sacrum, a section of several vertebrae which are linked to the hip bones.
Sagittal
Parallel to the medial (or middle) plane in an anatomical sense.
Saurian
Reptile, from the Greek, meaning 'lizard'.
Scapula
The shoulder blade.
Scapulocoracoid
A hook-like bone, the shoulder girdle in pterosaurs, composed of the scapula and the coracoid.
Sclerotic ring
A ring of flat bones lying in the eye-ball (the sclera or 'white' of the eye) providing additional strength.
Sedimentation
The process of forming and accumulating sediment in layers.
Shale
Consolidated silt or clay, showing parting parallel to the bedding plane.
Silt
Grains, or finely broken pieces of rock that slowly settle out of water.
Species
The basic unit of biological classification; a group of animals which can breed together – something which is impossible to prove in fossils. *Pterodactylus antiquus* is a species. There may be many species in a genus (*Pterodactylus* is the genus).
Stem reptiles
A group of early, primitive reptiles which lived in the Carboniferous Period and are considered to be the ancestors of most later reptiles.
Sternum
Breastbone.
Stratigraphy
The study of the pattern of rock layers (strata) and their chronological sequence.
Stratum (pl. strata)
A layer of (usually sedimentary) rock.
Supraoccipital
The upper bony element at the back of the skull in many reptiles.
Synapsids
A reptile group characterised by a single opening low down on the skull behind the eye socket; they include the pelycosaurs and the mammal-like reptiles of the Permian and Triassic Periods.
Synsacrum
The fused sacral and posterior dorsal vertebrae in birds and some pterosaurs.
Systematics
The arrangement of groups (taxa) of organisms (plants and animals) according to their phylogenetic relationships.

T

Tarsals
Small bones forming the ankle.
Tarsus
The ankle.
Taxon (pl. taxa)
A systematic unit of classification of plants or animals of any category, for example species, genus, family, order or class.
Taxonomy
Science of the grouping and subdivision of organisms according to their distinct categories (taxa).
Temporal opening
An opening in the temporal region of the skull behind the eye socket, also called temporal fenestra.
Tertiary
The first period in the Cenozoic Era, following the Cretaceous Period and followed by the Quaternary Period. It lasted from 65-2 million years ago and charts the rise of mammals and birds.
Tetrapods
Vertebrates with four limbs, e.g. amphibians, reptiles, mammals *and* birds.
Thecodonts
Early archosaurian reptiles of the Permian and Triassic Periods, with socketed teeth. The group from which most of the more advanced archosaurs evolved.
Thoracic
Pertaining to the thorax or chest.
Tibia
One of the two bones of the lower leg, the shin bone; usually the larger of the two, the other being the fibula.
Tibiotarsus
A single bone composed of the tibia and the proximal series of the tarsal bones of the ankle; a normal condition in birds and advanced pterosaurs.
Triassic
The first period of the Mesozoic Era, following the Permian and followed by the Jurassic Periods. It lasted from 248-213 million years ago. The pterosaurs, dinosaurs and many other reptilian groups appeared towards its close.
Trochlea
The pulley-shaped distal articular end of the upper arm (humerus) against the forearm (radius and ulna).
Tuatara
Lizard-like reptile *(Sphenodon punctatus)* of New Zealand, the only living representative of the rhynchocephalians, a group widely distributed in the Mesozoic Era.

U

Ulna
One of the two long bones in the forearm or front leg of a tetrapod; the other being the radius.
Uropatagium
A skin membrane stretched between the tail and hind legs of bats; probably not present in pterosaurs.

V

Varve
A sedimentary layer deposited in a body of still water within one year, usually by meltwater streams in a glacial lake in front of a glacier. A glacial varve includes a light-coloured summer layer and a dark-coloured winter layer. By counting the varves the ages of Quaternary glacial deposits can be measured.
Ventral
From beneath (opposite of dorsal).
Vertebra
An individual bone of the back (vertebral column=backbone). The backbone is made up of many vertebrae.
Vertebrates
Backboned animals, e.g. fish, amphibians, reptiles, birds and mammals.
Viscera
The internal organs, the intestines.

Z

Zoology
The science of the study of animals.
Zygapophyses
Processes on the vertebrae which articulate with the corresponding processes of the neighbouring vertebrae, and prevent them from slipping apart.

INDEX

Subjects mentioned in illustration captions are indicated by *italic* page numbers.

A

B

C

D

E

F

Picture Credits

The publishers wish to thank the many scientists, private collectors, museums and photographic libraries who have kindly supplied photographs for inclusion in this book, and by courtesy of whom they are reproduced. All photographs are here credited by page number.

S. K. H. Herzog Albrecht von Bayern: 22 upper
J. Augusta and Z. Burian, Artia Verlag, Prague: 8 both; 169 middle left
Bayerische Akademie der Wissenschaften, Munich: 24 lower
Bayerische Staatssammlung für Paläontologie und historische Geologie, Munich (Photographer: Franz Höck): 11 lower; 12-13 all; 14 upper right; 15 upper right; 17 middle; 23 upper; 25 both upper and bottom; 37 middle right and bottom left; 48; 49; 79 middle; 81; 82; 88; 89 upper; 92; 97 upper; 117 bottom; 124 all photographs; 125 top, middle left and lower right; 126 all photographs; 127 top left and right, bottom left and right; 149 upper; 171 top; 173 bottom right; 182 upper right; 185
C. Bennett: 36 bottom
J. F. Bonaparte: 131; 132; 173 middle right
China Ocean Press, Beijing: 80 top and bottom; 104 bottom; 117 top
City of Bristol Museum: 173 top left
Martyn Cowley: 6 upper; 39 lower right; 175 all photographs
R. Delun: 38 top right (Franz Höck)
Dong Zhiming: 105; 120 top; 121 top
K. Ehrenberg: 169 bottom
Foto-Wagner, Furth im Wald: 20-21
K. A. Frickhinger: 73 lower
Geological Institute of Basel: 170 middle
Dietrich Herm: 21 upper
F. Höck: 58 upper; 159 lower right; 162; 178 upper
H. Hofer: 38 top left (Franz Höck)
N. Hutchinson: 33 lower
Imitor: 28 top left; 29 top left; 35 upper; 36 top; 109; 113 top; 146 middle
Institut und Museum für Geologie und Paläontologie, Tübingen: 31 upper; 85, 96 bottom
G. Jakob: 15 lower
A. W. A. Kellner: 14 lower left; 122 bottom; 123 both; 125 bottom left and middle right skull; 127 middle right; 130 top and middle right
A. Kistner: 22 lower
E. Kuhn-Schnyder: 23 lower right
Frank Lane Picture Agency: 153 (Christiana Calvalho); 155 bottom (Bob Langrish); 157 bottom left (Chris Newton); 183 top right (Silvestris)
W. Langston Jr: 143
Maxberg Museum, Solnhofen: 91
Museum Hauff, Holzmaden: 73 top right
Museum of the Rockies, Bozeman: 173 middle left
Natural History Museum, London: 28 top right; 69 upper; 83 top; 108; 115 top right; 158 top left
Naturhistorisches Museum, Vienna: 150
F. Nopcsa: 58 lower (Franz Höck)
Nucolorvue Productions Pty Ltd: 172 top right
J. H. Ostrom: 134 top
Palaeontological Institute, Moscow: 101 upper; 102; 104 upper; 177 middle
Peabody Museum of Natural History, Yale University: 35 lower; 135 top, middle left and right; 137
Queensland Museum, Fortitude Valley: 122 top middle and right
Science Photo Library: 39 upper right (Martin Dohrn/Stephen Winkworth); 174 middle and bottom right (Martin Dohrn/Stephen Winkworth)
Senckenberg Museum, Frankfurt: 10 upper and middle right; 83 lower
Smithsonian Institution: 173 top right
Staatliches Museum für Naturkunde, Stuttgart: 73 top left (H. Lumpe); 76 top (H. Lumpe)
Stadtarchiv Munich: 24 upper
E. Stromer: 170 top
Texas Memorial Museum, Austin: 141 middle right and bottom
D. Unwin: 101 lower, 120 middle right; 164 middle and bottom left
G. Viohl: 15 upper middle
Peter Wellnhofer: 6 lower; 7 both; 11 upper; 14 upper left; 17 upper; 21 lower; 29 lower left; 30 upper; 46; 51 both; 52; 53; 54; 55; 56; 57; 77; 78; 79 top right; 84; 89 lower; 93; 95; 96 upper left and right; 97 middle right and lower right; 100 both; 107; 121 bottom; 130 bottom; 134 bottom; 135 bottom; 136; 140; 141 top left and right; 144 all photographs; 148; 149 bottom; 151; 152 all photographs; 156 all photographs; 158 bottom left and bottom middle; 159 middle left and bottom; 160 all photographs; 161 both photographs; 163 all photographs; 164 bottom right; 165; 167; 170 bottom; 171 middle and bottom; 172 middle; 177 top right; 178 lower; 179 all photographs; 182 upper left
R. Wild: 59; 62; 63; 66; 67

Artwork Credits

Listed here are the artists responsible for and sources of the artwork reproduced in this book. All the double-page colour restorations are the work of John Sibbick, while the accompanying maps, time charts and silhouette drawings were prepared by Richard Hawke from references supplied by Peter Wellnhofer.

O. Abel: 153 lower right; 157 lower right and bottom right
C. Bennett: 151 middle
J. Bonaparte: 42 upper; 42 lower (redrawn by John Sibbick); 131 (redrawn)
C. D. Bramwell and G. R. Whitfield: 38 middle left and right (Geoff Denney)
W. Buckland: 28 bottom (BMNH); 146 bottom (BMNH)
J. Buckman: 162 top left and top middle
Z. Burian: 8 both; 169 middle left
R. Carroll: 44 bottom (redrawn by John Sibbick)
E. H. Colbert: 105 lower right (redrawn); 106 bottom (redrawn)
G. Cuvier: 26 upper
De La Bèche: 169 top left (in F. J. Pictet, 1846)
Dong Zhiming: 105 upper left and right (redrawn)
C. F. Eaton: 36 middle (redrawn); 50 bottom (redrawn)
R. F. Ewer: 43 upper left and right
L. Figuier: 158 middle top
E. Frey and J. Riess: 148 top (redrawn)
P. Galton: 106 top (redrawn)
A. Goldfuss: 26 lower; 27 upper left
E. v. Holst: 39 top left (redrawn); 174 top (redrawn)
S. Howse: 113 bottom right (redrawn)
N. Hutchinson: 169 top right
F. v. Huene: 41 bottom (redrawn)
Illustrated London News (1853): 172 top left
J. Jensen and K. Padian: 106 middle (redrawn)
A. Kircher: 20 lower
G. Mantell: 29 upper right (BMNH)
O. C. Marsh: 37 middle left; 181 middle left and right (Franz Höck)
J. Martin: 168 lower (in Th. Hawkins, 1840)
H. v. Meyer: 31 lower; 32 upper and 33 upper (Franz Höck)
E. Newman: 27 bottom right
E. T. Newton: 32 lower (Franz Höck)
F. Nopcsa: 59 upper
R. Owen: 34 upper (Franz Höck); 68 lower; 69 lower; 110 top (redrawn); 112 middle and bottom (redrawn); 113 middle and bottom left (redrawn); 178 upper (Franz Höck)
K. Padian: 79 (redrawn)
N. Parker: 158 top left
C. J. Pennycuick: 148 middle (redrawn)
D. S. Peters and W. F. Gutmann: 44 top (redrawn)
S. Phillips: 172 bottom left
L. I. Price: 9 lower; 122 (redrawn)
M. Reichel: 137 lower left and right (redrawn); 169 middle right
R. Reisz: 40 lower (P. Wellnhofer)
Riou: 168 upper (in O. Fraas, 1866)
W. Schäfer: 13 lower left (redrawn)
H. G. Seeley: 34 lower
John Sibbick: 42 bottom; 44 bottom; 49 top, middle left and bottom; 67 middle right; 72 lower; 76; 116 bottom (after E. Thenius); 144; 154 (after C. Bramwell and G. Whitfield); 160 top (after P. Wellnhofer); 161 bottom (after R. T. Bakker); 162 (after P. Schouten); 176 top; 177 top left and bottom (after H. Haubold and E. Buffetaut); 180 top; 181 bottom
S. T. v. Soemmerring: 23 lower left; 25 middle left and right
C. Theodori: 30 lower
G. Wagler: 27 middle
M. Wellnhofer: 10 lower (Geoff Denney)
Peter Wellnhofer: 9 upper (Geoff Denney); 16 (Geoff Denney); 17 both; 18 (Geoff Denney); 19 (Geoff Denney); 40 upper (Geoff Denney); 41 upper; 42 middle (Geoff Denney); 43 lower (Geoff Denney); 45 (slightly changed after R. Wild); 46 all artwork; 47 all artwork; 49 middle right; 50 top and middle; 51 all artwork; 52 all artwork; 53 all artwork; 54 all artwork; 55 all artwork; 56 all artwork; 57 all artwork; 62 portrait; 67 both skulls; 70 portrait and skull; 72 upper; 74 skull; 75 portrait; 77 all artwork; 78; 80; 81; 84; 85 all artwork; 86 skulls; 87 portrait; 88 all artwork; 92 all artwork; 93 all artwork; 96 all artwork; 97; 99 portrait and skull; 100; 101; 104; 112 skull; 113 skull; 116 all skeletal drawings; 119 skull and portrait; 120 skull; 121 lower right; 124; 125; 126 both artworks; 127; 128 skull and portrait; 129 jaw; 130; 133 portraits; 136; 139 portrait and skull; 147; 148 bottom; 150 bottom; 151 top and bottom right; 152; 153 lower left; 155; 156 all artwork; 157 top left (after M. Reichel) and middle; 158 middle; 159 both artworks; 160 bottom left and right; 163 (after R. Wild); 165 all artworks; 166 (Geoff Denney); 174 middle sequence); 176 bottom; 177 middle left; 179 skulls; 180 skeletons; 183 bottom (after Novacek 1987)
R. Wild: 59 lower; 60 top; 62 all skeletal artwork; 63 all artwork; 66 all artwork; 67 top left and bottom left
S. W. Williston: 37 top left; 140 (redrawn)
C. C. Young: 120 middle left (redrawn); 121 upper left (redrawn)
K. A. Zittel: 37 upper right

Location of Illustrated Specimens

The fossil specimens figured in this book are housed in many different museums and collections around the world. The following listing, arranged in alphabetical order of the cities, is intended to help the interested reader to locate these specimens.

Austin, USA: Texas Memorial Museum, University of Texas, Austin, Texas; 141 upper left, 141 middle right.
Banz, Germany: Petrefaktensammlung Kloster Banz, Lichtenfels; 30 bottom.
Beijing, China: Institute of Vertebrate Palaeontology and Palaeoanthropology, Academia Sinica, Beijing; 105 upper left, 117 bottom, 120 top, 121 top left, 121 top right.
Bergamo, Italy: Museo Civico di Scienze Naturali, Bergamo; 59 lower right, 62 top, 62 bottom, 66 upper left, 66 bottom, 67 upper left.
Berlin, Germany: Museum für Naturkunde an der Humboldt-Universität, Berlin; 178 lower, 179 top left.
Bochum, Germany: Collection Helmut Leich, Bochum; 152 lower right, 163 upper right.
Bonn, Germany: Geologisch-Paläontologisches Institut der Universität, Bonn; 27 upper left, 92 top.
Brisbane, Australia: Queensland Museum, Fortitude Valley, Queensland; 122 upper middle, 122 upper right.
Cambridge, England: Sedgwick Museum, University of Cambridge; 110 upper right, 113 lower left.
Chicago, USA: Field Museum of Natural History, Chicago, Illinois; 140 upper right.
Curitiba, Brazil: Collection Guido Borgomanero, Curitiba; 127 middle right.
Dashanpu, China: Palaeontological Museum Zigong, Beipei Museum, Dashanpu, Sichuan; 80 upper right.
Eichstätt, Germany: Jura-Museum, Eichstätt; 84 middle left, 97 lower right, 100 middle left. Collection Karl Strobl, Eichstätt; 159 lower right.
Frankfurt, Germany: Naturmuseum Senckenberg, Frankfurt am Main; 10 top, 83 middle.
Haarlem, The Netherlands: Teyler's Museum, Haarlem; 11 bottom, 32 top, 33 top, 163 upper left.
Hays, USA: Sternberg Memorial Museum, Fort Hays State University, Hays, Kansas: 136 middle left.
Leningrad, USSR: Palaeontological Institute of the University, Leningrad; 121 lower right.
London, England: Natural History Museum, London; 15 upper right, 28 bottom, 32 bottom, 69 top, 79 middle, 83 top, 108 middle, 112 bottom, 112 middle right, 113 middle left, 113 bottom right, 115 upper right, 116 middle left, 165 lower left, 178 upper.
Milan, Italy: Museo Civico di Storia Naturale, Milano; 63 lower right.
Moscow, USSR: Palaeontological Museum, Moscow; 101 upper left, 101 lower right, 102 upper right, 104 upper left, 120 middle right, 164 lower left, 164 middle, 177 middle.
Munich, Germany: Bayerische Staatssammlung für Paläontologie und historische Geologie, München; 12 top, 12 bottom, 13 middle, 13 lower right, 14 upper right, 23 top, 25 upper right, 37 middle right, 48 top, 51 upper right, 51 middle, 52 upper left, 53 middle right, 54 middle, 55 middle right, 56 top, 57 middle left, 57 lower left, 82 top, 82 bottom, 84 middle right, 84 lower left, 88 lower right, 89 top, 93 upper left, 95 upper left, 96 top, 97 upper right, 97 middle right, 100 upper left, 124 top, 124 lower middle, 125 upper left, 127 upper right, 127 bottom left, 127 bottom right, 129 upper left, 148 middle, 149 middle, 149 bottom, 151 lower left, 152 top, 152 middle, 156 upper left, 156 middle, 158 lower left, 159 middle left, 159 middle right, 160 middle, 160 middle right, 160 lower middle, 161 upper left, 161 middle left, 163 middle, 164 lower right, 182 top right.
New Haven, USA: Peabody Museum of Natural History, Yale University, New Haven, Connecticut: 11 top, 36 bottom, 49 middle right, 105 lower right, 106 upper right, 135 middle left, 135 middle right, 137 top, 182 top left.
New York, USA: American Museum of Natural History, New York, N.Y.; 89 lower left, 106 lower right, 107 bottom, 125 top, 125 lower right, 126 upper right, 126 middle, 126 lower left, 127 upper left, 135 lower left, 136 top, 130 bottom left, 177 top right.
Paris, France: Musée National d'Histoire Naturelle, Paris; 121 lower middle.
Pittsburgh, USA: Carnegie Museum of Natural History, Pittsburgh, Pennsylvania; 77 top.
Provo, USA: Brigham Young University, Provo, Utah; 106 middle right.
Rio de Janeiro, Brazil: Departamento Nacional Produción Mines (DNPM), Rio de Janeiro; 9 bottom, 122 lower right, 123 lower right, 125 upper right, 125 lower middle left.
Salt Lake City, USA: Utah Museum of Natural History, University of Utah, Salt Lake City, Utah; 158 lower middle.
Solnhofen, Germany: Bürgermeister-Müller-Museum, Solnhofen; 179 middle left.
Maxberg-Museum, Solnhofen; 91 upper right.
Stuttgart, Germany: Staatliches Museum für Naturkunde, Stuttgart; 73 upper left, 76 top.
Tübingen, Germany: Institut für Geologie und Paläontologie der Universität, Tübingen; 31 top, 85 middle left.
Tucuman, Argentina: Instituto-Fundación Miguel Lillo, Universidad Nacional, Tucuman; 131 top, 134 top.
Udine, Italy: Museo Friuliano di Storia Naturale, Udine; 67 middle left, 67 bottom.
Uppsala, Sweden: Palaeontological Museum, Uppsala; 73 lower right.
Vienna, Austria: Geological Institute of the University, Wien; 116 upper right, 116 middle right.
Naturhistorisches Museum, Wien; 150 upper left.